IMRT · IGRT · SBRT
Advances in the Treatment Planning and Delivery of Radiotherapy
2nd, revised and extended edition

Frontiers of Radiation Therapy and Oncology

Vol. 43

Series Editors

J.L. Meyer San Francisco, Calif.
W. Hinkelbein Berlin

IMRT · IGRT · SBRT

Advances in the Treatment Planning and Delivery of Radiotherapy

2nd, revised and extended edition

Volume Editor

John L. Meyer San Francisco, Calif.

Advisory Editors

L.A. Dawson Toronto, Ont.
B.D. Kavanagh Aurora, Colo.
J.A. Purdy Sacramento, Calif.
R. Timmerman Dallas, Tex.

211 figures, 175 in color, and 41 tables, online supplementary material, 2011

Basel · Freiburg · Paris · London · New York · New Delhi ·
Bangkok · Beijing · Tokyo · Kuala Lumpur · Singapore · Sydney

Frontiers of Radiation Therapy and Oncology

Founded 1968 by J.M. Vaeth, San Francisco, Calif.

John L. Meyer, MD, FASTRO

Department of Radiation Oncology
Saint Francis Memorial Hospital
San Francisco, Calif., USA

Library of Congress Cataloging-in-Publication Data

IMRT, IGRT, SBRT: advances in the treatment planning and delivery of radiotherapy /
volume editor, John L. Meyer; advisory editors, L.A. Dawson ... [et al.]. -- 2nd, rev. and extended ed.
    p. ; cm. -- (Frontiers of radiation therapy and oncology, ISSN 0071-9676; v. 43)
 Includes bibliographical references and indexes.
 ISBN 978-3-8055-9680-0 (hard cover: alk. paper) -- ISBN 978-3-8055-9681-7 (e-ISBN)
 1. Cancer -- Radiotherapy. 2. Radiotherapy. I. Meyer, John, 1949– II. Series: Frontiers of radiation therapy and oncology; v. 43. 0071-9676
 [DNLM: 1. Radiotherapy, Computer-Assisted--Methods. W3 FR935 v.43 2011 / WN 250.5.R2]
 RC271.R3I435 2011
 616.99'40642--dc22
                                                                    2011000684

Bibliographic Indices. This publication is listed in bibliographic services, including Current Contents® and PubMed/MEDLINE.

Illustrations. The authors and the publisher have made every effort to obtain permission for all copyright-protected material. Any omissions are entirely unintentional. The publisher would be pleased to hear from anyone whose rights unwittingly have been infringed.

Disclaimer. The statements, opinions and data contained in this publication are solely those of the individual authors and contributors and not of the publisher and the editor(s). The appearance of advertisements in the book is not a warranty, endorsement, or approval of the products or services advertised or of their effectiveness, quality or safety. The publisher and the editor(s) disclaim responsibility for any injury to persons or property resulting from any ideas, methods, instructions or products referred to in the content or advertisements.

Drug Dosage. The authors and the publisher have exerted every effort to ensure that drug selection and dosage set forth in this text are in accord with current recommendations and practice at the time of publication. However, in view of ongoing research, changes in government regulations, and the constant flow of information relating to drug therapy and drug reactions, the reader is urged to check the package insert for each drug for any change in indications and dosage and for added warnings and precautions. This is particularly important when the recommended agent is a new and/or infrequently employed drug.

All rights reserved. No part of this publication may be translated into other languages, reproduced or utilized in any form or by any means electronic or mechanical, including photocopying, recording, microcopying, or by any information storage and retrieval system, without permission in writing from the publisher.

© Copyright 2011 by S. Karger AG, P.O. Box, CH–4009 Basel (Switzerland)
www.karger.com
Printed in Switzerland on acid-free and non-aging paper (ISO 9706) by Reinhardt Druck, Basel
ISSN 0071–9676
ISBN 978-3-8055-9680-0
e-ISBN 978-3-8055-9681-7

Contents

VIII **Frequently Used Abbreviations**

IX **Preface**
Meyer, J.L. (San Francisco, Calif.)

Introduction

1 **Advances in the Planning and Delivery of Radiotherapy: New Expectations, New Standards of Care**
Purdy, J.A. (Sacramento, Calif.)

General Concepts and Methods

29 **Advanced Technologies in the Radiotherapy Clinic: System Fundamentals**
Meyer, J.L. (San Francisco, Calif.); Sharpe, M.; Brock, K. (Toronto, Ont.); Deasy, J. (St. Louis, Mo.); Craig, T.; Moseley, D. (Toronto, Ont.); Alaly, J.; Zakaryan, K. (St. Louis, Mo.)

Economic Perspectives

60 **Controversies in the Adoption of New Healthcare Technologies**
Wallner, P.E. (Fort Myers, Fla.); Steinberg, M.L. (Los Angeles, Calif.); Konski, A.A. (Detroit, Mich.)

I. IMRT and IGRT: Intensity-Modulated and Image-Guided Radiotherapy

IMRT: Advances in Treatment Design and Delivery

80 **Clinical Implementation of Intensity-Modulated Arc Therapy**
Shepard, D.M.; Cao, D. (Seattle, Wash.)

IGRT: Advances in Targeting Therapy

99 **4D Imaging and 4D Radiation Therapy: A New Era of Therapy Design and Delivery**
Low, D. (Los Angeles, Calif.)

118 **Locating and Targeting Moving Tumors with Radiation Beams**
Keall, P. (Sydney, N.S.W.)

Integrating IMRT and IGRT into Treatment Delivery

132 **Technologies of Image Guidance and the Development of Advanced Linear Accelerator Systems for Radiotherapy**
Wu, V.W.C.; Law, M.Y.Y. (Hong Kong); Star-Lack, J. (Palo Alto, Calif.); Cheung, F.W.K. (Hong Kong); Ling, C.C. (New York, N.Y./Palo Alto, Calif./Hong Kong)

165 **Helical Tomotherapy: Image-Guided and Adaptive Radiotherapy**
Kupelian, P. (Los Angeles, Calif.); Langen, K. (Orlando, Fla.)

181 **The CyberKnife in Clinical Use: Current Roles, Future Expectations**
Dieterich, S.; Gibbs, I.C. (Stanford, Calif.)

II. IMRT and IGRT Clinical Treatment Programs

196 **Image Guidance and the New Practice of Radiotherapy: What to Know and Use from a Decade of Investigation**
Kim, J. (Toronto, Ont.); Meyer, J.L. (San Francisco, Calif.); Dawson, L.A. (Toronto, Ont.)

Head and Neck Cancers

217 **Intensity-Modulated and Image-Guided Radiation Therapy for Head and Neck Cancers**
Chu, K.P.-M.; Le, Q.-T. (Stanford, Calif.)

255 **Delineating Neck Targets for Intensity-Modulated Radiation Therapy of Head and Neck Cancer**
David, M.B.; Eisbruch, A. (Ann Arbor, Mich.)

Thoracic Cancers

271 **Motion Management and Image Guidance for Thoracic Tumor Radiotherapy: Clinical Treatment Programs**
Loo, B.W., Jr. (Stanford, Calif.); Kavanagh, B.D. (Aurora, Colo.); Meyer, J.L. (San Francisco, Calif.)

Breast Cancer

292 **Intensity-Modulated Radiotherapy for Breast Cancer: Advances in Whole and Partial Breast Treatment**
White, J.R. (Milwaukee, Wisc.); Meyer, J.L. (San Francisco, Calif.)

Abdominal Cancers

315 Image-Guided Radiotherapy Strategies in Upper Gastrointestinal Malignancies
Swaminath, A.; Dawson, L.A. (Toronto, Ont.)

Lymphomas

331 Radiotherapy Planning for the Lymphomas: Expanding Roles for Biologic Imaging
Hoppe, R. (Stanford, Calif.)

Prostate Cancer

344 Image-Guided, Adaptive Radiotherapy of Prostate Cancer: Toward New Standards of Radiotherapy Practice
Kupelian, P. (Los Angeles, Calif.); Meyer, J.L. (San Francisco, Calif.)

III. SBRT: Stereotactic Body Radiotherapy

SBRT Advances

370 The Expanding Roles of Stereotactic Body Radiation Therapy and Oligofractionation: Toward a New Practice of Radiotherapy
Kavanagh, B.D. (Aurora, Colo.); Timmerman, R. (Dallas, Tex.); Meyer, J.L. (San Francisco, Calif.)

382 Stereotactic Body Radiation Therapy: Normal Tissue and Tumor Control Effects with Large Dose per Fraction
Timmerman, R.; Bastasch, M.; Saha, D.; Abdulrahman, R.; Hittson, W.; Story, M. (Dallas, Tex.)

Thoracic Cancers

395 Stereotactic Body Radiation Therapy for Thoracic Cancers: Recommendations for Patient Selection, Setup and Therapy
Timmerman, R.; Heinzerling, J.; Abdulrahman, R.; Choy, H. (Dallas, Tex.); Meyer, J.L. (San Francisco, Calif.)

Gastrointestinal Cancers

412 Stereotactic Body Radiation Therapy for Gastrointestinal Malignancies
Minn, A.Y.; Koong, A.C.; Chang, D.T. (Stanford, Calif.)

Genitourinary Cancers

428 Stereotactic Body Radiotherapy for Prostate Cancer: Current Results of a Phase II Trial
King, C. (Los Angeles, Calif.)

IV. Proton Beam Radiotherapy

Techniques and Technologies

440 Proton Therapy: Clinical Gains through Current and Future Treatment Programs
Mohan, R. (Houston, Tex.); Bortfeld, T. (Boston, Mass.)

Medical Applications and Advances

465 Proton Therapy in the Clinic
DeLaney, T.F. (Boston, Mass.)

486 Author Index
487 Subject Index

 Online supplementary material: www.karger.com/FRATO43_suppl

---

Frequently Used Abbreviations

3DCRT	Three-dimensional conformal radiotherapy
CTV	Clinical target volume
DVH	Dose-volume histogram
EPID	Electronic portal imaging device
GTV	Gross tumor volume
IGRT	Image-guided radiotherapy
IMRT	Intensity-modulated radiotherapy
kV	Kilovoltage
MV	Megavoltage
PTV	Planning target volume
RTOG	Radiation Therapy Oncology Group
SBRT	Stereotactic body radiotherapy

Preface

This volume offers a guide to the techniques and technologies that can bring new, advanced radiotherapy capabilities into daily patient care. It is intended to be a readable and practical resource, presenting useful radiotherapy programs for clinical application, guidance on their efficient use and specific evidence for their selection from recent reports on clinical outcomes.

The volume is divided into four sections. The first offers explanations and discussions of the technologies themselves and technical methods for their implementation. The second section brings these technologies into the radiation clinic with presentations by noted physicians at major centers who have broad experience with these new treatment approaches. In each chapter, the authors give specific guidelines for current clinical practice. The third section explores the use of these high-precision technologies in the rapidly expanding field of stereotactic body radiotherapy. The final section discusses the important advances in proton therapy, which is adding a new dimension to the range of available treatment technologies.

This second edition takes a broad new vantage on the substantial changes that have recently occurred in the field of radiation oncology. For many cancer conditions, intensity-modulated, image-guided and stereotactic body treatment programs have become expected services of the modern, comprehensive radiotherapy clinic, and clear understanding of their use in now basic to practice. Despite their rapid adoption, many fundamental issues regarding their roles are still being defined at every step from tumor imaging and therapy planning to treatment delivery. Measures of clinical outcomes and their value based on cost/benefit analyses are just now being reported. Many new insights into their clinical indications, integration and efficient utilization are coming into focus, and their roles are redefining practice in new hypofractionated treatment programs. Further, the growing contributions of proton therapy add a new level of decision and treatment selection for clinicians. Central to all of these issues are the current economics and practical allocations of our medical resources, which may increasingly define the utilization of these technologies. These developments, and their relative merits, are the focus of this volume.

I have planned and developed the text based on presentations given at the 2009 San Francisco Radiation Oncology Conference, which was jointly sponsored by the Departments of Radiation Oncology of Stanford University, University of California at San Francisco, Saint Francis Memorial Hospital San Francisco and University of California at Davis. Drs. Richard Hoppe, Mack Roach III, James Purdy, Paul Keall and Jean Pouliot supported me in organizing the conference. I wish to thank each of them. Presentations have been expanded, updated, referenced and integrated for this volume, and several additional chapters were contributed to complete the scope of the text.

Advances in radiologic imaging are the foundation of much of the current work explored in this text. Throughout the volume, examples of this are often presented in more than one format. In addition to the printed illustrations, a website (www.karger.com/FRATO43_suppl) allows the reader to view a number of the important figures in time-elapse videos. This is especially useful in understanding the work on tumor motion and image guidance. Other illustrations are also posted on this website for greater clarity and dynamic visualization, and the website is an essential part of these presentations overall.

I wish to thank all of the authors for working with me on this text, especially Drs. James Purdy, Laura Dawson, Brian Kavanagh and Robert Timmerman for their excellent contributions and guidance. I wish to thank Dr. Catherine Burns for her expert assistance in the editing of the volume, and Mr. Josue Castellano for his expertise in organizing its illustrations and website materials. Finally I wish to thank Dr. Thomas Karger and the many associates of his fine publishing house.

John L. Meyer, MD, FASTRO
San Francisco, Calif., USA

Introduction

Advances in the Planning and Delivery of Radiotherapy: New Expectations, New Standards of Care

James A. Purdy

Department of Radiation Oncology, UC Davis Medical Center, Sacramento, Calif., USA

Abstract

The practice of radiation therapy continues to build on rapid advancements in treatment planning and delivery technology, which brings real potential for improving treatment outcomes. Manufacturers have employed advanced computer and imaging technology to produce treatment planning/delivery systems capable of precise shaping of dose distributions, conformal target volume coverage for even the most complex shapes and conformal avoidance of specified sensitive normal structures. However, these new systems have led to a more complex, less intuitive planning and treatment delivery process that presents great challenges for quality assurance/treatment verification. Advances in planning and delivery technologies will continue to occur at record paces, pushing the field toward even higher expectations for radiotherapy accuracy, reliability and applicability and leading the field to new standards of care. However, this optimism must be tempered with the realizations that for this to happen, progress is urgently needed in 3 areas, (1) accuracy in specification of gross tumor volume and clinical target volume, (2) radiation oncology informatics and (3) quality assurance, if we are to keep pace with these rapid planning and delivery developments. Copyright © 2011 S. Karger AG, Basel

The Expansion of Radiation Oncology Technologies

Radiation therapy treatment planning and delivery technology and workflow processes have changed dramatically since the introduction of 3D treatment planning in the 1980s [1, 2]. This expansion began in an era in which physicians used a conventional X-ray simulator to design beam portal apertures. Typically, these were based on 'class-solution' beam arrangement techniques and bone landmarks visualized on simulator-generated planar radiographs. This was an extremely efficient process, but one limited in ability to conform the dose to the target volume and shield normal tissues, thus limiting significant dose escalation in many disease sites. An image-based 3D treatment planning process in those early days required obtaining a volumetric CT imaging dataset of the patient in the treatment position, followed by tedious contouring of the target volume(s) and pertinent organs at risk. This planning was (and still is for some disease sites) a fairly inefficient process. It required learning new computer-based skill sets and understanding oncologic anatomy in new, 3D perspectives that were foreign from established approaches.

This increased effort was a key reason, in my opinion, that it took such a long time for the 3D planning approach to gain acceptance into clinical practice. The other key factor was the lengthy FDA 510(k) approval process which was in place at that time for 3D treatment planning systems. These 2 barriers caused delays that were actually beneficial to the profession: they allowed added time for universities to develop improved software tools, and for clinicians to gain experience in 3D treatment planning and conformal therapy (3DCRT) procedures. As a result, the technical and clinical processes were reasonably mature when commercial 3D planning systems became widely available and embraced by the profession.

In contrast, newer intensity-modulated radiation therapy (IMRT) techniques [3] spread relatively quickly throughout the radiotherapy community when the technology became commercially available. The rapid growth in the USA was no doubt aided when Medicare recognized IMRT as a billable service and described it by CPT planning and delivery codes that carried significantly greater valuations (then and now) than the corresponding 3DCRT service codes. This was certainly good news for those groups that were involved with the clinical development of IMRT, as it allowed recovery of some of their initial costs. However, in my view, the increased reimbursement spurred clinics to implement this technology before it had adequately matured.

The highly conformal doses and sharper dose gradients of IMRT permitted the use of smaller safety margins, decreasing the doses to the surrounding normal tissues. They also brought new concerns for our ability to account for intrafraction and interfraction variations in patient positioning, internal organ shapes and target locations. Was the IMRT actually delivered as accurately as the IMRT was planned? These fundamental concerns drove the development of treatment machines with integrated planar and volumetric imaging capabilities to guide and confirm the IMRT delivery [4], processes named image-guided IMRT or simply image-guided radiation therapy (IGRT) [5].

IGRT medical linear accelerators (linacs), which are rapidly being implemented in therapy clinics worldwide, can now obtain volumetric CT imaging before therapy. The radiation doses to be delivered can be projected on these images, enabling their inspection and approval before treatment. From day to day, the visualized organs and their corresponding radiation doses may change shapes somewhat. How can these images/doses be cumulated, since they will not quite match volumetrically? Technologies that will allow rapid deformable registration of such images are being developed and will allow continuing updated recalculation of the cumulative 'true doses' to the patient. Not only can the daily volumetric images be used for guidance, but they can also be used for the daily volumetric doses. If the delivered dose distributions are not the intended ones, as the therapy course progresses, the fields can be adapted for the remaining treatments to correct for these changes. Such a system will allow replanning when the treating physician deems it necessary, an exciting new process referred to as adaptive radiation therapy (ART) [6].

This quest to provide the most conformal radiation therapy possible is not confined just to external photon beam therapy, but it is also occurring in image-guided brachytherapy and proton radiation therapy, including intensity-modulated proton therapy (IMPT). It seems as if radiation oncology is on a 'technology super-highway' in which companies are bringing advances in planning and treatment delivery technologies to the market as fast as possible, while physicians and physicists are embracing these advances and using them as fast as possible. As a result, the rapid adoption of new technologies poses a significant challenge to keep quality assurance (QA) methodologies and instrumentation in front of the technology adoption curve.

This book series has played an important role in chronicling the profound changes that have occurred over the past several years. Today, we need to pause and critically assess how well we are using these technologies and then to identify the new advances on the horizon of this radiation oncology super-highway.

Definitions

We should first recognize that there are no universally understood definitions for the shortcut acronyms IMRT, IGRT and ART. I list here my interpretation of these terms.

> *Intensity-Modulated Radiotherapy (IMRT):* an advanced form of 3DCRT that uses nonuniform radiation beam intensities (incident on the patient) that have been determined using various computer-based optimization techniques [7].
>
> *Image-Guided Radiation Therapy (IGRT):* a process that integrates patient/tumor positioning and treatment machine image guidance tools and other motion management systems to better direct the radiation beam to the patient's tumor/target volume.
>
> *Adaptive Radiation Therapy (ART):* a process intended to improve radiation treatment by systematically monitoring patient/treatment positional and volumetric variations and incorporating them to reoptimize the treatment plan at appropriate intervals during the course of treatment.

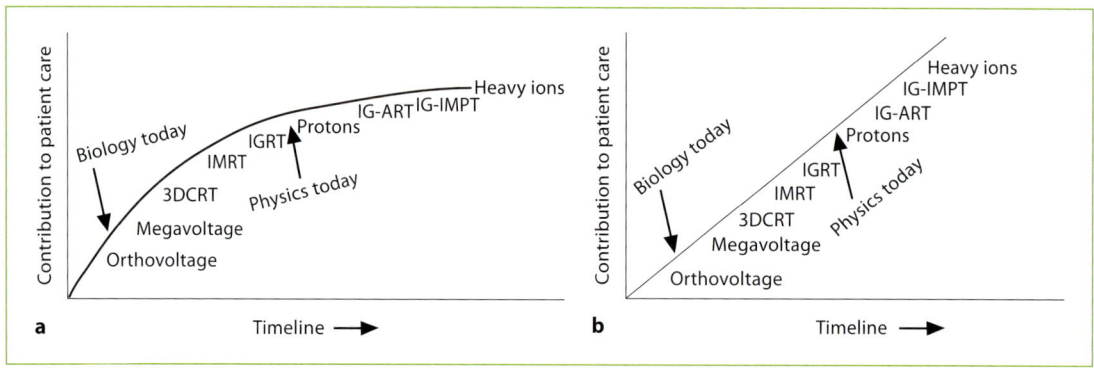

Fig. 1. Two distinctly different worldviews regarding biology and physics/engineering contributions to actual clinical practice. **a** One view predicts advances in physics/engineering will likely not significantly improve treatment outcomes over those currently achieved with IMRT/IGRT treatments. **b** The other predicts advances in physics/engineering will continue to improve treatment outcomes. The author is optimistic that the latter is more representative of the future. Note in both cases, there is significant upside for cancer biology to contribute to actual clinical practice.

IMRT is now the standard of practice for many disease sites. The definition that we established within the IMRT Collaborative Working Group (CWG) was a struggle, as several groups were already using a variety of beam-modulated techniques, such as 'field within a field' and compensating filters [7]. However, the CWG recognized that those older techniques were not really indicative of the change in practice occurring as a result of the new IMRT. The CWG finally agreed that the real change IMRT brought was the ability to create nonuniform beam fluences needed to achieve specified dose-volume constraints via a process referred to as 'inverse planning'.

Today, IGRT principally involves the frequent use of planar or volumetric imaging during a course of radiotherapy, coupled with software to allow registration with the original treatment plan to adjust the treatment table's x, y and z coordinates. This moves the patient's target volume back to the planned position. Other nonimaging repositioning/tracking systems are also used to accomplish this task and will be discussed later.

Adaptive radiotherapy is a process that is still in development and not well defined. But with the rapid adoption of IGRT, groups are well positioned to fully explore the integration of IMRT planning and delivery with on-line volumetric imaging. Clearly, the development of technology and efficient strategies for on-line plan adaptation and replanning will be a key area of interest moving forward.

Integration of Cancer Biology Discoveries into Clinical Practice

While there clearly has been an explosion of new discoveries in cancer biology, there has been relatively slow progress (in contrast to physics) in the clinical application of these cancer biology advances. In addition, some are now predicting that future physics and engineering developments in external beam radiotherapy will likely not significantly improve treatment outcomes over those currently achieved with 3DCRT and IMRT treatments [8]. This worldview of our profession can be illustrated as shown in figure 1a. It should be duly noted that I emphatically do not agree with the shape of that curve. While I do believe it is imperative that cancer biology discoveries move more rapidly into the clinic, I believe figure 1b is much more likely to depict the future as there are many more technological advances coming to radiation oncology. Our challenge will be to understand and use them properly and integrate them into the clinic safely, while maintaining and improving efficiencies.

Cost to the Health Care System

With the rapid pace of continuing technical developments, we must also appreciate their effects on the efficiencies of radiation oncology services and health care costs. The new technologies are not nearly as efficient as the pre-3DCRT techniques. In the past, 4 or 5 patients could be treated per hour. Today, the typical throughput is 3 patients per hour; for some IGRT delivery systems or treatment techniques such as SBRT, it may be as few as a single patient per hour. Granted the treatment quality today is vastly superior. However, an important message for all of us to hear loud and clear is this: going forward, we must address efficiency and cost effectiveness issues, while improving treatment quality and patient safety. This is a huge challenge, but one that we must meet.

Target Volume Specification and Treatment: Issues of Accuracy

Accuracy in Defining Tumor Volumes

The accurate specification of volumes remains the most critical issue in planning the patient's treatment, from 3DCRT to IMRT to IMPT. If we do not have these volumes defined as accurately as possible, then the advanced treatment delivery technology is wasted. Over the past 2 decades, the ICRU Reports 50/62 [9, 10] have served as a road map for improving target volume specification by providing nomenclature/methodology as listed below:

- Gross tumor volume (GTV) is the volume of the known tumor that is imaged or palpated.
- Clinical target volume (CTV) is a clinical-anatomical concept that represents a tissue volume that contains a demonstrable GTV and/or subclinical malignant disease that is not imaged.
- Internal target volume is a geometrical concept, introduced for treatment planning purposes, in which an appropriate margin is added around the CTV to account for uncertainties in size, shape and position of the CTV (defined in the patient coordinate system, use is optional in defining PTV).
- Planning target volume (PTV) is a geometrical concept, used for treatment planning and evaluation purposes, in which an appropriate margin is added around the CTV to ensure with a clinically acceptable probability that the prescribed dose will actually be delivered to all parts of the CTV despite geometrical uncertainties such as organ motion and setup variations (defined in treatment machine coordinate system).

Researchers have understood these concepts and focused on ways to improve the target volume accuracy. Continuing advances have been made in imaging to better identify the GTV, and in technologies to quantitatively determine and manage GTV motion and patient setup variations. Unfortunately, very limited attention has been directed to improving the accuracy of the CTV, the basis for treatment. Only a few publications and atlases for some clinical sites have been developed to establish consistent CTVs [11, 12]. This is another message that I hope physician-scientists, physicists, biologists and manufacturers hear clearly. We urgently need more effort focused on optimizing the specification of CTVs. In my opinion, the physician's decision about what to treat (delineating GTV/CTV) is likely to have a greater impact on optimal treatment than improving the technology for managing inter- and intrafraction motion.

Accuracy in Managing Motion

We have made significant progress in managing motion, reflected in the multiple technologies currently in use (see table 1). Many treatment machines now have an integrated CT system, allowing full volumetric imaging to be performed on a daily basis. Many treatment rooms are equipped with additional localization and tracking systems (see fig. 2). For instance, the Calypso® 4D System uses implanted Beacon® electromagnetic transponders to provide guidance on the position and movement of the target (usually the prostate) during therapy (Calypso Medical Technologies Inc., Seattle, Wash., USA) [13, 14], and the AlignRT System uses video-based 3D surface imaging to guide primarily breast radiotherapy (VisionRT Ltd., London, UK) [15]. Technologies are now available to allow the physician to

Table 1. Technologies currently in use for managing motion of patients undergoing radiation therapy

4D CT simulation
Electronic portal imaging with fiducials
Ultrasound-guided target localization
Body frame/compression
Active breathing control
In-room conventional CT
IGRT linacs kV-CBCT
MV-CBCT, MV helical CT
Video and surface imaging
2D planar X-ray tracking
2D fluoroscopic X-ray tracking
Electromagnetic tracking
Respiratory-gated radiation therapy

significantly reduce the PTV margin, representing expansion from the CTV. However, there is a real concern in reducing this margin too much. Since the boundary of the CTV may be an educated guess at best, is one willing to shrink the PTV boundary down to a guess?

Accuracy in Intensity-Modulated Radiation Therapy Delivery

Several investigators have shown that IMRT treatment delivery is very sensitive to multileaf collimator (MLC) leaf position errors. Sharpe et al. [16] showed that a rigorous QA program must be in place to ensure that leaf precision is tightly controlled, as uncertainties in leaf position can have significant effects on dosimetric accuracy. More recently, Mu et al. [17], as well as Rangel and Dunscombe [18], reported that dosimetric effects were insignificant for random MLC leaf position errors up to 2 mm. However, systematic leaf position errors caused significant dosimetric changes for both simple and complex IMRT plans. Both papers show that it is much more important to reduce systematic leaf position errors than random errors, and that simpler IMRT plans are less sensitive than more complex ones to leaf position errors in general. These publications reinforce the earlier statement that physicists must stay abreast of such issues, and insure that preventive maintenance/QA checks are in place to achieve and maintain strict tolerance standards [19].

There is an interplay effect between GTV/CTV motion and MLC motion that may be problematic when IMRT is used. Since each IMRT segment treats only a portion of the PTV at a time, there may be significant dosimetric consequences if the patient and/or the GTV/CTV move during treatment (intrafraction geometric uncertainties). In this situation, there are moving tumor cells and also moving

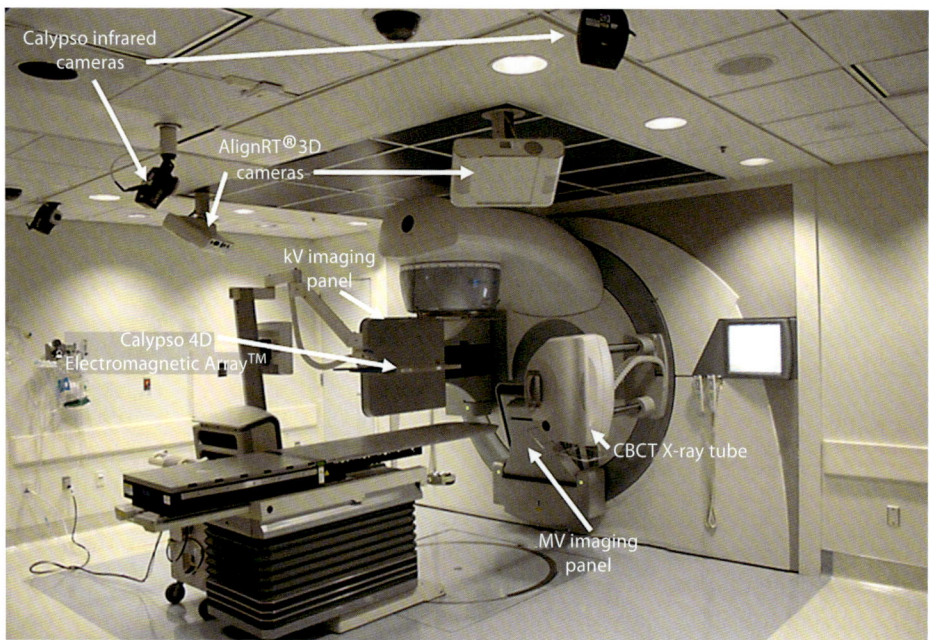

Fig. 2. Modern-day radiation therapy treatment room implemented at UC Davis with advanced multimodality medical linac with integrated kV cone-beam CT (CBCT) and MV EPID (Elekta Synergy-S; Elekta Ltd., Crawley, UK) allowing full volumetric imaging on a daily basis. The room also contains additional localization and tracking systems, i.e. the Calypso® System which uses implanted Beacon® electromagnetic transponders and 4D Tracking Station located at the linac console (Calypso Medical Technologies), and the video-based 3D surface imaging system (AlignRT, VisionRT).

MLC leaves, which could result in relative hot and cold spots in the dose delivered. Simulation studies have concluded that this interplay effect is not clinically significant in therapy courses using normal fractionation, but may be significant in courses using only a few fractions such as SBRT [20, 21]. In such cases, one must be very careful that GTV motion is negligible if IMRT is utilized.

Intensity-Modulated Radiation Therapy/Image-Guided Radiation Therapy Planning and Delivery Technologies: Current Challenges

Imaging for Planning

Even though the current generation of CT simulators (fig. 3) provides powerful 3D and 4D capabilities, several more improvements are likely to evolve over the next decade that will enhance the treatment planning process further. For example, a

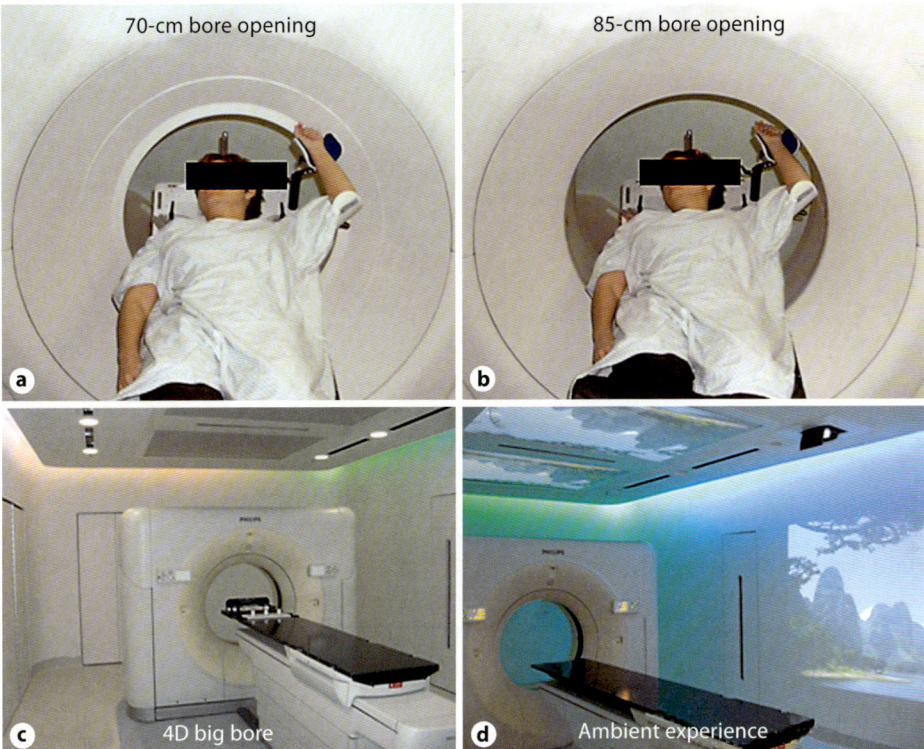

Fig. 3. a Small-bore CT simulator problematic for some treatment setups such as tangential breast. **b** Large-bore aperture (85 cm) which allows scanning patients in treatment position with their immobilization devices in place. **c** Modern-day CT simulator (Brilliance Big Bore Oncology), installed at UC Davis, with software tools for respiratory motion assessment, tumor localization for isocenter identification and patient marking. **d** Note new video and audio technology (Ambient Experience) on the ceilings and walls installed at the UC Davis facility to create a more comfortable experience for patients (Philips Healthcare, Andover, Mass., USA).

CT gantry bore larger than the currently available 85- to 90-cm diameter, which limits some treatment setups, will likely be developed. The CT reconstruction size will also be made larger (currently only a 60-cm diameter is provided). Continued improvements in the CT simulator couch will occur to more accurately mimic the geometry of the treatment delivery system. Patient registration and motion management systems will continue to evolve and become more integrated into the CT simulation process.

However, the most significant development now occurring is the continued integration of functional imaging with anatomical imaging in the simulation/planning process. Already there is a commercially available magnetic resonance (MR) simulator (fig. 4), and a large-bore positron emission tomography (PET)-CT simulator (fig. 5), both with features specifically designed for radiation oncology.

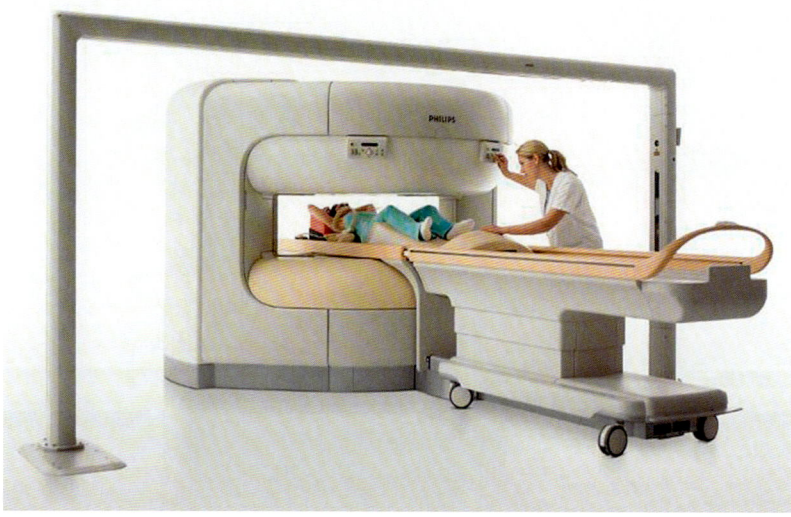

Fig. 4. Modern-day MR simulator (Panorama R/T) with open design, 1.5T system with rigid oncology table top mimicking linac table and a set of MR compatible immobilization devices allows for patient positioning and scanning in treatment position. Courtesy of Philips Healthcare.

While not yet commonplace in radiation oncology departments, it is very likely that multimodality imaging simulators and advanced simulation systems will become available within the next decade and include full 4D capabilities to address cases involving significant internal motion.

Intensity-Modulated Radiation Therapy Plan Optimization

IMRT treatment planning systems are designed to generate plans that meet the physician-specified constraints for the target volume coverage and the organs at risk. To achieve these dose distributions, nearly all planning systems use optimization engines with *dose-* and/or *dose-volume*-based objective functions. Efforts continue, though very slowly, to employ *dose-response* models, such as tumor control probabilities, normal tissue complication probabilities and equivalent uniform dose in the optimization process [22, 23]. This is an area of opportunity for researchers to improve the IMRT/IGRT planning process.

A more practical matter of concern is the slow progress made to establish national and/or international guidelines for dose prescription, planning and reporting of results for patients planned and treated with IMRT. This problem was highlighted by the publications by Das et al. [24] and Willins and Kachnic [25], who pointed out the large variability in IMRT planning and reporting among institutions.

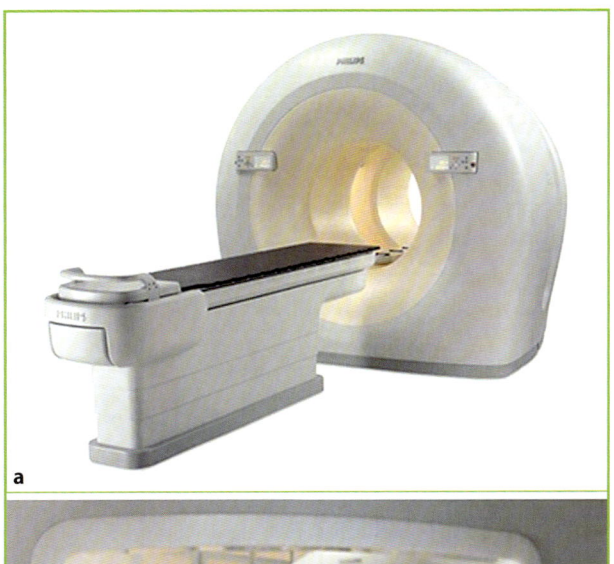
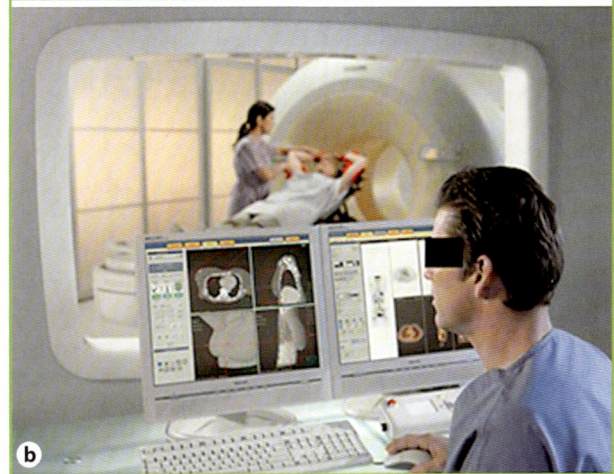

Fig. 5. a Modern day PET-CT simulator (GEMINI TF Big Bore PET/CT) with full 85-cm bore diameter for both CT and PET, thus allowing patients to be scanned with their immobilization devices in the treatment position. **b** Software tools for respiratory motion assessment, tumor localization for isocenter identification and patient marking provided with the simulator. Courtesy of Philips Healthcare.

In 2009, the American Society of Radiation Oncology recommended that specific details of the inverse treatment planning and image-guided treatment processes be recorded using (1) an IMRT treatment planning directive, (2) a treatment goal summary, (3) an image guidance summary and (4) a motion management summary [26]. Radiation oncology electronic medical record manufacturers (e.g. Elekta-Impac MOSAIQ and Varian ARIA) should quickly incorporate these templates into their user interfaces, allowing physicians to enter their IMRT prescriptions in a more robust and unambiguous manner. In addition, ICRU Report 83 [27] is now in press and hopefully will provide clear guidelines for dose prescription, planning and reporting of results for patients planned and treated with IMRT just as Report 50 did for 3DCRT.

Finally, it should be recognized that we still do not have tools that allow a physician to easily determine the 'best' plan when given a choice. Current technology

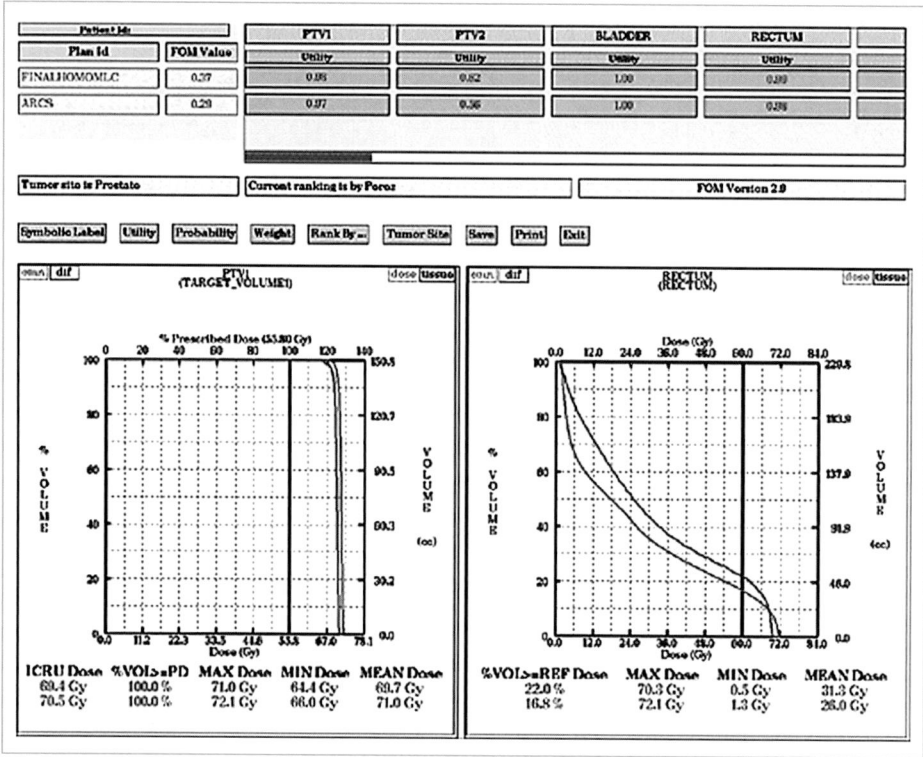

Fig. 6. Representative screen display of the FOM tool treatment plan for ranking competing plans for the same patient. In the upper left hand corner, indicators for 2 plans ranked by their FOM scores are shown. By scanning across the various plan issues at the top, the physician can also evaluate the merit of the plan on an issue-by-issue (organ at risk and target volumes) basis.

relies heavily on subjective evaluation by the physician using treatment planning system (TPS) displays showing dose-volume histograms, isodose lines or color washes superimposed on multiple gray scale CT images and/or 3D isodose clouds. Hence, IMRT planning today remains more of an art, by which the physician/planner adjusts the constraints and uses planning structures to drive the dose distribution decision toward one finally deemed acceptable.

The planning process could be greatly improved, I believe, if the TPS manufacturers provided a software tool similar to one we developed in the 3DCRT era called the Figure-of-Merit (FOM; fig. 6) [28]. The FOM tool is a generic plan-ranking model that can be refined to form a user-specific plan-ranking model. This refinement is accomplished by incorporating the treatment preferences of individual radiation oncologists and the clinical features of patients. A tool like FOM would likely improve the efficiency and quality of radiation therapy planning.

Dose Calculation

Advanced 3D dose calculation algorithms, such as the convolution/superposition algorithm and Monte Carlo, should now be considered the standard of practice [29]. These models provide accurate results even for complex heterogeneous geometries.

Unfortunately, some clinical practices and even cooperative groups have continued to ignore tissue heterogeneity effects and use archaic dose calculation algorithms, such as Bentley-Milan, Clarkson and pencil beam methodologies. The study by Frank et al. [30] provides a clear methodology for safely transitioning from planning using a homogeneous patient model to one using a heterogeneous patient model. We should work to ensure that the new level of accuracy provided by these advanced algorithms becomes the accepted standard worldwide.

Intensity-Modulated Radiation Therapy Delivery: New Technologies

Today, the majority of IMRT treatments are delivered with a traditional multimodality linac, equipped with a MLC originally designed for shaping field apertures but now greatly improved, and a computer control system. This form of IMRT is delivered at fixed gantry angles by (1) delivering multiple field segments (called segmental MLC or step-and-shoot IMRT) or (2) having the leaf pairs move across the field at a varying rate with the X-ray beam on (called dynamic MLC or sliding window IMRT) [7]. In the past decade, there have been continued engineering improvements in the various MLC systems, especially reducing leaf widths, to improve plan quality and improve reliability [31]. This is likely to continue over the next several years until the MLC is fully integrated into the treatment head design (no longer being a tertiary system, but a jaw replacement), with increased head shielding and thicker leafs to reduce transmission and lower the leakage radiation dose.

This past decade also saw the development of 2 specialized IMRT systems with integrated planning and image guided delivery. The TomoTherapy HI-ART System (TomoTherapy Inc, Madison, Wisc., USA) is a 6-MV (megavoltage) linac rotational approach using a binary collimator, based on the concept first proposed by Mackie et al. [32]. Another unique IGRT system is the CyberKnifeTM (Accuray Inc., Sunnyvale, Calif., USA) which consists of a 6-MV x-band linear accelerator mounted on a robotic arm which moves along predetermined, nonisocentric paths (or trajectories) around the patient [33]. Both of these IGRT systems have demonstrated the ability to deliver highly conformal dose distributions.

Most recently, manufacturers of medical linacs and their associated planning systems have introduced products for rotational IMRT, e.g. Elekta VMAT and Varian RapidArc [34]. The linac-based rotational IMRT concept was first pro-

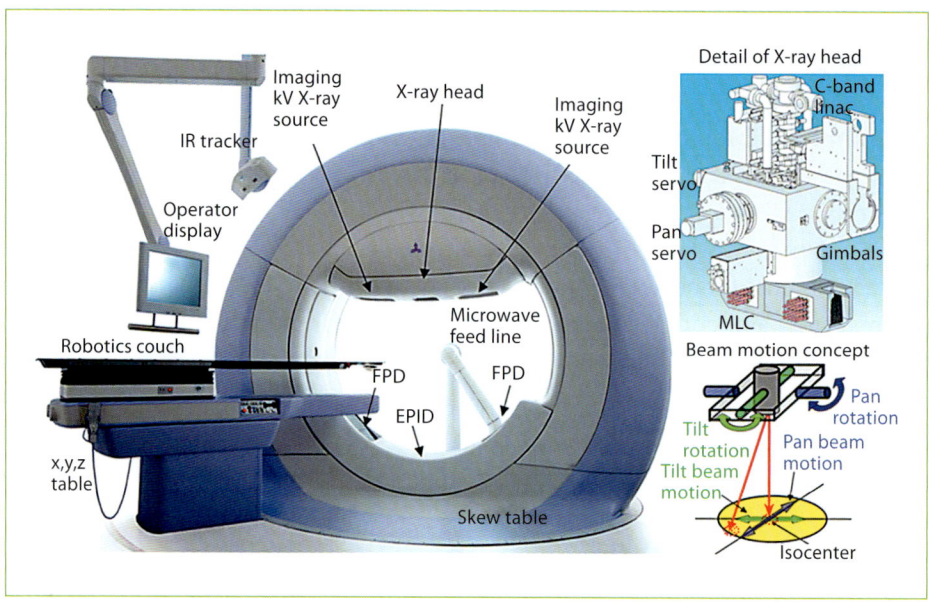

Fig. 7. Medical linac with gimbaled X-ray head mounted on an O-ring with 2 kV X-ray tubes, 2 flat panel detectors (FPDs), and an electronic portal imaging device (EPID). The O-ring can be rotated 360° around the isocenter and can be skewed +60° around its vertical axis. The couch has motion in x, y and z. The X-ray head has a MLC to control dose distribution. The pan and tilt rotations provide a precise and quick beam motion around the isocenter. Courtesy of Yuichiro Kamino, PhD.

posed by Yu [35] and called intensity-modulated arc therapy. Rotational IMRT approaches on conventional linacs may provide more conformal dose distributions, compared with segmental MLC- or dynamic MLC-IMRT approaches that use only a limited number of gantry directions. In addition, plan optimization is simpler since it eliminates the planner's iterative choices of beam number and direction. Compared with tomotherapy, the use of a conventional MLC for rotational IMRT is likely to improve delivery efficiency significantly [34], although this remains somewhat controversial at present [36, 37] and more users will need to report their intensity-modulated arc therapy experiences over the next few years.

There are several other image-guided IMRT approaches that also show significant promise. For example, the 4D IGRT system proposed by Kamino et al. [38] and shown in figure 7 has a unique, gimbaled X-ray head that allows the linear accelerator head to be pivoted. By easily allowing non-coplanar beams without couch rotations, new degrees of freedom are made available for IMRT optimization, and even more conformal dose distributions may be possible.

Another highly promising image-guided IMRT delivery system on the horizon, which has significant potential to improve our abilities to localize and track soft

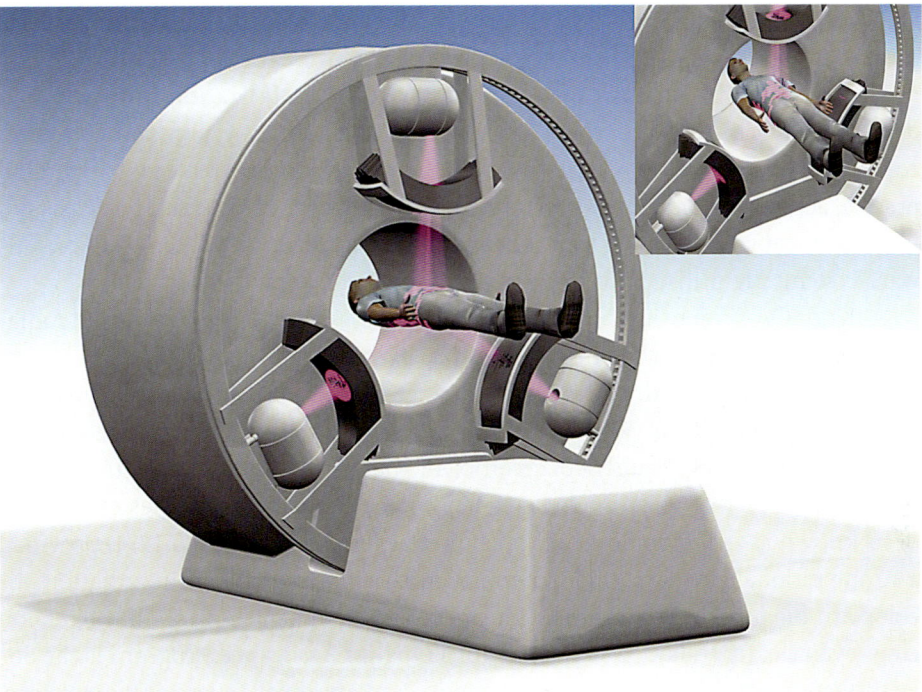

Fig. 8. Rendering of integrated MR-Co-60 IMRT system (RenaissanceTM System 1000) that will provide real-time beam-on imaging and targeting of the tumor and IMRT treatment delivery. Courtesy of ViewRay.

tissue tumors, is the RenaissanceTM System 1000 (ViewRay Inc., Cleveland, Ohio, USA). Depicted in figure 8, this is a multiheaded cobalt-60 rotational IMRT system with MR image guidance and the ability to image continually during treatment even while the beam is on. In addition, MRI-linac hybrid systems continue to be an important area of development [39].

Intensity-Modulated Radiation Therapy Physics Quality Assurance

Today's treatment planning and delivery processes have become much more complex and much less intuitive. These complexities, coupled with the inadequate informatics infrastructure (discussed in a later section) and outdated national QA guidelines, have created enormous QA challenges for modern radiation therapy. Communication among the planning team, including the physician, resident, simulation therapist, dosimetrist, physicist and treating therapist, is often cryptic and complicated by the hybrid charting systems (electronic and paper medical records) now commonly used. Multiple imaging modalities are often used to define target volumes, and 4D imaging is emerging in routine practice, all complicating the information processing further. In a typical clinic, over 40% of patients are now managed with IMRT plans using computer-assisted optimization software. Plan data, transferred over a network to the verify and record system and computer controlled linac systems, carry complex specifications for treatment including precise, accurate positioning of MLC leafs, variable dose rates, gantry angles and, in some instances, a moving treatment table. In many cases, different vendor software systems are utilized, and interoperability is not always robust.

A major QA symposium was held in 2007 to assess physical QA issues in radiation therapy [40], though little progress has been made in the ensuing 3 years to address the issues raised. With newer advanced treatment modalities continuing to be implemented, e.g. VMAT and ART, updated national QA guidelines/approaches and the development of an efficient and robust IMRT/IGRT QA verification system must be given higher priority by the profession. A series of publications known as the 'Blue Book', published over the period 1968 to 1991, provided a strong rationale for the development, purpose and need for QA in radiation oncology in the pre-3DCRT era [41]. The following statement from the 1991 version rings even clearer today: 'The purpose of a Quality Assurance Program is the objective, systematic monitoring of the quality and appropriateness of patient care. Such a program is essential for all activities in Radiation Oncology'.

The greater complexity of radiation therapy today requires a far stronger teamwork approach since no individual has all of the skills necessary to insure its maximum quality. Unfortunately, in my opinion, we are gradually losing control over important quality aspects of radiation oncology to hospital and financial administrators who may not fully understand the complexity of radiation therapy and may focus on cost-cutting efficiencies. I strongly believe that the American Society of Radiation Oncology and AAPM should return to the 'Blue Book' concept and publish a version current to our ongoing work. This should emphasize the needs for (1) metrics for proper staffing of clinics that provide IMRT, SBRT, SRS, brachytherapy and other advanced modalities, (2) proper staffing to allow continuing education and focused efforts in error prevention, and (3) continued assessment and updating of QA methodologies and procedures. Such a document would be a powerful tool in negotiating with administrators the funding needed to support QA adequately.

Image-Guided Radiation Therapy: Still Many Questions

While advanced IGRT treatment machines are now installed in many facilities, few sites have taken the steps to use their IGRT data to determine PTV margins specific for their clinic's work. These values are readily available from each clinic's IGRT practice data using well-established methodologies [42]. Instead, many (if not most) clinics continue to use margins that are literature-based. This first step in using IGRT data is an opportunity for many institutions to directly improve the quality of their practices. Beyond this, many practical questions need answering to establish the optimum clinical use of IGRT, including:

- What is the optimal use frequency of IGRT for specific disease sites? Indeed, daily imaging may be necessary, depending on the size of the PTV margin used.
- What type of IGRT [e.g. cone-beam CT (CBCT) vs. planar] is optimal for specific disease sites? In some cases, planar imaging may be all that is required.
- What is the additional radiation dose to the patient from IGRT? While there have been published reports on this, it is not routinely accounted for or documented in the individual treatment. Doses can be significant, typically on the order of 3 cGy per scan, but doses can be higher depending on the machine settings used. This increased dose, combined with the increase in leakage dose from IMRT, may prove problematic years from now with regard to secondary malignancies. The NCRP has recently formed a committee (SC 1–17: Second Cancers and Cardiopulmonary Effects after Radiotherapy) reviewing this matter. I believe it is important that the radiation-oncology community take the lead on this issue, and, at a minimum, ensure that there is adequate documentation in the patient's record to allow these doses to be accurately estimated at some future time. Also, treatment machine manufacturers should take steps now to reduce the dose from imaging and IMRT leakage.
- What are the most effective periodic technical QA methodologies and patient-specific QA methodologies, including weekly chart rounds, for IGRT? I do not think that our profession's QA programs are sufficiently developed for IMRT/IGRT.
- Does IGRT improve clinical outcomes? It is important that we acquire and analyze the data for evidenced-based medicine, though we may intuitively understand the benefit of our work.

Adaptive Radiotherapy: Understanding and Adjusting for Variation in Daily Organ Dose Distributions

Internal organ motion/deformation and daily setup errors can cause temporal variations in dose distributions through an organ. For each subvolume of the organ, these variations in dose with each fraction create a cumulative dose variation [43] (fig. 9). IGRT treatment machines can now obtain volumetric CT data sets that more accurately reflect these daily variations in patient anatomy and target volume configurations. In most cases, these data are used only to realign the patient at present. However, 3D deformable image registration algorithms are becoming available on advanced image processing workstations in radiation oncol-

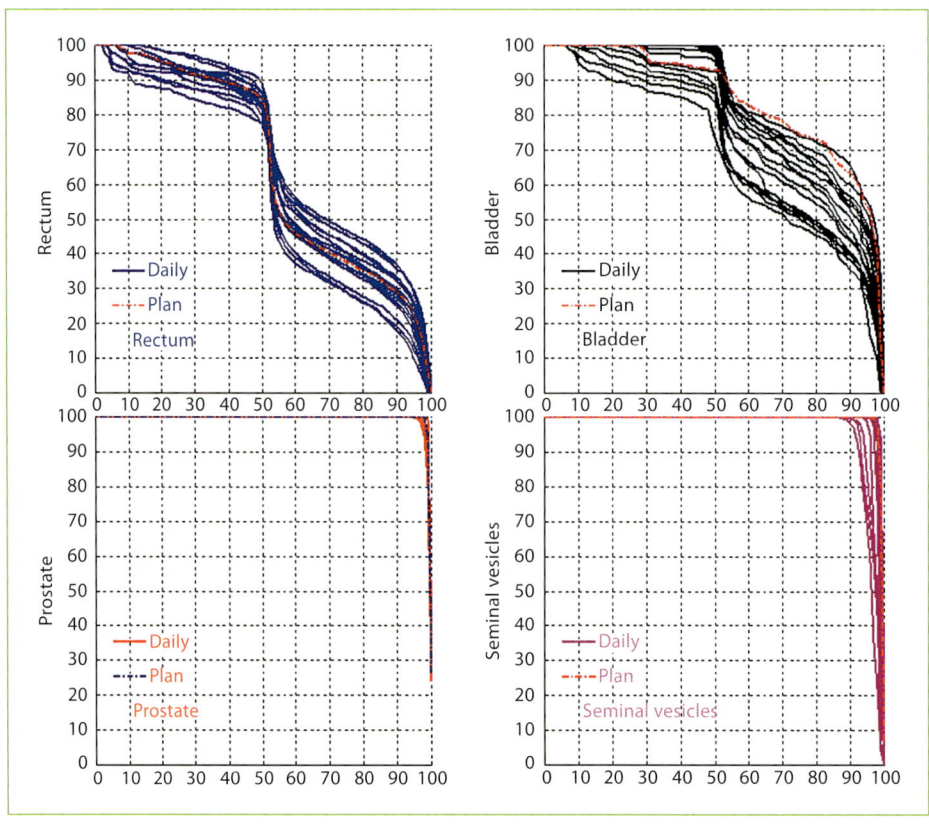

Fig. 9. Dose-volume histograms illustrating effect of temporal variation in organ dose distribution, consisting of both dose variation per treatment fraction and cumulative dose variation in each subvolume of organ. Shown are daily dose-volume histograms for the prostate cancer patient's rectum, bladder, prostate and seminal vesicles. Data includes both internal organ (bladder wall, rectal wall, prostate and seminal vesicles) motion and daily setup error. These types of data clearly point to the need for robust deformable registration tools to account for such temporal variations in organ dose distribution. Courtesy of Di Yan, DSc.

ogy. Major improvements in their accuracy and speed are needed before such tools can be used widely, but robust systems will likely become available within the decade. The daily images can then be used both to reposition the patient and to determine the daily accumulation of doses to the target volumes and organs at risk more accurately. Thus, replanning of the radiation therapy could be performed at various times (daily, weekly) to incorporate the impact of these variations in the dose distributions. Appropriate changes could be made in the delivery of the radiation therapy based on the actual doses already delivered to the patient. However, more research and development of robust deformable registration tools are still needed to make ART practical.

Proton and Heavy Ion Radiation Therapy

There is increasing worldwide interest in proton and heavy ion radiation beams because their depth-dose characteristics show advantages over photon beams [44]. In most proton facilities, cyclotrons are used to accelerate proton beams to sufficient energy (200–250 MeV) and beam intensity. Cyclotrons produce continuous beams with fixed energy, which makes their design quite simple. Beam-spreading mechanisms obtain suitable field sizes for radiation therapy by passive modulation (scattering foil) systems or by dynamic pencil beam spot-scanning systems [45], which allow dose conformation not only at the distal edge of the tumor, but also at the proximal edge. There are now 5 major proton facilities in the USA, and 5 more scheduled to open over the next 2 years – 10 facilities in operation in this country by 2012.

There are fewer clinical facilities for heavy ion radiation therapy, primarily carbon ion, which are currently active (in Japan, Germany and China) [44]. Heavy ion radiation therapy requires beams at much higher energy than proton therapy. For example, a proton beam of 150 MeV can penetrate 16 cm in water. To achieve the same penetration with carbon ions, energy of 3,000 MeV or 250 MeV/u (energy per nucleon) is needed. Currently, synchrotrons are the only available sources for high-energy heavy ion beams, and are large, complex machines requiring large facilities and similarly large capital expenditures. Interest in heavy ion radiation therapy is due to its combination of 2 important physical advantages: (1) its depth dose characteristics and (2) high linear energy transfer in the Bragg peak region of the beam. For its depth characteristics, the ratio of the Bragg peak dose to the entrance region dose is even larger than for protons. In addition, unlike proton beam therapy, there is a large increase in the radiation linear energy transfer in the Bragg peak region of the beam. The combination of these two characteristics results in a potentially unique advantage: a high linear energy transfer region that can be closely conformed to the target volume. It is likely that a heavy ion clinical facility will be built in the USA in the next decade, but these programs are more likely to remain in the realm of promising research endeavors than mainstream radiation therapy systems for the foreseeable future.

New technologies for the delivery of proton beam radiation therapy (PBRT) are now emerging. Already, single gantry systems are being designed and manufactured, including the Monarch 250TM PBRT system (Still River Systems, Littleton, Mass., USA) shown in figure 10 and scheduled to be installed at Washington University-St. Louis in 2010. Recently, Varian announced their single room PBRT system (not yet FDA-approved), which uses superconducting cyclotron technology and can be expanded into a multiroom facility when needed.

A more futuristic design concept, depicted in figure 11, has been reported by Caporaso et al. [46]. This system is being codeveloped by the Compact Particle

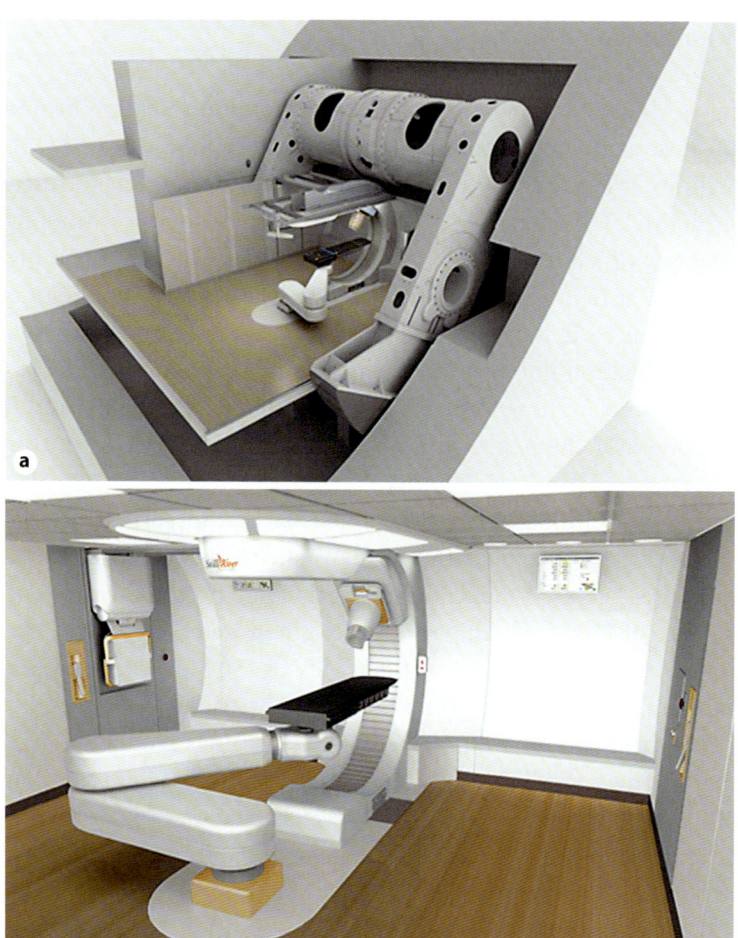

Fig. 10. Rendering of next generation proton therapy system (Monarch 250TM PBRT System). **a** The outer gantry holds the proton accelerator in the center, pointing directly to the isocenter. The accelerator is a 250-MeV cyclotron specifically designed (uses superconducting magnets) for proton therapy (with IMPT capabilities, high dose rate, high reliability and easy maintenance). **b** Treatment room depicting the inner gantry and 6-degree-of-freedom robotic couch. Courtesy of Still River Systems.

Acceleration Corp. and TomoTherapy Inc. It uses a dielectric wall accelerator technology (developed by the Lawrence Livermore National Laboratory [46]) which can produce individual pulses that can be varied in intensity, energy and spot width. While this is still largely at a research level of development, concept designs are already underway to develop a compact, CT-guided PBRT system that can provide rotational IMPT and can be installed in a treatment room not much larger than for a conventional linac.

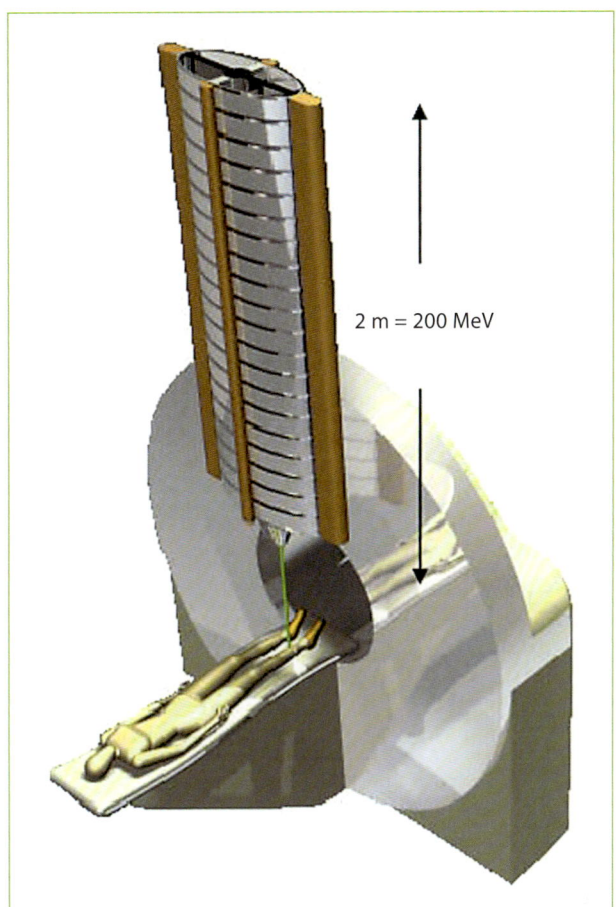

Fig. 11. Artist's rendition of a possible reduced size proton therapy system being codeveloped by the Compact Particle Acceleration Corp. and TomoTherapy using the dielectric wall accelerator technology developed by the Lawrence Livermore National Laboratory [46].

These are exciting developments, and are only the beginnings in expanding approaches for providing PBRT. I firmly believe that new proton beam technologies will be coming forth, and costs for units capable of IMPT will decrease, creating opportunities for their mainstream use.

Radiation Dose and Second Cancers

The additional dose given to the patient from IMRT and IGRT is an important issue for our profession, but we have done little to reduce this dose in practice. As discussed by Hall and Wuu [47], the transition from 2DRT to 3DCRT and IMRT has resulted in an increase in dose to the target volume, but an overall reduction in the volume of normal tissues receiving high doses, since tumors are more precisely treated [47]. This precision could lead to a decrease in the number of sarcomas induced, and a possible small decrease in the number of carcinomas as well. However, both Followill et al. [48, 49] and Hall and Wuu [47] express concern that the move from 2DRT to 3DCRT and IMRT may result in an increased rate of secondary malignancies because of the larger volumes of normal tissue being irradiated at lower doses. Also, compared with 2DRT and 3DCRT, IMRT requires a significantly larger number of monitor units to deliver a comparable prescribed dose. Typically, this monitor unit increase is by a factor of 2–3, but can be as large as 8–10 for serial tomotherapy. This results in an increase in the dose outside the boundary of the primary collimator as a result of leakage and scattered radiation [48–50]. Hence, the total body dose received is substantially increased.

Hall and Wuu [47] estimated that IMRT has the potential to nearly double the incidence of second malignancies compared with 2DRT – from about 1 to 1.75% for patients surviving 10 years. These numbers may be even larger for younger patients and longer survivals, though the ratio of increased incidence should remain the same. This unwanted dose can be decreased by technical redesigns of the treatment units, including increased shielding in the primary collimator and treatment head, and removal of the flattening filter [32]. In the treatment planning process, clinicians can also reduce exposure doses to sensitive organs by delineating normal structures with higher potentials for developing second malignancies, such as the thyroid and breasts, and using the optimization processes to limit these doses.

Dose due to Daily Imaging

While the patient dose from a single IGRT volumetric imaging acquisition is small, compared with the therapy dose, daily image guidance procedures can accumulate to potentially significant dose levels for normal tissues. Perks et al. [51] showed that peripheral dose contributed by daily CBCT imaging can be on the same order of magnitude as the IMRT peripheral doses. The dose to the patient from a CBCT depends significantly on the scanning parameters used. Additionally, the dose across the patient may vary, depending on scanning technique (i.e. some peripheral areas of the patient will be irradiated only during portions of a 360 degree scan, while the more central regions will be irradiated at all times). Islam et al. [52] reported on the dosimetric properties of the clinical CBCT system on an Elekta linear accelerator (Synergy®, XVI System). The dose at the center and surface of a body phantom were determined to be 1.6 and 2.3 cGy per scan for a typical imaging protocol. For the MV helical CT of the TomoTherapy HI-ART System, Forrest et al. [53] reported that images can be obtained at a dose level of about 3 cGy per scan.

Radiation Oncology Informatics

Many recent advances, including extensive laboratory investigations and technological developments, can be the basis for further improvements in radiation treatment and patient outcomes. These many advances are also providing vast amounts of data. Difficulty in accessing the available data in practical ways has become a critical limitation to investigators in our field and to its advancement. Unfortunately, the commercially available radiation oncology informatics infrastructures are not yet adequate to meet this challenge. The great volume and diversity of the data have rendered their storage, management and processing extremely problematic, resulting in real-world situations in which important data and information are inefficiently disseminated and sometimes lost. In the busy clinic, failure to manage critical information may compromise patient safety. Here, a major issue is the lack of integration of the radiation oncology electronic medical record, based in systems such as IMPAC MOSAIQ or Varian ARIA, with the electronic hospital record, based in systems such as EPIC. In most cases, we must resort to hybrid charting systems (ad hoc combinations of the electronic medical record, the electronic hospital record and paper charting), which is a dangerous situation. Immediate, focused efforts in achieving such integration and improving the radiation oncology informatics infrastructures are urgently needed.

The challenge to manage this diverse, increased data volume is also important for cooperative groups involved with radiation oncology standards, which are charged with producing clear, high-quality information to help guide community clinical practice. The Advanced Technology Consortium for Clinical Trials QA (ATC; http://atc.wustl.edu/), a consortium of QA centers in the USA, is striving to stay abreast of these challenges. The ATC has strongly advocated the use of advanced medical informatics to facilitate education, collaboration and QA review, providing an environment in which clinical investigators can receive, share and analyze volumetric multimodality treatment planning and verification digital data. It is currently using the paradigm first established by the pioneering efforts of the Washington University 3D QA Center, which is now referred to as the Image-Guided Therapy QA Center (ITC), and the Radiation Therapy Oncology Group (RTOG), which partnered to conduct RTOG 94–06, a prostate 3DCRT dose escalation study [54, 55]. To date, the ITC has archived over 10,000 protocol case digital data sets for multiple protocols and now provides support to several clinical trial groups in the USA, Europe and Japan. It should be noted that while we have made significant headway in digital data transfer and web-based remote QA review, we have thus far failed to truly harmonize among the various cooperative groups/QA centers in the US all the credentialing/QA tests which institutions must pass for participation in clinical trials. This difference is even more pronounced when we compare US credentialing requirements with those of other countries. I strongly believe that establishing common worldwide credentialing/QA methodology/criteria for clinical trials should be given the highest priority at this time.

It should be noted that in the 2007 edition of this series, I stated that 'the development of a robust radiation oncology PACS will be one of the most important developments for radiation oncology, as the use of information technology will be "mission critical" in order to make radiation oncology more effective and efficient.' Unfortunately, we have made little progress in developing an RO-PACS since then. The systems that are in place provide little capability for improving workflow efficiencies or real data mining. The systems remain mostly a large 'info file'. This issue remains a major challenge for our profession, and thus there is a great opportunity in this area. I think one of our mistakes was using the term 'PACS' because that is exactly the approach that has been taken – basically a big storage system that gives one viewing capabilities, but little else.

Obviously, we need much more than that. We need a system that is able to link all of the treatment planning data objects, electronic medical record data objects (e.g. MOSAIQ), the patient's biological data (if any is obtained) and the associated outcomes database. This would create a discovery data warehouse (see fig. 12) that would allow clinic workflow and the QA process to be much better managed, and would provide an enormous amount of data and data mining tools for research purposes.

Hence, I modify slightly what I said in 2007 dropping the reference to PACS, 'the development of a robust radiation oncology informatics infrastructure providing workflow efficiency tools and data mining tools will be one of the most important developments going forward for radiation oncology, and most importantly, for the cancer patient.'

Conclusions

The practice of radiation therapy continues to experience rapid advancements in treatment planning and delivery technologies that have the potential to significantly improve treatment outcomes. Advanced IMRT/IGRT systems are now available that allow physicians to significantly increase tumor doses and/or lower doses to normal tissues. Certainly, we will continue to see imaging systems using CT, US, MR, PET and perhaps other modalities fully integrated into radiation treatment delivery systems. However, further improvements in IMRT plan quality may need new approaches. Some are already on the horizon, including a gimbaled 4D IG/IMRT system, integrated MR imaging/IMRT delivery systems and new proton therapy technology.

Advances in planning and delivery technologies will continue to occur at record paces and push the field toward even higher expectations for radiotherapy

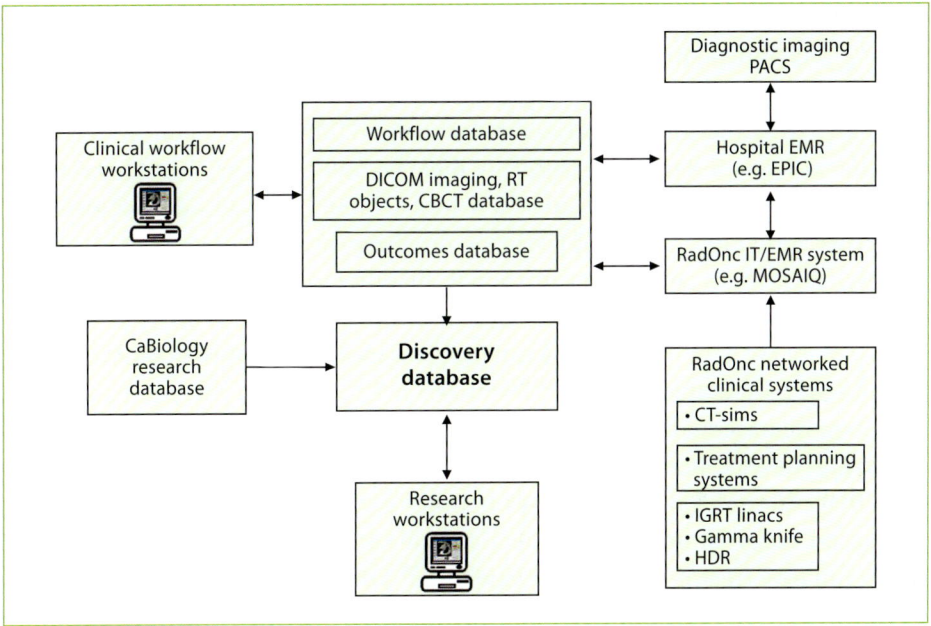

Fig. 12. Block diagram depicting a proposed radiation oncology informatics infrastructure for improving workflow efficiencies, QA processes and research (basic, translation and comparative effectiveness).

accuracy and reliability, all leading to new standards of care. However, this optimism must be tempered with the realization that, for this to happen, progress is urgently needed in 3 areas, (1) accuracy in specification of GTV/CTV, (2) radiation oncology informatics and (3) QA, if we are to keep pace with these rapid planning and delivery developments.

With reference to the first issue, advances in both anatomical and functional imaging provide improved ability to define the GTVs and organs at risk. However, there continue to be wide variations among physicians in delineating GTVs for many disease sites and we have made little progress in improving accuracy specifying CTVs.

Regarding the second issue, radiation oncology has seen an enormous increase and diversity in data, data types and information in this decade. The lack of integration of the electronic hospital record with the radiation oncology electronic medical record has become a major problem. Hence, a much more capable informatics infrastructure is needed to allow clinic workflow to be much better managed for improved QA and to provide data mining tools for the enormous amount of information that could be made available for research purposes.

Concerning the third issue, it is essential that updated national QA guidelines be put in place along with the development of more efficient and robust IMRT/

IGRT QA verification systems. In addition, the administrative framework and staffing recommendations described by the 'Blue Book', last published in 1991, must be updated now by the pertinent national organizations.

Finally, I remind the reader of what I wrote in 2000 for an editorial in the *International Journal Radiation Oncology Biology Physics,* suggesting where radiation oncology may be 2 or 3 decades in the future [56]: 'By 2035, advances in technology will provide a systems approach to radiation oncology treatment planning, dose delivery, and treatment verification. Capability to accurately define volumes containing both gross disease and subclinical disease will be available.' Over the past 10 years, we have made significant progress in some areas. In others, e.g. CTV definitions, radiation oncology informatics, QA/treatment verification and other concerns, we have made little progress and must increase the intensity of the research and development efforts.

In that editorial, I also wrote 'I strongly believe that this next generation of radiation oncology clinicians and scientists have a unique opportunity to significantly improve treatment outcomes and lower costs thus making high quality radiation therapy available the world over.' I continue to stand by that statement.

References

1 Purdy JA: 3-D radiation treatment planning: a new era: in Meyer JL, Purdy JA (eds): 3-D Conformal Radiotherapy: A New Era in the Irradiation of Cancer. Front Radiat Ther Oncol. Basel, Karger, 1996, pp 1–16.
2 Purdy JA: From new frontiers to new standards of practice: advances in radiotherapy planning and delivery: in Meyer JL (ed): IMRT, IGRT, SBRT – Advances in the Treatment Planning and Delivery of Radiotherapy. Front Radiat Ther Oncol. Basel, Karger, 2007, pp 18–39.
3 Webb S: Intensity-Modulated Radiation Therapy. Bristol, Institute of Physics Publishing, 2000.
4 Jaffray DA, Siewerdsen JH, Wong JW, Martinez AA: Flat-panel cone-beam computed tomography for image-guided radiation therapy. Int J Radiat Oncol Biol Phys 2002;53:1337–1349.
5 Bortfeld T, Schmidt-Ullrich R, De Neve W, Wazer DE: Image-Guided IMRT. Berlin, Springer, 2006.
6 Martinez AA, Yan D, Lockman D, Brabbins D, Kota K, Sharpe M, Jaffray DA, Vicini F, Wong J: Improvement in dose escalation using the process of adaptive radiotherapy combined with three-dimensional conformal or intensity-modulated beams for prostate cancer. Int J Radiat Oncol Biol Phys 2001;50:1226–1234.
7 Intensity Modulated Radiation Therapy Collaborative Working Group: Intensity-modulated radiation therapy: current status and issues of interest. Int J Radiat Oncol Biol Phys 2001;51: 880–914.
8 Schulz RJ, Verellen DL, Orton CG: Point/Counterpoint. Future developments in external beam radiotherapy will be unlikely to significantly improve treatment outcomes over those currently achieved with 3D-conformal and IMRT treatments. Med Phys 2007;34:3123–3126.
9 International Commission on Radiation Units and Measurements: ICRU Report 50: Prescribing, Recording, and Reporting Photon Beam Therapy. Bethesda, ICRU, 1993.
10 International Commission on Radiation Units and Measurements: ICRU Report 62: Prescribing, Recording, and Reporting Photon Beam Therapy (Supplement to ICRU Report 50). Bethesda, ICRU, 1999.
11 Gregoire V, Coche E, Cosnard G, Hamoir M, Reychler H: Selection and delineation of lymph node target volumes in head and neck conformal radiotherapy: proposal for standardizing terminology and procedure based on the surgical experience. Radiother Oncol 2000;56:135–150.

12 Martinez-Monge R, Fernades PS, Gupta N, Gahbauer R: Cross-sectional nodal atlas: a tool for the definition of clinical target volumes in three-dimensional radiation therapy planning. Radiol 1999;211:815–828.
13 Kupelian P, Willoughby T, Mahadevan A, Djemil T, Weinstein G, Jani S, Enke C, Solberg T, Flores N, Liu D, Beyer D, Levine L: Multi-institutional clinical experience with the Calypso System in localization and continuous, real-time monitoring of the prostate gland during external radiotherapy. Int J Radiat Oncol Biol Phys 2007;67: 1088–1098.
14 Willoughby TR, Kupelian PA, Pouliot J, Shinohara K, Aubin M, Roach M 3rd, Skrumeda LL, Balter JM, Litzenberg DW, Hadley SW, Wei JT, Sandler HM: Target localization and real-time tracking using the Calypso 4D localization system in patients with localized prostate cancer. Int J Radiat Oncol Biol Phys 2006;65(2):528–534.
15 Bert C, Metheany KG, Doppke KP, Taghian AG, Powell SN, Chen GTY: Clinical experience with a 3D surface patient setup system for alignment of partial-breast irradiation patients. Int J Radiat Oncol Biol Phys 2006;64:1265–1274.
16 Sharpe MB, Miller BM, Yan D, Wong JW: Monitor unit settings for intensity modulated beams delivered using a step-and-shoot approach. Med Phys 2000;27:2719–2725.
17 Mu G, Ludlum E, Xia P: Impact of MLC leaf position errors on simple and complex IMRT plans for head and neck cancer. Phys Med Biol 2008;53: 77–88.
18 Rangel A, Dunscombe P: Tolerances on MLC leaf position accuracy for IMRT delivery with a dynamic MLC. Med Phys 2009;36:3304–3309.
19 Klein EE, Hanley J, Bayouth J, Yin F-F, Simon W, Dresser S, Serago C, Aguirre F, Ma L, Arjomandy B, Liu C, Sandin C, Holmes T: Task Group 142 report: quality assurance of medical accelerators. Med Phys 2009;36:4197–4212.
20 Bortfeld T, Jokivarsi K, Goitein M, Kung J, Jiang SB: Effects of intra-fraction motion on IMRT dose delivery: statistical analysis and simulation. Phys Med Biol 2002;47:2203–2220.
21 Yu CX, Jaffray DA, Wong JW: The effects of intra-fraction organ motion on the delivery of dynamic intensity modulation. Phys Med Biol 1998;43: 91–104.
22 Niemierko A: A generalized concept of equivalent uniform dose (EUD) (abstract). Med Phys 1999;26:1100.
23 Wu Q, Mohan R, Niemierko A, Schmidt-Ullrich R: Optimization of intensity-modulated radiotherapy plans based on the equivalent uniform dose. Int J Radiat Oncol Biol Phys 2002;52:224–235.
24 Das IJ, Chang C-W, Chopra KL, Mitra RK, Srivastava SP, Glatstein E: Intensity-modulated radiation therapy dose prescription, recording, and delivery: patterns of variability among institutions and treatment planning systems. J Natl Cancer Inst 2008;100:300–307.
25 Willins J, Kachnic L: Clinically relevant standards for intensity-modulated radiation therapy dose prescription. J Natl Cancer Inst 2008;100: 288–290.
26 Holmes T, Das R, Low D, Yin F-F, Balter J, Palta J, Eifel P: American Society of Radiation Oncology recommendations for documenting intensity-modulated radiation therapy treatments. Int J Radiat Oncol Biol Phys 2009;74:1311–1318.
27 International Commission on Radiation Units and Measurements: ICRU Report 83: Prescribing, Recording, and Reporting Intensity-Modulated Photon-Beam Therapy (IMRT). Bethesda, ICRU, in press.
28 Jain NL, Kahn MG, Drzymala RE, Emami B, Purdy JA: Objective evaluation of 3-D radiation treatment plans: a decision-analytic tool incorporating treatment preferences of radiation oncologists. Int J Radiat Oncol Biol Phys 1993;26: 321–333.
29 Ahnesjö A, Aspradakis MM: Dose calculations for external photon beams in radiotherapy. Phys Med Biol 1999;44:R99–R155.
30 Frank SJ, Forster KM, Stevens CW, Cox JD, Komaki R, Liao Z, Tucker S, Wang X, Steadham RE, Brooks C, Starkschall G: Treatment planning for lung cancer: traditional homogeneous point-dose prescription compared with heterogeneity-corrected dose-volume prescription. Int J Radiat Oncol Biol Phys 2003;56:1308–1318.
31 Zhang G, Jiang Z, Shepard D, Earl M, Yu C: Effect of beamlet step-size on IMRT plan quality. Med Phys 2005;32:3448–3454.
32 Mackie TR, Holmes T, Swerdloff S, Reckwerdt P, Deasy JO, Yang J, Paliwal B, Kinsella T: Tomotherapy: a new concept for the delivery of dynamic conformal radiotherapy. Med Phys 1993;20: 1709–1719.
33 Adler JR, Chang SD, Murphy MJ, Doty J, Geis P, Hancock SL: The CyberKnife: a frameless robotic system for radiosurgery. Stereotact Funct Neurosurg 1997;69:124–128.
34 Ling CC, Zhang P, Archambault Y, Bocanek J, Tang G, LoSasso T: Commissioning and quality assurance of RapidArc radiotherapy delivery system. Int J Radiat Oncol Biol Phys 2008;72:575–581.
35 Yu CX: Intensity modulated arc therapy with dynamic multileaf collimation: an alternative to tomotherapy. Phys Med Biol 1995;40:1435–1449.

36 Ling CC, Archambault Y, Bocanek J, Zhang P, LoSasso T, Tang G: Scylla and Charybdis: longer beam-on time or lesser conformality – the dilemma of tomotherapy. Int J Radiat Oncol Biol Phys 2009;75:8–9.
37 Mehta M, Hoban P, Mackie TR: Commissioning and quality assurance of RapidArc radiotherapy delivery system: in regard to Ling et al. (Int J Radiat Oncol Biol Phys 2008;72;575–581): absence of data does not constitute proof; the proof is in tasting the pudding. Int J Radiat Oncol Biol Phys 2009;75(1):4–6.
38 Kamino Y, Takayama K, Kokubo M, Narita Y, Hirai E, Kawawda N, Mizowaki T, Nagata Y, Nishidai T, Hiraoka M: Development of a four-dimensional image-guided radiotherapy system with a gimbaled X-ray head. Int J Radiat Oncol Biol Phys 2006;66:271–278.
39 Kirkby C, Stanescu T, Rathee S, Carlone M, Murray B, Fallone BG: Patient dosimetry for hybrid MRI-radiotherapy systems. Med Phys 2008;35:1019–1027.
40 Williamson JF, Dunscombe PB, Sharpe MB, Thomadsen BR, Purdy JA, Deye JA: Quality assurance needs for modern image-based radiotherapy: recommendations from 2007 inter-organizational symposium on 'quality assurance of radiation therapy: challenges of advanced technology'. Int J Radiat Oncol Biol Phys 2008; 71(Suppl 1):S2–S12.
41 ISCRO: Radiation Oncology in Integrated Cancer Management: Report of the Inter-Society Council for Radiation Oncology. Philadelphia, ISCRO, 1991.
42 van Herk M: Errors and margins in radiotherapy. Semin Radiat Oncol 2004;14:52–64.
43 Yan D, Xu B, Lockman D, Kota K, Brabbins DS, Wong J, Martinez AA: The influence of interpatient and intrapatient rectum variation on external beam treatment of prostate cancer. Int J Radiat Oncol Biol Phys 2001;51:1111–1119.
44 Schulz-Ertner D, Jakel O, Schlegel W: Radiation therapy with charged particles. Sem Radiat Oncol 2006;16:249–259.
45 Coutrakon G, Bauman M, Lesyna D, Miller D, Nusbaum J, Slater J, Johanning J, Miranda J, DeLuca PM Jr, Siebers J, Ludewigt B: A prototype beam delivery system for the proton medical accelerator at Loma Linda. Med Phys 1991;18:1093–1099.
46 Caporaso GJ, Mackie TR, Sampayan S, Chen YJ, Blackfield D, Harris J, Hawkins S, Holmes C, Nelson S, Paul A, Poole B, Rhodes M, Sanders D, Sullivan J, Wang L, Watson J, Reckwerdt PJ, Schmidt R, Pearson D, Flynn RW, Matthews D, Purdy JA: A compact linac for intensity modulated proton therapy based on a dielectric wall accelerator. Phys Med 2008;24:98–101.
47 Hall EJ, Wuu C-S: Radiation-induced second cancers: the impact of 3D-CRT and IMRT. Int J Radiat Oncol Biol Phys 2003;56:83–88.
48 Followill D, Geis P, Boyer A: Estimates of whole-body dose equivalent produced by beam intensity modulated conformal therapy. Int J Radiat Oncol Biol Phys 1997;38:667–672.
49 Followill D, Geis P, Boyer A: Errata: estimates of whole-body dose equivalent produced by beam intensity modulated conformal therapy. Int J Radiat Oncol Biol Phys 1997;39:783.
50 Williams P, Hounsell R: X-ray leakage considerations for IMRT (correspondence). Br J Radiol 2001;74:98–102.
51 Perks JR, Lehmann J, Chen AM, Yang CC, Stern RL, Purdy JA: Comparison of peripheral dose from image-guided radiation therapy (IGRT) using kV cone beam CT to intensity-modulated radiation therapy (IMRT). Radiother Oncol 2008; 89:304–310.
52 Islam MK, Purdie TG, Norrlinger BD, Alasti HM, Douglas J, Sharpe MB, Siewerdsen JH, Jaffray DA: Patient dose from kilovoltage cone beam computed tomography imaging in radiation therapy. Med Phys 2006;33:1573–1582.
53 Forrest LJ, Mackie TR, Ruchala K, Turek M, Kapatoes J, Jaradat H, Hui S, Balog J, Vail DM, Mehta MP: The utility of megavoltage computed tomography images from a helical tomotherapy system for setup verification purposes. Int J Radiat Oncol Biol Phys 2004;60:1639–1644.
54 Purdy JA, Harms WB, Michalski J, Cox JD: Multi-institutional clinical trials: 3-D conformal radiotherapy quality assurance; in Meyer JL, Purdy JA (eds): 3-D Conformal Radiotherapy: A New Era in the Irradiation of Cancer. Basel, Karger, 1996, pp 255–263.
55 Purdy JA, Harms WB, Michalski JM, Bosch WR: Initial experience with quality assurance of multi-Institutional 3D radiotherapy clinical trials. Strahlenther Onkol 1998;174(Suppl 2):40–42.
56 Purdy JA: Future directions in 3-D treatment planning and delivery: a physicist's perspective. Int J Radiat Oncol Biol Phys 2000;46:3–6.

Dr. James A. Purdy, PhD, Professor and Vice Chairman
Department of Radiation Oncology, UC Davis Medical Center
4501 X Street, Suite G140, Sacramento, CA 95816 (USA)
Tel. +1 916 734 3932, Fax +1 916 454 4614, E-Mail james.purdy@ucdmc.ucdavis.edu

General Concepts and Methods

Advanced Technologies in the Radiotherapy Clinic: System Fundamentals

John L. Meyer[a] · Michael Sharpe[b] · Kristy Brock[b] · Joseph Deasy[c] · Tim Craig[b] · Douglas Moseley[b] · James Alaly[c] · Konstatin Zakaryan[c]

[a]Department of Radiation Oncology, Saint Francis Memorial Hospital, San Francisco, Calif., USA;
[b]Department of Radiation Oncology, University of Toronto and Radiation Medicine Program, Princess Margaret Hospital, Toronto, Ont., Canada; [c]Department of Radiation Oncology, Washington University, St. Louis., Mo., USA

Abstract

The radiotherapy treatment process is undergoing rapid development at every step from planning through delivery, and each step is increasingly automated and assisted by new imaging, positioning, contouring and treatment tools. Plan delivery and verification is now aided using an increasing range of image guidance technologies, and imaging at treatment now brings broad opportunities for dose guidance and adaptation for improving overall treatment quality. While these many tools bring exciting opportunities for exact, reliable and efficient targeting of radiation dose, a consistently high level of accuracy must be achieved at every step to achieve the desired results. This level of workmanship requires thorough understanding of the basic methods involved in each step, including the opportunities and limitations, by both the clinicians and the planning/delivery staff alike. These processes and their clinical implementation are discussed in depth throughout this volume. Here, we overview their integration and guiding background concepts, as well as a range of workday efficiencies for clinical practice.

Copyright © 2011 S. Karger AG, Basel

There are four basic elements to the clinically successful implementation of a modern intensity-modulated radiotherapy/image-guided radiotherapy (IMRT/IGRT) treatment course, often divided into its component planning and delivery operations. First, the therapy course can only be as successful as it is planned initially. This plan must encompass the full range of potential uncertainties involved with each step of the therapy sequence [1–3]. For the delivery of the plan, the daily patient setup for each fraction must reliably represent the patient model of the plan. To do this, the target must be accurately localized at treatment with 3D data sufficient for guiding therapy. This localization data must represent the initial staging and treatment planning data by being consistently and accurately registered at every point in space throughout the therapy process. Finally, variations from the original staging data – changes in tumor and tissue geometries during therapy – must be appreciated with consideration of adaptation to these changes as clinically required [4]. The four sections of this discussion present the essential concepts and quality concerns of IMRT/IGRT: plan development, target localization, image registration and dose adaptation.

A Typical Intensity-Modulated Radiotherapy Treatment Planning System

IMRT has been a dramatic change in the planning and delivery of radiotherapy. While the available software programs differ in many details, their basic elements are similar. It is easiest to view these from the point of view of the practical steps involved. In a typical IMRT commercial planning system, the planner is first asked to set certain mathematical statements (usually dose specifications) to always hold. These are called *constraints*. For example, 'No more than 45 Gy should be delivered to the spinal cord.' Other mathematical functions are directed to be made as small or as large as possible. For instance, 'The average square difference between the target prescription doses and the computer-predicted doses should be as small as possible.' Such mathematical statements are added together resulting in the *objective function*.

The objective function, or the *linear sum objective function*, includes terms for normal tissues as well as the target volume. Usually the different terms have different multiplying weights. This represents a way of specifying the relative importance of these terms. The state-of-the-art treatment planning system typically has other inputs called *dose-volume constraints*, where the planner does not want any more than X percent of an organ, by volume, to receive Y dose. The treatment planning system will try to match or exceed the goal dose-volume histogram parameters for target volumes and normal tissues. In IMRT treatment planning, the computer is driven to find the best beam weight solution, the one with a lowest defined mathematical objective function consistent with the constraints that have been set.

A mathematical equation that is often used in IMRT treatment planning as an objective function term is the *generalized equivalent uniform dose* (gEUD). The gEUD, as it has been named by Niemierko, is a volumetric power-law average of the dose distribution over a given structure:

$$\text{GEUD} = \left\{ \frac{1}{N} \sum_{i=1}^{N} d_i \times d_i^{(a-1)} \right\}^{1/a},$$

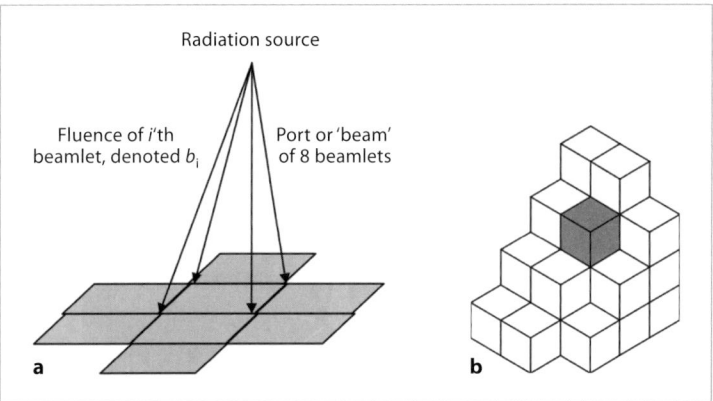

Fig. 1. a Radiation beams and beamlets. For IMRT, the incident radiation beams are typically idealized by mapping them into small discrete beamlets. The beamlets are represented as matrix lists of dose contributions to different voxels of the treatment volume. **b** Treatment volumes and voxels. The overall dose to a volume, for instance a target volume, will be the weighted contributions of all of the beamlets from all beams.

where a is the user-defined parameter. It is written this way to emphasize that it is just a power-law weighting of the dose distribution. This is very similar (in fact it is mathematically equivalent) to the dose volume histogram reduction step in what is called the Lyman-Kutcher-Burman NTCP model [5]. The advantage of this simple equation is that, depending upon what the value of the parameter a is chosen to be, the gEUD will tend toward the minimum dose, maximum dose or something between. For instance, if a is chosen to be negative and large in magnitude, the gEUD will tend toward (but does not quite reach) the minimum dose for that structure. If a is positive and large in magnitude, the gEUD will tend toward the maximum dose. This choice might be appropriate for some organs that are so-called serial structures (organs that transmit activities or products, such as the spinal cord). However, if a is chosen to be close to 1, gEUD is similar to the mean dose, which might be appropriate for parallel structures (organs that function with many similar subunits, such as the lung parenchyma) [6–8]. This makes the planning more specific to the tissues treated.

After the objective function is specified, the incident beams are typically idealized by mapping them into small discrete elements, called *beamlets*, shown in figure 1 [9, 10]. Each treatment beam is mathematically modeled, comprised of a sum of these beamlets. Typically, thousands of such beamlets are used in the first part of the optimization process. The aim is for the computer to determine the best or 'optimal' beamlet weights.

In the computer, beamlets are represented as dose contributions to different *voxels* (volumetric pixel) for a given level of beam monitor units (for example, dose for 1 MU). The overall dose to a volume, for instance a target volume, will be the weighted contributions of all of the beamlets from all beams.

Once all the terms are put together into an objective function and the beamlet matrices are computed, the computer is asked to iterate the beamlet weights until progress in optimizing the objective function between iterations falls to a specified low level. In actual practice, it is a clinical decision as to how many optimization iterations should be performed and when optimization is complete.

After the beam *fluence maps* have been derived, they must be broken up (decomposed) into deliverable segments for each beam that the multileaf collimator (MLC) or other device can deliver in a stepwise fashion. For some systems, the beam can be on when the segment is delivered, even when the MLC leaves are moving *(dynamic delivery)*. Otherwise, the beam is off when the MLC leaves are moving between segments *(static* or *step-and-shoot delivery)*. The many resulting segments together comprise the full delivery of the fluence map. An attempt is made to reconstruct that fluence map (as well as possible) by the superposition of these segments, but there is always some discretization error (usually, but not necessarily, small).

Most commercial systems take the approach wherein beamlet fluences are first optimized and afterwards MLC settings are derived as a deliverable sequence. This is the most common approach, although some companies are considering an approach that starts with apertures, and directly modifies or optimizes the aperture shapes.

Target Localization and Image Guidance

Target localization is the process of maintaining the planned relationship to the isocenter for each subsequent treatment session. Target localization, performed using appropriate technologies and frequency, is a critical component of any treatment quality assurance strategy. It assures that the accuracy and precision of each fraction meet predetermined error-tolerance criteria that are appropriate for the planning target volume (PTV) margin selected. There is a range of technologies to consider in the design of localization procedures [11, 12].

Sources of Geometric Uncertainty

In practice, a number of uncertainties must be factored into the radiotherapy planning process [11].

(a) *Clinical uncertainties*, such as the risk of undetected microscopic spread and metastases, are addressed during the treatment planning process by determining the clinical target volume and prescribing an appropriate dose and fractionation.

Table 1. Time scale of the processes leading to geometric uncertainties

Intrafraction	Seconds	Cardiac cycle breathing		
	Minutes	Organ-filling peristalsis	Patient movement	
Interfraction	Hours			
	Days		Setup variation	Anatomical changes
	Months			Normal tissue and tumor response

(b) *Physical uncertainties* are related to setup variation, organ movement and tissue deformation. Setup variations arise from gross inconsistencies in the patient's treatment position. Organ motion is related to physiological processes, such as breathing motion, organ filling and peristalsis; all of which can lead to shifts in organ and target position from fraction to fraction, or even within a treatment session. As the patient progresses through the treatment, target and normal tissue response can lead to volume changes and deformation. As a result, the relative position, size or shape of the tumor and also the normal tissues can deviate from the organ models defined in the planning phase of the process. Physical uncertainties are controlled using target localization systems.

The effects of these variations are classified as a *random* variation around a mean, and as a *systematic* variation manifested by an average shift in a landmark's position, relative to its position at simulation [13]. Systematic errors may accumulate and form a time trend, for example if the average position or volume of a structure drifts with time. Systematic and random errors can occur within a treatment and over longer time intervals, as summarized in table 1. Clearly the factors leading to random and systematic uncertainties vary in relative importance for a particular disease site, or even a particular patient. Effective implementation of conformal and IMRT treatment techniques requires an objective understanding of these sources of uncertainties, and the adoption of appropriate strategies to control them.

Choosing Target Localization Technologies

Several technologies are available to re-establish the patient setup with respect to the machine isocenter prior to each fraction, and to monitor the patient's position during treatment delivery (see online supplementary material). There is

an emerging shift from localization inferred from surface marks or radiographs, to the more direct use of implantable markers and soft tissue localized via volumetric imaging in the treatment suite.

Optical Tracking

Optical tracking has emerged as a means of monitoring surface landmarks objectively in real-time. Most systems track infrared-reflecting markers attached to the patient's external surface. The position of reflecting landmarks is determined within the field-of-view of a pair of stereoscopic cameras, which are calibrated with respect to the machine isocenter [14]. More recent systems track surface topography, e.g. breast and chest wall surface contours.

Target localization is achieved by establishing the positions of several optical markers relative to the target volume in a CT simulator. In the treatment room, the marker positions are measured relative to the isocenter to assess the need for setup correction, and to monitor with real-time feedback. Real-time tracking offers the possibility to detect setup changes during treatment delivery and interrupt treatments in the event of a gross misalignment of the treatment beam. Optical tracking systems can be integrated with other target localization technologies, such as bite plates used in stereotactic radiotherapy, or ultrasound or radiographic X-ray imaging, which are discussed below. In some body sites, the influence of organ motion caused by breathing, organ filling and interabdominal pressure can confound target localization using surface landmarks [15]. When dealing with mobile internal anatomy, great care must be taken to assure the external surface marks are a valid surrogate for target position.

Stereotaxy

Stereotaxy is the methodology involved in the 3D localization of structures using a mechanical frame and precise coordinate system to align and direct surgical instruments. Stereotactic radiosurgery uses stereotaxy adopted from neurosurgery applications to localize small brain tumors and certain benign brain disorders [16]. Recently, radiosurgery principles have been extended to extracranial disease, using a stereotactic body frame [17]. This approach incorporates a variety of technologies with the aim of controlling lung and other organ movement that would otherwise require large PTV margins to compensate for target motion. This approach allows a dramatic reduction of treatment volumes, which opens the avenue of hypofractionation with distinctly increased daily doses and very short courses of treatment.

Megavoltage Electronic Portal Imaging

This widely available localization technology has long been used to create radiographic images using the therapy X-ray beam. Decades of development have led to the current generation of electronic portal imaging devices (EPIDs) [18]. State-of-the-art portal imaging is based on amorphous silicon panels, which contain millions of photo-detectors coupled to switching transistors and fast read-out electronics. These devices produce good image quality, sufficient for detecting bone structures and implanted radiographic markers. Modern EPIDs offer advantages over film-based megavoltage (MV) radiography, mainly because of the capacity to adjust display contrast, and to assess target position and adjust the patient promptly.

While MV-EPID technologies have achieved a presence in commercial treatment systems, a number of factors have impeded the widespread adoption of frequent of portal imaging in clinical practice, including the inherently low contrast of radiographs made at MV energies, field-of-view limitations and a lack of tools and computer networking infrastructure to permit integration in the clinical environment. Using MV radiographs, it is difficult to infer field placement with respect to soft-tissue structures. The desire to further improve normal tissue sparing and shorten fractionation schemes is driving requirements for greater accuracy in target localization.

Radio-Opaque Markers

A number of studies have demonstrated a lack of correlation between prostate position and the localization of pelvic bone anatomy, and the ability to improve target localization using implanted markers as a surrogate [15, 19]. Many institutions implant multiple gold seeds into the prostate under transrectal ultrasound guidance, usually a few days prior to CT simulation. The seeds are identified on the planning CT, as well as on a pair of orthogonal digitally reconstructed radiographs. Radiation therapists acquire MV portal images on a daily basis, and align the seeds with the corresponding digitally reconstructed radiographs. Setup discrepancies exceeding a predefined threshold are corrected by adjusting the couch.

The practice of using radiographic markers may not be restricted to passive seeds or wires. Some innovative emerging technologies include a solid-state radiation dosimeter hermetically sealed in a glass tube [20]. The marker includes a MOSFET dosimeter, which integrates radiation exposure, and is visible on radiographs and on CT images. The integrated dose can be interrogated using an external radio frequency antenna to energize the sensor, which radiates back a signal reporting the dose information. The capability of confirming dose delivery in concert with target localization is intriguing. However, the benefits of implanting any form of marker for guidance should be weighed against the risk of infection or tumor seeding along the needle track [21].

Radiofrequency Transponders

A novel system has been introduced for real-time localization of radiofrequency transponders that are energized and monitored with an external antenna array

[22]. Each marker resonates a unique frequency, and its position is determined with respect to the array. The transponders are implanted prior to the CT simulation and are linked to the treatment isocenter in the planning phase. In the treatment room, the array is positioned over the target to monitor the positions of the transponders with respect to the array. The array is integrated with an optical tracking system, which in turn, links the transponder location to the treatment isocenter. This system offers the unique capability to monitor implanted markers in real-time, without fluoroscopy, within a 15-cm cubic volume, and further development may refine it for localization of larger volumes.

Kilovoltage Imaging
Because of the more pronounced photoelectric absorption in the lower energy range, radiographic imaging with kilovoltage (kV) X-rays offers higher contrast than MV imaging. This translates into greater visibility of bones or implanted markers at a lower imaging dose, and simplified interpretation of images. Several kV imaging systems are offered commercially. These systems frequently offer fluoroscopic imaging capability for real-time monitoring and include two or more imaging systems to permit real-time 3D localization of markers or bone anatomy. Dynamic tracking of radio-opaque markers in real-time gives the option to adopt treatment gating strategies and the possibility to limit the margins required to compensate for breathing motion. However, the benefit of real-time tracking must be weighed against the possibility of delivering an excessive skin dose due to the fluoroscopy procedure [23].

Ultrasound
In some situations, imaging of soft-tissue contrast can provide a better targeting surrogate than implanted markers, or skeletal landmarks on radiographs. Ultrasound systems are relatively inexpensive and offer soft-tissue visualization capability in some clinical situations. Transabdominal systems for guiding prostate cancer treatment have received a large amount of attention [24]. Images of the prostate and adjacent organs are acquired in the treatment position by placing a transducer on the patient's abdomen. The transducer is calibrated and coregistered with the isocenter using an additional localization technology, such as optical tracking, possibly in both the CT scanner and treatment rooms [14].

Contours from the planning system are displayed in the ultrasound context and aligned to position of the prostate correctly with respect to the isocenter. Placement of the transabdominal imaging transducer requires a degree of technical skill and training to acquire suitable images and to avoid displacing the prostate with extra-abdominal pressure [25]. While ultrasound systems have been employed successfully for prostate cancer and upper abdominal treatment, difficulties are encountered with sites where bone or large air cavities generate reflections that overwhelm the ultrasound signal from soft-tissue targets.

Computed Tomography Imaging in the Treatment Room

There is growing interest in the use of CT for the localization of soft-tissue targets. CT images are geometrically accurate and applicable to a broad range of anatomical sites. CT images are acquired without contact with the patient, but require exposure to additional ionizing radiation. Most CT systems provide an accurate measure of tissue attenuation, i.e. Hounsfield numbers, which is important for dose calculations in treatment planning. This may be an advantage in adaptive treatment strategies when a midcourse reoptimization of the treatment plan is required. Several implementations bring CT capability to the treatment suite, and are outlined briefly here.

1. Computed Tomography on Rails. One of the first systems to provide CT imaging in the treatment room was the so-called 'CT-on-rails' device [26–28]. A conventional scanner is located within the treatment room, adjacent to a linear accelerator. The treatment couch has been modified to minimize CT artifacts, and to transit patients between the CT scanner and accelerator. Images are acquired by translating the scanner along the rails and along the stationary patient. Images acquired prior to treatment are compared with the planning CT and couch movements are calibrated so that the patient can be correctly repositioned from the scanning position to the treatment isocenter following the evaluation of images. This affords a high-degree degree of accuracy and precision in setup correction based on soft-tissue anatomy.

Frequent CT imaging during the course of therapy has opened the door to assess how organs move and deform, how to properly assess the dose distribution in light of these anatomical changes, and when to replan treatment. CT-on-rails systems have demonstrated the implications of frequent volumetric imaging for image guidance, including dramatic changes in tumor shrinkage that can occur over the course of treatment, especially in head and neck cancer patients [29], and the capacity for frameless stereotactic treatment for early-stage lung cancer [30].

2. Tomotherapy Imaging. The Tomotherapy device (Madison, Wisc., USA) includes a MV CT imaging system based on a fan beam and helical acquisition paradigm [31, 32]. Volumetric MV CT images are acquired in the treatment position, just prior to treatment delivery, using a detuned linear accelerator producing 3.5-MV X-rays and a reduced photon output. An array of xenon ion chambers separated by thin tungsten septa forms the detector system, and photons interacting in the septa release electrons that ionize the xenon gas [33]. This process is very 'photon-efficient' compared with MV portal imaging devices. As a result, imaging doses can be acquired at doses in the range of 0.5–3 cGy, which is competitive with conventional CT systems. Because MV photons interact via Compton scattering, MV CT images are immune from artifacts produced by high atomic number materials, like dental amalgam or metal prostheses, as shown in figure 2.

3. Cone-Beam Computed Tomography. Cone-beam CT systems enable volumetric imaging capability on a conventional medical linear accelerator. Cone-beam

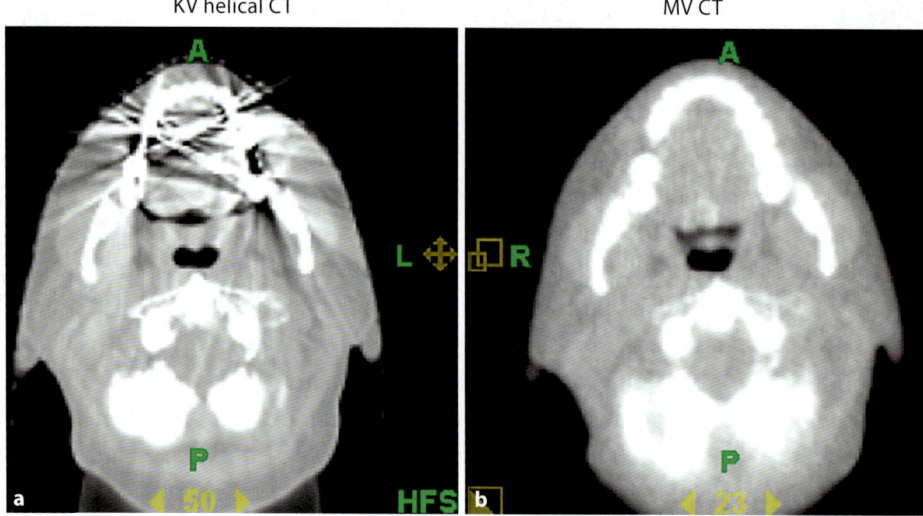

Fig. 2. a Artifacts can be seen in kV cone-beam CT images, including X-ray scatter and beam hardening in the vicinity of the shoulders. **b** MV CT images are immune from artifacts produced by high atomic number materials, like dental amalgam or metal prostheses. Courtesy of Ken Ruchala, Tomotherapy Inc.

CT differs in the source-detector geometry for image acquisition, compared with the conventional fan beam geometry. Like the Tomotherapy system, cone-beam CT reconstructions are acquired in the treatment position, just prior to treatment delivery. In a single gantry rotation, volumetric images are reconstructed by backprojection of hundreds of 2D radiographs acquired from using a large-area amorphous silicon detector. The field-of-view can approach a 50-cm diameter in the transverse plane, and 25 cm in the craniocaudal direction. Jaffray et al. [34] have reported on kV cone-beam CT systems, and clinical experience is emerging [35, 36]. The imaging dose associated with a volumetric image is typically less than 3 cGy with kV cone-beam CT systems. MV cone-beam CT systems have also been described [37, 38] and the dose used for volumetric imaging is comparable with portal imaging using contemporary MV EPIDs. Objective comparisons of imaging dose require consideration of the associated image contrast, resolution and noise characteristics. kV systems tend to be more flexible in terms of dose, and support planar radiographs and real-time fluoroscopy for patient monitoring over the course of treatment delivery. Most of the leading radiotherapy vendors offer cone-beam CT technology, all of which appear capable of producing high-resolution volumetric images of soft-tissue anatomy at reasonable doses (<5 cGy).

Transverse, coronal and sagittal images of a patient with head and neck cancer, shown in figure 3, illustrate the utility of kV cone-beam CT for image guidance

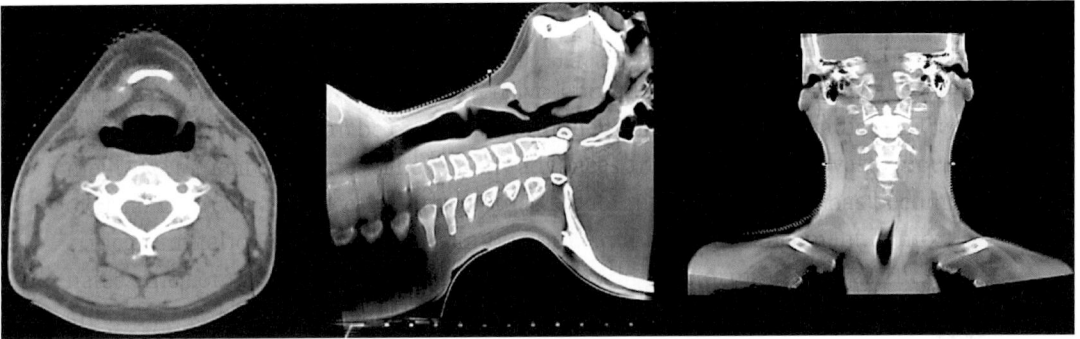

Fig. 3. kV cone-beam CT images of the head and neck.

based on soft tissues. Good contrast is seen between fat, vessels and muscle groups. It should also be pointed out that cone-beam CT systems do not yield proper calibration relating CT numbers to attenuation coefficients at this time, i.e. Hounsfield numbers.

Magnetic Resonance Imaging in the Treatment Room
Since MRI provides superior soft-tissue imaging of tumors and avoids radiation exposure, it offers real advantages for IGRT. In several organs such as the liver, breast and brain, intraparenchymal tumors or tumor resection sites may be poorly visualized except by MRI, and MRI opens a new realm of image guidance if it can be effectively integrated into the delivery process. In fact, MR-guided radiotherapy appears to be feasible, both from an engineering and a physics perspective [39], and MRI has recently been introduced into the therapy room at the University of Toronto, University Medical Center Utrecht [40], and elsewhere. Ultimately, it may provide high-quality, real-time MRI imaging at the same time radiation is being administered. Potential drawbacks include the effect of magnetic field on dose deposition, as well as its added costs for equipment and treatment room redesign. Specific research and development in this area is rapidly growing, clinical installations are being implemented and commercial enterprises are underway (Imris Inc., ViewRay Inc., and others).

Designing an Image-Guided Intervention Strategy

Target localization technologies permit frequent assessment of treatment to minimize patient setup uncertainties, and in some cases, provide a record of anatomical changes over a treatment course. In particular, the potential for frequent volumetric CT imaging during the course of therapy generates a number of opportunities and challenges. Fortunately, the experience gained in the MV EPID era has led to a framework for the continuing development of clinical guidance strategies [13]. Implementation of target localization falls broadly into *on-line* and *off-line* approaches for assessment and intervention. On-line approaches evaluate information acquired immediately prior to each fraction, and simple couch translations are applied to correct for observed deviations in treatment position that exceed a predefined threshold. An off-line approach refers to the frequent acquisition of setup information at the start of a treatment course (e.g. first 3 fractions), followed by an off-line statistical analysis to determine the patient's systematic (mean offset) and random (standard deviation) setup errors. A correction of the systematic error, the most important component of geometric uncertainties, is then made for future fractions.

On-line correction strategies tend to lead toward a larger reduction in geometric errors than an off-line approach, but action thresholds must be set to control the time and effort spent in 'chasing' random setup errors. On-line correction strategies appear to be most appropriate when the tumor is relatively unaffected by uncharacterized organ movement and in close proximity to critical normal tissues, where small errors in setup could compromise toxicity or local control. On-line approaches also seem appropriate when high-dose radiation is delivered in one or very few fractions, where there is little opportunity to obtain statistical information to detect systematic positioning errors. However, the impact of geometric uncertainties on tumor control may be small, depending on the size of the PTV margins [41]. An off-line analysis may also be useful to identify and address the random and systematic uncertainties due to breathing motion in hypofractionated treatment [42].

Table 2 compares the target localization systems discussed in this section. In-room CT and ultrasound solutions offer soft-tissue imaging. Some CT solutions can demonstrate breathing motion using breathing-correlated reconstructions, but real-time updates are not feasible with CT. Alternatively, volumetric CT images can be used in concert with fluoroscopy systems to monitor targets in real-time, but the formation of images in both systems requires the administration of additional dose. The extra dose delivered to the patient from more frequent imaging appears clinically insignificant in the radiotherapy context, but long-term follow-up of patients is not available. The benefit of imaging should be weighed against any perceived risk of second malignancies, especially in young patients. Alternatively, optical tracking of surface markers report data in real-time. This surrogate may be too remote to be used alone for targets influenced by internal organ movement, but this deficiency can be addressed by combining optical tracking with implanted transponders, ultrasound, or CT to establish a valid relationship with the target location. The real-time aspects of optical tracking

Table 2. Summary of the features and limitations of target localization systems (adapted from H. Sandle)

Method	Dimensions (x, y, z)	Evaluated features	Time frame	Cons
Radiography, MV/kV	2D/3D	bones/markers	snapshot	dose?
Fluoroscopy, kV	2D/3D	bones/markers	real-time	dose?
Tomography	3D	soft tissue	snapshot	dose
Ultrasound	2D/3D	soft tissue	snapshot/real-time?	expertise
Optical	1D/2D/3D	skin surface	real-time	remote surrogate
Implantable sensors	0D/3D	markers	real-time	invasive, FOV?

could then use surface makers to detect gross setup changes and breathing motion. A number of other scenarios can be considered. Clearly there are a number of trade-offs to consider in the implementation of a target localization procedure. To be safe and effective, the selection and commissioning of equipment must be done in the context of the clinical requirements for target and normal tissue localization.

Image Registration

Image registration is central to every step of the radiotherapy planning and delivery process. It can improve the ease and accuracy with which multimodality images can be incorporated into a single model of the patient by resolving the geometric discrepancies that exist between the images. This allows all the information gained from different imaging studies to be fully exploited for treatment definition. As the image registration becomes more accurate, comparisons between images can be made on a voxel-by-voxel basis. Image registration can also improve the accuracy and precision of treatment delivery by relating the volumetric images obtained just prior to treatment delivery with the images used for treatment planning. Registration of these images will allow the offset to be calculated, which can then be applied to the patient position to allow accurate delivery of the intended treatment.

During radiation planning, different forms of imaging provide unique information that allow differentiation of tissue density (CT, MR and ultrasound) and functionality (MRSI, PET and functional CT). Incorporating these images into one model of the patient can be challenging since the images are obtained on different devices, each requiring the patient to be repositioned for its imaging process [43].

During radiation delivery, delivery of a precise treatment plan requires that the patient be in the same position as the model that was used to develop that plan.

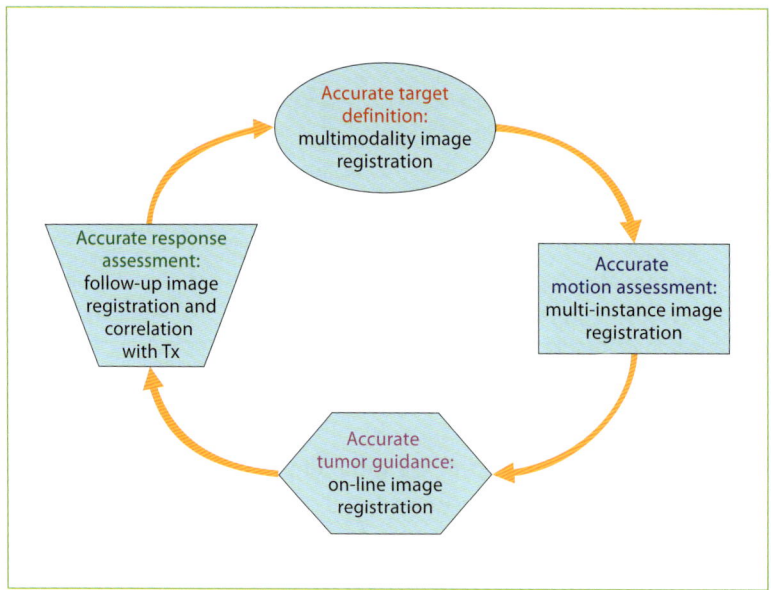

Fig. 4. The role of image registration to improve the integration of the radiotherapy process.

Advances in imaging at the time of treatment, including portal imagers [44, 45], in-room CT scanners [46, 47], ultrasound [48], tomographic imaging [32, 49] and cone-beam imaging [50, 51], allow a reduction in the expected deviations and the potential to eliminate unexpected deviations through daily corrections of the patient. These multiple sources of image data are useful only if they are accurately registered and fused during their evaluation.

Registration allows us to complete the loop in radiotherapy, potentially improving the accuracy of each step of the process, as shown in figure 4. The first is for *accurate target definition* through multimodality image registration. Target definition may be improved through deformable registration, as it will reduce the uncertainty in the correspondence of these multiple sources of information. The second is *accurate motion assessment*. When multiple examples of geometry of the patient are obtained, such as inhale and exhale scans, a more accurate deformation map between the two of these may be generated. The next step is treatment delivery; a*ccurate tumor guidance* for on-line image registration may be improved using deformable registration. Finally, in *accurate response follow-up*, the response assessment of the tumor and changes in the normal tissue can be improved using deformable registration to compare the pretreatment images with those obtained following the completion of treatment.

The Image Registration Process

The four components of registration are (a) determining common surrogates, (b) calculating the correspondence between surrogates, (c) interpolating between surrogates and (d) evaluating and displaying results. Several surrogates can be used depending on the application, including points, surfaces and image intensities. Calculating the correspondence depends on the surrogate and can be as simple as minimizing the difference for points and surfaces, or complex, such as mutual information for comparing image intensities. The appropriate surrogate and correspondence must be selected for each problem and the limitations of the surrogates must be understood and included in the PTV margins for the treatment delivery. Currently, most commercially available systems interpolate the image space between the surrogates using a linear translation, accounting for rotation and translation only. This is a limitation for soft tissue, which deforms due to physiological motion, flexibility in patient positioning and anatomical changes over the course of treatment. Deformable registration algorithms, which allow for more complex interpolations, are currently being developed. Local improvements in linear registration can be achieved in the presence of deformation by optimizing the registration to focus on the region of interest (ROI) that is most important. The use of clip boxes or ROI delineations can restrict the algorithm to the ROI of importance; however, caution should be used to ensure that the misregistrations outside of the ROI do not significantly violate planning or treatment constraints.

Common Surrogates and Correspondence between Surrogates

If the position of the tumor itself is not visible on the images, as is usually the case, surrogates for that position are needed. Choosing a surrogate closer to the actual tumor will likely increase the precision. Neighboring organs, such as a bone or diaphragm that are visible in the images, can also be used as a surrogate for alignment [52, 53]. Implanting markers or fiducials into the tumor itself can create a close surrogate for the tumor [54–56], indicating the position of the tumor, but potentially not indicating deformation that is occurring in the tumor. Guidance using the tumor-bearing organ can also increase precision; however, this still remains a surrogate, with the accuracy of its surrogacy dependent on the amount of deformation that the organ and tumor undergo, as well as any differences in coupling between the organ motion and the tumor motion. It is sometimes possible to identify the tumor boundary on volumetric images obtained at the time of treatment used for guidance; however, surrogates are still used in a majority of the cases. The degree of uncertainty between the surrogate and the target for treatment must be included in the uncertainty margins in the treatment design. These uncertainties can be determined from repeat images where the tumor and the surrogate are visible.

Once the surrogates are identified, the method of correspondence must be determined. This can be simple, by using points such as seed locations, or complex, by using the intensity information in the image. Three common methods for calculating correspondence of surrogates are described below: alignment by points, alignment by surfaces and alignment by intensity pattern.

Alignment by Points

For 3D alignment using points, 3 noncoplanar points must be identified. If no deformation exists between the position of the seeds on the initial image and the position on the secondary image, then an exact alignment of all three seeds can be performed using a rotation and translation. This does not happen for soft-tissue anatomy, however, so the residual error of all the seeds must be minimized. The residual error indicates the error remaining after the correction is carried out.

An example of a common *point* surrogate is implanted gold fiducials in the prostate, which are seen well on MV and KV portal images and cone-beam images. With seeds, a digitally reconstructed radiograph or reference CT is obtained from planning with the seed location, and this is compared with a MV, kV or volumetric image during treatment. A least squares difference can determine the difference between these seed positions, and this displacement can provide a *transformation* to align the patient to the intended position. Marker implants have been shown to be a very effective method of registering the prostate prior to treatment. It is important to note, however, that these seeds are only surrogates. A study performed at Princess Margaret Hospital, where repeat MR images were obtained of the prostate and seeds indicated the potential for substantial deformation to exist following registration of the prostate using translation only and translation and rotation [57]. 10% of the patients had greater than 50% of the prostate surface deform by more than 3 mm after alignment of the seeds, with the worst case exhibiting 1.3 cm of deformation following translation and 1.1 cm of deformation following translation and rotation. Figure 5 shows the worst case of the 29 patients examined. The mesh represents the prostate position at the first MR image and the colorwash represents the prostate at the second MR image; the colorwash indicates the difference, as a vector magnitude, in surface position between prostate on the second MR image and the prostate on the first MR image.

Alignment by Surfaces

Surfaces (represented in planar 2D or volumetric 3D data) can be used for alignment of the tumor, the tumor-bearing organ or a neighboring organ. Surface alignment is performed by dividing the surface into a series of points and minimizing the difference between these points, defined on both surfaces. This is com-

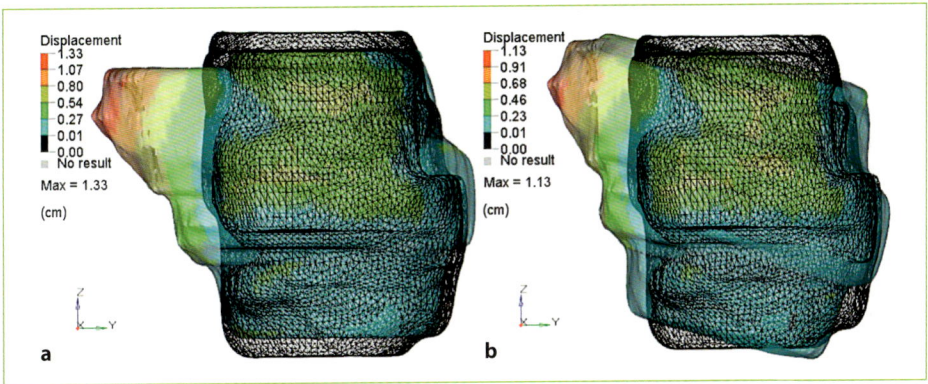

Fig. 5. Residual error of the prostate surface, due to deformation, following registration of the seeds using translation only (**a**) and translation and rotation (**b**).

monly referred to as chamfer matching [58]. The surfaces can be defined manually or by setting a threshold in the image to define the surface.

With volumetric imaging, registration of the surface of the organ can be performed, increasing the precision of registration by improving the relationship between the surrogate, the tumor-bearing organ and the tumor. The tumor-bearing organ will provide a closer surrogate for the tumor position than neighboring organs, such as bone. Registration can be performed manually by registering the contours from the planning scan onto the images from the pretreatment images or by identification of the entire organ surface, either through manual contouring or autosegmentation.

Pretreatment alignment of the liver provides an example of surface registration for neighboring organs and the tumor-bearing organ. The liver, contoured on the planning CT, can then be registered to the liver contoured on the kV cone-beam CT to match the entire organ volume. Even with good alignment, it is important to remember that this still remains only a surrogate. The organ may deform, prohibiting a rigid registration, even with rotation, to perfectly align these contours [59, 60]. Therefore, it is critical to know what part of the anatomy offers greatest accuracy for alignment. The degree of this uncertainty must be reflected in PTV margins, even with on-line image guidance every day.

In lung cancer therapy, registration benefits from the contrast between the lung and the tumor, allowing the tumor to be visualized on pretreatment volumetric images [61]. In this case, alignment of the tumor itself to the planning scan volumes at the time of treatment can be performed. In addition, some tumors encompass the entire organ, such as the prostate. Registration is therefore dependent on the tumor itself; residual errors, due to deformation and changes in anatomy over the course of treatment can be visualized after registration.

Alignment by Intensity
The information inherent in the image, i.e. the intensity of each voxel, can be used to drive the registration using automated registration methods. Rather than using anatomic or contour surface details, this approach simply uses the 'raw data' of the image itself. Alignment by intensity is performed by identifying a metric that describes the similarity of the voxel intensities between the two images. A transformation is then applied to the secondary image and the similarity metric is recalculated. This is repeated several times, through an optimization algorithm, to find the optimal transformation that leads to the best similarity between the images. The similarity can be calculated using several different techniques. Three of the most common include sum of the squared difference, cross correlation and mutual information [62–69].

Automated Registration: Linear Interpolation Methods

Results from an automated registration system will depend on its flexibility and accuracy in performing the interpolation and extrapolation steps required. The majority of commercially available registration algorithms use a linear interpolation method, allowing the secondary image to only translate and rotate. The major types are described below. Unfortunately, the human body does not behave in such a simplistic manner, since organ deformation occurs and would need to be accounted for by a more sophisticated interpolation method. A few methods of deformation registration are briefly described later in the chapter.

Clip Box
When an automated registration is performed using an intensity-based method described above, the registration will examine the intensity throughout the entire image. The particular linear optimization that the algorithm chose may not be the best one from a clinical viewpoint. For example, if a head and neck patient is imaged prior to treatment and the flexion in the neck is different than that in the CT scan, which is the reference image, the registration algorithm needs to know which area is most important for registration. The user can indicate this by using a *clip box*, or a box that encompasses the ROI of greatest interest. The registration algorithm then focuses only on the image intensities in this box.

An example is shown below for a head and neck case using the Synergy XVI R3.5 General Release software (Elekta Ltd., Crawley, UK) with a clip box to perform image guidance at the time of treatment. The algorithm sets a threshold for the bones and then performs a surface match on the segmented bone [58]. The reference image is a CT image and the secondary image is a kV cone-beam CT image. In figure 6a, the clip box is placed on the neck, resulting in good registration

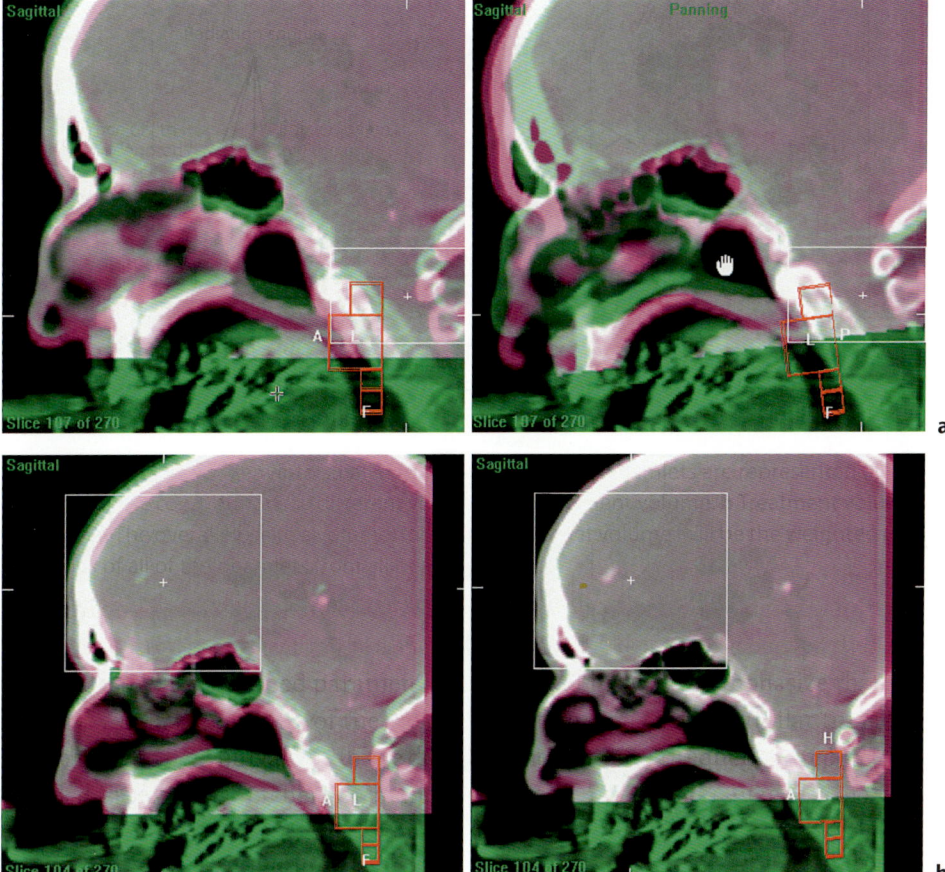

Fig. 6. a Head and neck image alignment using a clip box positioned on the neck (upper 2 panels). **b** Registration of the same head and neck, with a clip box placed on the front skull (lower 2 panels). A correct match will appear as gray by overlaying green and purple.

of the neck but poor registration of the skull, due to deformation resulting from neck flexion between the primary (green) and secondary (purple) images. In figure 6b, the clip box is placed on the front skull, resulting in good registration of the skull and poor registration of the neck.

Organ Limitation

A second optimization approach, similar to the clip box, is to use the contoured ROI to limit the registration region. This approach allows a more tailored ROI, but requires user intervention with contouring of the ROI. For treatment planning purposes, contouring the ROI is already required, so registration can take advantage of this more detailed type of clip box.

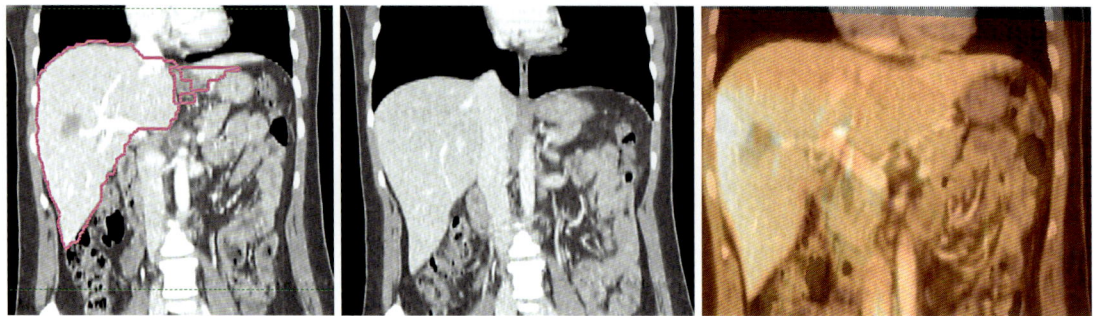

Fig. 7. Exhale breath hold CT image (left) with liver contour for limiting the ROI of registration, inhale breath hold CT image (middle), and results of the registration (right).

The example shown in figure 7 is an inhale and exhale breath hold image of the abdomen for a liver cancer patient. The primary image, on the left, is at the exhale breath hold. The liver has been contoured, indicated by the purple outline. The image in the middle is at inhale breath hold. The spine and ribs have not moved substantially between the two image sets and will cause problems for a rigid registration algorithm; however, if the liver contour is used to limit the registration, the results, shown on the right, are very good for the liver, which is the important ROI in this case.

Deformable Registration Methods

The human body does not behave in a linear fashion; deformation exists due to internal physiological motion and external patient positioning, and the tumor and normal tissue responses to them. Accounting for this deformation in the registration process is currently the subject of much research. Several different approaches are being investigated and although it is out of the scope of this chapter to describe them in detail, four of the primary algorithms are briefly described: fluid flow, optical flow, spline-based and biomechanical methods. Two recent publications compared the results for several deformable registration algorithms using phantoms [70] and clinical data [71].

Fluid Flow
In a fluid flow approach, the deformations in the images are modeled as a fluid. The algorithm optimizes a similarity metric, described above, while constraining the interpolation by the laws of continuum mechanics. This method is currently being investigated by several groups for 4D CT registration [72–74], intracavitary brachy-

therapy [75] and brain [76, 77]. The approach can be fully automatic, invertible and has shown accuracy of less than 4 mm in the lung. Two potential limitations include accommodating noncontinuous motion, such as the stationary spine next to the moving lung, and accommodating a lack of intensity correspondence, such as an image with a full rectum being registered to an image with an empty rectum.

Optical Flow
Optical flow uses the differences between the images and the gradient of the image as forces to drive the registration, while maximizing a similarity metric. This approach is often referred to as the 'demon's approach', which was first implemented by Thirion [78]. Additional forces, such as an active force, based on the gradient in the moving image, can also be added to improve the outcome of the registration [79–81]. This algorithm has been investigated in the head and neck and prostate [79]. It can also be fully automatic and has shown good efficiency. As with the fluid flow, accommodating noncontinuous motion may be a limitation in some instances as well as accommodating lack of intensity correspondence. These are only potential problems, and research is ongoing to solve these issues for this type of registration.

Thin-Plate Spline and B-Spline
A spline-based approach deforms the image using control points that are placed in specific locations (thin-plate spline) or on a regular grid (B-spline). The control points guide the deformation of the image as a similarity metric is optimized. In a thin-plate spline approach, each control point affects the deformation of the entire image, the extent depends on the distance to the control point, making this approach best for single organ registration. The control points in B-spline only affect a local area of the registration, allowing for a multiorgan registration to be performed. Spline approaches have been used in a variety of anatomical sites, including liver [66, 82, 83], lung [63, 84, 85] and prostate [86–88]. The algorithm can be fully automated and efficient and an accuracy of 1.0–3.5 mm has been shown. Potential limitations are the same as the fluid and optical flow – accommodating noncontinuous motion and lack of intensity correspondence.

Biomechanical Method
A biomechanical approach deforms the image according to the material properties of the tissue, i.e. if the material is hard or soft. The approach is typically implemented using a finite element model of the anatomy in the image, which represents each ROI as a series of connected nodes, forming tetrahedrons and solved using finite element analysis. Boundary conditions are required, which describe the motion and deformation of certain parts of the model. These are determined using a similarity metric, often by deforming the contoured ROI on one image to the con-

toured ROI on a secondary image. This approach has been used for several anatomical sites, including the thorax [89, 90], abdomen [91, [92] and pelvis [93–95]. Accuracy of 1.2–2.5 mm has been shown. Potential limitations include the dependence of contours on the registration and the uncertainty in defining the material properties of the anatomy that is modeled.

Adaptive Radiotherapy

Effective treatment planning now integrates data representing many characteristics of tumor and normal tissues, including their referenced positions, anatomical shapes, densities, 4D changes in shape and position and sometimes biological activities as measured on FDG-PET or related imaging. These characteristics may change during a treatment course, and information detecting such changes can be helpful in further refining and updating the operative treatment plan. The development of comprehensive and reliable approaches to this process is a major, current effort in radiation oncology. Systematic methods that monitor, detect and correct for variations in any data used for the original treatment planning are known as adaptive radiotherapy (ART).

Encompassing corrective mechanisms for tumor positional and anatomical changes, ART is a process intended to improve radiation treatment by systematically monitoring patient/treatment positional and volumetric variations and incorporating them to reoptimize the treatment plan at appropriate intervals during the course of treatment [96]. Introduced by the William Beaumont Hospital group in 1997 [97], the concept of ART includes the following key features: incorporating systematic measurements of treatment variations into a closed-loop radiation treatment process, providing feedback to reoptimize the treatment plan during the course of treatment, and delivering treatment that is customized to the daily patient target volumes.

Ideally, the perfect ART system would monitor *all* of the tumor/normal tissue characteristics used in the initial planning and rapidly adapt the treatment plan prior to *every* therapy delivery. Obviously there are many challenges to realizing this ideal, though important gains are rapidly being achieved. Already, imaging in the treatment room provides extensive data documenting setup inaccuracies, movement and anatomic changes of targets during and between therapy sessions that can be clinically significant and provide the basis for adapting therapy [98]. Immediately attainable goals of ART now include opportunities to reduce initial PTV expansion margins by correcting setup inaccuracies in a programmed way, to adapt PTV expansion margins by monitoring the movement characteristics of tumor/normal tissue anatomy during therapy, and to modify plans based on anatomical changes occurring during a treatment course. Monitoring of the effectiveness of motion management strategies and adapting them to daily therapy are being incorporated into many clinical programs. Methods to record and accumulate dose distributions actually delivered, and to correct for inaccuracies in subsequent treatments, are now under development and will be meaningful for increasing treatment dose accuracy and potentially therapeutic gain [99].

Currently, there are two practical aspects of ART now used: (a) the detection of systematic positional/motion changes and the correction of setup points and PTV expansions used in daily therapy [100], and (b) the detection of changing anatomy during a therapy course and the modification of plans based on these changes [101, 102]. Applications of ART are discussed throughout this volume, especially in the chapters by Kim et al. [pp. 196–216] introducing IGRT concepts and by Kupelian and Meyer [pp. 344–368] on adaptive therapy for prostate cancer; here, we introduce the essential concepts involved.

Planning Target Volume and Daily Treatment Uncertainties

Since patients will vary in their degree of tumor motion, deformation and other characteristics influencing the PTV selection, setting one PTV expansion for all patients is not optimal. This is illustrated in figure 8 for management of setup variations, but the same ART concepts can be applied for other PTV considerations. In figure 8a, the clusters of dots represent the actual daily setup positions for four patient treatment courses. In an ideal course, the patient would always be set up at the prescribed isocenter (the interception of the x- and y-axes in the figure) and the target volume would always be the same as originally planned. In practice, the actual daily setup positions are found to be distributed at distances away from this ideal position, and the target may move and deform. If one wants to adjust the PTV expansion margins to accommodate for all such daily variations found for an entire treatment group (for instance, all prostate cancers treated with IMRT), then a quite large treatment margin would need to be added (represented by the shaded circle). In each patient, two types of variation may occur: *systematic*, in which the mean of the observed changes are a given amount from the planned, or *random*, which can be measured as daily changes from this mean.

The systematic type of uncertainty is amenable to *off-line* analyses of imaging acquired at treatment. In figure 8b of the figure, Patient A can be effectively treated with a smaller margin if systematic errors are identified early on during the treatment course and corrected, as demonstrated by the shift with the arrow. Now a smaller margin can be prescribed since it only needs to account for the random variations specific to the patient. In general, a systematic error may be the more detrimental, because the dose may be consistently delivered to an unintended location if not encompassed in a large PTV margin. Random errors are more forgiv-

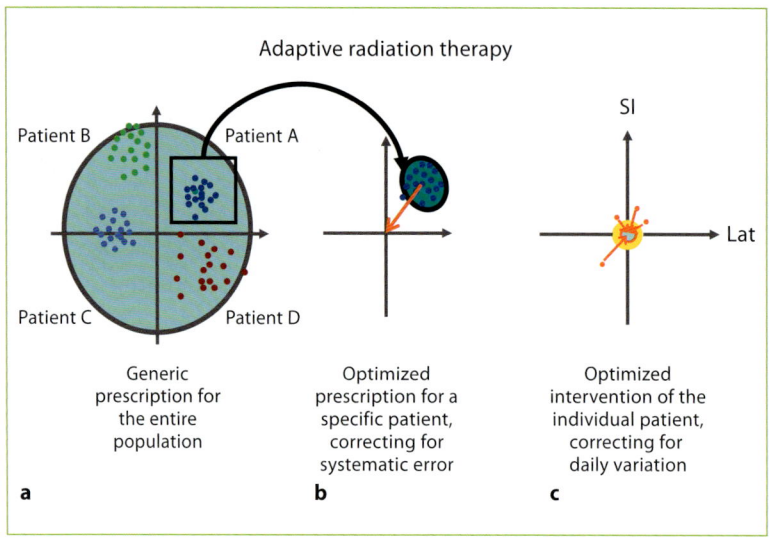

Fig. 8. ART strategies for correcting setup variations. **a** Conventional radiotherapy, enlarging PTV for all patient setup variations. **b** ART, correcting for systematic setup variations in a patient's treatment course. **c** ART, correcting for all systematic and random daily variations. Circle and ellipse show required PTV expansions to encompass uncertainties. Courtesy of J. Wong.

ing, since any effects on the final dose distribution tend to smooth out during the treatment course, especially if it involves many fractions. However, when a treatment course involves high doses delivered using a small number of fractions, large random variations become unacceptable. Random errors can only be addressed with daily *on-line* imaging and modification. Figure 8c illustrates the example of a treatment course where every daily setup variation is observed and corrected, permitting the smallest PTV margins to be used. It is estimated that the majority of treatment errors can be corrected through off-line analyses. These same ART concepts can be applied to adjustments in the treatment plan based on tumor movement and other anatomical changes to make each day's therapy plan highly specific for that day's treatment anatomy.

Adaptive Radiotherapy and Changing Anatomy during a Radiotherapy Course

In addition to daily anatomical and positional changes, target changes may occur during a therapy course and gradually introduce systematic error. Therapy adaptation may be critical to maintaining target dose objectives [103, 104]. Figure 9 shows a head and neck cancer example. Because of patient weight loss and tumor regression evident after 21 fractions, continued treatment with an unaltered plan

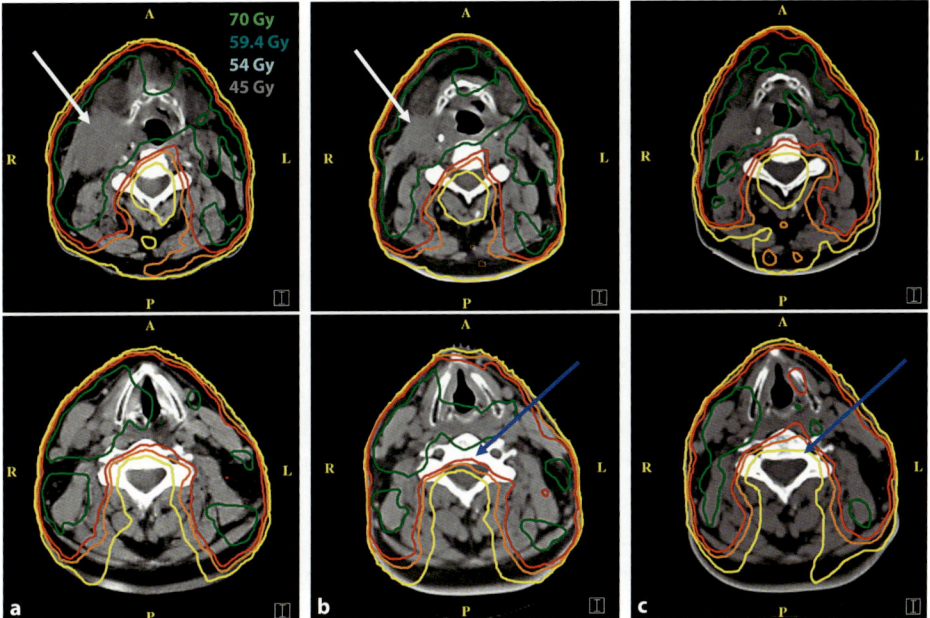

Fig. 9. IMRT treatment isodoses for a T4N2c base of tongue cancer patient. Top row: level of hyoid bone; bottom row: lower neck. **a** CT 1, initial plan before therapy. **b** Same plan after 21 fractions; tumor has regressed (white arrow) and patient has lost 5% of his body weight. Note that spinal cord dose (blue arrow) has become unacceptable, requiring plan reoptimization. **c** CT 2, plan reoptimized correcting spinal cord dose. Courtesy of L. Verhey and P. Xia (UCSF). See online supplementary material.

would result in significant overtreatment to the spinal cord. In other cases, such anatomic changes might result in unacceptable variations in dose to other important regions, including the tumor volumes themselves. Establishing triggers for re-evaluation of patient/tumor anatomy and replanning therapy can be essential for achieving the intended goals of therapy.

Adaptive Radiotherapy and the Future of Radiotherapy

Adaptive control principles are expected to have broad implications in the emerging relationship between target localization and the replanning of patients under treatment. Currently, replanning is infrequently conducted in clinical practice because it is labor and time intensive and appropriate anatomical information is often not available. Furthermore, the optimal frequency, or even the clinical need, for replanning is unknown in many disease sites. Ideally, when clinically indicated, temporal and volumetric imaging would be obtained while the treatment is delivered with the opportunity for efficient adaptation of therapy during and between radiation treatments. It may be tempting to embark on fully on-line adaptive replanning, but the desire for submillimeter technical precision should be balanced with the risk of chasing only modest clinical gains, as well as with the possibility of imposing an unacceptable workload on clinical planning, treatment delivery and review processes. The continuing development of tools and algorithms will assist in target and normal tissue contouring, planning, deformable registration of images and decision support for determining which corrections are required before replanning can be investigated properly in clinical practice [94, 105–107]. The computer networking and data storage involved in the management and visualization of large quantities of image-based information must be considered, and continuing technological developments must lead to practical solutions operating within the constraints imposed by time and treatment resources. Further development of decision support tools and robust correction algorithms will facilitate more complex adaptive guidance strategies.

The ART mechanism must identify treatment errors and determine whether they represent significant variation. Ultimately, ART involves recognizing the daily treatment delivery process and adjusting it as required by identified positional or volumetric variations of the target [98]. To achieve these goals, ART will need to integrate replanning and dose reconstruction tools [108]. It requires infrastructure for distributing information and tools for off-line decision-making to approve or disapprove actions [109]. New skills must be learned to make better use of these new technologies, and the radiation treatment process must be redeveloped as an informatics system for the efficient management of imaging data for treatment control.

ART is patient-, tumor site-, equipment- and institution-specific. Practical development of ART requires an understanding of how the PTV were developed for the individual patient or for a patient population as a class solution. These standards will be specific to each institution, which must then address fundamental questions. What should be the limits for on-line image-guided setup correction? How will these vary when imaging is directed by soft-tissue contours, implanted markers or other surrogates? How reproducible is the clinic's management of breathing motion by gating or other strategies? What is the appropriate PTV after the introduction of a new IMRT method, which may be different than the PTV used before this technology? These decisions must be part of a directed process within the clinic's self-assessments, and an effective ART method will depend on having a well-established and continuing process of radiotherapy quality assurance overall.

References

1 Urie MM, Goitein M, Doppke K, et al: The role of uncertainty analysis in treatment planning. Int J Rad Oncol Biol Phys 1991;21:91–107.
2 Pickett B, Kurhanewicz J, Coakley F, Shinohara K, Fein B, Roach M 3rd: Use of MRI and spectroscopy in evaluation of external beam radiotherapy for prostate cancer. Int J Radiat Oncol Biol Phys 2004;60:1047–1055.
3 Black QC, Grills IS, Kestin LL, et al: Defining a radiotherapy target with positron emission tomography. Int J Radiat Oncol Biol Phys 2004;60:1272–1282.
4 Galvin JM, Ezzell G, Eisbrauch A, et al: Implementing IMRT in clinical practice: a joint document of ASTRO and AAPM. Int J Radiat Oncol 20004;58:1616–1634.
5 Kutcher GJ, Burman C, Brewster L, et al: Histogram reduction method for calculating complication probabilities for three-dimensional treatment planning evaluations. Int J Radiat Oncol Biol Phys 1991;21:137–146.
6 Deasy JO, Fowler, JF: The radiobiology of intensity modulated radiation therapy; in Mundt AJ, Roeske JC (eds): Intensity Modulated Radiation Therapy: A Clinical Perspective. Hamilton, B. C. Decker, 2005.
7 Moiseenko V, Deasy JO, Van Dyk J: Radiobiological modeling for treatment planning; in Van Dyk J (ed): Modern Technology of Radiation Therapy. Madison, Medical Physics Publishing, 2005, vol 2.
8 Marks LB, Yorke ED, Jackson A, et al. Use of normal tissue complication probability models in the clinic. Int J Radiat Oncol Biol Phys 2010;76:S10–S19.
9 Langer M, Lee EK, Deasy JO, et al: Operations research applied to radiotherapy, an NCI-NSF-sponsored workshop February 7–9, 2002. Int J Radiat Oncol Biol Phys 2003;57:762–768.
10 Webb S: Intensity Modulated Radiation Therapy. Bristol, Institute of Physics Publishing, 2001.
11 Mageras GS: Introduction: management of target localization uncertainties in external-beam therapy. Semin Radiat Oncol 2005;15:133–135.
12 Langen KM, Jones DT: Organ motion and its management. Int J Radiat Oncol Biol Phys 2001;50:265–278.
13 Van Herk M: Errors and margins in radiotherapy. Semin Radiat Oncol 2004;14:52–64.
14 Meeks SL, Tome WA, Willoughby TR, et al: Optically guided patient positioning techniques. Semin Radiat Oncol 2005;15:192–201.
15 Schallenkamp JM, Herman MG, Kruse JJ, Pisansky TM: Prostate position relative to pelvic bony anatomy based on intraprostatic gold markers and electronic portal imaging. Int J Radiat Oncol Biol Phys 2005;63:800–811.
16 Podgorsak EB, Pike GB, Olivier A, et al: Radiosurgery with high energy photon beams: a comparison among techniques. Int J Radiat Oncol Biol Phys 1989;16:857–865.
17 Kavanagh BD, Timmerman RD: Stereotactic radiosurgery and stereotactic body radiation therapy: an overview of technical considerations and clinical applications. Hematol Oncol Clin North Am 2006;20:87–95.
18 Antonuk LE: Electronic portal imaging devices: a review and historical perspective of contemporary technologies and research. Phys Med Biol 2002;47:R31–R65.
19 Chung PW, Haycocks T, Brown T, et al: On-line aSi portal imaging of implanted fiducial markers for the reduction of interfraction error during conformal radiotherapy of prostate carcinoma. Int J Radiat Oncol Biol Phys 2004;60:127–133.
20 Briere TM, Beddar AS, Gillin MT: Evaluation of precalibrated implantable MOSFET radiation dosimeters for megavoltage photon beams. Med Phys 2005;32:3346–3349.
21 Haddad FS, Somsin AA: Seeding and perineal implantation of prostatic cancer in the track of the biopsy needle: three case reports and a review of the literature. J Surg Oncol 1987;35:184–191.
22 Balter JM, Wright JN, Newell LJ, et al: Accuracy of a wireless localization system for radiotherapy. Int J Radiat Oncol Biol Phys 2005;61:933–937.
23 Valentin J: Avoidance of radiation injuries from medical interventional procedures. Ann ICRP 2000;30:7–67.
24 Kuban DA, Dong L, Cheung R, et al: Ultrasound-based localization. Semin Radiat Oncol 2005;15:180–191.
25 Artignan X, Smitsmans MH, Lebesque JV, et al: Online ultrasound image guidance for radiotherapy of prostate cancer: impact of image acquisition on prostate displacement. Int J Radiat Oncol Biol Phys 2004;59:595–601.
26 Uematsu M, Fukui T, Shioda A, et al: A dual computed tomography linear accelerator unit for stereotactic radiation therapy: a new approach without cranially fixated stereotactic frames. Int J Radiat Oncol Biol Phys 1996;35:587–592.

27 Yenice KM, Lovelock DM, Hunt MA, et al: CT image-guided intensity-modulated therapy for paraspinal tumors using stereotactic immobilization. Int J Radiat Oncol Biol Phys 2003;55:583–593.

28 Shiu AS, Chang EL, Ye JS, et al: Near simultaneous computed tomography image-guided stereotactic spinal radiotherapy: an emerging paradigm for achieving true stereotaxy. Int J Radiat Oncol Biol Phys 2003;57:605–613.

29 Barker JL Jr, Garden AS, Ang KK, et al: Quantification of volumetric and geometric changes occurring during fractionated radiotherapy for head-and-neck cancer using an integrated CT/linear accelerator system. Int J Radiat Oncol 2004;59:960–970.

30 Uematsu M, Shioda A, Suda A, et al: Computed tomography-guided frameless stereotactic radiotherapy for stage I non-small cell lung cancer: a 5-year experience. Int J Radiat Oncol Biol Phys 2001;51:666–670.

31 Ruchala KJ, Olivera GH, Schloesser EA, Mackie TR: Megavoltage CT on a tomotherapy system. Phys Med Biol 1999;44:2597–2621.

32 Ruchala KJ, Olivera GH, Kapatoes JM, et al: Megavoltage CT image reconstruction during tomotherapy treatments. Phys Med Biol 2000;45:3545–3562.

33 Keller H, Glass M, Hinderer R, et al: Monte Carlo study of a highly efficient gas ionization detector for megavoltage imaging and image-guided radiotherapy. Med Phys 2002;29:165–175.

34 Jaffray DA: Emergent technologies for 3-dimensional image-guided radiation delivery. Semin Radiat Oncol 2005;15:208–216.

35 McBain CA, Henry AM, Sykes J, et al: X-ray volumetric imaging in image-guided radiotherapy: the new standard in on-treatment imaging. Int J Radiat Oncol Biol Phys 2006;64:625–634.

36 Oldham M, Letourneau D, Watt L, et al: Cone-beam-CT guided radiation therapy: a model for on-line application. Radiother Oncol 2005;75:271–278.

37 Pouliot J, Bani-Hashemi A, Chen J, et al: Low-dose megavoltage cone-beam CT for radiation therapy. Int J Radiat Oncol Biol Phys 2005;61:552–560.

38 Groh BA, Siewerdsen JH, Drake DG, et al: A performance comparison of flat-panel imager-based MV and kV cone-beam CT. Med Phys 2002;29:967–975.

39 Fallone, Carlone M, Marray B, et al: Investigations in the design of a novel linac-MRI system. Int J Radiat Oncol Biol Phys 2007;69:S19.

40 Lagendijk J, Raaijmakers AJ Overweg J, et al: MR-XRT at 1.5T, the UMC Utrecht hybrid MRI linac system. Med Phys 2009;36:2775.

41 Craig T, Moiseenko V, Battista J, Van Dyk J: The impact of geometric uncertainty on hypofractionated external beam radiation therapy of prostate cancer. Int J Radiat Oncol Biol Phys 2003;57:833–842.

42 Hugo G, Vargas C, Liang J, et al: Changes in the respiratory pattern during radiotherapy for cancer in the lung. Radiother Oncol 2006;78:326–331.

43 Brock KK, Hawkins M, Eccles C, et al: Improving image-guided target localization through deformable registration. Acta Oncol 2008;47:1279–1285.

44 McParland BJ, Kumaradas JC. Digital portal image registration by sequential anatomical matchpoint and image correlations for real-time continuous field alignment verification: Med Phys 1995;22:1063–1075.

45 Michalski JM, Graham MV, Bosch WR, et al: Prospective clinical evaluation of an electronic portal imaging device. Int J Radiat Oncol Biol Phys 1996;34:943–951.

46 Barker J, Garden A, Dong L, et al: Radiation-induced anatomic changes during fractionated head and neck radiotherapy: a pilot study using an integrated CT-LINAC system. Int J Radiat Oncol Biol Phys 2003;57:S304.

47 Cheng CW, Wong J, Grimm L, et al: Commissioning and clinical implementation of a sliding gantry CT scanner installed in an existing treatment room and early clinical experience for precise tumor localization. Am J Clin Oncol 2003;26:e28–e36.

48 Ghanei A, Soltanian-Zadeh H, Ratkewicz HA, et al: A three-dimensional deformable model for segmentation of human prostate from ultrasound images. Med Phys 2001;28:2147–2153.

49 Mackie TR, Holmes T, Swerdloff S, et al: Tomotherapy: a new concept for the delivery of dynamic conformal radiotherapy, Med Phys 1993;20:1709–1719.

50 Jaffray DA, Siewerdsen JH: Cone-beam computed tomography with a flat-panel imager: initial performance characterization. Med Phys 2000;27:1311–1323.

51 Ford EC, Chang J, Mueller K, et al: Cone-beam CT with megavoltage beams and an amorphous silicon electronic portal imaging device: potential for verification of radiotherapy of lung cancer. Med Phys 2002;29:2913–2924.

52 Dawson LA, Brock KK, Kazanjian S, et al: The reproducibility of organ position using active breathing control (ABC) during liver radiotherapy. Int J Radiat Oncol Biol Phys 2001;51:1410–1421.

53 Wong JW, Sharpe MB, Jaffray DA: The use of active breathing control (ABC) to reduce margin for breathing motion, Int J Radiat Oncol Biol Phys 1999;44:911–919.
54 Dehnad H, Nederveen AJ, van der Heide UA, et al: Clinical feasibility study for the use of implanted gold seeds in the prostate as reliable positioning markers during megavoltage irradiation, Radiother Oncol 2003;67:295–302.
55 Balter JM, Lam KL, Sandler HM, et al: Automated localization of the prostate at the time of treatment using implanted radiopaque markers: technical feasibility. Int J Radiat Oncol Biol Phys 1995;33:1281–1286.
56 Litzenberg D, Dawson LA, Sandler H, et al: Daily prostate targeting using implanted radiopaque markers. Int J Radiat Oncol Biol Phys 2002;52:699–703.
57 Jaffray DA, Brock KK, Nichol A, et al: An analysis of inter-fraction prostate deformation relative to implanted fiducial markers using finite element modeling. Int J Radiat Oncol Biol Phys 2004;60:S334–S335.
58 van Herk M, Kooy HM: Automatic three-dimensional correlation of CT-CT, CT-MRI, and CT-SPECT using chamfer matching. Med Phys 1994;21:1163–1178.
59 Dawson LA, Eccles C, Kirilova A, Brock KK: Three dimensional motion of liver Tumours using cine MRI compared to liver motion assessed at fluoroscopy. Radiat Oncology 2006;A483:S214.
60 Hawkins M, Brock KK, Eccles C, et al: Assessment of residual error in liver position using kV cone-beam CT for liver cancer high precision radiation therapy. Int J Radiat Oncol Biol Phys 2006;66:610–619.
61 Coselmon MM, Balter JM, McShan DL, Kessler ML: Mutual information based CT registration of the lung at exhale and inhale breathing states using thin-plate splines. Med Phys 2004;31:2942–2948.
62 Keall PJ, Joshi S, Vedam SS, et al: Four-dimensional radiotherapy planning for DMLC-based respiratory motion tracking. Med Phys 2005;32:942–951.
63 Rietzel E, Chen GT, Choi NC, Willet CG: Four-dimensional image-based treatment planning: target volume segmentation and dose calculation in the presence of respiratory motion. Int J Radiat Oncol Biol Phys 2005;61:1535–1550.
64 Dong L, Boyer AL: An image correlation procedure for digitally reconstructed radiographs and electronic portal images. Int J Radiat Oncol Biol Phys 1995;33:1053–1060.
65 Fitchard EE, Aldridge JS, Reckwerdt PJ, Mackie TR: Registration of synthetic tomographic projection data sets using cross-correlation. Phys Med Biol 1998;43:1645–1657.
66 Meyer CR, Boes JL, Kim B, et al: Demonstration of accuracy and clinical versatility of mutual information for automatic multimodality image fusion using affine and thin-plate spline warped geometric deformations. Med Image Anal 1997;1:195–206.
67 Maes F, Vandermeulen D, Suetens P: Comparative evaluation of multiresolution optimization strategies for multimodality image registration by maximization of mutual information. Med Image Anal 1999;3:373–386.
68 Kim J, Fessler JA, Lam KL, et al: A feasibility study of mutual information based setup error estimation for radiotherapy. Med Phys 2001;28:2507–2517.
69 D'Agostino E, Maes F, Vandermeulen D, Suetens P: A viscous fluid model for multimodal non-rigid image registration using mutual information. Med Image Anal 2003;7:565–575.
70 Kashani R, Hub M, Halter JM, et al: Objective assessment of deformable image registration in radiotherapy: a multi-insititution study. Med Phys 2008;35:5944–5953.
71 Brock KK, Deformable Registration Accuracy Consortium: Results of a multi-institution deformable registration accuracy study (MIDRAS). Int J Radiat Oncol Biol Phys 201;76:583–96.
72 Keall PJ, Starkschall G, Shukla H: Acquiring 4D thoracic CT scans using a multislice helical method. Phys Med Biol 2004;49:2053–2067.
73 Keall P: 4-dimensional computed tomography imaging and treatment planning,' Semin Radiat Oncol 2004;14:81–90.
74 Keall PJ, Siebers JV, Joshi S, Mohan R: Monte Carlo as a four-dimensional radiotherapy treatment-planning tool to account for respiratory motion. Phys Med Biol 2004;49:3639–3648.
75 Christensen GE, Carlson B, Chao KS, et al: Image-based dose planning of intracavitary brachytherapy: registration of serial-imaging studies using deformable anatomic templates. Int J Radiat Oncol Biol Phys 2001;51:227–243.
76 Christensen GE, Joshi SC, Miller MI: Volumetric transformation of brain anatomy. IEEE Trans Med Imaging 1997;16:864–877.
77 Haller JW, Banerjee A, Christensen GE, et al: Three-dimensional hippocampal MR morphometry with high-dimensional transformation of a neuroanatomic atlas. Radiology 1997;202:504–510.
78 Thirion JP: Image matching as a diffusion process: an analogy with Maxwell's demons. Med Image Anal 1998;2:243–260.

79 Wang H, Dong L, Lii MF, et al: Implementation and validation of a three-dimensional deformable registration algorithm for targeted prostate cancer radiotherapy. Int J Radiat Oncol Biol Phys 2005;61:725–735.

80 Wang H, Dong L, O'Daniel J, et al: Validation of an accelerated 'demons' algorithm for deformable image registration in radiation therapy. Phys Med Biol 2005;50:2887–2905.

81 Brock KK, McShan DL, Ten Haken RK, et al: Inclusion of organ deformation in dose calculations. Med Phys 2006;30:290–295.

82 Brock KK, Balter JM, Dawson LA, et al: Automated generation of a four-dimensional model of the liver using warping and mutual information. Med Phys 2003;30:1128–1133.

83 Rohlfing T, Maurer CR Jr, O'Dell WG, Zhong J: Modeling liver motion and deformation during the respiratory cycle using intensity-based nonrigid registration of gated MR images. Med Phys 2004;31:427–432.

84 Rosu M, Chetty IJ, Balter JM, et al: Dose reconstruction in deforming lung anatomy: dose grid size effects and clinical implications. Med Phys 2005;2:2487–2495.

85 Paganetti H, Jiang H, Adams JA, et al: Monte Carlo simulations with time-dependent geometries to investigate effects of organ motion with high temporal resolution. Int J Radiat Oncol Biol Phys 2004;60:942–950.

86 Schaly B, Bauman GS, Battista JJ, Van Dyk J: Validation of contour-driven thin-plate splines for tracking fraction-to-fraction changes in anatomy and radiation therapy dose mapping. Phys Med Biol 2005;50:459–475.

87 Venugopal N, McCurdy B, Hnatov A, Dubey A: A feasibility study to investigate the use of thin-plate splines to account for prostate deformation. Phys Med Biol 2005;50:2871–2885.

88 Schaly B, Kempem JA, Bauman GS, et al: Tracking the dose distribution in radiation therapy by accounting for variable anatomy. Phys Med Biol 49 2004;5:791–805.

89 Schnabel JA, Tanner C, Castellano-Smith AD, et al: Validation of nonrigid image registration using finite-element methods: application to breast MR images. IEEE Trans Med Imaging 2003;2:238–247.

90 Zhang T, Orton NP, Mackie TR, Paliwal BR: Technical note: a novel boundary condition using contact elements for finite element based deformable image registration. Med Phys 2004;31:2412–2415.

91 Brock KK, Hollister SJ, Dawson LA, Balter JM: Technical note: creating a four-dimensional model of the liver using finite element analysis. Med Phys 2002 29:1403–1405.

92 Brock KK, Dawson LA, Sharpe MB, et al: Feasibility of a novel deformable image registration technique to facilitate classification, targeting, and monitoring of tumor and normal tissue. Int J Radiat Oncol Biol Phys 2006;64:1245–1254.

93 Liang J, Yana D: Reducing uncertainties in volumetric image based deformable organ registration. Med Phys 2003;30:2116–21223.

94 Yan D, Jaffray DA, Wong JW: A model to accumulate fractionated dose in a deforming organ. Int J Radiat Oncol Biol Phys 1999;44:665–675.

95 Bharatha A, Hirose M, Hata N, et al: Evaluation of three-dimensional finite element-based deformable registration of pre- and intraoperative prostate imaging. Med Phys 2001;28:2551–2560.

96 Martinez AA, Yan D, Lockman D, et al: Improvement in dose escalation using the process of adaptive radiotherapy combined with three-dimensional conformal or intensity-modulated beams for prostate cancer. Int J Radiat Oncol Biol 2001;50:1226–1234.

97 Yan D, Vicini F, Wong J, Martinez A: Adaptive radiation therapy. Phys Med Biol 1997;42:123–132.

98 Mohan R, Zhang X, Wang H, et al: Use of deformed intensity distributions for on-line modification of image-guided IMRT to account for interfractional anatomic changes. Int J Radiat Oncol Biol Phys 2005;61:1258–1266.

99 Velec J, Moseley J, Dawson LA, et al: Accumulated treatment dose for image guided liver radiotherapy using deformable registration of 4D cone beam CT. Int J Radiat Oncol Biol Phys 2008;72:S1–49.

100 Yan D, Lockman, D, Martinez A, et al: Computed tomography guided management of interfractional patient variation. Semin Radiat Oncol 2005;15:168–179.

101 Barker JL Jr, Garden AS, Ang KK, et al: Quantification of volumetric and geometric changes occurring during fractionated radiotherapy for head and neck cancer using an integrated CT/linear accelerator system. Int J Radiat Oncol Biol Phys 2004;59:960–970.

102 Kuo YC, Wu TH, Chung TS, et al: Effect of regression of enlarged neck lymph nodes on radiation doses received by parotid glands during intensity-modulated radiotherapy for head and neck cancer. Am J Clin Oncol 2006;2:600–605.

103 Castadot P, Lee JA, Geets X, Gregoire V: Adaptive radiotherapy of head and neck cancer. Semin Radiat Oncol 2010;20:84–93.
104 Han C, Chen YJ, Liu A, et al: Actual dose variation of parotid glands and spinal cord for nasopharyngeal cancer patients during radiotherapy. Int J Radiat Oncol Biol Phys 2008;70: 1256–1262.
105 Pekar V, McNutt TR, Kaus MR: Automated model-based organ delineation for radiotherapy planning in prostatic region. Int J Radiat Oncol Biol Phys 2004;60:973–980.
106 Brock KK, McShan DL, Ten Haken RK, et al: Inclusion of organ deformation in dose calculations. Med Phys 2003;30:290–295.
107 Pevsner A, Davis B, Joshi S, et al: Evaluation of an automated deformable image matching method for quantifying lung motion in respiration-correlated CT images. Med Phys 2006;33: 369–376.
108 Castadot P, Lee JA, Parraga A, et al: Comparison of 12 deformable registration strategies in adaptive radiation therapy for the treatment of head and neck tumors. Radiother Oncol 2008;89:1–12.
109 de la Zerda A, Armbruster B, Xing L: Formulating adaptive radiation therapy (ART) treatment planning into a closed-loop control framework. Phys Med Biol 2007;52:4137–4153.

John L. Meyer, MD, FASTRO
Department of Radiation Oncology, Saint Francis Memorial Hospital
900 Hyde Street
San Francisco, CA 04109 (USA)
Tel. +1 415 353 6420, Fax +1 415 353 6428, E-Mail jmeyer1@chw.edu

Controversies in the Adoption of New Healthcare Technologies

Paul E. Wallner[a] · Michael L. Steinberg[b] · Andre A. Konski[c]

[a]21st Century Oncology, Fort Myers, Fla., [b]Department of Radiation Oncology, David Geffen School of Medicine at UCLA, Los Angeles, Calif., and [c]Department of Radiation Oncology, Wayne State University School of Medicine, Detroit, Mich., USA

Abstract

Healthcare economists generally agree that the development and rapid introduction of new technologies and the expanding utilization of existing ones in national healthcare systems have been significant factors in the dramatic and potentially unsustainable growth in healthcare spending. Creating a rational system for evaluation of emerging technologies in this country has been complicated by 3 broad issues: the often conflicting needs and expectations of the variety of stakeholders; an arcane and often illogical system of service valuation and payment; and the lack of a standardized, transparent and validated approach to the measurement of 'value.' Recent discussions on reforming the elements of healthcare delivery have increased focus on these systemic shortcomings and conflicts. As a specialty that is clinically wedded to modern and increasingly expensive technology, radiation oncology has often been singled out for scrutiny. A thorough examination and understanding of the various factors and controversies involved in technology development, implementation and valuation analysis is essential to rational growth and development of the specialty.

Copyright © 2011 S. Karger AG, Basel

Health Technology Assessment

The widespread adoption of clinical services employing new technologies is ultimately determined by whether public and/or commercial payers will recognize, authorize and reimburse the use of the technologies. There are a number of major stakeholders with disparate agendas involved in this cycle of assessment, approval, implementation, reimbursement and market penetration: (1) inventors, developers, vendors and investors; (2) individual and institutional healthcare providers; (3) patients, families, caregivers and advocacy groups; (4) public and commercial payers; and (5) policy-makers. Each of these groups approach the issues involved from very different vantage points, and from very different senses of urgency regarding their individual concerns. The first group is concerned that new products or devices are not being introduced quickly enough, or that approved and marketed devices have insufficient levels of reimbursement or market share. The second group may have similar concerns, but may also be interested in professional development or personal and institutional competitive advantages. Among the other groups, patients and their supporters are most concerned about rapid access to what they perceive to be advances in their care (despite the financial hazards related to increasing their own direct costs), and payers and policy-makers increasingly raise concerns regarding systemic cost, risk and value. Eventually, in an era of active dialogue regarding healthcare reform, each of these groups presumably must ameliorate hardcore, uncompromising positions to progress to a more orderly system of health technology adoption and utilization.

Medical technology development and introduction into the care process have accelerated over the past 50 years and intensified in the current era of sophisticated computer hardware, software and microcircuitry. Despite this explosion in development, methodologies for assessment of new medical technologies and treatments have not progressed at a similar pace [1]. Currently, the pathways to the adoption of new technologies that are truly of 'value' remain essentially market-driven, and a social paradigm has emerged characterized by the philosophy that new technology by definition equates to improvement in care. More often than not, 'proof' of benefit is not a requirement for adoption, and the various stakeholders cannot even agree on a clear definition of appropriate proof. In this milieu, 'efficacy by proclamation,' essentially weighing the testimonials of a variety of influential voices in the marketplace and our professional societies, has become the norm, absent a systematic analysis that would be appropriate [2].

Health Technology Assessment: Methods and Their Users

Because of the myriad of issues noted above, a serious attempt has been initiated to develop an academic regimen of health technology assessment (HTA):

> *Health technology assessment* encompasses the evaluation of medical technologies for evidence of safety, efficacy, cost, cost-effectiveness and risk-benefit, including ethical, legal and policy implications. Evaluations can be specific to a single technology, or comparative among technologies that are designed to provide similar processes [3].

HTA evaluations are not analogous or redundant to those performed by the US Food and Drug Administration (FDA), which is legislatively mandated to evaluate only safety and effectiveness of any device when used for its stated indications. The FDA is actually legislatively precluded from evaluating cost and cost-effectiveness – what might be termed *value* in the healthcare reform debate. Determination of safety and effectiveness, as judged by the FDA, essentially means only that a device can perform the generally limited indication that is requested for approval by a vendor. It must do so within a level determined to be safe, with the Agency's determination generally based on limited data often obtained under tightly controlled and monitored test situations. The process does not determine the *comparative effectiveness* of the device to existing ones already approved and marketed. Nor does it attempt to determine if the new device adds to the *process of care* for any particular disease, or improves the ultimate *outcome* of a therapeutic intervention. As an example, FDA evaluation of an implantable organ motion tracking device might limit its analysis to whether the product is biocompatible and positionally stable after implantation, whether it could cause electronic interference or be manipulated/moved by MRI, and whether it is indeed capable of externally tracking the movement of the implanted monitor or fiducial. The evaluation does not assess whether the tracking in and of itself adds to the process of care for a particular health problem or whether one device performs better or less expensively than another.

HTA initiatives, each comparing the effectiveness of medical services, devices, drugs, therapies and procedures, have been undertaken worldwide. In the United States, most of these evaluations have been performed in the private sector, carried out by vendors unilaterally, or by academic centers and 'independent' or payer-funded health policy organizations. In Europe and other areas of the developed world, this trend has been reversed, with the overwhelming majority of comparisons performed by, or for, public agencies. In some respects, this disparity is a reflection of the American free-market approach to health care delivery. In all of these venues, there are 3 basic methodologies used in HTA.

(1) The first methodology, and arguably the most widespread, is essentially limited to review and analysis of the available literature. The US Agency for Healthcare Research and Quality (AHRQ), a component agency of the US Department of Health and Human Services has been the primary public agency functioning in this manner to date. Other evaluation organizations, including among others the VA Technology Assessment Program, the Cochrane Collaborative [4] and the California Technology Assessment Forum, also function using literature review solely. Specialty societies have been somewhat hesitant to launch HTA efforts, often because of ongoing close and multilevel relationships with vendors, but the American Society for Radiation Oncology has created an Emerging Technology Committee, which currently carries out literature review and analysis on a limited scale.

The shortcomings of evaluations using literature review are easily understood. There is generally a limited or absent literature base for new and emerging technologies, the existing literature may provide inconclusive or conflicting results, and there may be significantly different interpretations of available data. Literature related to these technologies is often generated by vendors, vendor-funded entities or early investigators/adopters under idealized conditions, and it rarely offers findings in terms of meaningful health outcomes. Conflicts of interest in preparation of these reports are almost unavoidable and are often clearly apparent in the interpretation of results.

(2) A second methodology begins with literature review but adds selected elements of health policy judgment such as cost-effectiveness assessments. This type of assessment is seen with increasing frequency from academic and payer-funded organizations. Despite attempts to formalize the process and add layers of external advisers and reviewers, there continue to be a variety of issues regarding the quality and objectivity of the judgment processes. One of the more visible organizations involved in this type of assessment is the Institute for Clinical and Economic Review (ICER), which is based at Harvard University but includes individuals from other distinguished institutions and represents many academic disciplines. ICER has performed and published a number of radiation oncology evaluations including assessments of various modalities used for the treatment of early stage, low-intermediate risk prostate cancer. These have compared results of active surveillance with various treatment modalities by using a Markov cohort model methodology yielding cost-effectiveness conclusions [5].

(3) Over the past several years and primarily based on initiatives developed within the Centers for Medicare and Medicaid Services (CMS), there has been increased interest in a third method of HTA: actual clinical trials evaluating new, emerging or marketed devices and procedures. Funding can be provided through defined channels. Under a reimbursement concept termed *Conditional Coverage Strategies* (CCS), it is likely (unless there is congressional interference to reduce the efforts secondary to intense pressure from vendors) that these types of initiatives will steadily increase in number and scope.

One element of CCS is called *Coverage with Evidence Development* (CED), which entails payment for selected procedures for the predetermined length of an investigation and analysis period, followed by a final payment policy decision [6]. Comparisons can be performed in traditional randomized controlled trials (RCTs), or alternatively, from tightly controlled registries. A number of CED trials have been carried out successfully with high-impact results, including the National Emphysema Treatment Trial. In this trial, lung volume reduction surgery was studied. It had rapidly gained in popularity with providers and patients for the management of chronic obstructive pulmonary disease despite the absence of literature-based data demonstrating a clear outcome benefit. At its height of penetra-

Table 1. International technology assessment initiatives

National Institute for Clinical Effectiveness (NICE, UK)
Medical Services Advisory Committee (MSAC, Australia)
Institute for Quality and Efficiency in Health Care (IQWiG, Germany)
Committee on Evaluation and Diffusion of Innovative Technologies (CEDIT, France)
Swedish Council on Technology Assessment in Health (SBU, Sweden)
Canadian Agency for Drugs and Technologies in Health (CADTH)

Examples of publicly funded HTA agencies.

tion into the health care market, the procedures were costing Medicare alone approximately USD 300 million per year. At the study's completion, it became apparent that the surgical group had a *shorter* life expectancy than the control population, which had been managed medically including aggressive pulmonary physical therapy. Following circulation of the study results, use of lung volume reduction surgery fell dramatically, and Medicare payments for the procedure have nearly disappeared now.

Similar evidence-driven reductions in cost and improvements in care have been generated by CMS-sponsored CED trials, including evaluations of high-dose chemotherapy/bone marrow transplantation for high-risk breast cancer and implantable cardiac defibrillators for management and prevention of specific arrhythmias. In the latter implantable cardiac defibrillator study, rather than suggesting that the technology be abandoned entirely, there was appropriate identification of those patients that were most likely to benefit.

Unlike the mostly private and voluntary HTA initiatives in the USA, other nations have taken more aggressive, publicly-funded and legislatively-mandated approaches. The National Institute for Health and Clinical Excellence (NICE) was organized in the United Kingdom in 1999 and has received much publicity (and notoriety). NICE has performed a number of significant drug and device assessments, and also developed treatment guidelines for a variety of conditions, all of which have directly driven UK National Health Service payment policies. NICE currently employs a threshold value of GBP 20,000 per year for cost of a quality-adjusted life-year (QALY) as a maximum value for payment. The system is relatively transparent and functions by developing proposals for review by panels with broad stakeholder representation that ultimately make policy decisions. Beneficiaries have the right to appeal panel decisions, and decisions have been reversed on occasion [7]. Throughout the developed world, other public agencies are active participants in technology assessment (table 1), and most have policy-making or policy-recommending authority. All of these agencies espouse a similar mission of cost-control and improved value and safety.

Health Technology Assessment through Clinical Trials

Healthcare economists have estimated that new technologies, or new or increased uses of mature technologies, may be responsible for a 38–62% increase in healthcare spending in the United States [8–10]. By rigorous HTA standards, much of this technology is unproven. Well-designed, practical and affordable prospective trials are needed in relevant applied research. There could be many collaborators for these trials, such as large cooperative clinical trial groups that already exist for investigating many disease sites and treatment modalities and also professional specialty societies. The AHRQ is a 'Comparative Effectiveness Institute' mandated to service by current healthcare reform legislation. There are a variety of innovative public/private partnerships, and active discussions among many of these entities are ongoing. Finally, private not-for-profit organizations such as the Center for Medical Technology Policy have been active in generating prospective outcome evaluations, and the authors (P.W., A.K.) have been involved with this group to promote important trials [11].

Theoretically, the existing large cooperative clinical trial groups could be superb testers of comparative effectiveness, although the Radiation Therapy Oncology Group and other organizations would argue that all of their efforts are essentially comparative effectiveness studies already. However, the groups are neither organized nor funded to evaluate specific applied research initiatives for HTA, and they have not been encouraged or incentivized by their program managers at the National Institutes of Health (NIH) to do so. Changes in these policies and funding streams would require high-level redirection of funds and programs within the NIH and probably Executive Branch and Congressional mandates.

The AHRQ seems to be moving toward greater involvement in comparative effectiveness research (CER) through expediting and facilitating comparative clinical trials, partly because of its added mandate and funding embedded in the American Recovery and Reinvestment Act of 2009. This allocated USD 1.1 billion in direct funding for CER infrastructure and programs. In addition, prominent policy makers in the Obama administration have written and publicly espoused a commitment to strong, vibrant and meaningful CER efforts. Although iterations of healthcare reform measures under consideration at the time of this writing specifically preclude the results of CER from being used for payment policy decisions (at least by CMS), it is unlikely that these data would or could be ignored moving forward, both in the public and private payer sectors. One possible scenario often mentioned is that patients would have a greater financial stake in decision making regarding their own care, if health insurance copayments for 'unproven' devices or procedures were higher than those for measures that had been verified for 'value' by rigorous CER.

Future Challenges

There are many inherent problems and essential requirements in the establishment and maintenance of programs devoted to HTA. Similar to any large-scale, long-term research effort, a stable and consistent funding base is needed to assure that appropriate infrastructure is available and that an adequate stream of trained and highly motivated investigators is available. Entities performing and reporting research must be visible, transparent and representative of diverse stakeholder interests. Further, their findings must be subject to careful scrutiny to uphold the necessary trust of the stakeholders. Investigators and their conclusions must also be shielded from undue influence from either the public or private sectors.

Providers and patients offer additional challenges to meaningful CER in our current system. Providers may often be driven by personal and/or institutional development or competitive/financial motivations, while patients, caregivers and advocates may be locked onto unproven and sometimes illogical beliefs that newer (and usually more expensive) technologies will always be better. The public discourse regarding these important issues has not been facilitated by stoking fears regarding access and rationing of services. Vendors raise the fear that vibrant and meaningful CER will inhibit research and development in new technologies and procedures, which they argue have helped drive the American consumer-based economy. They and others suggest that CER will increase the specter of a wave of litigation secondary to reduced utilization and access to these 'life-saving' advances. Only a strong, unbiased and reproducible HTA base can adequately address these concerns.

Pragmatic political, social, professional and economic obstacles also stand in the way of objective technology assessment programs. At present, FDA HTA efforts are hampered by legislative restrictions on agency actions that would require a legislative amendment, which may ultimately be difficult to achieve in a highly partisan Congress. HTA studies are costly to perform, and the economic downturn has set back availability of necessary discretionary funds to launch such efforts. Standardized methodologies of research, data collection and data analysis must be developed, and individual providers, institutions and vendors may need to accept a level of personal risk in an attempt to move toward a stronger and ultimately more vibrant and sustainable healthcare system.

Payment Policy Development: Past and Present Realities, and Future Uncertainties

How are payments for new medical services determined? In the United States, the pathway to incorporation and payment coverage for new technologies is typically not scientifically driven. Often the science is trumped by politics, and the avenues that lead to coverage for a new technology may even belie medical evidence and cogent policy. The actors in this drama include not only physicians and their patients, but also advocacy groups, hospitals, device vendors, drug companies, organized medicine, the media, politicians and even paid lobbyists [12].

For a moment, consider the 1990s odyssey of bone marrow transplantation (BMT) for women with locally advanced breast cancer who faced the dismal circumstance of known low survival rates. BMT was enthusiastically offered with a rationale and a promise considered by many providers and desperate patients to be medically sound. It was thought of as being so much better than standard chemotherapy that many argued strenuously against RCTs to prove its efficacy. When the results of the RCTs came out, delayed by many years as a result of true believers not wanting patients to enter trials and patients themselves believing that participation in a trial would put their life in jeopardy, it was found that the treatment was ineffective and actually more toxic than conventional chemotherapy regimens. Although this episode is often cited as an example of the system's inadequacies and the need for evidence-based coverage, its ultimate effect on policy has remained limited.

The Medicare (CMS) Mandate and New Technologies

Medicare policy regarding the incorporation of new medical technologies into routine medical care has had significant effect on coverage and payment policies for the entire healthcare system. Medicare has statutory authority to cover services that are 'reasonable or necessary' for the diagnosis or treatment of illnesses or injury. In this regard, the process through which Medicare determines coverage and payments for medical services and technology is arcane, but continues to evolve. It comes as a surprise to many that within the statutes, Medicare does not have direct authority to consider cost and cost-effectiveness for individual treatments and devices [13].

In the last decade, CMS (the USA agency managing government-sponsored healthcare payments including Medicare and Medicaid) has started to consider clinical outcomes data related to quality of life, morbidity and mortality. In addition, CMS has begun to look at questions such as: 'Is the service or device as good or better than currently covered alternatives?', 'Does it have broad relevance to the larger Medicare population?' and 'Is the service of high cost with minimal or marginal benefit [13]?' In fact, the economic stimulus package Congress passed in February 2009 allocated over USD 400 million to CMS to study the *comparative effectiveness* of medical treatments.

To better address the emerging needs to evaluate new medical treatments and devices, CMS established a public committee, the Medical Evidence Development and Coverage Advisory Committee (MEDCAC), to provide the agency with independent external guidance and expert advice on new and emerging technologies and treatments, as well as address how to assimilate emerging technology into the system. To address the question of what evidence is necessary to deem a treatment or device ready for routine use and for Federal government payment, MEDCAC dedicated an entire meeting to what it termed 'evidentiary priorities' [14, 15]. The committee considered various methodologies to obtain evidence for new treatments including RCTs and the use of registries. Also considered were practical arguments for flexibility in the use of each methodology, since there are potential drawbacks of RCTs and both positive roles and limitations of registries. The committee concluded that although RTCs are valuable, they are not always appropriate. In fact, in this era of personalized medicine, some MEDCAC members suggested that RTCs may not always be an appropriate evaluation method for many of the new genetic and proteomic interventions. Some members suggested that evidence criteria must be applied on the basis of specific technologies and conditions. The committee also identified a clear need to recognize 'thresholds of evidence' for accepting new technologies and to define standards for determining those thresholds, but did not reach agreement on the use of any generic standards. Unlike the commercial insurance industry, MEDCAC avoided definitive criteria, in favor of flexibility in evaluating technology. The objective was to avoid politically unpopular noncoverage decisions according to testimony at the meeting [16].

Valuation and Payment for Physician Services and Technologies: The Roles of CPT©, RUC and CMS

Medicare and the commercial insurance industry look to the Common Procedural Terminology (CPT©) nomenclature, governed by the CPT Editorial Panel [a committee of the American Medical Association (AMA)], for *descriptions* of medical procedures and services. Once the code is established by the panel, it is submitted to the AMA/Specialty Society Relative Value Scale Update Committee (RUC) for the *valuation* of the professional work and the technical aspects (practice expense) of the service.

Services that are in routine use and have established RUC valuations are designated CPT Category I. To qualify as Category I, a service must demonstrate: (1) FDA approval, (2) a distinct medical activity provided by many physicians in the United States and (3) clinical efficacy well established and documented in the scientific literature (currently interpreted as 5 articles published in the peer-reviewed literature of the United States) [17].

After the CPT Editorial Panel approves a new code, it forwards the code to the RUC for valuation of the physician work and practice expense associated with the delivery of the service. In this regard, the RUC solicits interested specialty societies to perform a rigorous survey to determine the relative value of the physician work compared with other procedures in the nomenclature, as well as to gather practical expense information about the clinical use of this service. With this information, the RUC adjudicates the relative value units for the physician work, and relative value units are multiplied by a CMS established dollar conversion factor to realize a dollar-based valuation for the procedure. The RUC also gathers the direct practice expense information associated with the service, including the cost of the equipment, the labor costs and supplies necessary to fulfill the service. The RUC partially validates this information and then passes it directly on to CMS. The agency uses this information to determine the technical value of the procedure and accepts greater than 90% of the RUC's physician work valuations without any alteration [17].

The notion of 'common use' of a new technology is subjectively defined by the CPT rules as 'many physicians'; but, in practice, the RUC requires 30 survey responses by individual medical users to determine a procedure's valuation. In the experience of ASTRO and other specialty societies, about 100 surveys must be tendered in order to receive the required 30 completed and valid survey responses. As far as the authors can determine, from a policy perspective, the implications of these numbers on technology assessment and evidence development have not been validated.

The CPT review process has become much more stringent in recent years. In the past several years the CPT Editorial Panel has substantially raised the standards for evidence required to approve Category I applications. It has done so by more closely scrutinizing the supporting peer-reviewed literature, asking that the literature demonstrate improvement in an actual health outcome as opposed to simple technical enhancement, and by investigating the veracity of claims that new procedures have been widely accepted within the medical community. Many of the purveyors of new and emerging technologies, particularly small start-up companies, find these elevated standards to be daunting and at times insurmountable. Members of the venture capital community have increasingly questioned their ability to back new interventions because of these higher requirements. The problem is further complicated by the fact that, although the bar has been raised, the AMA and CMS have not provided specific, precisely defined standards of evidence for new technology. At the extremes of evidence analysis and evaluation, there is often common agreement among policy makers (i.e. circumstances of minimal or overwhelming evidence of efficacy), but in the middle, the standard is similar to one Supreme Court Justice's view of obscenity – 'they can't define it, but they know it when they see it'.

In 2002, CPT Category III codes were introduced for the purpose of defining and tracking the use of new and emerging technologies. Initially intended only for categorizing data collection on investigational services, recently some of these codes have attained reimbursement status by Medicare and non-Medicare payers on a region-by-region basis [17].

Medicare Payment to Hospitals: Ambulatory Payment Classifications/Hospital Outpatient Prospective Payment System

The Hospital Outpatient Prospective Payment System (HOPPS) uses the Ambulatory Payment Classifications (APCs) to determine Medicare payment of the technical component of hospital-based outpatient procedures. The system parallels the fee schedule for the CPT although there is no physician work included as part of the valuation. The APC reimbursement rates are developed by averaging and bundling procedures described by CPT, Healthcare Common Procedure Coding System (HCPCS) Level II and other revenue codes. The method then groups the codes based on similarity in clinical resources and established payment rates. CMS gathers cost and charge information from individual hospital providers on the particular procedures, and then employs a complex 'cost-to-charge ratio' methodology to set reimbursement values each year. Unlike the RUC valuation methodology for nonfacility (nonhospital-based) services, the APC methodology employs an advisory committee composed of various stakeholders, though their recommendations carry limited weight and final decisions on payment rates reside within CMS.

The HOPPS/APCs provide for technical aspects of service only and not for physician work. Emerging technologies may be covered with a G-code or so-called 'new technology APC'. These codes and their valuation have in the past been established without input from professional societies or from experts in the field. Instead, vendors employ individuals, typically attorneys, to lobby CMS for assignment of these codes. Granting a code for a new technology in the HOPPS does not require a precedent CPT code. The rules of evidence appear to be consistent with the CMS statutory language of 'reasonable and necessary' and 'safe and effective', but represent lower thresholds when compared with the CPT process. Comparative cost or clinical effectiveness data are not mandated. This circumstance, therefore, can create powerful economic incentives on the technical side, which can drive equipment utilization even for devices whose efficacy has not been validated.

CMS and the Business of Proton Beam Therapy

There are currently 6 active, full clinical proton facilities and one limited clinical site in the United States, and more than 20 in development or planned over the next few years. The pricing of protons is not incorporated in the CPT-based Medicare fee schedule. Proton beam therapy (PBT) treatment delivery has never been valued through the RUC process for the purpose of gathering valid practice expense costs to pass on to the CMS for its evaluation. However, the technical aspects have been priced through the APC process for more than 10 years. Using the cost-to-charge methodology for setting the price of APCs, PBT suffered a 30% decrease in 2008 and a 16% decrease in 2009. However, due to the pricing methodology as new high-cost facilities go on-line, this APC pricing could increase in the next few years.

The high and rapidly increasing costs of PBT are getting the attention of CMS. In 2009, PBT was listed among procedures to be considered for a Medicare National Coverage Determination. This procedural 'red flag' is prompted by policy makers when a new treatment or technology raises concerns; that is, its rapid diffusion could have a significant impact on the financial stability of the Medicare program and is coupled with significant uncertainties about its health benefit, patient selection or appropriate facility and staffing requirements. CMS highlighted specific concerns regarding PBT for prostate cancer, stating that it is: 'Proposed as means to concentrate radiation therapy and reduce side effects. Very high upfront cost to build these facilities and thus only at very few facilities. For prostate cancer treatment, no current comparative trials comparing usual therapy [15].'

Current Socioeconomic Context

The paths that emerging technologies take to be incorporated into the United States healthcare delivery system are products of the social, economic and political milieu, which effectively modulate the rapidity of market diffusion of these technologies. However, given the current environment of dramatically rising healthcare costs and growing numbers of uninsured persons, US health policy makers have begun to embrace policies aimed at improving healthcare value and addressing cost, as well as recognize cost-effectiveness and comparative effectiveness analyses. As a result, the payment system is currently in transition. At present, the effectiveness of emerging, expensive medical technologies can only be determined after (a) substantial capital has been invested for product development, and (b) significant operating revenue is made available for the clinical activity needed to evaluate its treatment efficacy. Policies going forward will need to address the economic realities of healthcare delivery, and must effectively embrace the use of evidence-based assessment of technology to ensure the solvency of the system. In addition, such polices must define ways to pay for the development and clinical testing of new and emerging medical advancements that are similarly cost effective, or development efforts will stagnate [12].

In the meantime, health financing inconsistently covers the cost of many new technologies, and there are few mechanisms for evaluating their effectiveness. Their purveyors (vendors) and the healthcare providers using the technologies have little to gain by submitting to evaluations of their comparative effectiveness. With providers incentivized to use more technology (some of questionable or marginal comparative benefit), and with patients seeing their healthcare as a free good (with little moral hazard in demanding more medical care for themselves), our system is inefficient in controlling the costs to society.

Cost-Effectiveness Methodologies, Implications in an Era of Value-Based Purchasing

To objectively evaluate the benefit of new technologies compared with their costs, one must develop reliable measures of their outcomes that are transparent, trusted and useable. These results can then be the basis for value-based purchasing. Cost-effectiveness methodologies, although limited in their influence on coverage decisions for new technologies or medications currently, will certainly influence health policy in the future. We must be aware of their specific applications to radiation oncology, while the percent of the gross domestic product (GDP) devoted to healthcare progressively increases and the needs to control costs increase equally.

Cost-Effectiveness Methodologies

All cost-effectiveness analyses are based on incremental assessment methods: the new intervention is compared with a standard or base case intervention, in terms of their relative cost and effect. The cost of the old or standard intervention is subtracted from the cost of the new intervention, giving an incremental *cost*. The effect of the standard or comparator intervention would likewise be subtracted from the new intervention, resulting in an incremental *effect*. The incremental cost is divided by the incremental effect, giving the cost-effectiveness ratio *(cost/effect)*.

There are several cost-effectiveness methodologies currently used in health economics:

(a) A *cost-minimization analysis* assumes no difference in outcome between interventions, and the lowest cost item is selected.

(b) The *cost-benefit analysis* applies a dollar value to a specific outcome, such as a life-year or number of new cancers detected. This methodology has some associated problems, as it is very difficult to assign a dollar value to a year of life or to a new cancer case detected or prevented.

(c) The *cost-effectiveness* analysis is the most widely used methodology. The incremental cost is divided by a unit of outcome of the intervention. The unit of outcome could be a year of life saved or the number of cancers prevented. The resultant ratio is dollar/life-year or dollar/new cancer case diagnosed.

(d) The *cost-utility* analysis results in a dollar value per life-year gained or lost-adjusted based on utility or patient preference for a health state, varying from 1.0 for perfect health down to 0 for death. For example, if a person lives a year in a health state that they value as 0.5, the resultant quality-adjustment would be calculated as 1 year × 0.5 = 0.5 QALY.

Value is best defined as outcome divided by cost, e.g. something is of better value if it gives you greater or improved outcome at the same cost, or the same outcome at lower cost. Value can be thought of as the inverse of a cost-effectiveness ratio, where one evaluates cost divided by outcome. In comparison with other countries, the American healthcare system scores very low in value, i.e. the quality of our healthcare provided divided by our costs of delivering that healthcare. Most other industrialized countries achieve much greater value for their healthcare dollar. However, our policy-making Federal agencies are now beginning to evaluate the efficiency of our healthcare expenses and the value of the healthcare we are providing.

Value-Based Purchasing

In value-based purchasing, buyers hold providers of healthcare accountable for both the cost and quality of care. The decision making process combines information on quality of healthcare, including patient outcomes and health status, with data on the dollar outlays going towards health. It focuses on managing the use of the healthcare system to reduce inappropriate care and to identify and reward the best-performing providers. This strategy can be contrasted with more limited efforts to simply negotiate price discounts, which reduce costs but do little to ensure that quality of care is improved [18]. Unfortunately, the moral hazard in our society is that the end-consumers of the healthcare are not the actual buyers of the healthcare. Most of the buyers are government agencies, Medicaid and Medicare, or employers for work-based healthcare. In some cases, buyers may be influenced by total costs to the plan, regardless of outcomes, simply to lower costs to the plan. Unquestionably, the purchasers of healthcare should combine objectively measured information on the quality of healthcare and patient outcomes, which are critical factors in the cost analyses.

The development of a plan to transition to a Medicare value-based purchasing program for physician and other professional services is a key objective of the current United States administration under President Obama [19].

Why is value-based purchasing such an important objective? Healthcare expenditures are becoming an increasing percentage of the United States GDP and more importantly, business expenses. American manufacturers are finding it increasingly difficult to compete in the global economy and must face the potential for loss of jobs to foreign competitors. At the same time, healthcare expenditures are adding to our country's massive budget deficits. Peter Orszag, Director of the Office of Management and Budget (OMB), testified to the United States Senate on January 13, 2009 that 'the principal driver of our long-term deficits is rising health care costs'. At that time, he specifically cited the need for expanding research on comparative effectiveness of different options for treating a given medical condition, which could provide information on both medical benefits as well as costs.

1.1 billion dollars has been allocated for comparative effectiveness studies with USD 300 million distributed to the AHRQ, USD 400 million to the NIH and USD 400 million to the Office of the Secretary of the US Department of Health and Human Services. The funding is to be used to conduct or support research to evaluate and compare clinical outcomes, effectiveness, risk and benefits of 2 or more medical treatments or services that address a particular medical condition. In this regard, a 'one-size-fits-all' approach to treatment is inappropriate and subpopulations must be considered when research is conducted or supported with these funds.

Cost-Effectiveness Analysis in Healthcare Policy

How can we incorporate cost-effectiveness analysis (CEA) into healthcare policy, and is it being done currently? The types and scope of health economic analyses are diverse, and there are obstacles in the use of CEA in setting health policy. There are (a) difficulties in trying to compare different types of analyses (as previously mentioned, there are four types). The results of a *cost-benefit* analysis are not easily compared with a *cost-utility* analysis, for example. There can even be difficulties in trying to compare the results of a cost-utility analysis performed from a payer's perspective with the results of a cost-utility analysis performed from a societal perspective. Currently, there are (b) no standard criteria for determining when economic factors are relevant, or how to use them in deciding patient access and payment policy. There is (c) no uniformly accepted standard for what information is to be included in CEAs. There is (d) no national, standardized process for setting priorities among health issues that could merit CEA. Consider the processes in other countries, for instance. CEA studies are less common in the USA than in Australia, Canada and the UK. Also, these countries use CEA and other healthcare cost/benefit information quite differently, and there are differences in who has the authority to use this information.

In the USA, the FDA is legislatively limited in the development or use of CEA evidence, and the FDA has no statutory authority or mechanism for evaluating the economic impact of guidance. In fact, it has been suggested that consideration of CEA or other cost-health tradeoff evidence during market approval or postmarket surveillance could compromise or distract from the FDA's core mission of ensuring safety and effectiveness of regulated health care products.

Are there any current examples of healthcare CEA use in health policy? This is certainly under active discussion and planning. The US Preventive Services Task Force has presented the uses of CEA to inform evidence-based recommendations for preventive services. CEA can be used to quantify differences between 2 or more effective services for the same condition, and illustrate the impact of delivering a given intervention at different intervals, different ages or to different risk groups. In addition, CEA can be used to evaluate the potential role of new technologies, identify key conditions that must be met to achieve the intended benefit of an intervention and incorporate preferences for intervention outcomes. But CEA will not be used for developing a ranking of services in order of their costs and expected benefits, at least at this time [20].

Already some other countries use CEA for public policy development. Australia has used CEA in the economic evaluations required for new drugs since 1993. In Europe, economic evaluations have been used in a number of countries as an input into decisions about drug reimbursement. In the UK, NICE has been established as an independent, government-funded organization that advises the British National Health Service. Assessing cost-effectiveness is the most controversial of all of NICE's work. NICE considers an intervention or drug to be cost-effective if the CEA ratio is GBP 20,000 ($\approx$ USD 34,000)/QALY. The highest cost-effective ratio that was adopted or ever accepted by NICE was USD 84,000/QALY for imatinib mesylate used in the blast-cell phase of CML. If one looks at other work performed by NICE, CEA ratios vary from USD 122,000/QALY for sunitinib to USD 294,638 for bevacizumab for advanced renal cell cancer.

Potential Uses of Cost-Effectiveness Analysis in Radiation Oncology: Prostate Cancer

How can we apply CEA in radiation oncology, and how might it affect health policy or coverage decisions? Consider its applications to prostate cancer treatment. For a patient with early-stage, low-risk prostate cancer, there are several approaches using radiotherapy with the costs of these treatments varying considerably. If one considers the expected Medicare payments, these range from USD 8,500 for low-dose-rate brachytherapy to USD 33,000 for intensity-modulated radiation therapy (IMRT) with daily cone-beam CT guidance image-guided radiation therapy (CBCT IGRT) for 80 Gy given in 2-Gy fractions. PBT exceeds even this, costing more than USD 38,000 before a recent 16% reduction in 2009. This cost will likely rise in 2010, given the increase in reimbursement for the daily proton beam delivery. If one considers CBCT IGRT and adds to this adaptive radiation therapy with replanning the patient weekly, the IMRT costs may approach those of proton therapy.

Table 2. Results of CEA comparing IMRT to different forms of IG-IMRT

Modality	Expected mean cost, USD	Outcome in QALY	Cost-effectiveness ratio, USD/QALY
IMRT	30,361	6.76	
Ultrasound IG-IMRT	32,995	7.14	6,931
Stereoscopic IG-IMRT	34,492	7.14	14,244
CBCTIG-IMRT	38,073	7.14	23,369

The results of the cost-utility analysis show the expected mean results of IMRT without image guidance the lowest, and IMRT with CBCT the highest. All of the image-guided (IG) techniques have higher quality-adjusted survival. The cost-utility ratios are all below USD 50,000/QALY. CBCTIG = Cone-beam computed tomography image guidance.

The Markov model that we have used in the past for CEA in prostate cancer radiotherapy was modified to compare IMRT with different forms of IGRT. The model encompassed treatment costs of initial therapy as well as outcomes, and was modified to include certain transition costs such as resulting GI toxicity in a percentage of cases, and treatment failure requiring systemic management in others [21–23]. We established utility values for hormone therapy and chemotherapy, and estimated the probability of GI toxicity after IMRT or IGRT. The utility value for IMRT was determined to be 0.909 from a group of men with prostate cancer treated with IMRT, and the addition of image guidance (with ultrasound, stereoscopic imaging or CBCT) was assumed to increase this value to 0.95 because of reductions in toxicity from using smaller margins. However, we have no data to corroborate the assumption that by adding image guidance one will actually decrease the risk of significant rectal toxicity by half. In IMRT, the yearly probability of rectal toxicity is only 0.02, so by adding image guidance this only decreases to 0.01.

Processing these values in the model gives the results shown in table 2. This lists the expected mean costs for IMRT and the image-guided additions, with the outcomes in QALY and the cost-effectiveness ratios in dollars/QALY. All of the CEA ratios are well below USD 50,000/QALY. However, the *probability* of cost-effectiveness for image guidance is only 30%, since the differences in QALY between the interventions are small (only 0.38 QALY). Figure 1 shows that the 95% confidence ellipse crosses the y-axis on the cost-effectiveness plane, suggesting no statistical difference between the tested interventions.

As a specific example of the application of CEA in prostate cancer, consider the question of whether a 50-year-old man with early-stage, low-risk prostate cancer should be treated the same way as an 80-year-old man. Already, the AHRQ has recognized that the 'one-size-fits-all' approach to treatment is inappropriate and that subpopulations must be considered. Accordingly, one might

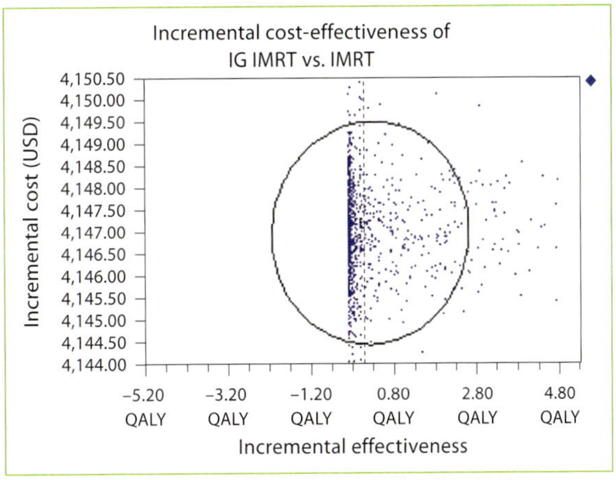

Fig. 1. The incremental cost-effectiveness scatterplot shows the 95% confidence ellipse crossing the y-axis. This can be interpreted as no statistical difference between the tested interventions.

consider whether the CEA methodology, using both cost and outcome data, could be used to stratify treatments by age groups to more efficiently use our resources. For a 50-year-old patient, possibly proton beam radiotherapy might be a more appropriate consideration, because one believes that there is a longer lifetime risk of secondary malignancies from whole-body (integral) radiation exposure and that this is reduced by proton therapy. For a 60-year-old, stereoscopic image guidance of IMRT might be a better treatment as it avoids the higher imaging radiation dose of the CBCT, but the lower integral radiation dose of PBT is no longer cost-effective. For the 70-year-old, CBCT IMRT might be preferred, while for the 80-year old, perhaps 3D CRT provides an optimal balance of cost and outcome factors.

Conclusions

It should be clear that CMS will continue to monitor radiation therapy costs closely. Medicare Part B utilization and expenditure data document the levels of concern. In 2007, IMRT treatment delivery CPT code 77418 was ranked as number 18 in total cost to the Medicare system for the year. In 2004, it was only number 38 in the cost rank of codes. Radiation oncologists currently represent only 0.3% of all Medicare service providers, yet account for 3% of their total costs. As a professional group, radiation oncologists will continue to be scrutinized for their increasing costs to the system. Payment for those costs will remain under active evaluation, especially if increasing outcome benefits for patients are not clearly documented. If new radiation technologies are to be accepted into our payment system, their value to the system must be established through clinical trials showing objective improvements in patient outcomes.

References

1 McIsaac ML, Goeree R, Brophy JM: Primary data collection in health technology assessment. Int J Tech Assess Health Care 2007;23:24–29.
2 Peters LJ: Through a glass darkly: predicting the future of radiation oncology (presidential address). 36th Annual Meeting of the American Society for Therapeutic Radiology and Oncology, San Francisco, Calif., 1994.
3 Banta HD, Luce BR: Health Care Technology and Its Assessment: An International Perspective. New York, Oxford University Press, 1993.
4 The Cochrane Collaboration: http://www.cochrane.org/ (accessed April 7, 2010).
5 Institute for Clinical and Economic Review: http://icer-review.org/ (accessed April 7, 2010).
6 Pearson SD, Miller FG, Emanuel EJ: Medicare's requirement for research participation as a condition of coverage. JAMA 2006;296:988–990.
7 Harris G: British balance gain versus cost of latest drugs. http://www.nytimes.com/2008/12/03/health/03/health/03.nice.html?pagewanted=print (accessed December 3, 2008).
8 McClellan MB: Technology and innovation: their effects on cost growth of healthcare. Before Joint Economics Committee, US Congress. July 9, 2003.
9 Beever C, Burns H, Karbe M: US health care's technology cost crisis. http://www.strategy-business.com/press/enewsarticle/enews033104 (accessed July 20, 2007).
10 Congress of the United States Congressional Budget Office: Technological change and the growth of health care spending. Washington, D.C., 2008. http://www.cbo.gov/ftpdocs/89xx/doc8947/01-31-TechHealth.pdf (accessed April 20, 2010).
11 Pear R: U.S. to compare treatments. http://nytimes.com/2009/02/16/health/policy/16health.html?_r=1&hp=&pagewanted (accessed February 16, 2009).
12 Steinberg M, Konski A: Proton beam therapy and the convoluted pathway to incorporating emerging technology into routine medical care in the United States. Cancer J 2009;15:333–338.
13 Tunis S: Medicare coverage of new technology: PowerPoint presentation. Centers for Medicare and Medicaid Meeting, Baltimore, Md., September 3, 2003.
14 Tunis S: Testimony to Medicare Evidence Development and Coverage Advisory Committee. April 30, 2008.
15 15Neumann PJ, Kamae MS, Palmer JA: Medicare's national coverage decisions for technology, 1999–2007. Health Aff (Millwood) 2008;27:1620–1631.
16 Straube B: Testimony to Medicare Evidence Development and Coverage Advisory Committee. April 30, 2008.
17 American Medical Association: CPT process – how a code becomes a code. http://www.ama-assn.org/ama/no-index/physician-resources/3882.shtml (accessed March 31, 2009).
18 Meyer J, Rybowski L, Eichler R: Theory and Reality of Value-Based Purchasing: Lessons from the Pioneers. AHCPR Publ No 98–0004. Rockville, Agency for Health Care Policy and Research, 1997.
19 Centers for Medicare and Medicaid: http://www.cms.hhs.gov/PhysicianFeeSched/downloads/PhysicianVBP-Plan- (accessed February 16, 2009).
20 Agency for Healthcare Research and Quality: http://www.ahrq.gov/clinic/ajpmsuppl/costsum.pdf (accessed February 16, 2009).
21 Konski A, Sherman E, Krahn M, et al: Economic analysis of a phase III clinical trial evaluating the addition of total androgen suppression to radiation versus radiation alone for locally advanced prostate cancer (Radiation Therapy Oncology Group protocol 86–10). Int J Radiat Oncol Biol Phys 2005;63:788–794.
22 Konski A, Speier W, Hanlon A, et al: Is proton beam therapy cost effective in the treatment of adenocarcinoma of the prostate? J Clin Oncol 2007;25:3603–3608.
23 Konski A, Watkins-Bruner D, Brereton H, et al: Long-term hormone therapy and radiation is cost-effective for patients with locally advanced prostate carcinoma. Cancer 2006;106:51–57.

Paul E. Wallner, DO
5013 Cedar Croft Lane, Bethesda, MD 20814 (USA)
Tel. +1 301 897 22091, Fax +1 301 897 2092, E-Mail pwallner@rtsx.com

I. IMRT and IGRT: Intensity-Modulated and Image-Guided Radiotherapy

Clinical Implementation of Intensity-Modulated Arc Therapy

David M. Shepard · Daliang Cao

Swedish Cancer Institute, Seattle, Wash., USA

Abstract

Intensity-modulated arc therapy (IMAT) is a rotational approach to radiation therapy delivered on a conventional linear accelerator using a conventional multileaf collimator. There are 2 key advantages of IMAT. First, the rotational nature of the delivery provides great flexibility in shaping each dose distribution. As a result, IMAT can provide dosimetric advantages relative to fixed-field intensity-modulated radiation therapy (IMRT). The second advantage is the highly efficient nature of the delivery. For centers with an active IMRT program, the clinical implementation of IMAT should be relatively straightforward. For clinical implementation of IMAT, it is important to fully characterize the accuracy of the dose model used, and the performance of the quality assurance equipment.

Copyright © 2011 S. Karger AG, Basel

Intensity-modulated arc therapy (IMAT) is an arc-based approach to delivery of radiation therapy that combines the dose painting capabilities of intensity-modulated radiation therapy (IMRT) with the dosimetric advantages of rotational delivery. Additionally, IMAT provides highly efficient delivery capabilities. This chapter provides an overview of the IMAT delivery technique including comparisons with other IMRT delivery techniques, considerations for clinical implementation of IMAT, and IMAT quality assurance (QA).

Intensity-Modulated Arc Therapy: Basic Concepts

IMAT is a rotational approach to IMRT that can be delivered on a conventional linear accelerator equipped with a conventional multileaf collimator (MLC) [1]. During each arc of an IMAT delivery, the leaves of the MLC move continuously as the gantry rotates. The degree of intensity modulation is related to the number of beam shapes per arc and the total number of arcs.

The IMAT delivery technique has 2 key advantages. The first is that the rotational nature of the delivery provides more flexibility in terms of shaping the distribution. The second advantage is that IMAT is a highly efficient delivery technique due to the continuous nature of the delivery.

Background

In 1993, an article appeared in *Medical Physics* entitled, 'Tomotherapy: a new concept for the delivery of dynamic conformal radiotherapy' [2]. This article, authored by Rock Mackie, described a novel delivery technique called tomotherapy that involved the delivery of IMRT using the rotational delivery of a fan beam of radiation [3, 4].

In 1995, Cedric Yu [1] introduced an alternative approach to the delivery of rotational IMRT in a paper entitled 'Intensity-modulated arc therapy with dynamic multileaf collimation: an alternative to tomotherapy'. In this paper, Yu described the delivery of rotational IMRT using a cone beam of radiation. A key difference from tomotherapy is that IMAT can be delivered using a conventional linear accelerator (linac) and a conventional (nonbinary) MLC. Between 1995 and 2007, researchers published a number of manuscripts on the topic of IMAT [5–15]. As a clinical tool, however, the IMAT technique did not mature into medical application. There were 2 main reasons for this. First, the linac manufacturers did not offer delivery control systems that were capable of taking full advantage of the IMAT delivery technique. Second, there were no robust inverse planning solutions for IMAT that were commercially available.

In 2008, both Elekta and Varian introduced new delivery control systems that were capable of delivering IMAT. The critical innovation was that the new control systems provided the ability to vary the dose rate, the gantry speed and the MLC leaf positions dynamically during rotational beam delivery.

Another key development was the introduction of the first robust commercial inverse planning solutions for IMAT [16, 17]. As compared with fixed-field IMRT, IMAT poses much more complex inverse planning problems mathematically. In an IMAT plan, the algorithm must account for the 'interconnectedness' of the aperture shapes between adjacent control points. That is, from one beam angle to the next within an arc, there are restrictions on the magnitude of the changes that can be made to the beam shapes. These are mainly due to the limits on the speed of the MLC leaves. These constraints must be met to ensure the deliverability of the plan. Another reason to restrict the MLC leaf motion during IMAT planning is to ensure that the delivered dose matches the planned one [18].

The availability of more advanced delivery control systems and robust inverse planning solutions have made it possible to realize the full potential of IMAT as a delivery technique. Consequently, IMAT is now becoming a widely adopted clinical tool for the delivery of IMRT.

Terminology

The growing number of commercial and scientific terms used to describe rotational IMRT delivery techniques can create confusion. In this chapter, the term *intensity-modulated arc therapy* will be used to describe all techniques where rotational IMRT is delivered using a cone beam of radiation.

A brief glossary of rotational IMRT terminology is as follows:
- *Tomotherapy*: rotational IMRT delivered using a fan beam of radiation.
- *Helical tomotherapy (HT)*: tomotherapy in which the patient is moving through the fan beam during radiation delivery, analogous to spiral CT.
- *Serial tomotherapy*: tomotherapy in which the patient is stationary during radiation delivery and is moved by integer multiples of the fan beam width between radiation deliveries, analogous to transverse CT.
- *IMAT*: rotational IMRT delivered using a cone beam of radiation.
- *Volumetric-Modulated Arc Therapy (VMAT)*: a term originally introduced to describe IMAT delivered using a single arc [6]. More recently, Elekta adopted the term VMAT to describe their commercial IMAT delivery solution.
- *RapidArc*: a VMAT planning and delivery solution from Varian Medical Systems.
- *SmartArc*: an IMAT planning solution from Philips Medical that is incorporated into the Pinnacle^3 treatment planning system.
- *Cone Beam Therapy*: an IMAT solution from Siemens.

Why Rotational Intensity-Modulated Radiation Therapy?

Rotational IMRT provides the advantage of selectively delivering the radiation dose from the beam angles that are optimal. As the number of beam angles increase, there is greater flexibility in shaping the dose distribution. This was demonstrated in the paper entitled, 'A simple model for examining issues in radiotherapy optimization' [19]. In this paper, simulations were run comparing plan quality as the number of beam angles increased. The results demonstrated that plan quality continued to improve as the number of beam angles increased, even as they increased beyond 21 angles.

A common misconception is that the total dose delivered to the patient increases as the number of beam angles increases. Studies have demonstrated, however, that the integral dose delivered is essentially independent of the number of angles [20–22]. Generally, rotational delivery delivers low doses to a larger volume and high doses to a smaller volume as compared with fixed-field IMRT. The total dose delivered to the patient, however, is essentially unchanged.

Comparison of Intensity-Modulated Arc Therapy with Other Intensity-Modulated Radiation Therapy Delivery Techniques

With the development of commercial IMAT solutions, researchers are now exploring how IMAT compares with fixed-field IMRT in terms of plan quality and delivery efficiency. For example, clinical planning studies have been published that compare Varian's RapidArc solution with conventional IMRT [23, 25–27]. In the study by Verbakel et al. [25], planning and dosimetry of RapidArc were compared with conventional IMRT plans for head and neck patients. They found that RapidArc is able to significantly reduce treatment times and the total number of monitor units (MUs) while maintaining similar dose distributions to those available with a 7-field sliding window IMRT technique. Fogliata et al. [26] compared treatment plans for RapidArc, HT and standard IMRT for 12 patients with benign brain tumors. Their study demonstrated that all 3 techniques yield comparable plan quality in the cases studied. They also concluded, however, that further investigations with more complex cases should be performed. Palma et al. [27] provided a planning study comparing RapidArc with 5-field IMRT and 3D conformal radiotherapy for the treatment of prostate cancer. They found that RapidArc resulted in more favorable dose distributions than fixed-field IMRT and reduced the MUs significantly.

A study comparing step-and-shoot IMRT and IMAT plans was carried out by Cao et al. [6]. This study indicated that IMAT can provide similar or better plan quality in terms of target dose coverage and critical structure sparing with fewer MUs per fraction as compared with step-and-shoot IMRT. IMAT also provided an average reduction in the delivery time of 65%.

Researchers at the Swedish Cancer Institute (Seattle, Wash., USA) and the University of Virginia (Charlottesville, Va., USA) performed a study where comparisons were made between Elekta VMAT, HT and fixed-field IMRT in terms of plan quality, delivery efficiency and accuracy. Comparisons were made for 18 clinical

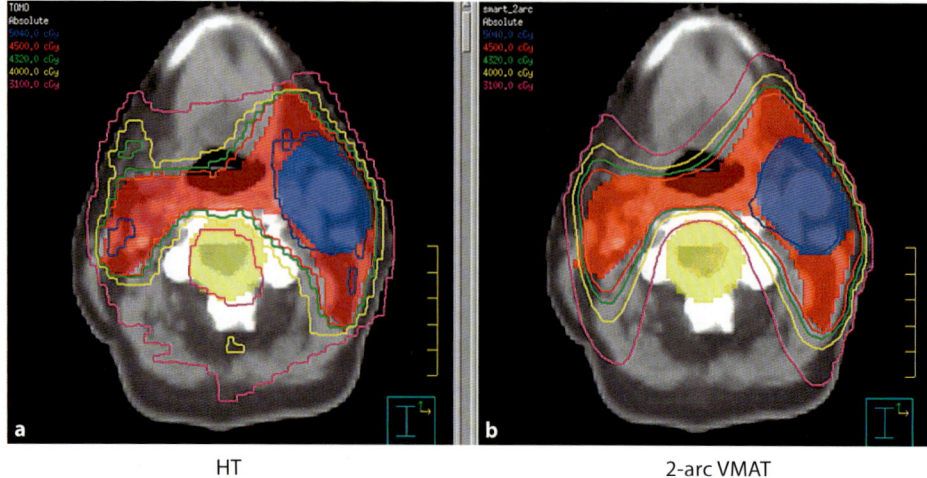

Fig. 1. Comparison of plans using HT (**a**) and 2-arc VMAT (**b**) for a head and neck cancer case with 2 targets at different prescription levels. The targets are plotted in red and blue.

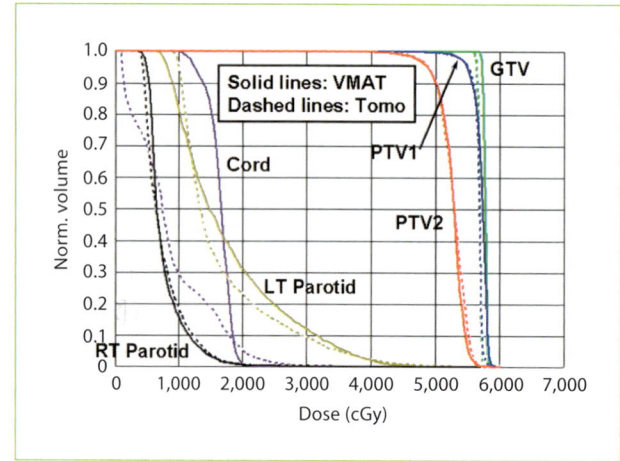

Fig. 2. DVH comparison for the HT and VMAT cases shown in figure 1. Tomo = Tomotherapy.

cases covering a range of treatment sites. In each case, the original treatment planner had a second opportunity to improve the plan after reviewing the quality of the plans obtained using the alternative delivery techniques.

Figure 1 shows a 2D isodose distribution comparison for a head and neck cancer case between a HT plan and a 2-arc IMAT plan. It is clear that both plans provide highly conformal dose to the targets. The corresponding DVHs are shown in figure 2. Just as observed in figure 1, both plans provide very similar target dose coverage. The average V_{95}, defined as the target volume receiving at least 95% of the prescribed dose, had values of 98.4 and 98.6% for the HT and IMAT plans, respectively. Both HT and IMAT plans also provide excellent sparing of critical structures, also shown in figure 2. The mean dose to the parotid glands is slightly less in the HT plan, while the IMAT plan has a smaller maximum dose to the spinal cord. While both HT and IMAT plans have comparable plan quality, the major difference between these 2 plans is the delivery efficiency. In fact, the delivery time for the IMAT plan is 4.4 min, while this value increases to 9.1 min for the HT plan.

This study concluded that VMAT can provide highly conformal dose distributions that are comparable to those of HT even in complex cases. However, the VMAT technique provides a more efficient delivery than HT. Detailed results were summarized in the article by Rao et al. [24] titled, 'Comparison of Elekta VMAT with helical tomotherapy and fixed field IMRT: plan quality, delivery efficiency and accuracy'. Shorter treatment delivery times not only improve the clinical throughput, but also reduce the chance of intrafractional patient motion.

Commercial Intensity-Modulated Arc Therapy Solutions

Delivery Solutions

Both Varian and Elekta offer commercial IMAT delivery solutions. The term *RapidArc* is used exclusively to describe Varian's complete IMAT solution that includes both their delivery control system and their RapidArc planning module in the Eclipse treatment planning solution [23, 28–31]. Varian RapidArc has been marketed primarily as a single arc solution under the tagline 'One revolution is all it takes'. More recently, however, Varian has added support for multi-arc IMAT deliveries. Elekta has adopted the term *volumetric modulated arc therapy* to describe their commercial implementation [32, 33].

Treatment Planning Solutions

Varian's Eclipse treatment planning system includes an IMAT planning module. The planning is performed using a technique called Direct Aperture Optimization [10, 34–36], an inverse planning algorithm first described by Shepard et al. [36] whereby all of the machine delivery constraints are included in the plan optimization process. In his paper entitled, 'Volumetric modulated arc therapy: IMRT in a single gantry arc', Karl Otto [37] described how the use of Direct Aperture Optimization in combination with progressive sampling can serve as a powerful planning tool for rotational IMRT. His work focused on treatment delivery using a large number of apertures delivered in a single arc.

IMAT plans produced using Eclipse typically use 1 arc, with 177 control points or beam shapes within that arc. In some cases, users are taking advantage of using multiple arcs to improve the plan quality or provide adequate coverage for large targets.

The SmartArc module in the Pinnacle^3 treatment planning system (Philips Medical, Bothell, Wash., USA) can be used to create IMAT plans for either Varian or Elekta linear accelerators. The SmartModule was developed by RaySearch (Stockholm, Sweden). The algorithm is described in detail in an article entitled, 'Development and evaluation of an efficient approach to volumetric arc therapy planning' by Bzdusek et al. [17].

The basic IMAT planning workflow in the Pinnacle^3 system is quite similar to that used for fixed-field IMRT. The dose objectives are defined and the optimization type is set to SmartArc. During the inverse planning, the algorithm completes the following steps:

(1) Beams are generated at the start and the stop angles and at 24° increments from the start angle.
(2) A fluence map optimization is performed.
(3) The fluence maps are sequenced and filtered so that there are only 2 control points per initial beam angle.
(4) These control points are distributed to adjacent gantry angles and additional control points are added to achieve the desired final gantry spacing.
(5) All control points are processed to comply with the motion constraints of IMAT.
(6) The direct machine parameter optimization [38, 39] algorithm is applied with an aperture-based optimization that takes into account all of the IMAT delivery constraints.
(7) The jaws are conformed to the segments based on the characteristics of the linac.

A sample SmartArc plan is shown in figure 3 for a head and neck cancer case. This plan was created using two 360° arcs with the final gantry angle spacing of 4°. It is clear that this plan has a dose distribution that is highly conformal to the targets, and also sparing to the critical structures such as the parotid glands and spinal cord.

A VMAT module has been introduced for Nucletron's Oncentra treatment planning system. The IMAT module was developed by RaySearch Laboratories and shares the underlying optimization engine used in the Philips SmartArc. The Nucletron Oncentra IMAT planning tools can be used with either Elekta or Varian linacs.

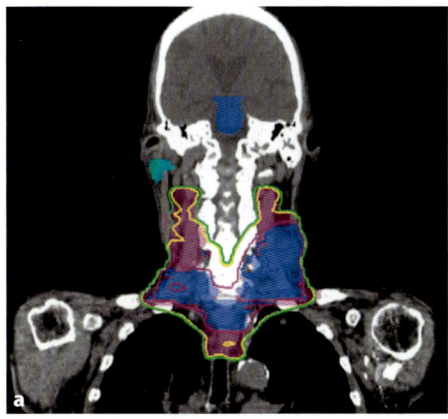 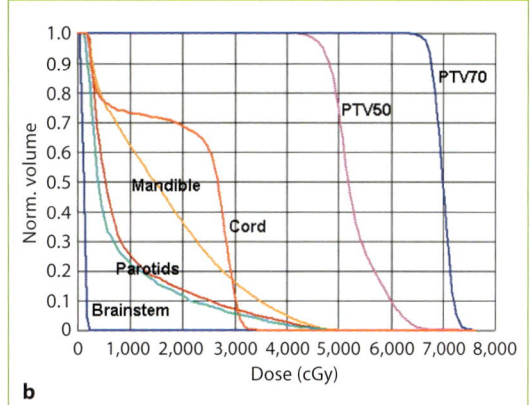 

Fig. 3. SmartArc head and neck plan (**a**) and corresponding DVH (**b**). See online supplementary material for video display of a head and neck plan delivered using SmartArc.

Elekta has introduced 2 commercial IMAT planning solutions. The first was incorporated into the Ergo++ treatment planning system. The Ergo++ IMAT planning solution uses anatomy-based semi-inverse planning, i.e. the aperture shapes are defined based on the patient's anatomy. For example, for a prostate IMAT delivery, one could deliver 2 arcs: the first arc tracking the shape of the prostate and the second arc tracking the shape of the prostate minus the rectum. The optimizer would then optimize the weights of the individual segments. This approach to IMAT more closely approximates dynamic conformal arc delivery. The key difference is that the weights can vary from one beam angle to the next within an arc, making it possible to deliver a higher percentage of the radiation from beam angles that are more preferred.

The second commercial solution from Elekta is an IMAT planning algorithm in the Monaco treatment planning system. Two key features of the Monaco system are that it uses (1) a Monte Carlo-based dose calculation and (2) biology-based IMRT optimization. At this time, Monaco is only capable of planning IMRT (not 3D conformal therapy). Rather than performing an aperture-based optimization, Monaco employs a 2-step process. First, optimized fluence maps are produced at a series of discrete beam angles. Then these optimized fluence maps are converted into deliverable IMAT arcs. Figure 4 shows a prostate IMAT case planned using the Monaco system.

Monaco produces plans using a 'sweeping leaf sequencer' where the leaves move unidirectionally across the field. The leaf movement continues to alternate between sectors of the arc. This delivery approach was first suggested by Cameron [40] in his 2005 manuscript, 'Sweeping-window arc therapy'.

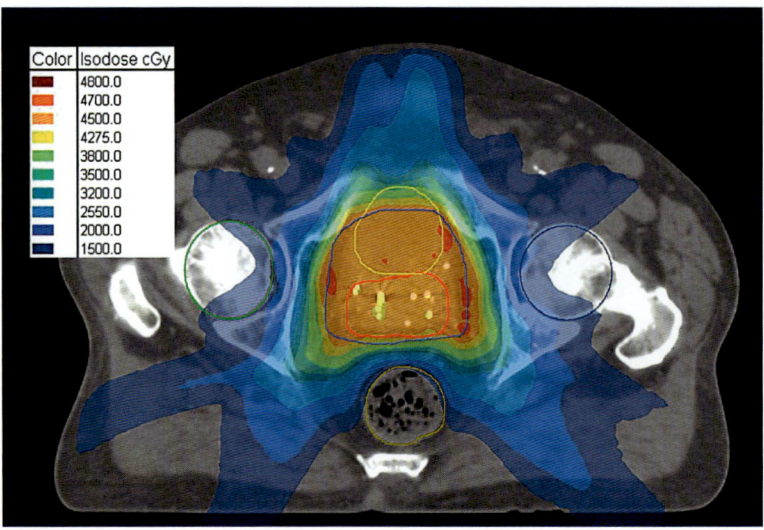

Fig. 4. A VMAT prostate plan optimized using the Monaco treatment planning system.

Clinical Implementation of Intensity-Modulated Arc Therapy

Intensity-Modulated Arc Therapy Treatment Planning

The basic treatment planning process for IMAT is very similar to that for fixed-field IMRT. The planner adds an IMAT beam and specifies the couch, collimator and beam angles. Next, the dose objectives are defined and an IMAT optimization is performed.

In IMAT planning, one also needs to specify IMAT-specific parameters. The key parameter is the number of arcs. Additionally, one may need to define parameters such as the allowable MLC leaf motion per degree of gantry rotation. For example, figure 5a plots the impact this leaf motion constraint has on plan quality for a head and neck patient using the Pinnacle SmartArc solution. Note that constraining the leaf motion to 1 mm per degree of gantry rotation resulted in a significant degradation in the plan quality. Beyond 3 mm per degree rotation, little improvement was observed. The delivery time is plotted as a function of the leaf-motion constraint in figure 5b. It can be seen that the delivery time increases with each increase in the allowable leaf motion. Trial and error may be required to find the best balance between plan quality and delivery efficiency.

It is important to keep in mind that IMAT is simply another form of IMRT. Due to the rotational nature of the delivery, IMAT can provide dosimetric advantages over fixed-field IMRT. As the number of beam angles increases, however, the differences between fixed-field IMRT and IMAT are minimized. An example is

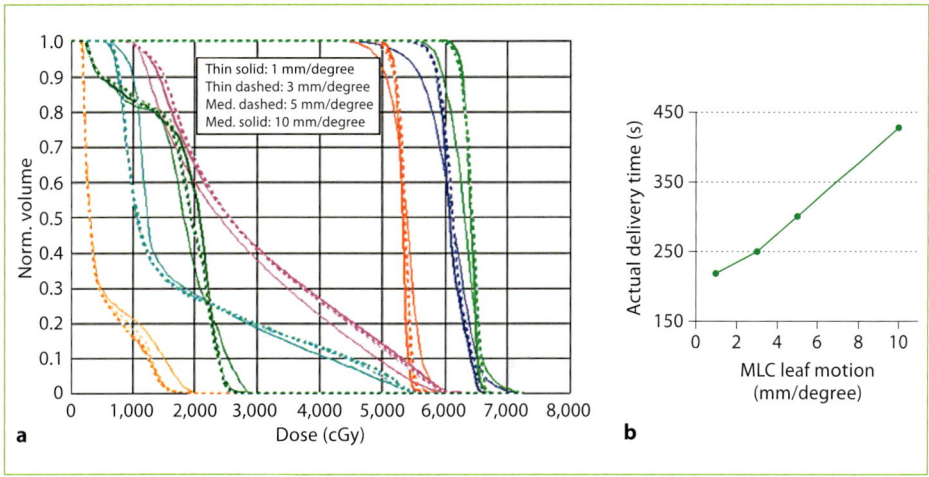

Fig. 5. a DVH overlay demonstrating the relationship between plan quality and the allowable leaf motion per degree of gantry rotation. **b** Plot demonstrating degree to which delivery time increases as one allows an increased leaf motion from one beam angle to the next in a VMAT plan.

shown in figure 6, which plots a DVH comparison between a 9-field IMRT and a multi-arc IMAT plan. In this rather complex head and neck case, it can be seen that IMAT and 9-field IMRT provided very similar plan quality.

Single Arc versus Multiple Arcs

In his original paper on IMAT, Yu [1] described the use of multiple overlapping arcs to deliver IMAT. A key concept in the paper was that the use of multiple overlapping arcs creates a modulated intensity pattern from each beam direction. Consequently, IMAT is clearly classified as an IMRT delivery technique. For example, if 3 coplanar arcs were utilized, each beam angle would see 3 aperture shapes, each with its own assigned MU value. By adding up the aperture shapes along with their weights, one would obtain a modulated intensity pattern from each beam direction.

The delivery of IMAT using a single arc has been described by a number of authors [23, 29, 31, 37, 41, 42]. In the single arc delivery approach, a large number of beam apertures are included in a single arc. A potential advantage of delivering with a single arc is the highly efficient nature of the delivery. For example, figure 7 plots a prostate case planned with a single arc. The total delivery time in this case was only 2.0 min.

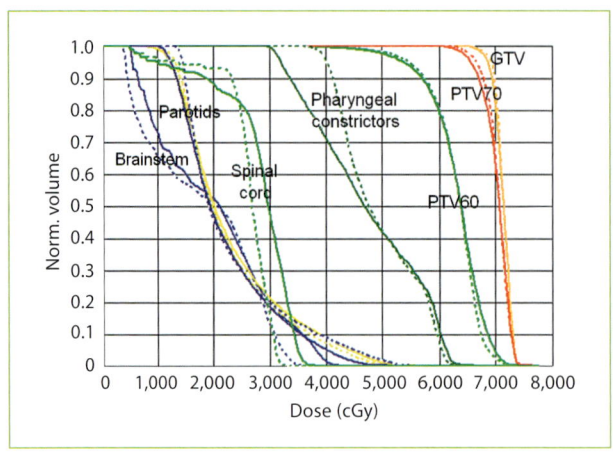

Fig. 6. DVH overlay for IMAT (solid lines) and 9-field IMRT (dashed lines) for a complex head and neck case. Note that IMAT and 9-field IMRT provide very similar plan quality.

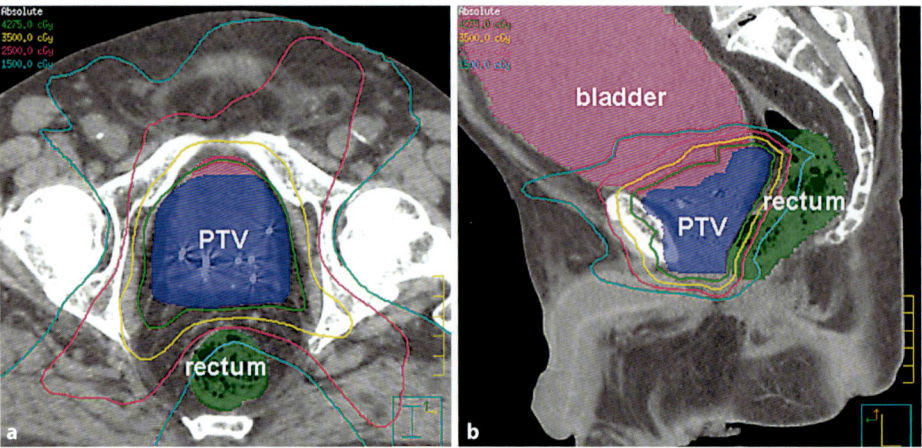

Fig. 7. Optimized single arc prostate IMAT plan.

In developing an IMAT treatment plan, a key question is how many arcs to use. This was explored by Cao et al. [43]. In general, multiple arcs can provide a better plan quality, but at a cost of compromising the delivery efficiency. For a simple case, such as prostate, partial brain or pancreas, the target volume is relatively small and has a relatively regular shape. Using multiple arcs in IMAT planning for relatively simple target shapes will not give a significant dosimetric advantage over that provided by single arc plans. On the other hand, the delivery time can increase

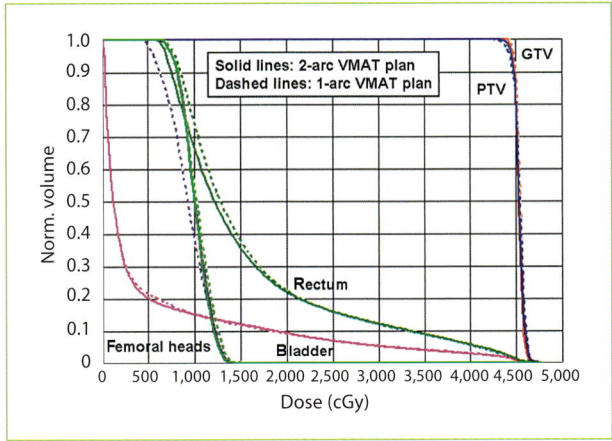

Fig. 8. Comparison of single arc and 2-arc VMAT plans for the prostate case shown in figure 7. Note that the 2 sets of DVH curves are nearly identical.

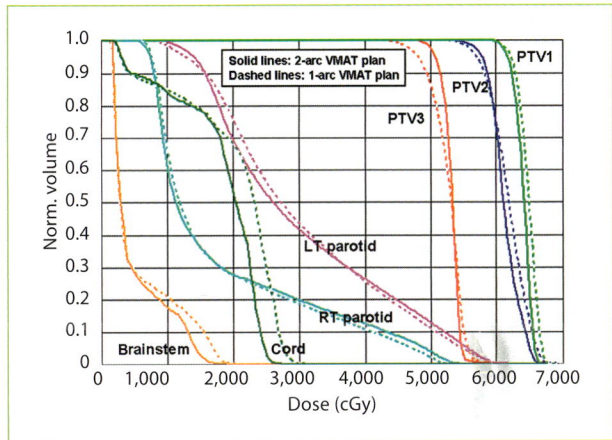

Fig. 9. Comparison of 1-arc and 2-arc VMAT plans for a complex head and neck case with 3 prescription levels. Note that there is a noticeable improvement when a second arc is added.

dramatically when the number of arcs increases from 1 to 2 or 3. For example, figure 8 shows a DVH comparison between IMAT plans using 1 and 2 arcs for the prostate case shown in figure 7. Both plans were created using the SmartArc inverse planning module with the same set of optimization objectives. With careful inspection, no significant differences can be found between the 2 plans. However, the 2-arc plan takes 3.6 min to deliver, which is 80% more than that of the single arc plan.

For more complicated cases, such as treatment of pelvic or head and neck sites, the improvements in target dose coverage and uniformity are more pronounced when more than 1 arc is used. Otherwise, achieving acceptable plan quality using a single arc usually requires more intensity modulation. This is accomplished by adding more control points or allowing more MLC leaf motion between adjacent control points. This, in turn, leads to longer delivery times for the single arc plans. Therefore, for the more complicated cases, the difference in the delivery time between single and multiple arc IMAT plans is less dramatic. As shown in figure 9, a DVH comparison of single and 2-arc IMAT plans for a head and neck case, the average V_{95} and σ_{PTV} are 98.8% and 5.2 cGy/fraction for the 2-arc plan, which are reduced from 95.6% and 7.3 cGy/fraction for the single arc plan. The delivery time only increased from 2.1 to 3.0 min when the number of arcs increased from 1 to 2.

In daily clinical practice, a balance must be struck between plan quality and delivery efficiency that determines how many arcs should be used in IMAT planning. Our experience suggests that 1 arc can provide high-quality plans with much better delivery efficiency for less challenging cases, such as prostate and pancreas cases. For more complex cases, such as pelvis and head and neck cases, multiple arcs are preferable because the resulting plans provide better target dose coverage without significantly prolonging the delivery time.

Intensity-Modulated Arc Therapy Commissioning and Quality Assurance

When initiating an IMAT program, personnel should carefully review the report from the American Association of Medical Physicists (AAPM) entitled, 'Task Group 142 report: quality assurance of medical accelerators' [44]. It details the key elements of a quality assurance (QA) program. Although this report was written before fully dynamic IMAT control systems were available, the tests in the report serve as an excellent base for an IMAT QA program.

As compared with fixed-field IMRT, IMAT necessitates additional QA measurements. This is because of the dynamic nature of IMAT, where delivery includes MLC leaf movement, gantry rotation and dose rate variations. There are 2 key components to an IMAT QA program: (a) machine-based QA and (b) patient-specific QA.

Commissioning and Machine Quality Assurance

It is anticipated that the American Association of Medical Physicists will provide specific QA recommendations for all VMAT platforms. To date, publications provide guidance on commissioning and QA for Varian RapidArc and Elekta VMAT. The tests examining the accuracy of MLC leaf positioning, the dose rate and gantry angles were performed and recommended for Varian RapidArc by Ling et al.

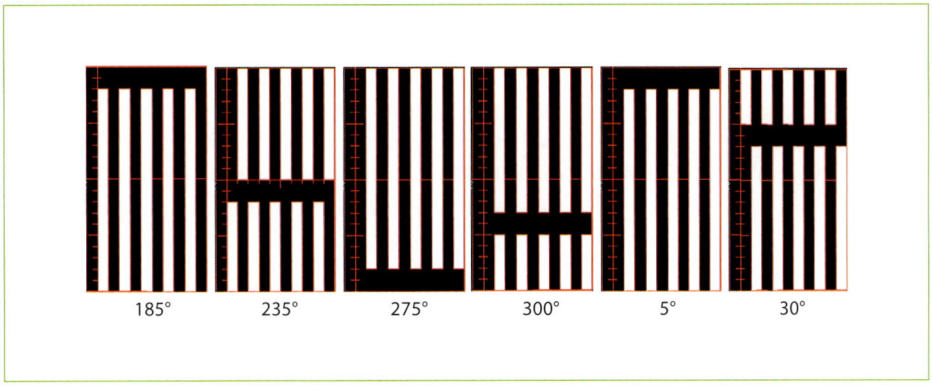

Fig. 10. MLC test pattern used at the Swedish Cancer Institute to test the machine performance when delivering using VMAT.

[29] For Elekta VMAT, tests on beam flatness and symmetry, MLC leaf calibration, sliding window dose, as well as rotational accuracy were considered necessary by Bedford et al. [33].

At the Swedish Cancer Institute, a specific MLC moving pattern along with a varied dose rate was introduced to test the accuracy of MLC leaf positioning, dose rate and gantry angle. Figure 10 shows the aperture shapes at different gantry angles to show the patterns of MLC leaf motion. A key feature of this test is that all 3 major variables during IMAT delivery can be tested within one 360° gantry rotation, which takes approximately 90 s. The detecting device used in this test is an IBA MatriXX 2D ion chamber inserted in a MULTICube solid water phantom.

A sample machine QA result for an Elekta Synergy linear accelerator is shown in figure 11. The results suggest that a maximum gantry angle error of 2.6° can be observed. The maximum leaf position error can reach up to 3.1 mm with a standard deviation of 0.2 mm. As to the dose rate, the mean error is 3.2 MU/min.

Patient-Specific Quality Assurance

One of the most widely adopted techniques for patient-specific IMRT QA has been the measurement of fluence maps for each beam angle. The fluence maps are each compared versus the predicted fluence map as a means to verify the accuracy of the delivery.

The dynamic nature of IMAT delivery, however, makes the fluence measurement undesirable. Simple fluence map measurements can fail to capture errors in the delivery introduced by the rotational delivery of the dose. As a result, a preferred option for IMAT patient-specific QA is the measurement of a composite dose. This can be accomplished using a number of available tools including:

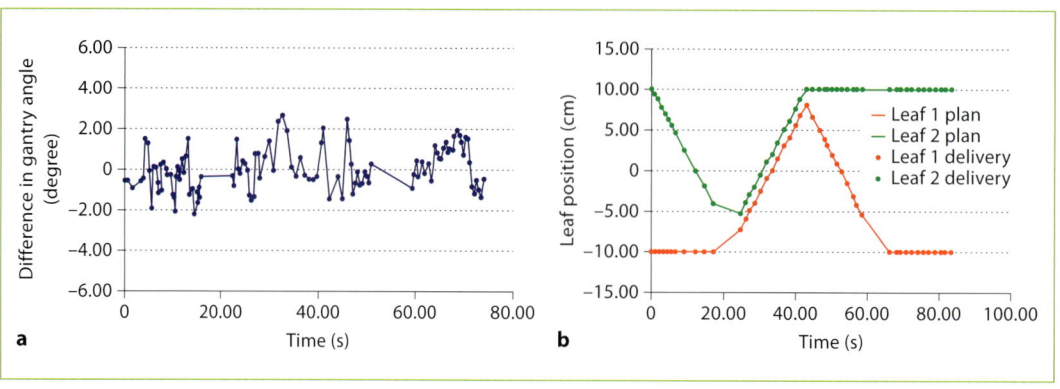

Fig. 11. Sample machine QA results for VMAT. **a** Gantry angle error. **b** MLC leaf position.

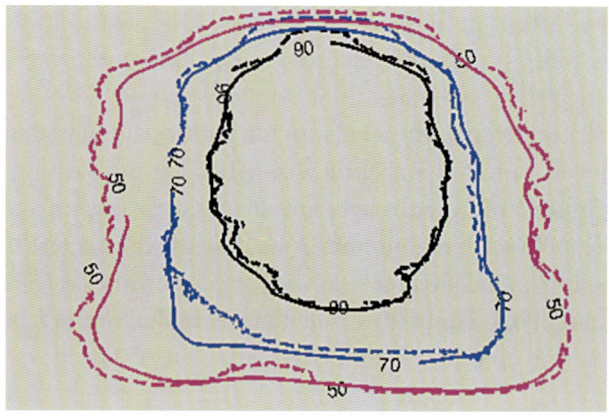

Fig. 12. Overlay of planned versus measured doses using a film measurement for a prostate case delivered using VMAT.

(1) Film and ion chamber
(2) 2D diode array
(3) 2D ion chamber array
(4) Other 2D/3D diode systems
(5) Dose reconstruction

A film and ion chamber combination has been widely adopted for conventional fixed field IMRT plan QA for many years. Typically an ion chamber is used for an absolute point dose measurement, while a film is used for 2D relative dose measurement. As shown in figure 12, in a 2D isodose overlay measured for a single arc prostate IMAT plan, the measured film results matched very well to the planned

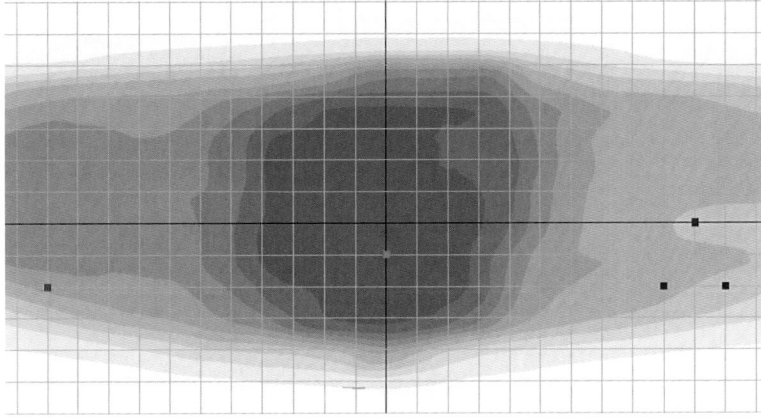

Fig. 13. Measured result for VMAT QA for a pancreatic case. With a gamma analysis of 3%/3 mm, the blue dots show diodes that were cold and the red spot indicates a measurement that was hot.

dose. For this case, the measured ion chamber dose was 2.0% lower than the planned dose. Despite the excellent resolution provided by the film, the prolonged data analysis (including film processing, scanning and registration) makes film QA less popular as compared to the systems that have instantaneous dose read out, such as the MapCheck and MatriXX systems.

A 2D diode array inserted into a solid water phantom is another option for IMAT plan QA. Figure 13 shows such a QA result using MapCheck and MapPhan (SunNuclear) for a pancreas case. With 3% and 3 mm passing criteria, the gamma analysis for the case is 98.6%. For a 2D diode array system consisting of diodes that are very small in size, an angular dose response typically exists. Such an angular dose response can be cancelled out in most IMAT plan QAs due to the rotational nature of IMAT delivery. However, this does not always happen, especially for cases with partial arcs or cases with asymmetric dose delivery. A measurement with built-in correction for the angular dose response can help improve the performance of the 2D diode array system.

A 2D ion chamber array inserted into a solid water phantom can also be used for IMAT plan QA. Commercial ion chamber arrays include the PTW 729 and IBA MatriXX systems. A 2D isodose overlay of a sample QA result for a head and neck case using the IBA MatriXX system inserted in a MULTICube phantom is shown in figure 14. The gamma analysis provided a passing rate of 99.8% with 3%/3 mm criteria for this case. Although an angular dose response also exists in MatriXX ion chamber array system, the magnitude is smaller compared with that of the MapCheck diode system. With an inclinometer, a real-time correction of this angular dose response can be achieved.

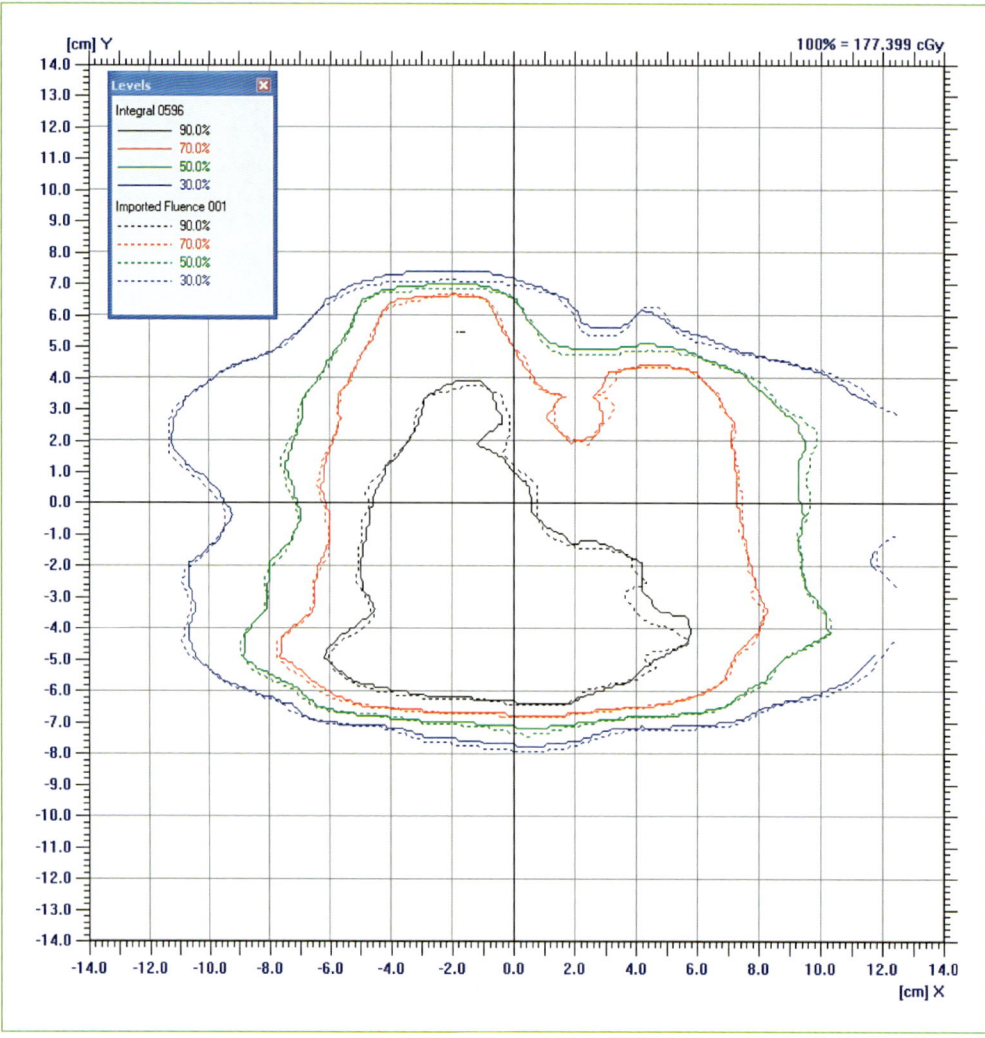

Fig. 14. Overlay of planned and measured isodose curves for head and neck VMAT plan verified using a 2D ion chamber array.

Other 2D/3D diode array systems, such as Scandidose Delta[4] and SunNuclear ArcCheck, can also provide excellent IMAT plan QA. A head and neck IMAT plan QA result using ArcCheck is shown in figure 15 as an example. The gamma passing rate for this case is 98.7% with 3%/3 mm criteria.

In considering a QA solution for patient-specific QA, it is important to understand the performance of your QA tool as a measurement system for rotational IMRT. The users need to know the limit of each system in order to obtain high-quality IMAT plan QA.

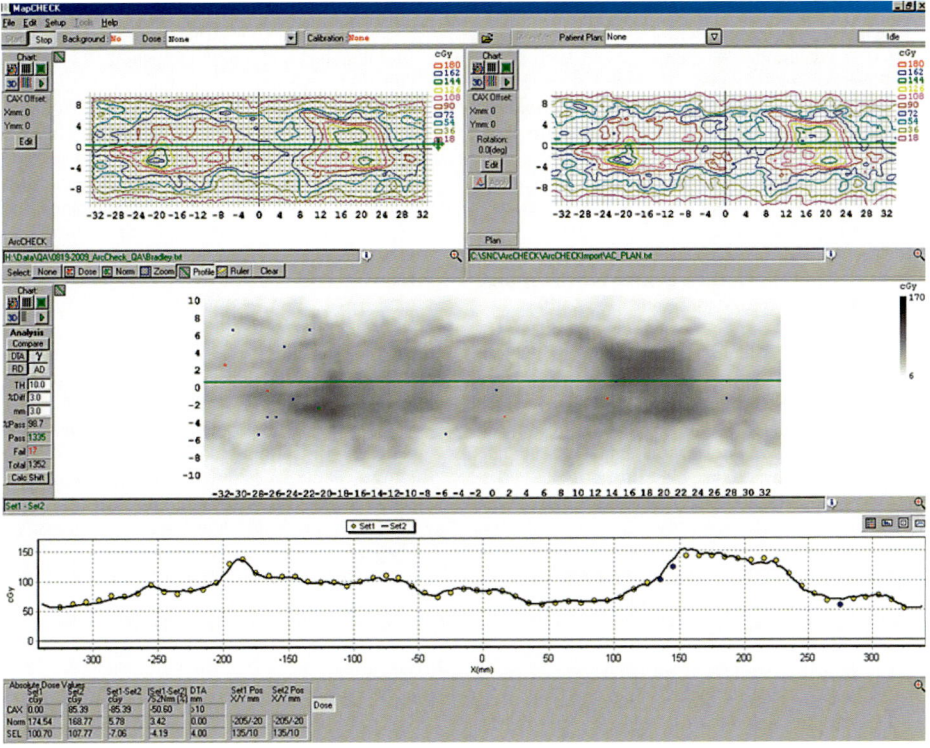

Fig. 15. VMAT QA results for a head and neck plan verified using a 3D cylindrical diode array. See online supplementary material.

References

1 Yu CX: Intensity-modulated arc therapy with dynamic multileaf collimation: an alternative to tomotherapy. Phys Med Biol 1995;40:1435–1449.
2 Mackie TR: et al: Tomotherapy: a new concept for the delivery of dynamic conformal radiotherapy. Med Phys 1993;20:1709–1719.
3 Mackie TR: History of tomotherapy. Phys Med Biol 2006;51:R427–R453.
4 Mackie TR, et al: Tomotherapy. Semin Radiat Oncol 1999;9:108–117.
5 Iori M, et al: IMAT-SIM: a new method for the clinical dosimetry of intensity-modulated arc therapy (IMAT). Med Phys 2007;34:2759–2773.
6 Cao D, et al: Comparison of plan quality provided by intensity-modulated arc therapy and helical tomotherapy. Int J Radiat Oncol Biol Phys 2007; 69:240–250.
7 Shepard DM, et al: An arc-sequencing algorithm for intensity modulated arc therapy. Med Phys 2007;34:464–470.
8 Duthoy W, et al: Clinical implementation of intensity-modulated arc therapy (IMAT) for rectal cancer. Int J Radiat Oncol Biol Phys 2004;60: 794–806.
9 Duthoy W, et al: Whole abdominopelvic radiotherapy (WAPRT) using intensity-modulated arc therapy (IMAT): first clinical experience. Int J Radiat Oncol Biol Phys 2003;57:1019–1032.
10 Earl MA, et al: Inverse planning for intensity-modulated arc therapy using direct aperture optimization. Phys Med Biol 2003;48:1075–1089.
11 Yu CX, et al: Clinical implementation of intensity-modulated arc therapy. Int J Radiat Oncol Biol Phys 2002;53:453–463.
12 Wong E, Chen JZ, Greenland J: Intensity-modulated arc therapy simplified. Int J Radiat Oncol Biol Phys 2002;53:222–235.
13 Ma L, et al: Optimized intensity-modulated arc therapy for prostate cancer treatment. Int J Cancer 2001;96:379–384.
14 Bratengeier K: Applications of two-step intensity modulated arc therapy. Strahlenther Onkol 2001; 177:394–403.

15 Cotrutz C, Kappas C, Webb S: Intensity modulated arc therapy (IMAT) with centrally blocked rotational fields. Phys Med Biol 2000;45:2185–2206.
16 Oliver M, et al: Analysis of RapidArc optimization strategies using objective function values and dose-volume histograms. J Appl Clin Med Phys 2009;11:3114.
17 Bzdusek K, et al: Development and evaluation of an efficient approach to volumetric arc therapy planning. Med Phys 2009;36:2328–2339.
18 Chen F, et al: Impact of Leaf Motion Constraint on VMAT Plan Quality, Deliver Efficiency and Accuracy. Submitted for publication, 2010.
19 Shepard DM, et al: A simple model for examining issues in radiotherapy optimization. Med Phys 1999;26:1212–1221.
20 Penagaricano JA: Integral radiation dose to normal structures with conformal external beam radiation: in regards to Aoyama et al. (Int J Radiat Oncol Biol Phys 2006;64:962–967). Int J Radiat Oncol Biol Phys 2006;65:1274, author reply 1274–1275.
21 Aoyama H, et al: Integral radiation dose to normal structures with conformal external beam radiation. Int J Radiat Oncol Biol Phys 2006;64:962–967.
22 D'Souza WD, Rosen II: Nontumor integral dose variation in conventional radiotherapy treatment planning. Med Phys 2003;30:2065–2071.
23 Kjaer-Kristoffersen F, et al: RapidArc volumetric modulated therapy planning for prostate cancer patients. Acta Oncol 2009;48:227–232.
24 Rao M, et al: Comparison of Elekta VMAT with helical tomotherapy and fixed field IMRT: plan quality, delivery efficiency and accuracy. Med Phys 2010;37:1350–1359.
25 Verbakel W, et al: Volumetric intensity-modulated arc therapy vs. conventional IMRT in head-and-neck cancer: a comparative planning and dosimetric study. Int J Radiat Oncol Biol Phys 2009;74:252–259.
26 Fogliata A, et al: Intensity modulation with photons for benign intracranial tumours: a planning comparison of volumetric single arc, helical arc and fixed gantry techniques. Radiother Oncol 2008;89:254–262.
27 Palma D, et al: Volumetric modulated arc therapy for delivery of prostate radiotherapy: comparison with intensity-modulated radiotherapy and three-dimensional conformal radiotherapy. Int J Radiat Oncol Biol Phys 2008;72:996–1001.
28 Korreman S, Medin J, Kjaer-Kristoffersen F: Dosimetric verification of RapidArc treatment delivery. Acta Oncol 2009;48:185–191.
29 Ling CC, et al: Commissioning and quality assurance of RapidArc radiotherapy delivery system. Int J Radiat Oncol Biol Phys 2008;72:575–581.
30 Oliver M, Ansbacher W, Beckham WA: Comparing planning time, delivery time and plan quality for IMRT, RapidArc and Tomotherapy. J Appl Clin Med Phys 2009;10:3068.
31 Yoo S, et al: Radiotherapy treatment plans with RapidArc for prostate cancer involving seminal vesicles and lymph nodes. Int J Radiat Oncol Biol Phys 2009;76:935–942.
32 Haga A, et al: Quality assurance of volumetric modulated arc therapy using Elekta Synergy. Acta Oncol 2009;48:1193–1197.
33 Bedford JL, Warrington AP: Commissioning of volumetric modulated arc therapy (VMAT). Int J Radiat Oncol Biol Phys 2009;73:537–545.
34 Earl MA, et al: Jaws-only IMRT using direct aperture optimization. Med Phys 2007;34:307–314.
35 Jiang Z, et al: An examination of the number of required apertures for step-and-shoot IMRT. Phys Med Biol 2005;50:5653–5663.
36 Shepard DM, et al: Direct aperture optimization: a turnkey solution for step-and-shoot IMRT. Med Phys 2002;29:1007–1018.
37 Otto K: Volumetric modulated arc therapy: IMRT in a single gantry arc. Med Phys 2008;35:310–317.
38 Jones S, Williams M: Clinical evaluation of direct aperture optimization when applied to head-and-neck IMRT. Med Dosim 2008;33:86–92.
39 Dobler B, et al: Comparison of direct machine parameter optimization versus fluence optimization with sequential sequencing in IMRT of hypopharyngeal carcinoma. Radiat Oncol 2007;2:33.
40 Cameron C: Sweeping-window arc therapy: an implementation of rotational IMRT with automatic beam-weight calculation. Phys Med Biol 2005;50:4317–4336.
41 Cozzi L, et al: A treatment planning study comparing volumetric arc modulation with RapidArc and fixed field IMRT for cervix uteri radiotherapy. Radiother Oncol 2008;89:180–191.
42 Palma D, et al: Volumetric modulated arc therapy for delivery of prostate radiotherapy: reduction in treatment time and monitor unit requirements compared to intensity modulated radiotherapy. Int J Radiat Oncol Biol Phys 2008;75:S472–S473.
43 Cao D, et al: Comparison of single-arc and multiple-arc VMAT plans. Submitted for publication, 2010.
44 Klein EE, et al: Task Group 142 report: quality assurance of medical accelerators. Med Phys 2009;36:4197–4212.

Dr. David M. Shepard
Swedish Cancer Institute, 1221 Madison Street, First Floor
Seattle, WA 98104 (USA)
Tel. +1 206 215 3306, Fax +1 206 215 6150, E-Mail david.shepard@swedish.org

4D Imaging and 4D Radiation Therapy: A New Era of Therapy Design and Delivery

Daniel Low

Department of Radiation Oncology, University of California, Los Angeles, Calif., USA

Abstract

Recently developed 4D CT imaging technologies have shown that significant organ motion can occur within radiotherapy fields during treatment. Most often a result of respiration, this motion can cause dose delivery errors that are clinically significant when unmanaged, as demonstrated in many recent investigations. Motion during the regular breathing cycling is important, but day-to-day breathing variations, as may be caused by changes in residual tidal volume, can cause systematic shifts in tumor position. These may cause delivery misalignments because the tumor is not in the same average location at each treatment. Approaches to management of this motion may involve motion-inclusive planning, gating or tracking. 4D CT has been instrumental in most of these approaches. Given the state of treatment planning software, it is not possible to preplan whether a specific patient would benefit from one or another of these methods. Daily imaging (or use of a nonimage-based system such as Calypso) is necessary to locate the tumor, and the location must be correlated with measurements from a system that tracks breathing motion during treatment delivery. This is typically done using an independent metric that characterizes the breathing cycle (e.g. the height of the abdomen). Only then can the treatment plan be accurately implemented. There are many methods to manage tumor motion, though most are challenging to implement and remain poorly supported by vendors. When determining which system to use, an important distinction between competing approaches is whether they are amplitude- or phase-based. Some implementations may use different approaches for different parts of the treatment planning and delivery process, potentially introducing errors in the characterization of breathing motion. While many advances have been achieved and are discussed here, the development of solid, stable and robust processes to effectively manage breathing motion remains a foremost and continuing challenge in radiotherapy.

Copyright © 2011 S. Karger AG, Basel

4D imaging technologies offer the opportunity to understand and manage organ motion throughout the radiotherapy planning and delivery processes. In particular, the management of breathing-induced motion is a clinical problem that must be overcome before many prescribed treatment courses can be reliably delivered. What is the fundamental challenge? Strategies need to be created that can be broadly applied to patients that have different breathing pattern characteristics. One patient may breathe regularly while another quite irregularly, but the management solutions need to be robust for both. Breathing patterns of patients are often unpredictable, and the clinical tactics have to deal with these real-time variations. This chapter first discusses the impact of breathing motion on CT images, including the relevant measurement and characterization methods used to acquire 4D CT images and model breathing motion; second, the impact of breathing motion on treatments; and third, the primary methods to manage this motion.

Impact of Breathing Motion on Images

Breathing motion will affect CT imaging results. This was shown clearly by Rietzel et al. [1], who imaged a simple sphere that was moving under simulated breathing conditions. When no gaiting was performed, the spherical shape became distorted and broken up into multiple smaller shapes as shown in figure 1a. Further, the specific shapes were not predictable. This work used a single-slice CT scanner, but even when a multiple-slice helical CT scanner was used, craniocaudal movement as from breathing could cause significant distortion of the images. Figure 1b compares CT scans of a doll obtained first with the object stationary and then moving in a pattern that simulated breathing. The distortion of the doll was evident in the dynamic images, and each image shows that the distortion was unpredictable.

There are 4 basic methodologies for managing breathing motion in 4D CT imaging. These use *axial* or *helical* image acquisition with either *prospective* or *retrospective* reconstruction or binning of the images (table 1) [1–4].

(1) Helical scanning refers to a technique where the patient couch moves as the scanner rotates (fig. 2). The speed of this couch motion is characterized in the pitch, defined as the ratio of the couch motion (during a 360 degree rotation) to the width of the CT scanner beam. Low pitches move the couch slower, acquiring multiple scans in each location. The pitch in 4D helical scanning is very low, typically a ratio of 0.1–0.2, allowing the patient to be scanned multiple times as the couch moves so that images are acquired throughout the breathing cycle.

(2) Axial scanning involves acquisition of the scans without any motion of the couch. The scanner repeatedly images in the same couch location and acquires scan data throughout the breathing cycle. When the scanning is finished, the couch is moved to an abutting position and the scan cycle is repeated.

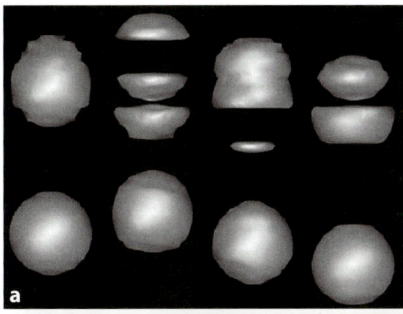

Fig. 1. Effects of motion on CT images of solid objects during simulated breathing: **a** sphere, CT obtained with a single slice CT scanner without gating [1]. **b** Doll, CT obtained with multi-slice helical scanner without gating; **1**: doll static; **2** and **3**: doll scanned while moving in a craniocaudal direction.

Table 1. Respiratory-correlated 4D CT techniques

Scanning method	4D imaging data		
	management	acquisition process	result
Axial	prospective	selected (gating) with manual or automatic triggering	data only for selected acquisition times
	retrospective	unselected	initial oversampling of data for later sorting
Helical	prospective	selected (gating) with manual or automatic triggering	data only for selected acquisition times
	retrospective	unselected	initial oversampling of data for later sorting

(3) In prospective scanning, the scanner is set to acquire images only during user-selected breathing phases.

(4) In retrospective scanning, the image data are analyzed after scan acquisition. The timing of the reconstructed scans is correlated to the breathing cycle and the CT images are sorted into breathing phases.

Within the available technologies, General Electric uses axial retrospective and Siemens and Philips use helical retrospective methods. One of the challenges

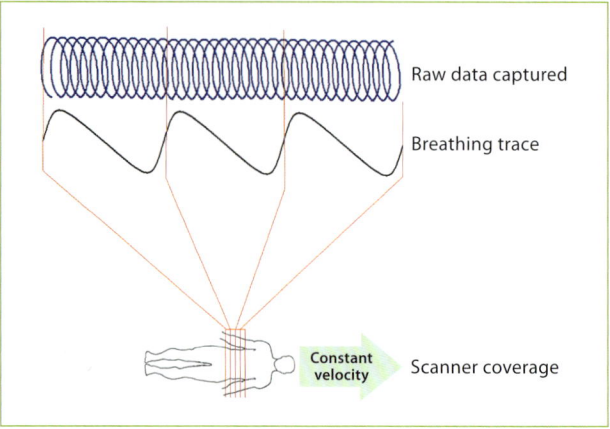

Fig. 2. Helical retrospective scanning. Pitch is set to gather an entire breath worth of raw data at every anatomical position. The pitch remains constant, so if an entire breath is not captured, the motion from the missed section will not be represented in the scan. With unpredictable breathing, it is best to set a lower pitch than needed to allow for longer breaths.

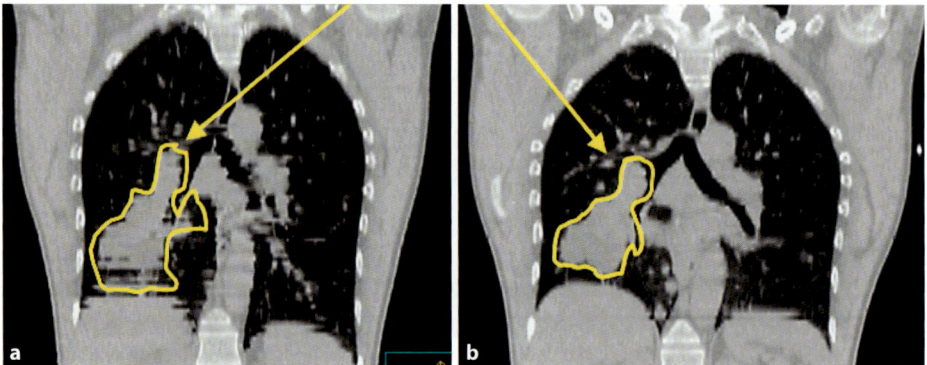

Fig. 3. Respiratory correlated CT. **a** CT obtained without gating. Inferior aspect of tumor image blurs with diaphragm contour and tumor delineation is less accurate. **b** CT obtained with gating, giving clearer definition of anatomy and smaller tumor volume. Courtesy of Medical College of Virginia.

of 4D CT helical scanning is that the scanner cannot be stopped if the patient breathes irregularly. If the couch moves too fast, or the patient's breathing is irregular, then the images will not be acquired though an entire breathing cycle and the subsequent images will have artifacts. These are sometimes severe and sometimes subtle, and can include tumors or organs that appear to split during some phase of the breathing cycle, or organ motion that appears less than the actual motion.

If the 4D CT process is done well, the tumor will be more accurately imaged on a CT scanner than when conventional scanning is used (fig. 3). This is also true for cone-beam CT (CBCT). A number of groups are gaining experience with 4D

CBCT [5–9], typically using retrospective gating to sort images according to the breathing cycle; this implies that they highly oversample the projection data before sorting. This 4D CBCT processing improves clarity significantly, but images are still not as clear as images of static anatomy. When gating is not used in CBCT of the thorax, there are significant motion artifacts and blurring.

The Breathing Cycle

There are 2 basic descriptions of the breathing cycle used by the radiation therapy community; one is based on the *amplitude* and the other on the *phase* of respiration. (a) With amplitude-based descriptions, the breathing cycle is defined by the depth of breathing, which can be thought of as proportional to the tidal volume: the volume of air inspired at a given time. (b) With phase-based descriptions, the breathing cycle is defined by the fraction of time between successive breaths. The point in time that defines the start of the breathing cycle is typically either the peak of inhalation or exhalation. The amount of time between successive peaks is broken down to define the phases only by the fraction of time between these peaks and not by anatomical changes. For instance, if 10 phases are being designated, each would be defined simply as one tenth of the time between successive peak inhalations.

Since 4D CT images are sorted by their designated point in the breathing cycle, the particular method of segmenting the breathing cycle affects how the images are sorted and ultimately used especially for gating. When the patient breathes regularly, i.e. consistent in both respiration amplitude and period, either the amplitude or phase-based breathing cycle definition will provide an effective basis for generating and using 4D CT images. Challenges arise when the breathing pattern is irregular, something that is common among lung cancer patients. Under these conditions, the two different approaches to breathing cycle definition will yield different results and potentially introduce systematic errors in treatment.

Lu et al. [2] compared gating results based on amplitude and phase interpretations of the breathing cycle. Figure 4 shows the breathing traces of a patient taken while being scanned. The breathing trace was acquired using a spirometer, so the breathing cycle was defined by tidal volume (other systems use different cycle metrics, but the results would be equivalent). In this case, the patient breathed regularly for the first 25 s of the measurement, then began to spend more time in exhalation. These investigators asked: what happens to a definition of 'mid-inspiration' in this patient using both methods, for example if a 'mid-inspiration CT scan' was requested? Using the amplitude-based approach, the patient's overall mid-inhalation tidal volume was found to be 350 ml. In each breathing cycle, the times at which the patient was near 350 ml are readily identified (as shown in blue, upper panel, fig. 4), even while the patient breathed irregularly. Presum-

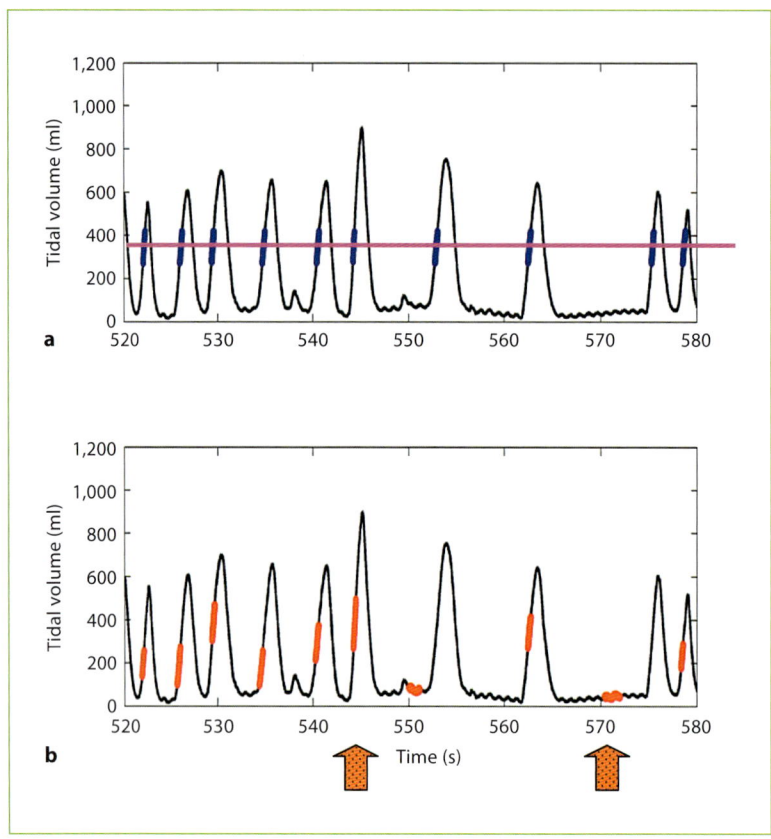

Fig. 4. a Mid-inspiration defined by amplitude sorting of 4D CT (blue line segments), based on the percentile of the tidal volume (purple line). **b** Mid-inspiration defined by phase sorting (orange line segments), based on time between exhalation and inhalation peaks. Mid-inhalation is accurately identified during regular breathing (left arrow), but not during irregular breathing (right arrow) [2].

ably, the internal tumor positions would be consistent at these times, and the resulting CT scan would accurately represent the patient's mid-inhalation geometry. The result is not the same when employing a phase-based approach to define mid-inspiration. The lower panel of figure 4 (arrows) shows 2 examples of defining the mid-inspiration breathing phase as the time halfway between peak exhalation and peak inhalation (in this case, the breathing cycles had been subdivided into the inhalation and exhalation). During the time the patient breathed regularly, mid-inhalation time corresponded to a tidal volume of approximately 400 ml and mid-inspiration was correctly identified (first arrow). However, by the time of the second arrow (572 s), the breathing pattern had changed significantly, and rather than identify a time when the patient was at mid-inspiration,

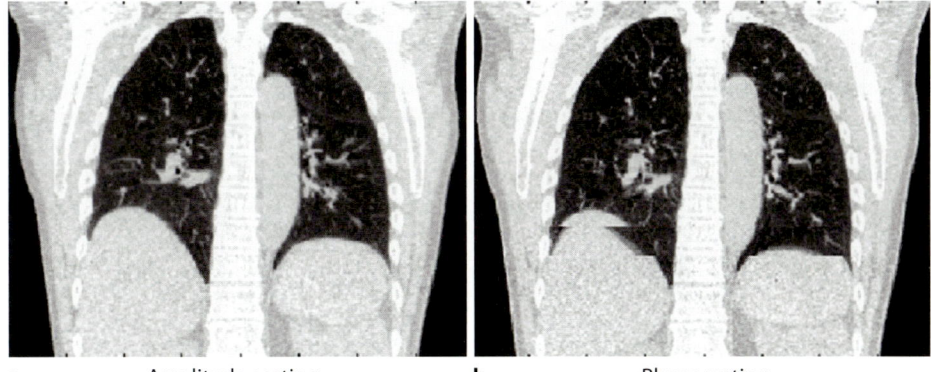

a   Amplitude sorting   b   Phase sorting

Fig. 5. CT sorting based on amplitude (**a**) or phase (**b**; can be better seen on online supplementary videos). In **b**, image acquisition was obtained during slightly irregular breathing and some of the image data were incorrectly associated with breathing cycle based on phase sorting.

the patient was still at exhalation. There is no doubt that the tumor, lungs and other tissues were *not* in the anticipated mid-inhalation position at that time. When breathing is irregular, there are challenges to using the phase angle definition of the breathing cycle.

The selection of amplitude or phase-angle-based gating is a fundamental issue with commercial 4D CT algorithms, since most do not automatically account for irregular breathing in the patient [4]. One example is seen in figure 5. It is important to understand that image acquisition during an unusually deep or shallow breath can cause the image dataset to inaccurately reflect motion and introduce a systematic error in planning; however, most commercial 4D CT systems do not allow for identification of irregular breathing patterns during acquisition. Some commercial software is beginning to allow the users to review the recorded respiratory cycles and identify the irregular breaths. For example, Mutaf et al. [10] were able to examine breathing patterns and appropriately reassign breathing phases, which allowed the reconstruction algorithm to ignore unreliable image data acquired at times when the breathing was sufficiently irregular. At Washington University in St. Louis, all associated breathing cycle patterns can be reviewed for the 4D CT planning. The CT scanner software places tags when assigning each peak inhalation, and inaccurately placed tags can be manually edited by the user. Regardless of the system that is used, none is sufficiently automated to be able to identify and compensate for significant patient breathing irregularity.

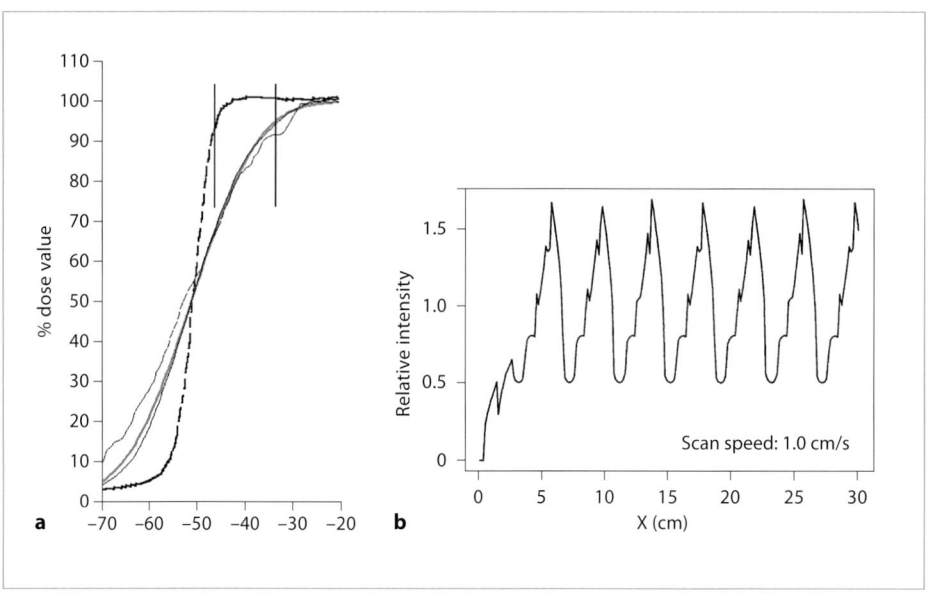

Fig. 6. Potential impact of breathing motion on treatments. **a** Edge penumbra without breathing (steep dashed sigmoid curve) and with breathing (solid curves). Note difference of up to 20% in delivered dose value between vertical lines [11]. **b** Sliding window dose delivery with a narrow sliding window. Dose errors as large as 50% were shown to occur if the breathing motion was in the same direction as the sliding window [13].

Impact of Breathing Motion during Treatments

Dose Gradients

Breathing motion also affects radiation delivery. There are 2 basic concerns. First, breathing motion blurs the dose gradients. McCarter and Beckham [11] showed the effect of craniocaudal breathing motion on the edge penumbra of the radiation beam, shown in figure 6a. The ordinarily steep, high-resolution dose distribution gradient becomes much shallower under the influence of breathing motion. What does this mean operationally? If one wishes to deliver a high radiation dose to the tumor, then the target volume margin must be shifted in order to account for the breathing motion. It turns out that it does not need to be moved as much as the maximum breathing motion itself. Mexner et al. [12] have shown that because tumors do not spend their entire time at the extremes of the breathing cycle, the margin that needs to be added is significantly smaller than the total breathing motion. Still, the dosimetric blurring effect is not trivial in lower lobe lesions or in patients that have large breathing amplitudes.

Interplay Effects and Multileaf Collimator

The second aspect of breathing motion to consider is its interplay effects. A decade ago, Yu et al. [13] published a fairly alarming paper, showing that for sliding window dose delivery with a narrow sliding window, one could have dose errors as large as 50% for any one delivery if the breathing motion was in the same direction as the sliding window (fig. 6b). If the tumor is moving along the same direction as the sliding window, some portions of the tumor move parallel to the window and stay under the window too long, while other portions move antiparallel to the window and are not irradiated long enough. Based on film measurements, it was demonstrated that very large errors could occur even with quite reasonable multileaf collimator (MLC) leaf and breathing motion magnitude parameters. Fortunately, breathing motion is typically perpendicular to the MLC motion, and there are a large number of beams used to deliver any one treatment. Both factors reduce the chances that these errors will occur to any meaningful extent.

The clinical significance of these interplay effects has been further questioned based on specific investigations. In 2002, Bortfeld et al. [14] accurately pointed out that these errors will not occur in the same manner in every fraction. If one applies a random association between where the maximum and the minimum errors occur, then the errors go from being large to averaging out. It does not take too many fractions before even the larger errors average out and the total dose error becomes negligible. More practically, Duan et al. [15] performed a 3D study, moving a phantom in a sinusoidal pattern to evaluate the effect of multiple fractions for 5-, 7-, 9- and 10-field IMRT treatment plans. With a single fraction, they found the greatest errors for the 7-field IMRT plan, with 1 beam having dose errors greater than 45%. For simulated treatments of 5 fractions, the random distribution of errors reduced the dose errors to about 2%. While, in theory, breathing-motion-induced errors average out over multiple fractions, the question arises of how large a dose error in a single fraction is the radiation community willing to tolerate? This is an issue that needs resolution in the near future.

Tomotherapy and Breathing

Tomotherapy differs from conventional therapy in the way it interacts with breathing motion, since it is fundamentally a narrow field with jaw settings of only 1.0-, 2.5- or 5.0-cm width. This narrow field moves very slowly, at a magnitude of tenths of a millimeter per second, as the patient is being irradiated typically from head to feet. This is nothing like the speed of a typical MLC, and there is no equivalent of MLC interplay effects in helical tomotherapy.

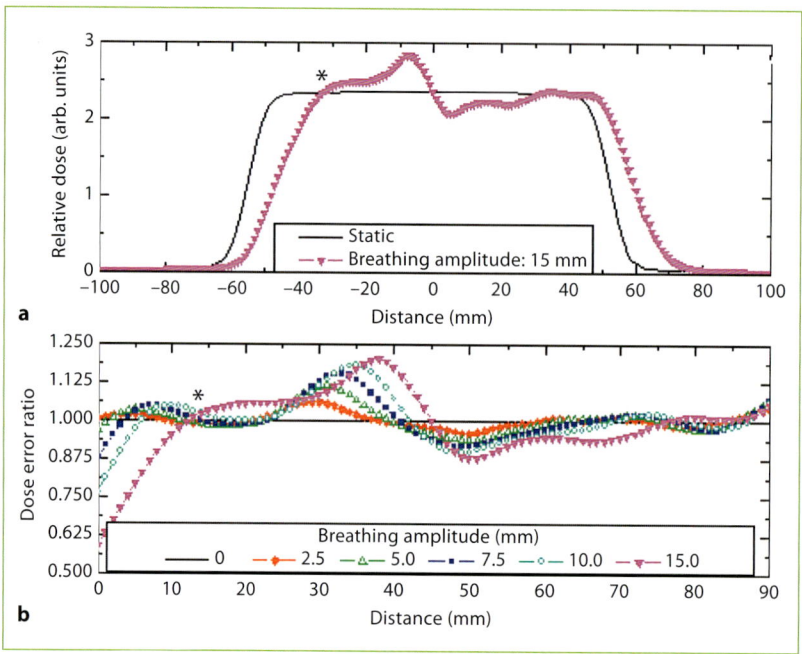

Fig. 7. Effect of breathing motion on helical tomotherapy treatments [16]. **a** Black line = craniocaudal radiation beam profile with no breathing motion. Pink line = radiation beam profile with 15 mm of breathing motion. **b** Same patient with breathing pattern with motion magnitudes 2.5–15 mm.

One might imagine that the tumor, which has faster respiration motion than the very slow motion of the tomotherapy unit, would be blurred by breathing motion. This may be true, but the relatively slow and regular motion of the tomotherapy delivery system may create other interactions. We recently conducted a study of the dosimetric impact of breathing motion on tomotherapy treatments [16]. A large number of patient-measured breathing patterns were used, and a model was developed to simulate the impact of breathing motion on tomotherapy-delivered dose distributions. We studied the relationship of couch speed, breathing motion magnitude and field size. While we expected blurring of the dose distributions with breathing, we also found nontrivial dose delivery errors and examined the reasons for these. In figure 7, the craniocaudal radiation beam profile with no breathing motion is graphed in black. It has the typical features of being uniform in the middle and has sigmoidally shaped penumbrae at both ends. The pink line is the radiation beam profile with 15 mm of breathing motion, demonstrating a maximum dose error of about 15%. The second part of the figure shows profiles of the same patient's breathing pattern with motion magnitudes from 2.5 to 15 mm. The peak error increases steadily as the breathing motion magnitude increases. In a summary of all 52 patients, patients had between 10 and 20% peak dose errors for even a 10-mm breathing motion amplitude.

What caused these errors? They were not caused by the interplay between the MLCs and motion, but rather by the interplay between the narrow field length and breathing motion. As patients breathe, their breathing often slowly drifts so that the average breathing volume slowly increases or decreases. This slow drift is accompanied by a slow shift in the average position of the tumor such that it can travel along or against the slowly moving tomotherapy radiation field. When the slow drift follows with or against the moving field, the tumor stays in the beam longer or less than intended, yielding an overdose or underdose. There are drifts in everyone's breathing patterns, some of which cause effectively no dose delivery error, while others cause larger errors.

Conclusion

In summary, we understand that [1] breathing motion changes margin requirements [2], margin requirements need to be known (the motion needs to be measured) to minimize the added margin, and [3] further reductions in margins may be possible if breathing is regulated or effectively reduced (gating).

Methods to Account for Breathing Motion

When treatment planning is based on conventional free-breathing CT scans, the contoured target volume may be distorted due to the breathing motion and may therefore be of limited use [17]. When accurate imaging of the moving target is conducted, treatment planning options open up that can improve the quality of the delivered dose. There are different approaches to account for the motion in therapy planning. These include the development of an internal target volume (ITV), linear accelerator gating, or mid-ventilation target positioning [17]. There are important clinical considerations for each approach. In the clinical chapters of this volume, additional methods that are patient-based (patient coaching, breath-hold and abdominal compression) are added.

Encompassing Tumor Motion: Internal Target Volume and Maximum Intensity Projection

In this approach, motion is accounted for by enlarging the radiation field and quantitatively shaping it to agree with the tumor motion. To do this, an accurate-

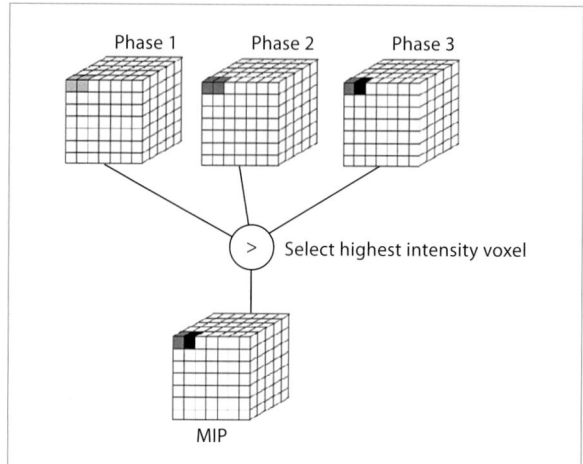

Fig. 8. MIP is a CT image based on selecting the greatest intensity voxels from all phases of the 4D CT dataset.

ly shaped contour must first be defined that reflects the full extent of the tumor motion envelope, designated the *ITV*. Then, radiation beams can be planned to encompass the tumor throughout the breathing cycle. To be as accurate as possible, this process requires a 4D CT dataset that has been divided into multiple phase CT images and the target contoured in each phase. Because of the large amount of work that this may require, automated approaches have been investigated. These take advantage of the fact that tumors generally have greater CT values than the surrounding tissues, which allows tumor volumes to be traced throughout the breathing phases. Automatic processing of the image data can take advantage of this finding to create a cumulative image of the ITV.

One such automated process is the maximum intensity projection (MIP; fig. 8, 9) [18, 19]. The MIP is generated using a 4D CT image dataset. The software examines each image voxel by voxel, looking at the CT numbers of each voxel location in the 4D CT dataset and assigning the greatest CT number value. Tumors typically have greater CT numbers than the surrounding lung tissues, so selecting the maximum CT numbers creates an image that represents the entire tumor trajectory though the breathing cycle. This can provide a very convenient image for contouring.

There are limitations to this technique, however. First, the images are acquired during a single CT session, and planning based on this session assumes that the patient's breathing will be consistently similar at each of the treatment fractions. A particular concern would arise if the patient's breathing became irregular as the scanner imaged the tumor region, since the motion would be inaccurately reflected for the planning of the entire treatment series. Second, if the tumor lies near a major bronchial structure, it may be difficult to visualize the border between the

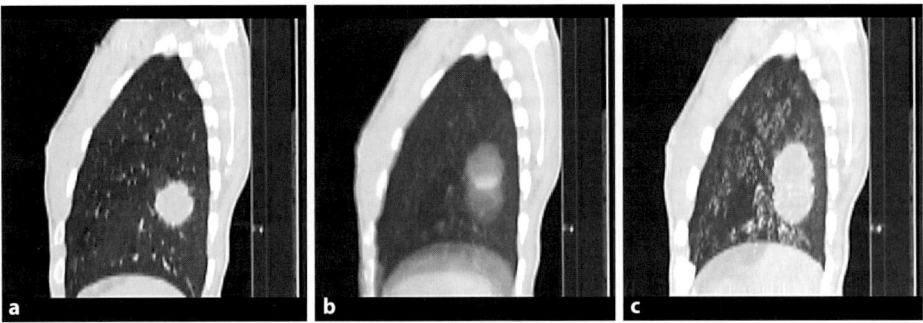

Fig. 9. Clinical example of MIP. **a** Original 4D CT dataset (single phase shown). **b** Mean image (as from 'slow CT' acquisition); note the blurred tumor margins and normal anatomy. **c** MIP; note the greater clarity of anatomy showing full range of tumor motion. Courtesy of Marcel van Herk.

tumor and that structure. Finally, if the tumor lies near the diaphragm, their contours may merge and the inferior portions of the tumor become masked by the diaphragm. We have observed cases with extensions of tumor both inferiorly and laterally that were eclipsed by the diaphragm. In such cases, it is important to realize that the MIP process cannot be used. *It is therefore critical to inspect single-phase images when employing the MIP process for ITV generation to assure that the tumor motion and size are adequately modeled by the MIP.*

Figure 9 shows a clinical example of a MIP. The sagittal view shows a single phase of a 4D CT dataset with the tumor, measuring 12.5 mm and clearly separated from the diaphragm. On the right, the third window shows the MIP of the same tumor and the impact of breathing motion. The MIP process has looked at the maximum intensity of all phases, and one can see that the tumor image becomes elongated to about 21 mm in the direction that would be expected, in the direction of breathing motion. Contouring the MIP is more efficient than contouring a series of single phase images.

Gating

Whether for imaging or therapy, gating results in reduction of motion effects by excluding times when the tumor is not in a defined physical location. In therapy, gating turns on the radiation beam only when the tumor is aligned with it. Since the beam effectively ignores much or most of the tumor motion, the beam portals can be more conformal. Accurate linear accelerator gating requires (a) a correct assessment of tumor motion and (b) a method for determining the tumor position during therapy. This is typically done using a noninvasive breathing surrogate

[20–24]. The surrogate records the breathing waveform, and the tumor is assumed to be in a location previously associated to that surrogate record in the planning 4D CT dataset.

One challenge to this approach is that the algorithm used to gate the 4D CT dataset often differs substantially from the algorithm used to gate the linear accelerator. For example, many 4D CT scanners use *phase-based* gating to retrospectively sort the image data. The 4D CT image datasets are presented as fractions of the breathing cycle as a function of time, so variations in breathing depth are not represented. In linear accelerator gating, phase-based approaches are challenging because the gating is done in real-time. In order to know the phase at a given point in time, one needs to know when the next breath would start. A predictive algorithm would be required for this, but the nature of human breathing makes predicting the breathing cycle very difficult. As a result, most linear accelerator gating algorithms operate based on breathing *amplitude* not phase.

How does the treatment planner rectify this discrepancy? This is solved by determining the amount of motion that the gating should tolerate and selecting the 4D CT phases that should be included in the gate. Once the tumor motion has been determined, the amplitude at the linear accelerator can be set to allow the tumor to move the required amount during the gated beam. This requires that the tumor motion be measured at the linear accelerator. For accurate gating, it is critically important that the tumor be visualized using a real-time imaging modality (e.g. fluoroscopy or 4D CBCT) [9, 25–30]. If the tumor cannot be visualized, then an internal surrogate such as a radio-opaque clip that can be visualized should be used [26, 27]. In that case, the relative motion of the tumor and surrogate will need to be determined from the 4D CT dataset; the final gating window will correspond to the motion of the internal surrogate rather than the tumor or tumor shadow.

Another challenge is that external breathing surrogates may not have a consistent relationship between their signal and the tumor motion [25]. The tumor may move different amounts for the same amount of motion of the surrogate, and the tumor may be in a different position on a day-to-day basis even for the same external surrogate position. While the source of this error is not fully understood, it may be a reflection of a changing residual tidal capacity. This is another reason why we recommend that direct imaging approaches be used to verify the amount of tumor motion present in the treatment gate.

The gating window needs to be defined. If a full window is used, there is effectively no gating. How does one decide or quantify the dosimetric impact of the window selection? In fact, no commercial tools to do this are available as yet. One can approach this manually by creating both gated and ungated plans, but this is not automated or easily guided. Most groups use some form of rule-based system. In other words, gating is arbitrarily undertaken if the motion is greater than 5 mm, or 1 cm, and often based on anecdotal data to make that determination.

Table 2. Tumor tracking methods

Imaging + surrogate
   OBI CBCT
   Soft-tissue contrast of CBCT used to identify tumor in 3D
   Surrogate maps motion and that map used to determine position throughout treatment (monitor surrogate)

Implants
   Implants + surrogate
      Fluoroscopic Imaging Systems
   Implants alone
      Calypso Medical

Although less common, there is the possibility that a lung tumor will systematically shift between fractions, as may occur with the resolution of atelectasis. With careful daily alignment to the tumor, the user may be allowing a critical structure to enter the high-dose regions. This concern is especially important in tumor tracking methods, but may be an issue in all motion management approaches. It is very important to relate the imaged anatomy that can be rigidly associated with the neighboring critical structures and the irradiated field. An example would be monitoring the spine position on fluoroscopy or portal images to determine if the spine has moved closer to the radiation fields. If the planned distance between the high-dose region and the spine is less than 1–2 cm, a systematic shift in the target position may place the cord within the high-dose region.

Tracking

Tracking means that the delivery device is following the moving tumor so that the dose is consistently delivered regardless of the breathing phase. Several vendors are actively working on practical tracking strategies. The CyberKnife system, the only radiotherapy tracking device commercially available, robotically adjusts the position of the linear accelerator to match the motion of the tracked tumor [31]. Other technologies include moving the couch to match the motion and adjusting the MLCs to move the aperture to match the motion, although these are not yet commercially available [32–34]. There are two general methods to track the tumor in real-time and both are similar to the technologies used for gating (table 2).

(1) One can perform an imaging sequence that is correlated to a breathing surrogate: the imaging sequence is used to determine the tumor position as a function

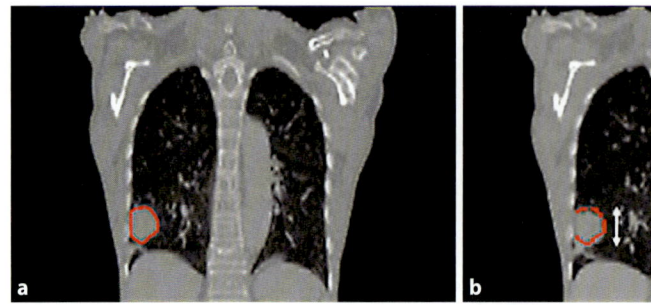
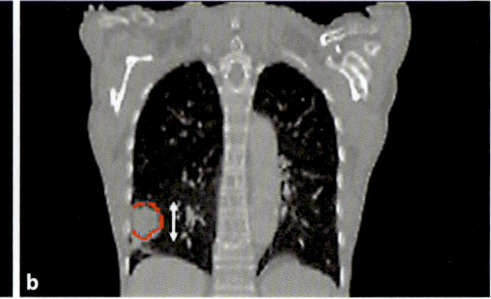

Fig. 10. a Mid-ventilation CT. **b** 4D CT. Mid-ventilation CT is simple and its clinical use is expanding. It eliminates systematic error due to imaging (except hysteresis), and approaches a full 4D plan geometrically and dosimetrically [37]. See online supplementary material for conceptual display. Courtesy of Marcel van Herk.

of the breathing surrogate. Once the relationship between the surrogate and tumor position has been determined, the surrogate is used to determine the tumor position during treatment. This is essentially the method that the CyberKnife uses.

(2) A second method for tumor tracking is based on implanted markers. If the implants can be tracked in real-time, their relationship to the tumor position can be measured and they can be used to directly aim the tracking algorithm. The Calypso system has the potential for doing this [35, 36]. It is composed of a flat board that contains a series of antennae that emit radiofrequency (RF) energy. The implanted markers are glass-enclosed RF transponders that absorb the RF energy. When the antennae stop emitting the RF power, the transponders re-emit their energy, the antennae measure their relative power and the transponder positions are determined. While the system was designed and built to localize the prostate gland for irradiation, it is being adapted to perform gating and tumor tracking. The system has excellent temporal resolution. For instance, it should be able to track a lung tumor in real-time if implanted into or near that lesion.

While tracking may adequately match the beam and tumor position, when one tracks the tumor, one 'untracks' everything else. This may be significant, such as when a tumor is moving towards a critical normal structure, unexpectedly increasing the dose to that structure. It should always be remembered that critical structures define dose limits, not tumors. When the tumor is made the point of tracking, doses to critical normal structures may be less well defined, depending on the magnitude and direction of the tumor movement. It will be important for any planning system that supports tracking to provide predictions of normal organ doses based on that tracking.

Gating and Tracking Alternatives

Marcel van Herk and coworkers [17, 37] suggest a simpler approach to managing tumor motion (fig. 10). They examined 4D CTs for a large number of patients and considered the dosimetry when the beam is simply positioned accurately in the middle of the tumor trajectory. They found that for most patients, the additional margins required for dosimetrically correcting for the positional breathing changes over time are not very large. This is because the tissues spend most of their time (and receive most of their radiation dose) in phases other than the extremes of inhalation or exhalation. For instance, the superior portion of a tumor might be in the beam penumbra only during maximum exhalation, when it would momentarily receive somewhat less than the prescribed dose. It spends most of its time well inside the radiation portal during phases other than exhalation, when it receives most of the delivered dose. Only a small additional margin is needed to achieve a therapeutic dose range in this superior portion. Van Herk and coworkers advocate using only small margins that provide sufficient dosimetric coverage, coupled with quantitative position verification to be certain that the treatment is planned and delivered daily at the tumor's mid-position as determined by the 4D CT ITV. In other words, the main issue is not 'is the tumor moving,' but rather, 'where is it centered on a daily basis?' They promote using gated CBCT [37] to identify the tumor position and positioning the patient so that the daily mean tumor position corresponds to the planned mean tumor position from 4D CT ITV. Figure 10 compares images created using the mid-ventilation CT and an image created using an ITV that covers all motion as defined by the 4D CT dataset.

Future Directions

The development of 4D CT has allowed an expanded understanding of the significance of organ motion in treatment planning and delivery. From 4D CT analyses, it is now recognized that breathing often causes significant errors in the expected delivery of treatment plans. Most importantly, breathing may cause delivery misalignment at therapy. Effective dynamic imaging methods at treatment are needed to confirm that the alignment between the tumor and the beam is correct, and also that physiologic changes have not occurred within the patient that alter the treatment plan. Breathing may cause dosimetric errors at delivery due to several types of interactions with existing delivery technologies. Some of these are minor and may be counterbalanced by the use of multiple fields and/or multiple fractions.

4D CT has provided the basis for several types of approaches to managing motion, though most will benefit from further development. Despite many worthwhile advances, CT, treatment planning and linear accelerator technologies need continued development toward robust, integrated processes for motion management. Care must be taken in how each commercial component specifically defines and accounts for breathing motion. It is particularly important that patients who are undergoing gated radiotherapy have direct validation of their therapy's accuracy at the time of daily delivery.

Acknowledgements

The author wishes to acknowledge the valuable contributions of Parag Parikh, Wei Lu, Sasa Mutic, Parag Parikh, Wei Lu, James Hubenschmidt, Camille Noel, Murty Goddu, Deshan Yang and Tianyu Zhao.

References

1 Rietzel E, Pan T, Chen GT: Four-dimensional computed tomography: image formation and clinical protocol. Med Phys 2005;32:874–889.
2 Lu W, Parikh PJ, Hubenschmidt JP, et al: A comparison between amplitude sorting and phase-angle sorting using external respiratory measurement for 4D CT. Med Phys 2006;33:2964–2974.
3 Pan T: Comparison of helical and cine acquisitions for 4D-CT imaging with multislice CT. Med Phys 2005;32:627–634.
4 Rietzel E, Chen GT: Improving retrospective sorting of 4D computed tomography data. Med Phys 2006;33:377–379.
5 Chang J, Sillanpaa J, Ling CC, et al: Integrating respiratory gating into a megavoltage cone-beam CT system. Med Phys 2006;33:2354–2361.
6 Dietrich L, Jetter S, Tucking T, et al: Linac-integrated 4D cone beam CT: first experimental results. Phys Med Biol 2006;51:2939–2952.
7 Kriminski S, Mitschke M, Sorensen S, et al: Respiratory correlated cone-beam computed tomography on an isocentric C-arm. Phys Med Biol 2005;50:5263–5280.
8 Li T, Xing L, Munro P, et al: Four-dimensional cone-beam computed tomography using an on-board imager. Med Phys 2006;33:3825–3833.
9 Sonke JJ, Zijp L, Remeijer P, van Herk M: Respiratory correlated cone beam CT. Med Phys 2005;32:1176–1186.
10 Mutaf YD, Antolak JA, Brinkmann DH: The impact of temporal inaccuracies on 4DCT image quality. Med Phys 2007;34:1615–1622.
11 McCarter SD, Beckham WA: Evaluation of the validity of a convolution method for incorporating tumour movement and set-up variations into the radiotherapy treatment planning system. Phys Med Biol 2000;45, 923–931.
12 Mexner V, Wolthaus JW, van Herk M, et al: Effects of respiration-induced density variations on dose distributions in radiotherapy of lung cancer. Int J Radiat Oncol Biol Phys 2009;74:1266–1275.
13 Yu CX, Jaffray DA, Wong JW: The effects of intra-fraction organ motion on the delivery of dynamic intensity modulation. Phys Med Biol 1998;43:91–104.
14 Bortfeld T, Jokivarsi K, Goitein M, et al: Effects of intra-fraction motion on IMRT dose delivery: statistical analysis and simulation. Phys Med Biol 2002;47:2203–2220.
15 Duan J, Shen S, Fiveash JB, et al: Dosimetric and radiobiological impact of dose fractionation on respiratory motion induced IMRT delivery errors: a volumetric dose measurement study. Med Phys 2006;33:1380–1387.
16 Chaudhari SR, Goddu SM, Rangaraj D, et al: Dosimetric variances anticipated from breathing-induced tumor motion during tomotherapy treatment delivery. Phys Med Biol 2009;54:2541–2555.
17 Wolthaus JW, Sonke JJ, van Herk M, et al: Comparison of different strategies to use four-dimensional computed tomography in treatment planning for lung cancer patients. Int J Radiat Oncol Biol Phys 2008;70:1229–1238.

18 Bradley JD, Nofal AN, El Naqa IM, et al: Comparison of helical, maximum intensity projection (MIP), and averaged intensity (AI) 4DCT imaging for stereotactic body radiation therapy (SBRT) planning in lung cancer. Radiother Oncol 2006;81:264–268.
19 Underberg RW, Lagerwaard FJ, Slotman BJ, et al: Use of maximum intensity projections (MIP) for target volume generation in 4DCT scans for lung cancer. Int J Radiat Oncol Biol Phys 2005;63:253–260.
20 Chi PC, Balter P, Luo D, et al: Relation of external surface to internal tumor motion studied with cine CT. Med Phys 2006;33:3116–3123.
21 Hunjan S, Starkschall G, Prado K, et al: Lack of correlation between external fiducial positions and internal tumor positions during breath-hold CT. Int J Radiat Oncol Biol Phys 2009;76:1586–1591.
22 Ionascu D, Jiang SB, Nishioka S, et al: Internal-external correlation investigations of respiratory induced motion of lung tumors. Med Phys 2007;34:3893–3903.
23 Korreman S, Mostafavi H, Le QT, Boyer A: Comparison of respiratory surrogates for gated lung radiotherapy without internal fiducials. Acta Oncol 2006;45:935–942.
24 Lu W, Low DA, Parikh PJ, et al: Comparison of spirometry and abdominal height as four-dimensional computed tomography metrics in lung. Med Phys 2005;32:2351–2357.
25 Lin T, Cervino LI, Tang X, et al: Fluoroscopic tumor tracking for image-guided lung cancer radiotherapy. Phys Med Biol 2009;54:981–992.
26 Shirato H, Harada T, Harabayashi T, et al: Feasibility of insertion/implantation of 2.0-mm-diameter gold internal fiducial markers for precise setup and real-time tumor tracking in radiotherapy. Int J Radiat Oncol Biol Phys 2003;56:240–247.
27 Tang X, Sharp GC, Jiang SB: Fluoroscopic tracking of multiple implanted fiducial markers using multiple object tracking. Phys Med Biol 2007;52:4081–4098.
28 Cui Y, Dy JG, Sharp GC, et al: Multiple template-based fluoroscopic tracking of lung tumor mass without implanted fiducial markers. Phys Med Biol 2007;52, 6229–6242.
29 Xu Q, Hamilton RJ, Schowengerdt RA, et al: Lung tumor tracking in fluoroscopic video based on optical flow. Med Phys 2008;35:5351–5359.
30 Xu Q, Hamilton RJ, Schowengerdt RA, Jiang SB: A deformable lung tumor tracking method in fluoroscopic video using active shape models: a feasibility study. Phys Med Biol 2007;52:5277–5293.
31 Brown WT, Wu X, Amendola B, et al: Treatment of early non-small cell lung cancer, stage IA, by image-guided robotic stereotactic radioablation – CyberKnife. Cancer J 2007;13:87–94.
32 Keall PJ, Cattell H, Pokhrel D, et al: Geometric accuracy of a real-time target tracking system with dynamic multileaf collimator tracking system. Int J Radiat Oncol Biol Phys 2006;65:1579–1584.
33 Neicu T, Shirato H, Seppenwoolde Y, Jiang SB: Synchronized moving aperture radiation therapy (SMART): average tumour trajectory for lung patients. Phys Med Biol 2003;48:587–598.
34 Shirato H, Oita M, Fujita K, et al: Feasibility of synchronization of real-time tumor-tracking radiotherapy and intensity-modulated radiotherapy from viewpoint of excessive dose from fluoroscopy. Int J Radiat Oncol Biol Phys 2004;60, 335–341.
35 Smith RL, Lechleiter K, Malinowski K, et al: Evaluation of linear accelerator gating with real-time electromagnetic tracking. Int J Radiat Oncol Biol Phys 2009;74:920–927.
36 Smith RL, Sawant A, Santanam L, et al: Integration of real-time internal electromagnetic position monitoring coupled with dynamic multileaf collimator tracking: an intensity-modulated radiation therapy feasibility study. Int J Radiat Oncol Biol Phys 2009;74:868–875.
37 Wolthaus JW, Schneider C, Sonke JJ, et al: Mid-ventilation CT scan construction from four-dimensional respiration-correlated CT scans for radiotherapy planning of lung cancer patients. Int J Radiat Oncol Biol Phys 2006;65:1560–1571.

Prof. Daniel Low, PhD
Department of Radiation Oncology, University of California
200 Medical Plaza, Suite B265
Los Angeles, CA 90095 (USA)
Tel. +1 310 983 3205, E-Mail DLow@mednet.ucla.edu

Locating and Targeting Moving Tumors with Radiation Beams

Paul Keall

Radiation Physics Laboratory, Sydney Medical School, The University of Sydney, Sydney, N.S.W., Australia

Abstract

3D knowledge of the tumor position during abdominal and thoracic radiotherapy is an important component of motion management in radiation therapy. A wide variety of real-time position monitoring systems are available or under development. These are based on a diversity of modalities including radiofrequency, radioisotopes, ultrasound and MRI in addition to the optical, kilovoltage and megavoltage imaging systems available on conventional accelerators. These systems are also providing new insights into the magnitude and complexity of target and normal tissue motion during a course of therapy, and are driving the development of real-time targeting systems. Real-time targeting devices to align the tumor and the radiation beam have built upon technologies of robots, multileaf collimators, and couch-based and gimbaled positioning systems. The integration and widespread dissemination of systems that locate and target moving tumors are ongoing developments in the early 21st century, and future systems are likely to include the functionality of targeting temporally changing tumors and normal tissue physiology as well as anatomy.

Copyright © 2011 S. Karger AG, Basel

This chapter will present the array of technologies that are available to determine where the target is located in real-time, and then review the systems available to move the beam in real-time. Finally, we will discuss how to put the *locating* and the *targeting* together into an integrated system, as shown in figure 1, and then be able to characterize the uncertainties of the system during radiation therapy.

In the clinic, we are observing substantial anatomic and potentially physiologic changes during radiotherapy. We are currently developing tools that are mov-

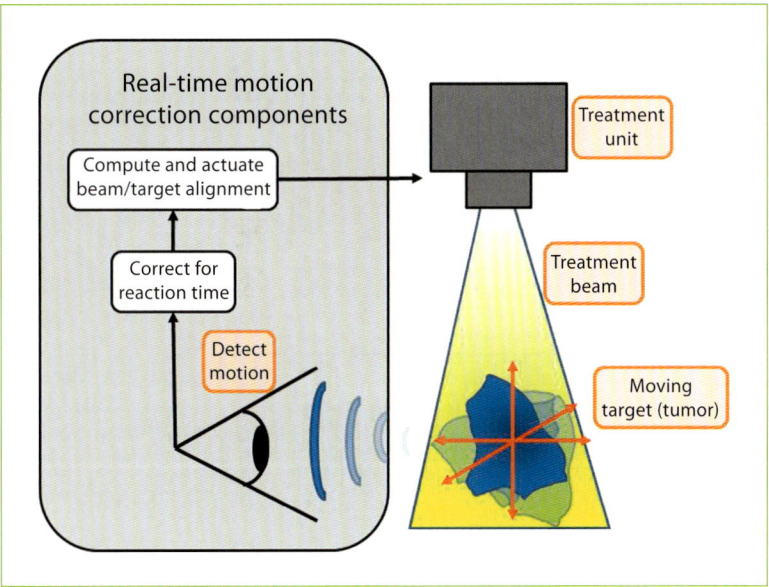

Fig. 1. A generalized flowchart showing the main components of a real-time motion correction system. A system is needed to locate moving targets within the patient in real-time. The system reaction time should be corrected for with prediction algorithms, and a targeting system developed to align the radiation beam and the moving target [52].

ing out of the research realm into the mainstream that are going to allow us to manage anatomic changes, both within a fraction and between fractions of radiotherapy. Figure 2 shows an example of some of the challenges faced in the clinic relating to tumor motion in all 3 dimensions. There is (a) variability in the range of motion of the tumor, (b) baseline drift in its position and (c) a period and pattern of its motion that is changing with time. If guidance was based on the current practice of imaging before radiotherapy, then, depending on the particular moment that the imaging occurred, this may or may not be representative of the motion during treatment. It is likely that there will be differences between the anatomic changes imaged during treatment and what is imaged prior to the treatment.

The intrafractional changes were exemplified by a stereotactic body case in which repeat cone-beam measurements were taken during the treatment (fig. 3). At the start of the treatment session, it was noticed that there was a misalignment between the target position and the skeletal anatomy. To correct for this, a change in position was made (the patient was in a BodyFix System). After 13 min, the patient was reimaged, showing a 10-mm drift in tumor position; we corrected for this and treated 2 beams. Twenty-three minutes later there was another 8-mm

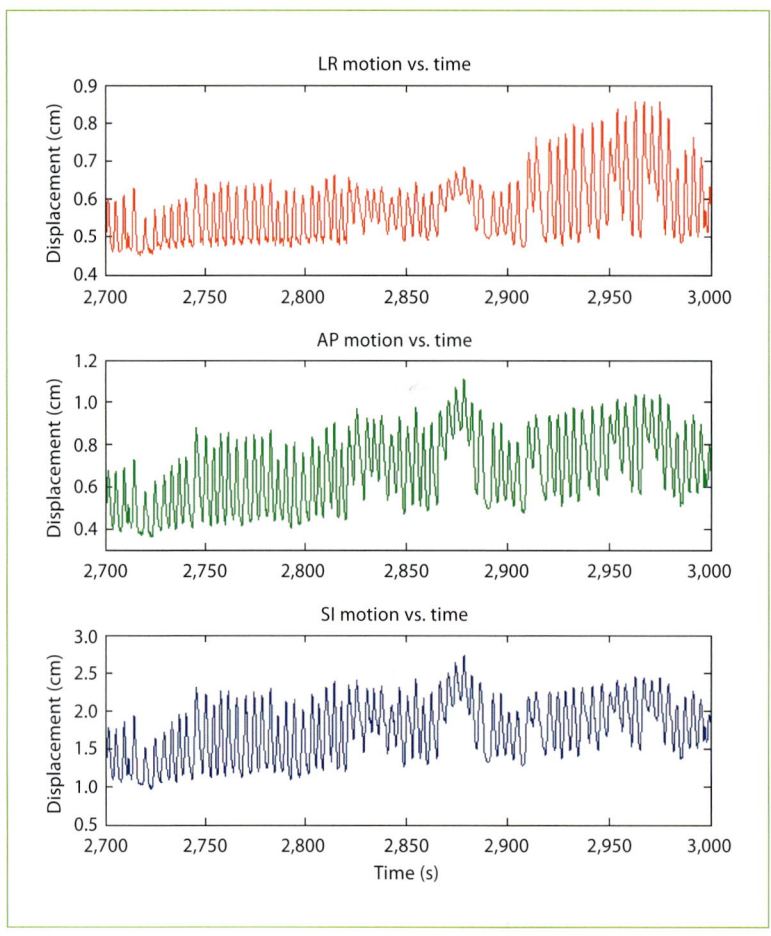

Fig. 2. Observed variation in lung tumor motion with time in 3 dimensions from CyberKnife Synchrony data. Notable is that there is significant motion in 3 dimensions, there are variations in the baseline position of the tumor in all 3 dimensions and variations in the cycle-to-cycle period, respiratory cycle shape and range of motion [52].

drift in tumor position. At completion of treatment, a final cone-beam CT showed an 18-mm shift from the time the treatment was initiated (28 mm from the time the patient was first imaged on the table). This patient was part of a single fraction non-small cell lung cancer protocol for definitive management of T1 tumors with the goal of 100% local control. Unless imaging is performed during the treatment itself, there is no opportunity to correct for such errors that are being observed during treatment.

Tumor targeting matters in 2 closely coupled ways: the delivery of radiation dose to the tumor and avoidance of dose to normal tissues. The importance of tu-

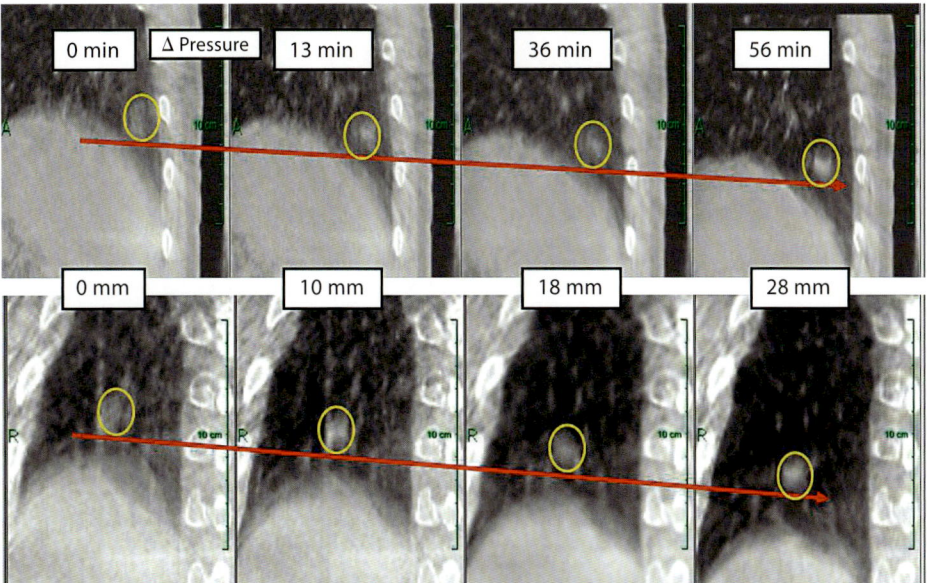

Fig. 3. Observed variation in average tumor position for a stereotactic body lung cancer radiotherapy patient from repeat cone-beam CT scans before and during treatment. Note the systematic drift of 28 mm from the originally measured position within an hour. Courtesy of Drs. Loo and Hristov, Stanford University.

mor dose is shown in a study by McCammon et al. [1], who reported on 141 consecutive patients with lung and liver lesions treated with 3-fraction stereotactic body radiation therapy. The doses prescribed were based on their evolving experience in observing local failures. A clear dose-response relationship was observed. Local control was strongly correlated to the nominal tumor dose ($\geq$54, 36–54 or $\leq$36 Gy) and there was a large difference in the rate of tumor control according to dose. With 54 Gy or higher, more than 90% of the cases achieved control, but this percentage dropped very quickly as the dose decreased. The lesson for radiotherapy is clear: the delivery of adequate dose is critical, and it is imperative to align the radiation beam with the target to achieve local control. On the other hand, there are data for many sites correlating the amount of dose given to normal organ structures with the rates of medical complications observed. If the radiation beam is not aligned with the tumor, then it is irradiating normal structures. By accurate targeting, we are achieving both goals of reducing normal tissue exposure and increasing tumor dose.

In the presence of tumor motion there are 3 challenges: (a) to develop methods to locate the moving target, (b) to develop methods to hit the moving target and (c) to combine these methods into an integrated treatment system.

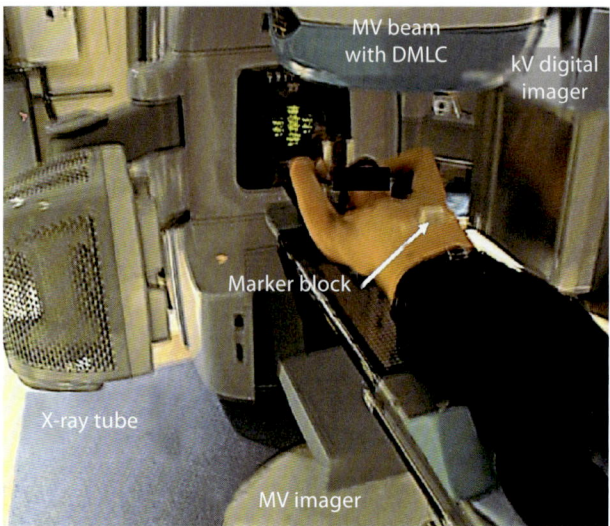

Fig. 4. Three imaging systems are available on conventional modern linear accelerators: optical, kV and MV. See online supplementary material for conceptual display.

How Do We Locate the Moving Target?

If one were designing an ideal 4D in-room imaging system from scratch, it might have many specific characteristics.

(1) It should give volumetric information with very high spatial resolution, and it should be able to report this information quickly.

(2) The information should accurately represent the anatomy that is being imaged.

(3) One should be able to calculate dose on the transformed contours, and have the ability to do adaptive radiotherapy in near real-time.

(4) There should not be any interference with the delivery system, either physical or otherwise.

(5) It should not be invasive or give imaging dose.

(6) One would like it to be integrated into an efficient workflow process so as to reduce treatment time.

(7) Finally, it should be inexpensive to buy and operate.

Of course, such a system is not currently available, but it is helpful to at least think about the characteristics of an ideal system, and then to consider what compromises are made with the present technology. Currently, there are a number of technologies available that meet some of these ideal criteria.

In the modern linear accelerator, one may have access to 3 different data streams that could be used for guidance (fig. 4). One is an optical system, which can be based on monitoring either points or surfaces. The second is a kilovoltage (kV) beam, which potentially can be turned on and off as desired during treatment.

There is also a megavoltage (MV) beam with a MV imager, which can provide some anatomic information some of the time. All of this information may be supplemented with data acquired during prior sessions or other evaluations to cumulate a model of the patient's internal motion, which can be used to better predict tumor position in the future.

Optical Systems

The advantage of optical systems is that real-time information of the external anatomy is obtained. In some cases, the external anatomy may be a very good surrogate for the internal target position, for example the cranium, or for surface lesions such as on the skin. From a dose point of view, use of these optical systems is essentially radiation-free. As a result, there is no clinical 'cost' to acquiring the data. The systems can be used during treatment to provide some degree of intrafraction information. When surface anatomy is not sufficient for guidance, optical systems can be combined with other imaging modalities that give information about the internal anatomy.

Kilovoltage Imaging

The beginning of kV imaging is tied with the introduction of the medical linear accelerator in the 1950s. One of the first linear accelerator rooms at Stanford Hospital incorporated a fluoroscopy unit that could provide a kV X-ray beam of the patient anatomy. Thus, even in the earliest days of linear accelerator radiotherapy, the idea of having an imaging beam incorporated into the delivery room was considered critical. Later, due to workflow, imaging systems were moved out of the delivery room and the traditional simulator was developed. Now that some of the work flow issues are being solved, the simulator is being integrated with the accelerator – analogous to what was available over 5 decades ago, though nowadays with far more advanced technology.

Pioneering scientific and clinical work in the application of real-time kV localization has been performed by Dr. Shirato and colleagues at Hokkaido University [2–5]. Their system consists of 4 kV systems, any 2 of which offer an unobstructed view of the patient at a given time. The markers are inserted in or near the tumor using image-guided implantation, and the system at 30 Hz will recognize their positions. The linear accelerator is gated to irradiate the tumor only when the markers are within a given tolerance from their planned coordinates relative to the isocenter. This system has been used for liver, lung and prostate tumor cases.

The CyberKnife System offers intermittent (every 30–60 s) stereoscopic imaging during treatment that allows the correction of translational and rotation tumor drifts during treatment, but cannot at present be used for real-time applications. It is discussed later in this volume.

There is now a trend toward using rotational delivery or intensity-modulated arc therapy [6–8]. This rotation allows the kV imager to locate targets in 3D during treatment. A limitation of a single X-ray is its lack of depth information, since it is very difficult to determine along the X-ray plane the depth of radio-opaque structures, such as skeletal anatomy or implanted markers. But by rotating the imager, one can use information acquired previously to estimate the current target 3D position. For example, we have used cone-beam CT reconstructions of a pancreas cancer case to determine the tumor trajectory during the cone-beam CT imaging itself. This 2D to 3D method has been extended to allow us, during radiation treatment delivery, to estimate in real-time where the 3D target position is using only a single X-ray imager. This method of real-time targeting has been investigated in phantom studies for rotational delivery to representative prostate and respiratory motion targets, and has shown errors of less than 1–2 mm in being able to follow the target motion as a function of time during the rotational delivery [9, 10]. More recent work has extended this method to fixed-field treatments such as those for conventional IMRT [11].

An issue with continuous kV imaging is the additional dose to the patient, which is approximately 0.18 mGy per image for detecting implanted markers [3]. This dose concern has prompted the development of systems integrated with respiratory monitoring systems, which may be useful to maintain accuracy with lower imaging doses.

Megavoltage Information

Most conventional linear accelerators are now purchased with a MV imaging panel, which will give some useful tracking information, but only some of the time. This is especially true when considering intensity-modulated radiation therapy, when much of the anatomy can be obscured due to the small treatment beam apertures. As yet, the use of these systems for real-time monitoring has been limited, though studies have been performed on preclinical and clinical applications [12–15].

If one combines the kV and the MV beams, then an accurate 3D position of the target can be obtained by triangulation. If one can see some of the markers most of the time when the beam is on, then very accurate therapy delivery can be achieved using this process. Further, if all 3 of these data streams are combined – MV, kV and optical – and one establishes correlations between the motions observed on the surface (optical) with the model that we are building internally (MV, kV), quite good information about the 3D target position throughout treatment delivery can be gained.

Electromagnetic Monitoring

Another method to follow tumor position is through electromagnetic monitoring, which involves the placement of wireless or wired transponders in or near the tumor. Tumor localization is performed by relating the measured position of the transponders during treatment with their position relative to the isocenter during CT treatment planning. In this way, the Calypso System yields high accuracy and precision, with a total system accuracy of less than 1 mm [16]. Clinical investigations at a number of institutions have found that the extent and frequency of prostate motion during radiotherapy delivery can be monitored and used for motion management [17, 18]. Recent work has investigated the preclinical integration of electromagnetic monitoring for lung applications [19] and also the integration of the real-time position monitoring system with the dynamic multileaf collimator (MLF) to correct for the measured translational offsets [20, 21].

MRI Monitoring

An intriguing concept under development at several institutions is integrating MRI with the linear accelerator [22–24]. This approach is compelling as it is possible to obtain volumetric data with an acceptable frame rate, with many free parameters to optimize the image acquisition. It is applicable to many sites.

The advantage of integrating MRI is that one can obtain high-quality anatomic images that can be used to guide radiation beams. With current MRI systems and pulse sequences, it is possible to obtain volumetric information at about 2 frames per second, and the trade off of spatial and temporal information can be used, along with various k-space sampling schemes, to accelerate the image acquisition process. An example of cine MRI imaging is shown in figure 5 and is better seen in the accompanying online supplementary video.

Monitoring Tumor Position with Radioactive Implanted Markers

Radioactive tumor position monitoring can be performed using 2 emission techniques: single photon or positron emission. A positron system was conceptually studied in which planar detectors were placed around the patient and a small radioactive implanted marker provided the signal. A new company, Navotek, is developing an implanted marker-based emission system in which directional dependent detectors are placed on the treatment head of the linear accelerator, allowing fast and precise internal target positioning.

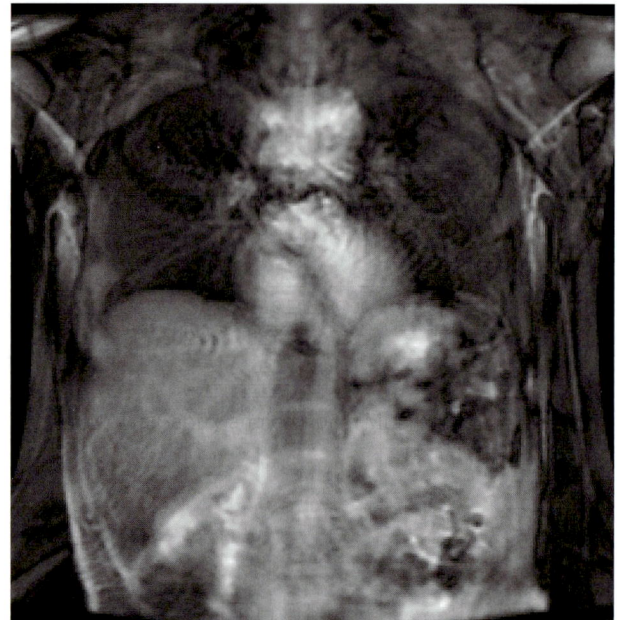

Fig. 5. Cine MRI (Amit Sawant, Stanford University). See online supplementary video for active display.

Ultrasound Monitoring

Ultrasound position monitoring systems [25–28] offer many desirable features. They are relatively cheap and the ultrasound system operation is compatible with the linear accelerator operation [25]. Over the past few years, there has been significant development in ultrasound technology, particularly of 3D probes, resolution, image quality and acquisition frequency. Additionally, advances in robots allow for remote probe placement, a requirement for operation in a linear accelerator vault.

How Do We Hit a Moving Target?

As described above, there are many methods to locate a moving target. What characteristics would be desirable in a system for hitting a moving target with a radiation beam? It would be desirable to be able to integrate all sources of imaging information that we have about the patient and about the patient motion. It would also be desirable to integrate all sources of uncertainty, as we would really like to know not only how well we are doing, but also how well we are not doing in order to understand the residual uncertainties which should feed back into determining the treatment margins. We would like to know the dose that has been actually delivered, and be able to modify the dose that is yet to be delivered, thereby optimizing the delivery in real-time. Systems with these characteristic are not currently available, though many systems are under development.

CyberKnife Linear Accelerator

The CyberKnife robotic treatment system [29–32] is the only commercially available system for radiotherapy tracking. The immediate difference between the CyberKnife and other linac technology is the use of a Kuka robot with very precise 6-degrees of freedom control. An x-band linear accelerator is attached to the robot; 2 fixed X-ray imaging systems and a real-time respiratory motion tracking camera provide the data needed to control the robot.

Movable Couch

The treatment couch is controlled by programmable motors, and in principle the couch can be used to compensate for detected target motion, by shifting the patient and continuously aligning the target with the radiation beam. Several projects are underway for motion compensation using the couch [33–36]. When the tumor moves within the patient, the detection system triggers a compensating movement of the treatment couch, including the patient and the tumor, to maintain the tumor/beam alignment.

Multileaf Collimator

The MLC is now widely available for radiation therapy delivery, and facilitates intensity-modulated radiation therapy delivery through predetermined leaf sequences. Another application, though technically challenging, is to use the MLC to continuously realign the radiation beam to the target during therapy, dynamically compensating for any detected target motion (see online supplementary material). One can consider each leaf of a MLC as a degree of freedom to solve the problem of moving the fluence that we want to deliver in real time. Guided by tumor monitoring/tracking, the shift in terms of translation in the beam view and also along the beam view may be accounted for in the MLC sequences – and potentially deformation and rotation as well in real-time because we have control over each of the different leaves in real-time. To make this happen, once we have the shape that we want to conform the beam to, the leaves are placed based on the incoming tumor position signals [37–49].

Experimentally, the use of the MLC for tracking has been demonstrated on 4 different vendor platforms, Accuknife [50], Siemens [51], TomoTherapy (personal communication, Gustavo Olivera, TomoTherapy, July 2008) and Varian [45, 46, 48, 49], though none are clinically available as yet.

Gimbaled Linac

Another method has been recently developed to align the radiation beam with the patient using rotational motors to steer the linac itself. Thus, the treatment beam moves [52] to follow the target. In the Mitsubishi Heavy Industries implementation, the linac head and rotational motors are attached to a rotating o-ring gantry, which also contains X-ray images for target localization. The system has been tested for dynamic tracking with submillimeter accuracy observed.

How Do We Create Integrated Systems to Locate and Hit Moving Targets?

Given a system that detects motion and a system that corrects for motion, the 2 systems need to be integrated into a single platform. If one has a moving target, and a system to observe this motion, there will always be latency – a delay – between the motion detection and motion correction steps. This latency can be somewhat accounted for through the use of prediction algorithms; however, these algorithms cannot predict respiratory motion perfectly and errors will be introduced. The errors can be minimized by reducing the total system latency, and by developing systems in which lowest achievable latency is included in the basic design specifications.

Conclusions

Unless treating a lesion within an immobilized cranium, clinicians must anticipate and strategize for skeletal, respiratory, gastrointestinal and other sources of motion that will occur during most tumor treatments. Many techniques are available or under development to image this temporally changing anatomy during therapy delivery, and several techniques are being developed to be able to react to and realign the beam with the changing anatomy. Technically, submillimeter accuracy in targeting has become an achievable reality. Clinically, there are many other sources of uncertainty in radiation therapy, particularly in target delineation. These may be on the order of several millimeters and must be taken into account when determining appropriate treatment margins.

What does the future of motion management hold for the field of radiation oncology? First, for anatomic targeting, it promises to move beyond accounting for translational motion, and include rotation and deformation. Second, it aims to study temporal physiologic changes and be able to measure physiologic changes in the treatment room and target radioresistant parts of tumors during the treatment itself.

References

1 McCammon R, Schefter TE, Gaspar LE, Zaemisch R, Gravdahl D, Kavanagh B: Observation of a dose-control relationship for lung and liver tumors after stereotactic body radiation therapy. Int J Radiat Oncol Biol Phys 2009;73:112–118.

2 Shirato H, Shimizu S, Kitamura K, Nishioka T, Kagei K, Hashimoto S, Aoyama H, Kunieda T, Shinohara N, Dosaka-Akita H, Miyasaka K: Four-dimensional treatment planning and fluoroscopic real-time tumor tracking radiotherapy for moving tumor. Int J Radiat Oncol Biol Phys 2000;48:435–442.

3 Shirato H, Shimizu S, Kunieda T, Kitamura K, van Herk M, Kagei K, Nishioka T, Hashimoto S, Fujita K, Aoyama H, Tsuchiya K, Kudo K, Miyasaka K: Physical aspects of a real-time tumor-tracking system for gated radiotherapy. Int J Radiat Oncol Biol Phys 2000;48:1187–1195.

4 Shirato H, Shimizu S, Shimizu T, Nishioka T, Miyasaka K: Real-time tumour-tracking radiotherapy. Lancet 1999;353:1331–1332.

5 Seppenwoolde Y, Shirato H, Kitamura K, Shimizu S, van Herk M, Lebesque JV, Miyasaka K: Precise and real-time measurement of 3D tumor motion in lung due to breathing and heartbeat, measured during radiotherapy. Int J Radiat Oncol Biol Phys 2002;53:822–834.

6 Yu CX: Intensity-modulated arc therapy with dynamic multileaf collimation: an alternative to tomotherapy. Phys Med Biol 1995;40:1435–1449.

7 Otto K: Volumetric modulated arc therapy: IMRT in a single gantry arc. Med Phys 2008;35:310–317.

8 Wang C, Luan S, Tang G, Chen DZ, Earl MA, Yu CX: Arc-modulated radiation therapy (AMRT): a single-arc form of intensity-modulated arc therapy. Phys Med Biol 2008;53:6291–6303.

9 Poulsen PR, Cho B, Ruan D, Sawant A, Keall PJ: Dynamic multileaf collimator tracking of respiratory target motion based on a single kilovoltage imager during arc radiotherapy. Int J Radiat Oncol Biol Phys 2010;77:600–607.

10 Poulsen PR, Cho B, Sawant A, Keall PJ: Implementation of a new method for dynamic multileaf collimator tracking of prostate motion in arc radiotherapy using a single kV imager. Int J Radiat Oncol Biol Phys 2010;76:914–923.

11 Poulsen PR, Cho B, Sawant A, Ruan D, Keall PJ: Dynamic MLC tracking of moving targets with a single kV imager for 3D conformal and IMRT treatments. Acta Oncol 2010;49:1092–1100.

12 Keall PJ, Todor AD, Vedam SS, Bartee CL, Siebers JV, Kini VR, Mohan R: On the use of EPID-based implanted marker tracking for 4D radiotherapy. Med Phys 2004;31:3492–3499.

13 Berbeco RI, Hacker F, Ionascu D, Mamon HJ: Clinical feasibility of using an EPID in CINE mode for image-guided verification of stereotactic body radiotherapy. Int J Radiat Oncol Biol Phys 2007;69:258–266.

14 Berbeco RI, Hacker F, Zatwarnicki C, Park SJ, Ionascu D, O'Farrell D, Mamon HJ: A novel method for estimating SBRT delivered dose with beam's-eye-view images. Med Phys 2008;35:3225–3231.

15 Park SJ, Ionascu D, Hacker F, Mamon H, Berbeco R: Automatic marker detection and 3D position reconstruction using cine EPID images for SBRT verification. Med Phys 2009;36:4536–4546.

16 Balter JM, Wright JN, Newell LJ, Friemel B, Dimmer S, Cheng Y, Wong J, Vertatschitsch E, Mate TP: Accuracy of a wireless localization system for radiotherapy. Int J Radiat Oncol Biol Phys 2005;61:933–937.

17 Kupelian P, Willoughby T, Mahadevan A, Djemil T, Weinstein G, Jani S, Enke C, Solberg T, Flores N, Liu D, Beyer D, Levine L: Multi-institutional clinical experience with the Calypso system in localization and continuous, real-time monitoring of the prostate gland during external radiotherapy. Int J Radiat Oncol Biol Phys 2007;67:1088–1098.

18 Willoughby TR, Kupelian PA, Pouliot J, Shinohara K, Aubin M, Roach M 3rd, Skrumeda LL, Balter JM, Litzenberg DW, Hadley SW, Wei JT, Sandler HM: Target localization and real-time tracking using the Calypso 4D localization system in patients with localized prostate cancer. Int J Radiat Oncol Biol Phys 2006;65:528–534.

19 Mayse ML, Parikh PJ, Lechleiter KM, Dimmer S, Park M, Chaudhari A, Talcott M, Low DA, Bradley JD: Bronchoscopic implantation of a novel wireless electromagnetic transponder in the canine lung: a feasibility study. Int J Radiat Oncol Biol Phys 2008;72:93–98.

20 Sawant A, Smith RL, Venkat RB, Santanam L, Cho B, Poulsen P, Cattell H, Newell LJ, Parikh P, Keall PJ: Toward submillimeter accuracy in the management of intrafraction motion: the integration of real-time internal position monitoring and multileaf collimator target tracking. Int J Radiat Oncol Biol Phys 2009;74:575–582.

21 Smith RL, Sawant A, Santanam L, Venkat RB, Newell LJ, Cho BC, Poulsen P, Catell H, Keall PJ, Parikh PJ: Integration of real-time internal electromagnetic position monitoring coupled with dynamic multileaf collimator tracking: an intensity-modulated radiation therapy feasibility study. Int J Radiat Oncol Biol Phys 2009;74:868–875.

22 Dempsey J, Dionne B, Fitzsimmons J, Haghigat A, Li J, Low D, Mutic S, Palta J, Romeijn H, Sjoden G: A real-time MRI guided external beam radiotherapy delivery system. Med Phys 2006;33:2254.

23 Fallone G, Carlone M, Murray B, Rathee S, Steciw S: Investigations in the design of a novel linac-MRI system. Int J Radiat Oncol Biol Phys 2007;69:S19.

24 Lagendijk JJ, Raaymakers BW, Raaijmakers AJ, Overweg J, Brown KJ, Kerkhof EM, van der Put RW, Hardemark B, van Vutpen M, van der Heide UA: MRI/linac integration. Radiother Oncol 2008;86:25.

25 Hsu A, Miller NR, Evans PM, Bamber JC, Webb S: Feasibility of using ultrasound for real-time tracking during radiotherapy. Med Phys 2005;32:1500–1512.

26 Wu J, Dandekar O, Nazareth D, Lei P, D'Souza W, Shekhar R: Effect of ultrasound probe on dose delivery during real-time ultrasound-guided tumor tracking. Conf Proc IEEE Eng Med Biol Soc 2006;1:3799–3802.

27 Sawada A, Yoda K, Kokubo M, Kunieda T, Nagata Y, Hiraoka M: A technique for noninvasive respiratory gated radiation treatment system based on a real time 3D ultrasound image correlation: a phantom study. Med Phys 2004;31:245–250.

28 Fuss M, Boda-Heggemann J, Papanikolau N, Salter BJ: Image-guidance for stereotactic body radiation therapy. Med Dosim 2007;32:102–110.

29 Murphy MJ: Tracking moving organs in real time. Semin Radiat Oncol 2004;14:91–100.

30 Schweikard A, Shiomi H, Adler J: Respiration tracking in radiosurgery. Med Phys 2004;31:2738–2741.

31 Ozhasoglu C, Murphy MJ, Glosser G, Bodduluri M, Schweikard A, Forster K, Martin DP, Adler JR: Real-time tracking of the tumor volume in precision radiotherapy and body radiosurgery – a novel approach to compensate for respiratory motion: Computer Assisted Radiology and Surgery. San Francisco, 2000, pp 691–696.

32 Seppenwoolde Y, Berbeco RI, Nishioka S, Shirato H, Heijmen B: Accuracy of tumor motion compensation algorithm from a robotic respiratory tracking system: a simulation study. Med Phys 2007;34:2774–2784.

33 D'Souza WD, McAvoy TJ: An analysis of the treatment couch and control system dynamics for respiration-induced motion compensation. Med Phys 2006;33:4701–4709.

34 D'Souza WD, Naqvi SA, Yu CX: Real-time intrafraction-motion tracking using the treatment couch: a feasibility study. Phys Med Biol 2005;50:4021–4033.

35 Qiu P, D'Souza WD, McAvoy TJ, Ray Liu KJ: Inferential modeling and predictive feedback control in real-time motion compensation using the treatment couch during radiotherapy. Phys Med Biol 2007;52:5831–5854.

36 Wilbert J, Meyer J, Baier K, Guckenberger M, Herrmann C, Hess R, Janka C, Ma L, Mersebach T, Richter A, Roth M, Schilling K, Flentje M: Tumor tracking and motion compensation with an adaptive tumor tracking system (ATTS): system description and prototype testing. Med Phys 2008;35:3911–3921.

37 McClelland JR, Webb S, McQuaid D, Binnie DM, Hawkes DJ: Tracking 'differential organ motion' with a 'breathing' multileaf collimator: magnitude of problem assessed using 4D CT data and a motion-compensation strategy. Phys Med Biol 2007;52:4805–4826.

38 McQuaid D, Webb S: IMRT delivery to a moving target by dynamic MLC tracking: delivery for targets moving in two dimensions in the beam's eye view. Phys Med Biol 2006;51:4819–4839.

39 Webb S, Binnie DM: A strategy to minimize errors from differential intrafraction organ motion using a single configuration for a 'breathing' multileaf collimator. Phys Med Biol 2006;51:4517–4531.

40 Webb S: Quantification of the fluence error in the motion-compensated dynamic MLC (DMLC) technique for delivering intensity-modulated radiotherapy (IMRT). Phys Med Biol 2006;51:L17–L21.

41 Papiez L, Rangaraj D: DMLC leaf-pair optimal control for mobile, deforming target. Med Phys 2005;32:275–285.

42 Papiez L, Rangaraj D, Keall P: Real-time DMLC IMRT delivery for mobile and deforming targets. Med Phys 2005;32:3037–3048.

43 Rangaraj D, Papiez L: Synchronized delivery of DMLC intensity modulated radiation therapy for stationary and moving targets. Med Phys 2005;32:1802–1817.

44 McMahon R, Papiez L, Rangaraj D: Dynamic-MLC leaf control utilizing on-flight intensity calculations: a robust method for real-time IMRT delivery over moving rigid targets. Med Phys 2007;34:3211–3223.

45 Keall PJ, Cattell H, Pokhrel D, Dieterich S, Wong KH, Murphy MJ, Vedam SS, Wijesooriya K, Mohan R: Geometric accuracy of a real-time target tracking system with dynamic multileaf collimator tracking system. Int J Radiat Oncol Biol Phys 2006;65:1579–1584.

46 Keall PJ, Kini VR, Vedam SS, Mohan R: Motion adaptive X-ray therapy: a feasibility study. Phys Med Biol 2001;46:1–10.

47 Neicu T, Shirato H, Seppenwoolde Y, Jiang SB: Synchronized moving aperture radiation therapy (SMART): average tumour trajectory for lung patients. Phys Med Biol 2003;48:587–598.

48 Sawant A, Venkat R, Srivastava V, Carlson D, Povzner S, Cattell H, Keall P: Management of three-dimensional intrafraction motion through real-time DMLC tracking. Med Phys 2008;35:2050–2061.

49 Zimmerman J, Korreman S, Persson G, Cattell H, Svatos M, Sawant A, Venkat R, Carlson D, Keall P: DMLC motion tracking of moving targets for intensity modulated arc therapy treatment – a feasibility study. Acta Oncol 2008:1–6.

50 Liu Y, Shi C, Lin B, Ha CS, Papanikolaou N: Delivery of four-dimensional radiotherapy with trackbeam for moving target using an AccuKnife dual-layer MLC: dynamic phantoms study. J Appl Clin Med Phys 2009;10:2926.

51 Tacke M, Nill S, Oelfke U: Real-time tracking of tumor motions and deformations along the leaf travel direction with the aid of a synchronized dynamic MLC leaf sequencer. Phys Med Biol 2007;52:N505–512.

52 Kamino Y, Takayama K, Kokubo M, Narita Y, Hirai E, Kawawda N, Mizowaki T, Nagata Y, Nishidai T, Hiraoka M: Development of a four-dimensional image-guided radiotherapy system with a gimbaled X-ray head. Int J Radiat Oncol Biol Phys 2006;66:271–278.

Prof. Paul Keall, NHMRC Australia Fellow
Radiation Physics Laboratory, Sydney Medical School, The University of Sydney
Room 474, Blackburn Building D06
Sydney, NSW 2006 (Australia)
Tel. +61 2 9351 3590, Fax +61 2 9351 4018, E-Mail radphyslab@sydney.edu.au

Technologies of Image Guidance and the Development of Advanced Linear Accelerator Systems for Radiotherapy

Vincent W.C. Wu[c] · Maria Y.Y. Law[c] · Josh Star-Lack[b] · Fion W.K. Cheung[c,d] · C. Clifton Ling[a–c]

[a]Department of Medical Physics, Memorial Sloan-Kettering Cancer Center, New York, N.Y., [b]Varian Medical Systems, Palo Alto, Calif., USA; [c]Department of Health Technology and Informatics, Hong Kong Polytechnic University, and [d]Department of Clinical Oncology, Queen Elizabeth Hospital, Hong Kong, China

Abstract

As advanced radiotherapy approaches for targeting the tumor and sparing the normal tissues have been developed, the image guidance of therapy has become essential to directing and confirming treatment accuracy. To approach these goals, image guidance devices now include kV on-board imagers, kV/MV cone-beam CT systems, CT-on-rails, and mobile and in-room radiographic/fluoroscopic systems. Nonionizing sources, such as ultrasound and optical systems, and electromagnetic devices have been introduced to monitor or track the patient and/or tumor positions during treatment. In addition, devices have been designed specifically for monitoring and/or controlling respiratory motion. Optimally, image-guided radiation therapy systems should possess 3 essential elements: (1) 3D imaging of soft tissues and tumors, (2) efficient acquisition and comparison of the 3D images, and (3) an efficacious process for clinically meaningful intervention. Understanding and using these tools effectively is central to current radiotherapy practice. The implementation and integration of these devices continue to carry practical challenges, which emphasize the need for further development of the technologies and their clinical applications.

Copyright © 2011 S. Karger AG, Basel

The evolution of technologies for image-guided radiation therapy (IGRT) has been rapid over the last 2 decades. During the 1980s, computer control of the linear accelerator (linac) was implemented, which was an important step for subsequent developments. Subsequently, electronic portal imaging devices (fig. 1) were intro-

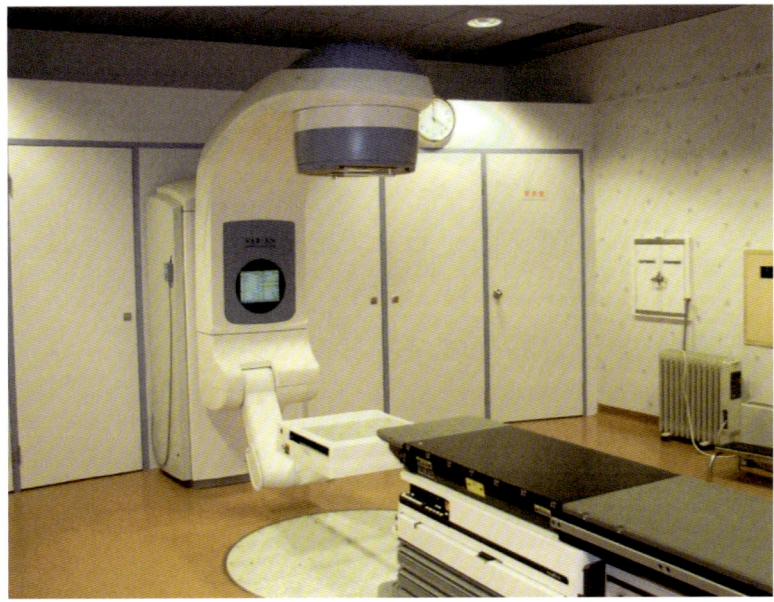

Fig. 1. Linear accelerator equipped with electronic portal imaging device.

duced, enabling digital portal imaging – the first step in modern IGRT. More recently, imaging devices capable of generating 3D images of soft tissues have been incorporated into radiation treatment devices, ushering in a new era of IGRT [1]. Of course, image guidance (IG) for radiation oncology has broader meanings [2], and perhaps the current process should really be called IG during radiation treatment delivery. In the past decade, we have seen a plethora of other IG devices such as kV on-board imagers, kV/MV cone-beam CT (CBCT) systems, CT-on-rails, and mobile and in-room radiographic/fluoroscopic systems. Nonionizing sources, such as ultrasound and optical systems, and electromagnetic devices, were introduced to set up, monitor, or track the patient and/or tumor positions during treatment. In addition, devices were designed specifically for monitoring and/or controlling respiratory motion [e.g. the Varian Real-Time Position Management (RPM) System; fig. 2].

Since many of these systems serve different functions, well-equipped facilities may have a multitude of such devices, each serving a specific function but not interfacing well with the other devices, or even with the linac. The delivery process may be much less seamlessly integrated. In addition, each new device arrives with a dedicated computer, requiring its own interfaces, networks and workarounds for practical implementation. As shown in figure 3, the control area of a radiation therapy machine can become crowded with computer monitors, increasing the burden and complexities confronted by the radiotherapists and staff.

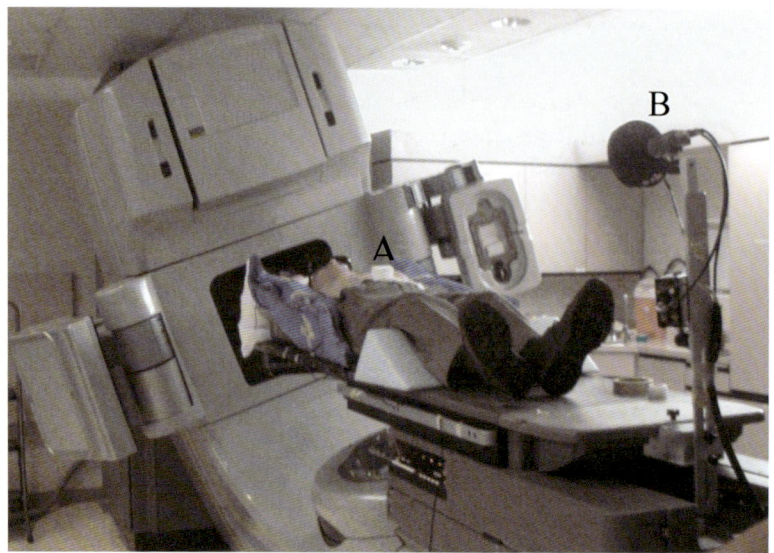

Fig. 2. RPM system by Varian Medical Systems: infrared reflector (A), infrared camera (B).

Fig. 3. The display workstations in the control room of a linear accelerator. Each workstation displays one function during treatment delivery.

Technologies to Meet the Goals of Image-Guided Radiation Therapy

The goal of IGRT is to reduce treatment uncertainties, which can occur between fractions (interfraction) or during the treatment (intrafraction). Imaging can provide guidance of treatment in several ways. To address interfractional errors, repositioning or replanning can be performed. To reduce intrafractional uncertainties, imaging during treatment can monitor the positions of internal structures and provide the basis for correction. In developing technologies for IGRT, the key issues are whether the imaging systems and processes are efficient and accurate enough. Ling et al. [1] and Greco and Ling [2], suggested that an ideal IGRT system should possess 3 essential elements: (1) 3D imaging of soft tissues and tumors, (2) efficient acquisition and comparison of the 3D images, and (3) an efficacious process for clinically meaningful intervention.

Huntzinger et al. [3] described 4 technical criteria for the assessment of successful technologies for image-guided stereotactic radiosurgery: (1) a treatment planning system that generates dose distributions, with rapid dose fall-off from the tumor in all 3 dimensions, through the use of beams converging from many directions; (2) imaging systems with submillimeter accuracy that can account for target motion; (3) a high-resolution, accurate and efficient delivery method; and (4) a system that is fully integrated to enable perfect geometric alignment between imaging, planning and targeting of tumor volumes.

Others have commented on the cost-benefit analysis of the new technologies [1, 4, 5]. In terms of economics, there is significant cost in initiating and implementing IGRT. Other than the investment in the in-room imaging and motion control facilities, human resources and machine time are also important considerations. The increased machine time per patient means an overall reduced throughput of patients per machine. Questions can be raised if the increased cost is not balanced by the benefits that IGRT will bring.

If our goal is to improve clinical outcomes in a cost-effective manner, then a major challenge becomes how to best apply the various technologies that have been developed. The addition of new, emerging technologies will only intensify this concern. For simplicity, perhaps we can consider 2 steps in meeting this challenge. The first is a vetting process in terms of what individual devices will be used, when to use them and how to use them. The second is how best to integrate the selected devices, systems and processes with linac-based radiation delivery in a coordinated approach.

Validation and Vetting of Image-Guided Radiation Therapy Devices

Ideally, the evaluation of the clinical efficacy of a technology should involve the quantification of its benefits in objective clinical outcome measures. To be reliable, this requires comparative phase III clinical trials with sufficient follow-up. There are many obstacles for such trials. The time required for their planning, approval and organization, reticence to submit cases to the randomization process, and clinical imperatives to individualize treatments based on the specifics of medical cases lead to difficulties in patient accrual and render such trials highly impractical. Even if conducted, due to the lengthy nature of clinical trials and the rapid evolution of the technologies, the conclusions may be outdated and irrelevant before they are even derived [4, 5]. A recent perusal of a database on IGRT publications revealed little data reporting objective clinical outcomes.

In the absence of clinical outcome data, we are left with surrogates for documenting clinical advances. This approach retreats to the dogma that reductions of setup uncertainties and minimizations of organ motion effects, carefully measured, lead to outcome improvements. Even with these data, there will be many challenges in the vetting process of devices. This is partly because there are many competing and an ever increasing variety of technologies and products, and partly because the standards and metrics for their evaluation are still evolving. Some of the metrics may be easy to define, such as spatial accuracy, acquisition time, invasiveness and soft tissue contrast/discrimination. Others are more nebulous, with some being device-specific and/or institution-specific in terms of the usefulness of an approach. The surrogates and metrics used in various studies on IGRT will be described in more detail in later sections.

Other factors in the evaluation of a device or system are its applicability and ease of use. In general, devices that can be applied to many major disease sites will be more widely accepted. As to ease of use, Joel Tepper recently said 'the less direct physician physical intervention, the more likely newer techniques will be accepted', highlighting the importance of physicians' acceptance of their involvement in the workflow of IGRT. This precept could be broadened to 'the less direct user (clinicians, physicists and radiotherapists) physical intervention, the more likely newer techniques will be accepted'.

The ease of use of a device also depends on how well it is integrated into the treatment process, with the other devices and, most importantly, the linac. Thus, while we separate the steps of 'vetting' and 'integration', they are in fact intertwined, and integration of devices and streamlining of the treatment process in connection to the linac, is one key to future development.

Integration

The purposes of integration are not only to improve efficiency, ease of use and streamlining of workflow, but also to prevent errors in communication between the devices, systems and computers. Integration will also facilitate automation in quality assurance procedures. The integration process has different levels: some direct and some indirect. By direct integration, we mean that the device is incorporated into the treatment linac, e.g. kV or MV CBCT. Examples of indirect integration include ultrasound guidance, CT-on-rails, and ceiling- or floor-mounted kV fluoroscopy units. Linac manufacturers have an advantage in developing integrated systems simply because they are in a better position to do so. However, integration is not easy, even when the components are from the same manufacturer and amenable to direct integration. Integration of complex imaging devices into treatment machines from different vendors is even more difficult.

Another area of integration is between the linac and treatment planning – not only 'pre-treatment' planning but also 'post-treatment' evaluation. Such integration should allow the display of planning CT images on the control console of the linac for comparison with the on-line CBCT images acquired before, during and after treatment. In addition, the CBCT data taken with the patient on the treatment couch can be overlaid with isodose distributions of the original plan. The comparative data can also be used for replanning when there is a significant deviation between the planning and treatment CT images, in accordance to the concept of adaptive radiotherapy [6–8]. However, for the 'post-treatment' treatment plan evaluation, it would be necessary to capture the actual patient anatomy 'during' radiation delivery, which is still an unsolved problem.

Image Guidance Systems and Procedures

There is a gathering body of literature on the various IG systems and their clinical applications, including the previous and present conference proceedings of the San Francisco Radiation Oncology Conference [9]. Yet many of the questions asked in our 2006 article, 'IMRT to IGRT: frontierland or neverland?' [1], remain unanswered and unresolved in a routine setting. For example, what are the relative roles of kV and MV CBCT, 2D versus 3D imaging, center-of-mass or complete boundary-based shifts, optimum frequencies of imaging, thresholds for intervention, and tradeoffs between tumor margin and organ-at-risk? In the following sections we will briefly describe some of the IG systems currently available and comment on some of the studies associated with their clinical evaluation. Clearly, our comments will not be comprehensive due to the large volume of information on this subject. Given our belief that gantry-mounted CBCT, particularly with kV X-rays, will be the most useful image-guidance system during radiation delivery, a separate section with a more detailed description will be on kV CBCT. Some systems, such as the CyberKnife, have special characteristics that are more appropriately addressed in the chapters dedicated to them.

In-Room Helical CT Scanners

The installation of helical CT scanners into treatment rooms represented the earliest attempt to avail oncologists with 3D soft tissue images of patients in treatment position. To start with, bringing a CT scanner to the treatment room is the easiest solution, but only if the treatment room is sufficiently large enough. The obvious advantage is that one can match the treatment simulation/planning CT images to similar images obtained in the radiation therapy treatment room. The CT scanner is usually mounted on rails in the treatment room and moves over the treatment couch of the linear accelerator. Patients can remain in the same position for imaging and treatment. The close proximity of the imaging CT and the treatment unit enables fine adjustments for changes in the size, shape or location of tumors and surrounding tissues by using CT images taken just before treatments are delivered. The first CT-on-rails was installed in Japan in 1996. Nowadays, commercial products such as PRIMATOM™ (Siemens Medical Solutions Inc.; fig. 4) and EXaCT Targeting™ (Varian Medical Systems Inc. and GE Medical Systems) are available.

Surface Topology Imaging Systems

Many investigators have studied the use of optical imaging methods to characterize the outer surface of patients undergoing radiotherapy. Recently a 3D stereo-imaging system (Align RT; Vision RT Ltd., London, UK) that provides a solution to capture the surface topology of the target area for patient setup and real-time surveillance has become commercially available [10, 11]. Coupled with gray-scale images of skin features, topological congruence can be evaluated by visual inspection of the texture-mapped images [12, 13]. Some authors have suggested that fusion of 3D surface data with in-room volumetric imaging will be useful for real-time image-based planning [14]. Evaluation of its use in setting up patients with cancer of the breast is discussed below.

The use of the breast surface for setting up patients undergoing accelerated partial breast irradiation has proven to be superior to the use of bone anatomy in terms of localization accuracy [15, 16]. Taking advantage of the distinctive shape of the breast, the constrained surface fitting algorithm has been developed to provide an efficient and robust solution that avoids local minima, and works even for patients with large and flaccid breasts [17]. Even so, there are hurdles that a 3D surface mapping system must overcome. One concern is the impact of morphological changes on the breast size and shape during the radiotherapy course due to a variety of causes, including inflammation, postoperative seroma formation, thoracic motion and ipsilateral arm repositioning [18, 19]. To compensate for the effect of respiratory motion, a respiration-gated acquisition of 3D surface models

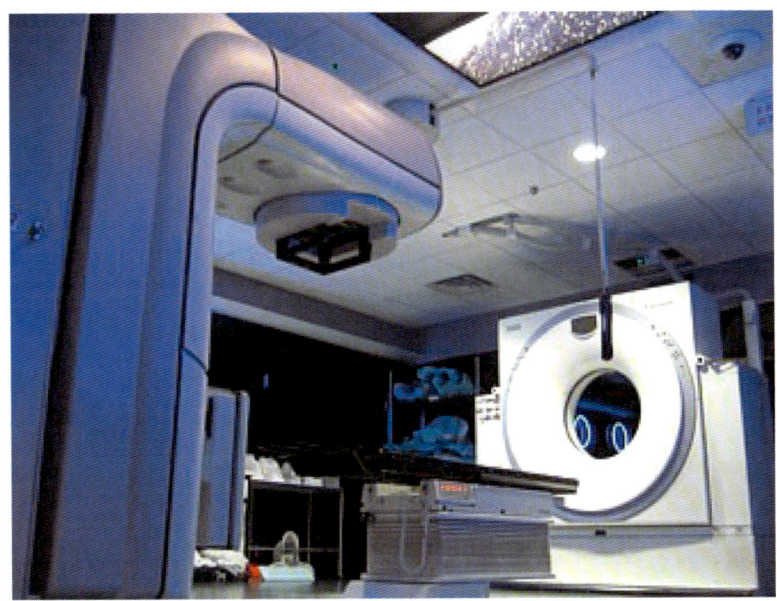

Fig. 4. CT-on-rails, Primatom. Courtesy of Siemens Medical Solutions.

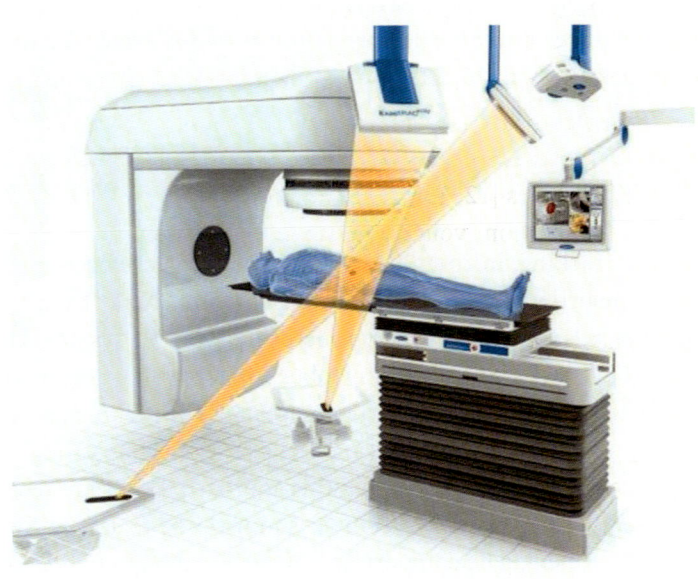

Fig. 5. Novalis™ equipped with a stereo fluoroscopic imaging device (ExacTrac®) enables continuous tracking of patient movement for IGRT (Varian Medical Systems and BrainLAB).

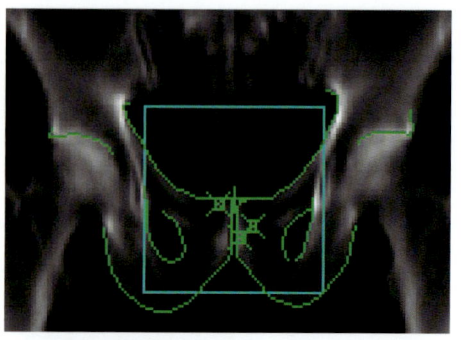
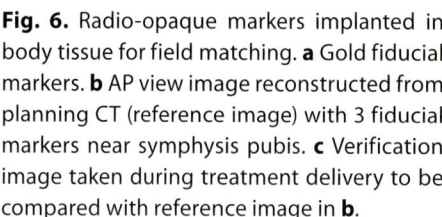
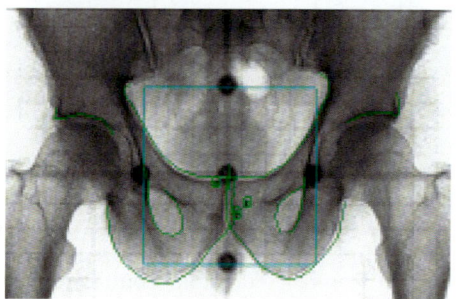

Fig. 6. Radio-opaque markers implanted in body tissue for field matching. **a** Gold fiducial markers. **b** AP view image reconstructed from planning CT (reference image) with 3 fiducial markers near symphysis pubis. **c** Verification image taken during treatment delivery to be compared with reference image in **b**.

may be incorporated [20, 21]. Another concern is the absence of internal anatomical information, and the uncertainty in the correlation between external patient surface and internal target position [15, 22].

In-Room Fluoroscopic/Radiographic Units

For real-time monitoring/tracking, in-room fluoroscopy systems have been implemented [23–25] to generate orthogonal pairs of kV images from 2 X-ray tubes and corresponding imaging panels. A similar and commercially available system is the ExacTrac® (fig. 5) X-Ray 6D which uses 2 kV X-ray tubes with opposing amorphous silicon panel imaging detectors to acquire 2 orthogonal planar X-ray images. The CyberKnife (Accuray, Sunnyvale, Calif., USA) is another example using orthogonal X-ray imaging [26, 27] for monitoring/tracking. In this approach, DRRs are reconstructed from planning CT for computerized comparison with the real-time X-ray images acquired during treatment. The comparison can be based on implanted radio-opaque markers (fig. 6) or high-contrast structures (e.g. of the thorax). The translation and rotation errors can then be calculated, and if adjustment is judged to be necessary, the patient is repositioned by moving the treatment couch. For monitoring and/or tracking during radiation delivery, computer-controlled processes can be designed for image acquisition and comparison, with treatment interruption/adjustment if necessary. However, the cost/benefit of the extra radiation dose delivered to the non-target regions in continued surveillance is a consideration.

Ultrasound Systems

Unlike other IGRT modalities, the benefits of ultrasound-based solutions include portability/availability, non-invasiveness and absence of ionizing radiation. The first commercial systems (e.g. BAT; formerly NOMOS Corp. and now North American Scientific, Cranberry Township, Pa., USA) could only acquire images in 2 cross planes, offering limited information for manual patient alignment [28, 29]. The newer systems (e.g. SonArray; Zmed Inc., Ashland, Mass., USA; and the newer BAT system, North American Scientific) hold promise for real-time 3D reconstruction of ultrasound images. With the geometrical relation between the ultrasound image and the linac isocenter being predetermined, the anatomic contours derived from planning CT are manually mapped to the ultrasound scans for daily position verification [30]. Their common drawback remains the need for human interaction, and the considerable interobserver uncertainty that can become even worse when there is significant interfraction prostate motion [31]. Despite the promising results of some feasibility studies, the accuracy of ultrasound-based alignment has been a subject of long-standing debate [32, 33]. Previous studies have reported good agreement between repeated CT and ultrasound prostate registration to within approximately 2 mm [34, 35]. Other studies, however, found substantial systematic errors with ultrasound, relative to that of CT-based or marker-based methods [36–38]. As a consequence, the daily prostate motion may be overestimated [39]. One major source for these systematic uncertainties was interuser variation in the contour alignment process, possibly because of difficulty in delineating the exact prostate boundary on ultrasound images [38, 40].

There is another inherent shortcoming of ultrasound-guided targeting systems: appropriate probe pressure is crucial to obtain a reasonably clear image, but such pressure could also induce prostate deformation [35, 41]. It also has been criticized for providing subjective information that requires expertise for image acquisition and interpretation. Under certain circumstances, ultrasound images could suffer from poor sonographic visibility. Factors attributing to low-quality images include inadequate bladder filling, large abdominal girth and limited prostatic volume superior to the pubic symphysis in an anteroposterior projection [33]. Special attention should be paid to the calibration and quality assurance programs for defect prevention and detection [42]. To better appreciate the extent of prostate motion, proper imaging protocols, adequate staff training and better patient preparation are of importance.

Electromagnetic Systems

Another approach that does not use ionizing radiation is the proprietary electromagnetic technology of the Calypso 4D localization system (Calypso System; Calypso Medical, Seattle, Wash., USA) for continuous, real-time target tracking during treatment delivery [43, 44]. In place of radio-opaque markers, electromagnetic transponders are implanted in or near the target. Promoted as 'GPS for the Body®' technology, the system can continuously monitor the 3D position of the transponders at a frequency of 10 Hz using an electromagnetic planar array [45]. The design is currently FDA-approved for use in the treatment of prostate cancer, while applications for lung cancers are also promising [46]. Nonetheless, there are at present several technical issues. Firstly, the electromagnetic planar array must be within 27 cm of the transponders, introducing limits on the patients' anteroposterior thickness [47]. Similar to gold markers, invasiveness and risk of transponder migration are of concern. Any conductive component should always be kept far away from transponders for accurate target localization. As a result, the Calypso System is contraindicated in patients with hip replacements or large metal implants that are in the vicinity of the prostate [48, 49]. For cardiac pacemaker recipients, extreme caution and close monitoring are essential as magnetic fields can cause the pacemaker to malfunction. An ideal IGRT system should provide anatomical details, but Calypso can only monitor a few points based on a RF readout without direct target visualization. To expand its applicability, work is under way to explore and develop synergies between the Calypso System and fluoroscopy [50].

Helical MV CT

Tomotherapy is delivered by a 6-MV linear accelerator that uses a ring gantry geometry like that of a CT scanner (fig. 7). With the slip ring technology of a diagnostic scanner, the unit is capable of continuous rotation around the patient. Intensity-modulated radiation therapy (IMRT) is given through a fan beam multileaf collimator from all angles around the patient slice by slice. MV CT imaging technology is incorporated within the gantry for precise localization of the tumor. The patient's anatomy can be reviewed prior to treatment and the radiation beam adjusted if necessary. In the Hi·Art TomoTherapy System (TomoTherapy Inc., Middleton, Wisc., USA), the radiation dose delivered to the patient can be computed and superimposed on the pretreatment CT and compared with the original treatment plan for possible dose-guided adaptation. However, MV imaging has the disadvantages of low inherent soft tissue contrast and poorer detection efficiency [51]. Quantitative analysis of contour variations has shown it to be inferior

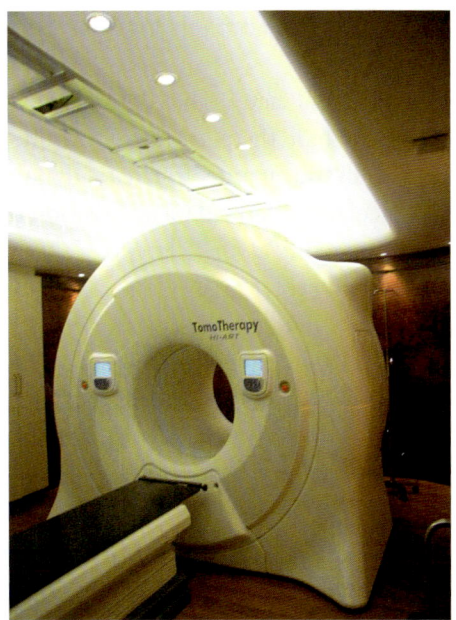

Fig. 7. Hi·Art TomoTherapy System.

to kV CBCT for prostate delineation [52]. The Hi·Art II currently lacks a treatment delivery model that can handle target motion and non-coplanar delivery [53]. The effort needed to develop a 4D delivery model and integrate it into the existing system is significant.

Cone-Beam CT for Linac-Based Image-Guided Radiation Therapy

CBCT is a volumetric imaging technique that holds great promise for IGRT. CBCT offers good contrast and spatial resolution and may enable several promising applications including adaptive radiotherapy. Only a single gantry rotation is required for volumetric data acquisition, which makes it particularly well suited for integration into linacs. While both MV and kV CBCT systems are now in clinical use, there is a general consensus that the superior soft tissue image quality and lower dose of kV CBCT makes it more broadly applicable to linac-based radiotherapy. While kV CBCT image quality is generally inferior to that provided by diagnostic (helical) CT systems, it is nevertheless proving to meet many IGRT imaging tasks sufficiently well. In this section, we provide a general description of CBCT, including technical and clinical considerations, and discuss future improvements.

Technical Considerations for Cone-Beam CT

The technical factors that determine whether a given set of goals for imaging can be met include the performance of the detector (quantum efficiency, size, spatial resolution, frame rate, linearity, dynamic range), the X-ray source (beam energy, spot size, power), the data acquisition time, imaging dose, and reconstruction and data processing algorithms.

Detector Technology
Large area digital X-ray detectors utilizing amorphous silicon technology have become widely available over the past decade and have found numerous clinical applications, including electronic portal imaging at MV beam energies, and diagnostic radiography, fluoroscopy and mammography at kV energies [54–56]. The availability of this technology, coupled with a suitable X-ray source and a rotating gantry, enables CBCT imaging. Compared with diagnostic CT detectors, which use crystalline silicon technology, amorphous silicon flat panel imagers (AMFPIs) are less costly to manufacture and more immune to radiation damage. But they also have several limitations, one of which is reduced dynamic range (AMPFI dynamic ranges are between 14–17 bits vs. 18–22 bits for conventional CT detectors), which can add quantization noise and/or may require limiting flux in order to prevent saturation of the detector. Fortunately, this weakness can be mitigated in part in kV energies by the use of a bowtie filter which reduces signal intensities in the unattenuated regions outside the patient, thus preventing saturation problems. Another mitigating factor is the smaller pixel sizes of AMFPIs which tend to reduce the overall dynamic range requirements if binning and filtering are performed after analog-to-digital conversion. Other AMPFI limitations include lower quantum efficiency, reduced frame rates and susceptibility to charge trapping effects including lag and gain changes, all of which can negatively affect image quality. These issues are discussed further below.

Beam Energy: kV Cone-Beam CT versus MV Cone-Beam CT
The MVision™ Megavoltage Cone Beam System (Siemens AG, Munich, Germany) uses the linac as an imaging source. Reasonable image quality for tissues of high contrast (e.g. bone and lung), or of materials with a high atomic number (e.g. dental implants or hip prostheses), can be achieved [57]. However, generally speaking, even for relatively high-dose levels of 50 mGy, image quality of soft tissue may not suffice for the accuracy desired for IG. Research into improving the image quality of MV CBCT includes using thick scintillators to increase detector DQE (detective quantum efficiency) [58–60] and alternative targets to tungsten, such as carbon or aluminum, to produce a softer Bremsstrahlung spectrum [61].

Standard diagnostic imaging is performed using an X-ray tube as a photon source with a peak voltage generally ranging from 70–120 kVp. Compared with the 'MV' photon spectra produced by linacs, these 'kV' photon spectra have a higher probability of interaction with the objects of interest, resulting in higher contrast images being produced at reduced doses. Also, kV images generally have higher spatial resolution than MV images because X-ray tube beam spot sizes are often smaller than those of linacs. These advantages have motivated the incorporation of a kV X-ray tube and an AMFPI detector into the linac gantry. The system can be used not only for CBCT, but also for fluoroscopy and radiography. The On-Board Imager™ system (OBI; Varian Medical Systems, Palo Alto, Calif., USA) consists of a kV source and a large area AMFPI that is mounted on 2 independent robotic arms orthogonal to the treatment beam. The arms can be retracted during treatment to shield the susceptible electronic components of the detector from scattered MV radiation. The original electronic portal imaging device, which is positioned opposite to the MV beam, still remains in the system, and the same center of rotation is shared by both the kV and MV beams. Other similar systems include the X-ray volume imaging system (Synergy®; Elekta Oncology, Stockholm, Sweden) and ARTISTE™ system (Siemens Medical Solutions, Malvern, Pa., USA).

Multiple studies confirm the superiority of kV CBCT image quality. It has been shown that kV CBCT is 10–100× more dose efficient than MV CBCT owing to kV CBCT's softer beam and higher DQE detector [51]. In a phantom study comparing MV planar, kV planar and CBCT imaging systems available on the Varian linear accelerators, it was reported that the kV planar system has the highest spatial resolution, followed by the MV planar and CBCT imaging systems [62]. A similar study showed that the uniformity and spatial resolution of MV CBCT images were comparable to kV CBCT images, but MV CBCT low-contrast resolution was significantly worse [63]. However, kV CBCT is more subject to streaking artifacts from metal implants, such as dental fillings and hip prosthesis where photoelectric cross sections are very high [64]. These artifacts are not evident in MV CBCT images where contrast is dominated by Compton interactions.

Imaging Volumes, Times and Doses
The size of the flat panel imager and the CBCT system geometry determines the maximum axial (cranio-caudal) and transaxial (lateral) imaging fields of view (FOVs). Commercial AMFPIs used for IGRT are approximately 40 cm wide and either 30 or 40 cm long depending on the manufacturer. Since RT systems generally have a 1.5× detector magnification (approx. 150-cm source-to-imager distance and approx. 100-cm source-to-isocenter distance), the maximum obtainable axial FOV is approximately 17 or 23 cm depending on the system. When the detector is centered over the source, the maximum transaxial FOV that can be re-

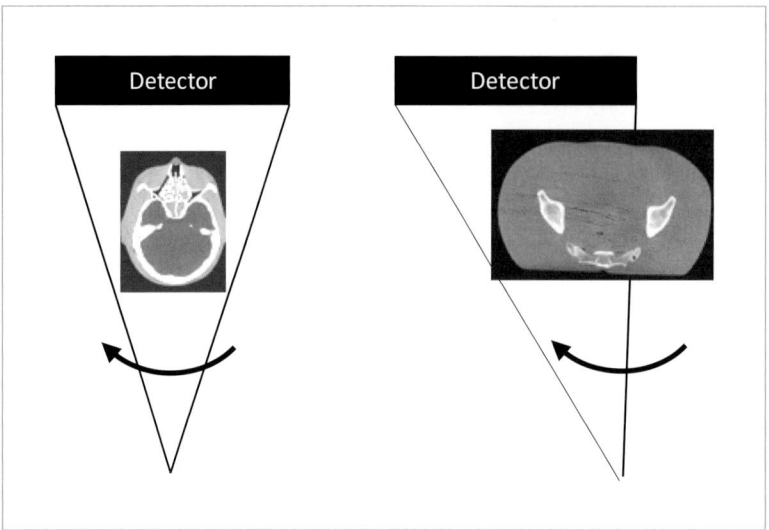

Fig. 8. The centered-detector geometry (left) is used for data acquisition of smaller FOVs and is compatible with either a '180 degree +fan angle' scan or a 360 degree scan. The offset-detector geometry provides a larger FOV but requires a 360 degree scan to acquire a complete data set.

constructed is approximately 26 cm. To increase the FOV for body scanning, the detector is shifted laterally and a 360 degree rotation is performed. This allows for reconstruction of up to a 50-cm transaxial FOV. Figure 8 shows the centered-detector and offset-detector geometries.

For a preset gantry speed, projections must be acquired at a sufficiently fast rate to avoid view aliasing artifacts, which are often manifested as streaking from bone tissue and air tissue interfaces [65]. Depending on the reconstruction parameters used, upwards of 650 projections are desired for a 360 degree rotation. Thus, for a gantry rotation speed of 1 rotation/min, which is the maximum currently allowed by International Electrotechnical Commission rules, AMFPI frame rates should be at least 11 Hz. At this time, however, some panels do not meet this goal. As a result, depending on the system, imaging times range from 1 to 2 min for a 360 degree scan, and from 35 s to 1 min for a '180 degree +fan angle' scan.

The dose delivered by CBCT depends on the beam energy and current, beam filtration, scan time, the part of the anatomy being imaged and the size of the patient. All manufacturers offer different protocols geared toward achieving a desired contrast-to-noise ratio for a certain tissue type. For example, for bone imaging, a much lower dose is required than for soft tissue imaging. Dose is generally calculated using a weighted average approach such as a CT dose index [66]. A typical kV CBCT CT dose index ranges from 1–25 mGy, depending on the protocol, while MV CBCT doses are generally above 40 mGy [67].

Reconstruction and Data Correction Methods

The CBCT volume is reconstructed in one operation and can produce high-resolution isotropic voxels instead of the conventional sliced images that are stacked after reconstruction [68]. Currently, the Feldkamp-Davis-Kress (FDK) convolution-backprojection reconstruction algorithm [69] is used for all systems. It has the advantage of operating sequentially on individual projections so that processing can begin as soon as data acquisition starts. Using multicore central processing unit technology or graphics processing units for computation, it is now possible for reconstruction to keep up with the data acquisition rate so that the final images are ready for review almost immediately after data acquisition stops. Software tools are then provided for registering the CBCT images to the planning CT images for estimation of setup errors.

Image Quality

CBCT image quality improvement is an active area of research both in industry and academia. Outlined below are areas where significant progress has been made and where research is ongoing.

Scattered Radiation. HU (Hounsfield unit) accuracy is a key element to enabling treatment replanning. Due to the large area of irradiation, CBCT projection data may possess very high scatter-to-primary ratios creating cupping artifacts and other non-uniformities in the reconstructed images [70, 71]. Scattered radiation also adds Poisson noise to the raw data that cannot be removed [72, 73]. Thus, it is best to minimize scatter as much as possible before it is detected. This can be achieved by reducing the extent of the exposed volume and using appropriate hardware such as antiscatter grids and bowtie filters when possible. Bowtie filters flatten the beam profile across the detector by increasing the beam intensity directed at the thicker part of the body relative to the intensity at the periphery which is a source of a significant amount of scatter. Phantom studies have demonstrated that bowtie filters can lead to significantly improved performance. In one instance, spatial non-uniformities decreased from 9.8 to 2.1% (across the image) and almost 100% CT number accuracy was restored at the periphery with the use of the bowtie filter [74, 75]. Antiscatter grids are effective in reducing scatter, but also decrease primary radiation, which lowers the overall image signal-to-noise ratio when the scatter-to-primary ratio is low [76].

Even when bowtie filters and grids are used, the remaining scatter-to-primary ratios may be very high (>2) for thick objects, which can result in large CT number errors that are on the order of 200 HU. Therefore, software corrections are still required, which is a very active area of research. Several methods have been proposed for estimating and correcting scatter in raw projection data. These include scatter deconvolution [77, 78], scatter modeling [79], Monte Carlo estimations [80, 81], intensity modulators and beam stop arrays [82–84], 2 scans [85] and detector

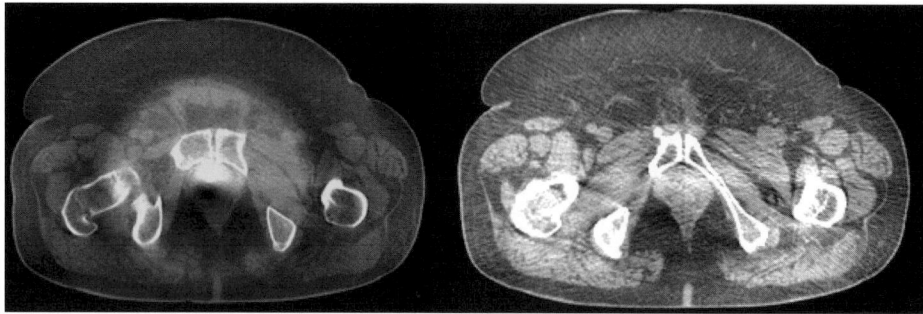

Fig. 9. Axial pelvis image without (left) and with (right) scatter correction using a deconvolution method. CT number accuracy is significantly improved and anatomic features are made more visible. The display window width is 500 HU. Data were acquired on the Varian Acuity Simulation CBCT system in offset-detector mode at 125 kVp.

shadowing [86]. The scatter deconvolution method, also known as scatter kernel superposition, is gaining favor as a practical, computationally efficient approach that requires no extra hardware and provides accurate results [87–91]. Figure 9 shows results from a new scatter kernel superposition method employing asymmetric kernels [92]. The in vivo pelvis data were acquired on the Varian Acuity Simulator, which has the same geometry as the on-board imager. The system had a 10:1 antiscatter grid and 7:1 bowtie filter and was operated at 125 kVp in offset-detector mode. A significant improvement in CT number accuracy and image uniformity was achieved.

Scatter correction research is expected to continue at an active pace. As computational capabilities increase, interest is expected to rise in the types of techniques that use Monte Carlo methods to estimate scatter from a preliminary reconstruction to then refine the raw data correction for a subsequent reconstruction [80, 81].

Charge Trapping. Due to the non-crystalline structure of the amorphous silicon detector, electrons and holes produced in the photodiode can be caught between the valence and conduction energy bands, causing gain changes and lag effects from frame to frame as charge is stored and released over the course of a scan [93, 94]. In CBCT reconstructions, several artifacts have been noticed, including streaking [95] and the so-called 'radar artifact' that is seen in non-circular objects and which occurs as a result of a detector pixel experiencing a transition from a high to low intensity during the course of a scan [96, 97]. The magnitude of these artifacts can range from 10–40 HU depending on the shape of the object being scanned. Several approaches have been proposed for ameliorating charge-trapping effects. Hardware solutions involve trying to fill the traps as much as possible before exposure to X-rays. These involve using either backlighting [94] or forward

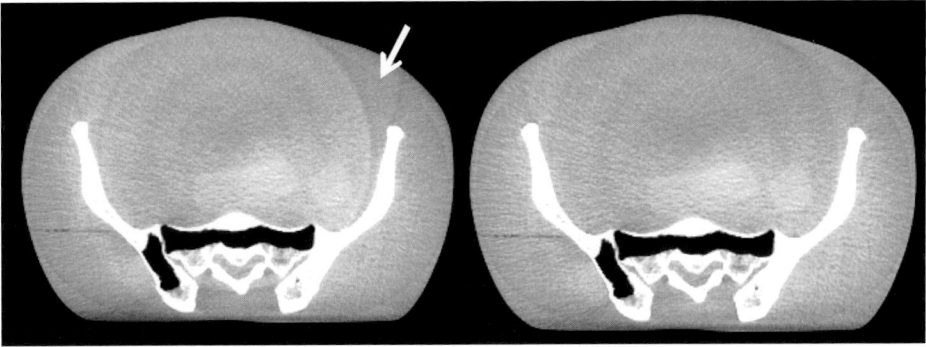

Fig. 10. 'Radar' artifact in a pelvis phantom scan caused by charge trapping effects (left). Correction of artifact using a charge state mapping approach (right).

biasing techniques [98]. Multiple software corrections have been proposed including deconvolution and charge-trap state modeling [96, 97, 99]. Figure 10 shows the application of the charge-trap state modeling solution [96] to reduce the radar artifact in a pelvis phantom scan.

Motion. Patient motion can lead to streaking artifacts, image blurring and distortion in the reconstructed images and is an especially important consideration in linac-based CBCT due to the long acquisition times of 0.5–2.0 min [100, 101]. While random motion (e.g. due to peristalsis) is difficult to correct, respiratory motion is in many ways addressable. One approach is to use respiratory control methods such as deep inspiration breath hold or the active breathing control system and gate the CBCT image acquisition [102]. In this case, an image at a single respiratory phase, is reconstructed. Alternatively, 4D CBCT techniques can be employed to create volume images at multiple respiratory phases. The most straightforward 4D approach involves slowing down the gantry to rotation times on the order of 4–6 min, thus enabling the acquisition of a sufficient number of projections for each respiratory phase to minimize view aliasing artifacts [103–106]. For reconstruction, the projections are first sorted into individual phase bins either using an internal signal derived from the projection data (e.g. the diaphragm) and/or an external signal such as the RPM. Each group of projections, corresponding to a single phase is then reconstructed using the FDK method to create a 4D image. This approach has the undesirable effect of significantly increasing scan times and may increase patient dose depending on the total mAs prescribed.

There is active research into finding means of reconstructing 4D CBCT images using shorter acquisition times on the order of 1–2 min. One approach uses deformable registration techniques to determine where rays should be backpro-

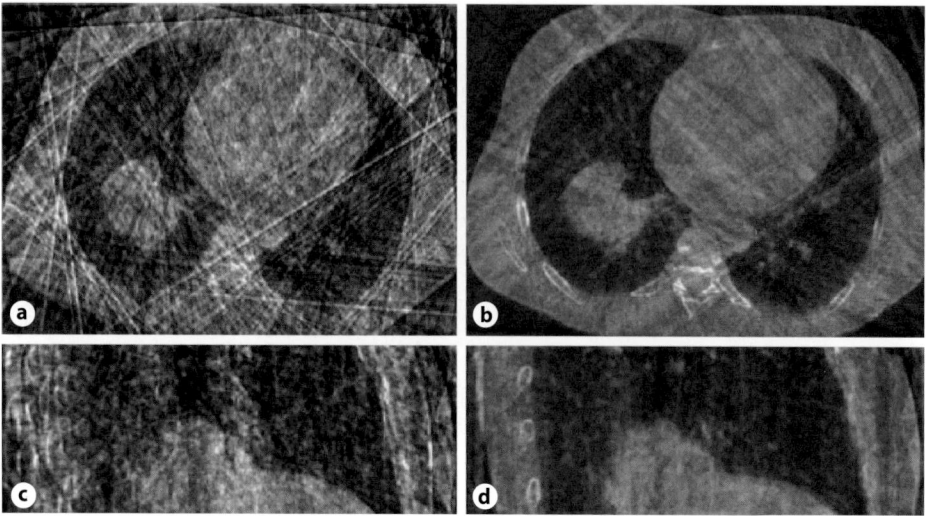

Fig. 11. Use of the McKinnon-Bates algorithm to reduce streaking artifacts in 4D CBCT scans. The acquisition time was 1 min. **a**, **c** Axial and sagittal images of a single phase using a conventional FDK reconstruction algorithm. **b**, **d** After application of the McKinnon-Bates algorithm. Reprinted from Leng et al. [112].

jected [107, 108]. The registration vectors may be predetermined from the 4D planning CT. A single phase is reconstructed using all projections and which then can be deformed to other phases. Another approach uses the FDK algorithm to reconstruct a 3D image of each phase and then uses a patient-specific motion model to deform the different phase images into a single phase [109]. Other reconstruction methods are based on treating stationary tissue differently from moving tissue during the reconstruction process to try to maximize the signal-to-noise ratio and minimize view aliasing artifacts. The McKinnon-Bates algorithm [110, 111], which was originally developed for cardiac imaging in the 1980s was recently adapted to create 4D CBCT lung images [112]. Good results were obtained for a 1-min scan as shown in figure 11. Compressed sensing techniques [113] may form the basis for other promising solutions for reconstructing under-sampled 4D CBCT images [114]. A recent development incorporates the use of prior information along with compressed sensing to improve image quality [115].

Cone-Beam Geometry. The circular trajectory used for CBCT acquisition does not completely satisfy Tuy's sampling condition [116]. Thus, some objects (a stack of disks, known as the Defrise phantom, is commonly used as the most extreme example) can never be fully and faithfully reconstructed, especially for slices for which the cone angle is largest. However, there is no evidence yet that this is a serious issue for IGRT applications. If it eventually turns out to be a significant problem, then more sophisticated gantry trajectories, such as 'circle + line' and 'saddle' trajectories that meet Tuy's sampling condition can be explored [117]. In some ways, a more significant concern is blooming and shading artifacts in 180 degree + fan centered-detector scans or in 360 degree offset-detector scans resulting from inconsistencies in the FDK algorithm. These 'cone-beam' artifacts are usually manifest at bone-tissue interfaces and also are most severe for slices where the cone angle is largest. Several processing techniques have been proposed for reducing these artifacts including iterative reconstruction [118] and alternatives to Parker weighting [117, 119].

Noise Reduction. For a given dose and spatial resolution, CBCT image noise may be higher than that of diagnostic CT reconstructions. The is partly due to the somewhat reduced quantum efficiency of AMFPIs compared with diagnostic detectors, and partly due to the presence of scattered radiation which adds Poisson noise to the raw projections. Therefore, an active area of research is in maximizing the signal to noise of reconstructions using noise estimation methods and non-linear filters [120]. Much research that has previously been applied to CT is directly applicable to CBCT, including iterative techniques and post-processing methods [66, 121, 122].

Metal Artifact Reduction. Metal artifacts from dental amalgams and artificial hips may limit the utility of CBCT for IGRT. Research is ongoing into the development of correction algorithms for these problems [66, 123, 124]. On the linac, there is the unique possibility of combining kV and MV imaging to achieve the best of both worlds [124, 125].

The use of CBCT in the clinical environment is a recent development and efforts are underway to determine how best to utilize the technology for IGRT applications.

Digital Tomosynthesis
The availability of CBCT capabilities on linac gantries has led to the exploration of digital tomosynthesis (DTS). In contrast to CBCT, DTS requires a more limited gantry rotation (e.g. 40 degrees) and therefore can reduce image acquisition times and patient doses with the caveat that spatial resolution of the final images is not isotropic and CT numbers are not accurate, thus prohibiting its use for treatment planning [126]. Similar to kV CBCT, DTS images can be reconstructed using the FDK algorithm [69] or with other approaches based on iterative or compressed sensing methods [114]. Alternative trajectories include the linear translation of the kV source and detection panel [127]. The DTS technique has been clinically investigated for accelerated partial breast irradiation [127], but is also applicable to many other disease sites. The speed of image acquisition may be desirable for imaging tumors in chest and abdominal regions [128].

Cone-Beam CT in the Clinic

Frequency of Imaging

IGRT allows reduction of patient positioning errors before and during treatment. However, if there is an increase in 'patient-on-couch' time because of the imaging process, it may add to patient setup instability and intrafractional errors. Also, the accumulated imaging and peripheral dose, and the extra human and facility resources consumed should be considered. These factors can influence the frequency of imaging as discussed by several investigators [129, 130].

Sheng et al. [131] reported that daily IG might not be necessary for well-immobilized head and neck cancer patients. Without daily IG, the setup errors (3.6 mm) and its dosimetric impact on planning target volume (2.1% dose reduction) and OARs (1.8 Gy added dose) were judged acceptable, and could be further reduced by half if weekly IG was used. Others [131–133] reported that kV 2D imaging could be an acceptable substitute for kV CBCT imaging for head and neck cases owing to the rigid immobilization. However, one should note that the radiation-induced organ shrinkage during a treatment course may require 3D imaging of soft tissues to account for anatomical changes. A recent study by Tanyi et al. [134] of 150 RT brain, head and neck, abdomen, and pelvis patients showed that 23.3% required treatment replanning at least once during the course of fractionated radiotherapy. In this study, a conventional helical CT scanner was used for replanning and CBCT was used only for determining when the new planning CT should be performed.

In disease sites such as the prostate and the lung, daily IG may improve treatment accuracy because of the relatively unpredictable and large target position variations [131–133]. Early pilot studies show that significant margin reduction in the prostate may be made possible by the improved positioning accuracy provided by soft tissue CBCT imaging [135, 136].

Clearly, disease-specific IG protocols need to be developed, including hybrid processes, e.g. combining weekly CBCT and daily verification (with kV radiographs or Calypso). Such hybrids may be desirable both in terms of operational efficiency, imaging dose and completeness of 3D anatomical information.

Criteria and Protocols for Image-Guided Intervention

Clinical criteria and protocols for IG during radiotherapy delivery are still evolving. The most common correction protocols in clinical practice are based on the matching of either bone landmarks or radio-opaque markers between treatment and reference planar images. In this approach, which emphasizes point-to-point matching, the changes in tumor shape, the relative geometry of the tumor and normal tissues, etc., are lost in the comparison. However, the comparison of CBCT and reference CT images is gaining momentum as tools become more user-friendly. In this regard, whether rigid body registration suffices, or whether deformable object matching is needed, is a subject for discussion.

Whatever the metrics for comparison in IGRT, it is often desirable to define a threshold for intervention so as to improve operational efficiency, i.e. to avoid the correction of small displacements that may be clinically irrelevant. Such thresholds may vary for different treatment regions, techniques and approaches, and should take into account estimations of institution-specific systematic and random errors for the particular disease sites.

The comparison of 'session' images with the reference is usually facilitated by software tools provided by the manufacturer. Continued improvement in such tools is necessary to make image registration more accurate, less tedious and less time-consuming so as not to deter the wide implementation of IGRT [5, 58]. In particular, improved speed in volumetric image comparison is needed, as well as the ability to register deformable objects accurately. In a study using rigid phantoms, the Varian image registration software for MV planar, KV planar and kV CBCT was found to be accurate, with kV planar image matching being the most accurate, to approximately 1 mm [62]. These authors noted that 'the attainable accuracy for rigid body translation… will more likely be limited by the resolution of the couch readouts than by inherent limitations in the imaging systems and image registration software.'

As CBCT becomes more widely used, the question of how to speed up the review and approval process becomes more important. In some institutions, the wait for physician's approval could become a bottle neck. The most effective solution is to train and delegate the duty to therapists, since they are the frontline users of the system. For treatment courses of 4–5 weeks, the oncologists can carry out off-line checks at regular intervals to monitor/approve the image-guided results, and to make adjustments if necessary. For hypofractionated treatment and stereotactic radiotherapy, real-time review and approval by the oncologists is necessary, and could be optimized by the development of efficient and reliable telecommunication systems.

Lastly, at present, most imaging protocols generate images before treatment, and less frequently after treatment, serving rather obvious purposes. What is lacking is the capability of acquiring images during radiation delivery with a view to yield actual anatomical data on which the delivered dose distribution can be superimposed and examined.

Intrafraction Motion Management
CBCT may help to reduce intrafraction dosimetric uncertainties stemming from involuntary patient movement or respiration-induced organ motion [88]. In the treatment of lung cancer, several schools of thought exist for how best to treat the disease and how best to use CBCT. For ITV-based treatments, the ITV position

can be determined daily from either a blurred 3D CBCT image [137] or from the appropriate phase of a 4D reconstruction [138]. For more advanced treatment approaches, including gating with the Varian RPM [139, 140] or, in the future, tracking with a dynamic MLC [141, 142], a 4D CBCT scan can be performed to find the 'trajectory-of-the-day' to synchronize tumor position with an external signal such as the RPM. During the course of a given treatment session, especially for hypofractionated cases, it may be desirable to recalibrate the trajectory. This can be achieved by performing another 4D CBCT scan, or by using 4D tomosynthesis [128] or kV fluoroscopy (with or without markers). How CBCT fits in with other motion monitoring technologies such as Calypso is still to be explored [143–145]. Clearly, the whole motion monitoring field is under rapid development and studies will be required to address technical problems, and obtain experimental and clinical validation before full utilization.

Image Guidance: Meeting Future Needs

It is important that IGRT systems meet the needs of current and future developments of radiotherapy. We expect to see much technical development, evaluation and implementation of methods and strategies in the above areas in the near future. For emphasis, we note again that process integration and streamlining of workflow are key to the success of clinical implementation. Some issues are specific to the IGRT platform being used, but many are common to all of the approaches. Given the general use of the linac in clinical practice, and often as a first treatment unit among alternatives, the goals and expectations for the development of new linac-based IGRT technologies are summarized in table 1.

Efficiency, Workflow and Practical Integrations

Radiotherapy delivery using linac-based and other technologies now involve many components, including IG. It is important that a highly functional, logical and efficient treatment process be achieved, and this will be a continuing goal among equipment vendors. Recently, Varian Medical Systems introduced the TrueBeam™ radiotherapy system in April 2010. An important design feature of this new device, relevant to the subject matter of this chapter, is the integration of TrueBeam's many functional components under one computer control system, in which a single 'supervisor' is in command of all the functional components, or 'nodes'. As illustrated in figure 12, the supervisor communicates with and controls the various functioning components, including beam generation, motions of the gantry, couch and MLC, kV and MV image acquisition, etc. The supervisor operates on a 10-ms

Table 1. Objectives for the future of linac-based IGRT

1	A highly integrated, flexible and efficient linac-based IGRT radiotherapy delivery system, not only direct integration of components actually on the linac, but upstream relative to treatment planning, and downstream in providing delivered dose distribution on images acquired during radiation delivery
2	All radiation delivery components will be dynamic (including spatial and temporal dose modulation, movement of MLC, collimator, gantry, couch) with collision-avoidance planning and anticollision features – with these features, non-coplanar and extended SSD treatments will become more common
3	Automated patient setup and image-guided verification of target position and configuration will be incorporated into the treatment delivery process – the enabling technologies and devices will either be part of the linac, or fully-integrated with it
4	Based on image-guided determination of target position and configuration, on-line adaptation of treatment plans may be triggered; disease site-specific criteria for such adaptation will be developed and refined
5	Task-assignment and workflow for physicians/physicists/radiotherapists will evolve with process automation. Monitoring/approval during treatment by clinicians/physicists, if necessary, of the above activities will be in near real-time via remote access devices
6	Motion management will be routine, with a combination of marker and marker-less techniques, gating and/or tracking, and new methods will be developed
7	Sometime in the more distant future, the integration of imaging of biological endpoints – including the use of regional imaging devices – should take place

heartbeat, i.e. receiving status reports from all the nodes and issues commands to them for implementation every 10 ms. The supervisor is the key to true integration in the TrueBeam™ linac-based IGRT system, it is the brain that controls all the nodes, so that the various functions are fully coordinated.

The integration of the various functional nodes under one computer control system facilitates process automation and the streamlining of IGRT workflow. Specifically, various combination/sequence of treatment delivery and image acquisition can be automated, such that all the steps in an IGRT procedure can be seamlessly coordinated and efficiently implemented. In a sense, TrueBeam™ provides the tools to achieve much of the objectives of table 1. With these capabilities, new clinical protocols can be developed and validated that improve the efficacy of IGRT.

Other improvements offered by the TrueBeam™ system are that (1) a geometric calibration tool has been implemented so that the kV imaging and MV imaging/treatment isocenters are coincident with each other, and (2) new scatter correction and beam-hardening correction algorithms [92] have been implemented to improve HU accuracy in kV CBCT images.

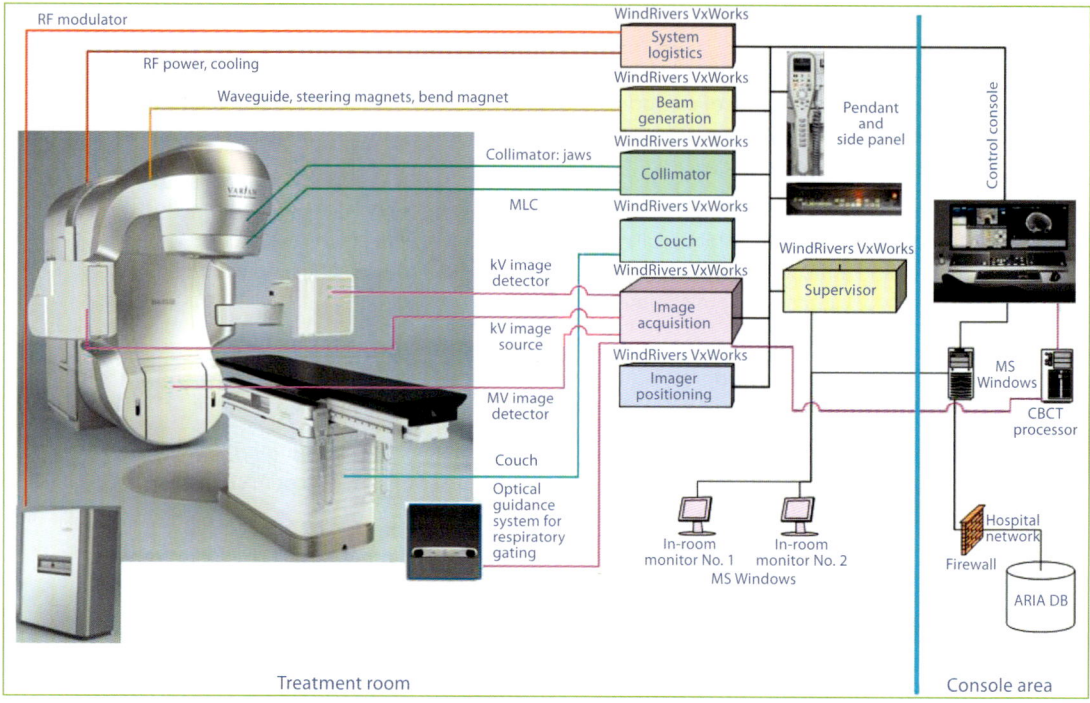

Fig. 12. A schematic illustrating the computer control system of the TrueBeam™. A single 'supervisor' is in command of all the functional components, or 'nodes', including beam generation, motions of the gantry, couch and MLC, kV and MV image acquisition, etc. The supervisor operates on a 10-ms heartbeat, i.e. receiving status reports from all the nodes and issues commands to them for implementation every 10 ms.

Oligofractionation and Stereotactic Body Radiotherapy

An important current trend is towards smaller number of fractions in a treatment course, using either hypofractionation, or oligofractionation in the stereotactic body radiotherapy (SBRT) approach (we define hypofractionation, relative to conventional fractionation, as having a somewhat reduced number of fractions coupled with a somewhat increase in dose per fraction; and oligofractionation as the use of one to a few fractions of much higher doses, as a simile to the definition oligometastasis). Such a trend seems at first contrary to almost all previously established radiobiological principles, but clinical results from oligofractionated radiotherapy for early lung cancer and paraspinal metastasis have produced promising treatment outcomes [146]. Regardless whether these are examples of good physics making up for bad biology, or a new paradigm of radiobiology, they are new frontiers for exploration. As mentioned in our article in 2006 [1], for treatment with a single or a few fractions, IG would probably be mandatory.

Radiotherapy with oligofractionation and SBRT involves large doses per fraction (even up to 24 Gy), and therefore longer treatment beam-on time than that encountered in conventional fractionation. Here technological improvements in treatment planning and delivery can make an important contribution by significantly reducing the radiation delivery or beam-on time. For example, the arc IMRT method introduced by Varian Medical Systems as RapidArc™ can deliver a 10-Gy treatment in 2–3 min, with the potential to increase patient comfort and reduce treatment uncertainty [147]. Future linac development will likely offer the feature of flattening-filter-free [148] beams to increase the radiation dose rate by factors of 2–5, further decreasing treatment and beam-on time. Aside from reducing patient discomfort and treatment uncertainty, a shorter treatment time will make motion management less challenging. Thus, for SBRT, improvement in the speed of treatment would be important as well as improvement in imaging during the treatment.

In view of the large doses and long treatment times associated with SBRT, consideration should be given to the monitoring of intrafraction target position. In this regard, one approach is the continuous localization of implanted markers using ionizing or non-ionizing radiation [43, 47, 149, 150], and another method is to interlace radiation delivery and 3D CBCT imaging of soft tissues. In situations where organ motion affects target and normal tissue positions, it is also important to incorporate strategies and techniques to minimize motion-induced treatment uncertainties.

Some of the developments described above have led, or will lead, to changes in the treatment times of a session/fraction, and in the radiation dose rate. For example, IMRT can increase beam-on time relative to 3D CRT. RapidArc™ can significantly reduce SBRT treatment time, and much higher average and instantaneous dose rates will be associated with flattening-filter-free linacs. Whether and how these physical changes impact biological results have been the subject of some studies and discussions [151].

Recently, Ling et al. [152] considered these factors in a review paper, so we will only remark on several basic points here. The effect of dose rate is primarily due to the phenomenon of sublethal damage repair (SLDR), and the kinetics of such repair appears to be different for tumors and normal tissues. The SLDR of tumor cells can be described with a single exponential function with a median T1/2 (the time to repair half of the damage) of approximately 60 min; however, that of most normal tissues involves 2 components with T1/2 of approximately 20 min and 2–3 h [152]. In general, treatment times significantly less than T1/2 will be associated with little repair and will not be affected by the dose rate effect, whereas sessions requiring time comparable to or longer than T1/2 will be affected, with decreasing biological effectiveness and increase in overall treatment time. For conventional fractionation, there is another factor. In the shoulder region of the survival curve, a large part of cell kill is due to the linear component not subjected to repair – further diminishing a dose rate effect – although, experimentally, the last point is difficult to prove. Damage repair and the dose rate effect are likely more important in oligofractionation because of the higher dose per fraction and the associated increase in treatment time. Thus, if technological advances result in a decrease in treatment time, biological efficacy will likely be increased. However, as tumor cell kill and normal tissue morbidity are both affected, any estimation of therapeutic advantage must take into account both sets of endpoints [152].

Another point, which is purely a curiosity, is that the instantaneous dose rate encountered in linac-based radiotherapy is very high. For a nominal dose rate of several hundred MU/min, the instantaneous dose rate is actually 10^6 MU/min. This is because the radiation is delivered in pulses of 2 µs, and 1/30 MU given in that duration yields an instantaneous dose rate of approximately 1 million MU/min. That may sound alarming, but in terms of radiochemical events, it is many orders of magnitude below that which would incur a dose rate effect [152].

Conclusions

Of the many important current topics discussed in this volume, medical economics casts long shadows of uncertainty. The political climate, global economic challenges and the untenable trend of healthcare costs are such that medical practices including radiation oncology will be affected, and with it technological development. Going forward, there will be increasing emphasis on flexibility, efficiency and economy, and the integration and streamlining of processes will be the key. In the words of an executive administrator at Memorial Sloan Kettering, everything has to be 'faster, better and cheaper'. But another infamous saying is that, 'when the going gets tough, the tough get going', so technical development will continue, albeit with redirected emphasis and perhaps a change of pace.

References

1 Ling C, Yorke E, Fuks Z: From IMRT to IGRT: frontierland or neverland? Radiother Oncol 2006;78:119–122.
2 Greco C, Clifton Ling C: Broadening the scope of image-guided radiotherapy (IGRT). Acta Oncol 2008;47:1193–1200.
3 Huntzinger C, Friedman W, Bova F, Fox T, Bouchet L, Boeh L: Trilogy image-guided stereotactic radiosurgery. Med Dosim 2007;32:121–133.
4 Baumann M, Holscher T, Zips D: The future of IGRT – cost benefit analysis. Acta Oncol 2008;47:1188–1192.
5 Amols H, Jaffray D, Orton C: Point/counterpoint. Image-guided radiotherapy is being overvalued as a clinical tool in radiation oncology. Med Phys 2006;33:3583–3586.
6 Foroudi F, Wong J, Haworth A, et al: Offline adaptive radiotherapy for bladder cancer using cone beam computed tomography. J Med Imaging Radiat Oncol 2009;53:226–233.
7 Yoo S, Yin F: Dosimetric feasibility of cone-beam CT-based treatment planning compared to CT-based treatment planning. Int J Radiat Oncol Biol Phys 2006;66:1553–1561.

8 Yan D, Vicini F, Wong J, Martinez A: Adaptive radiation therapy. Phys Med Biol 1997;42:123–132.
9 Meyer, JL: IMRT, IGRT, SBRT – Advances in the Treatment Planning and Delivery of Radiotherapy. Basel, Karger, 2007.
10 Noel C, Klein E, Moore K: A surface-based respiratory surrogate for 4D imaging. Med Phys 2008;35:2682.
11 Schöffel P, Harms W, Sroka-Perez G, Schlegel W, Karger C: Accuracy of a commercial optical 3D surface imaging system for realignment of patients for radiotherapy of the thorax. Phys Med Biol 2007;52:3949–3963.
12 Schoeffel P, Harms W, Karger C: Evaluation of repositioning accuracy of patients with breast cancer using a 3D surface imaging system. Radiother Oncol 2006;81:S195.
13 Krengli M, Gaiano S, Mones E, et al: Reproducibility of patient setup by surface image registration system in conformal radiotherapy of prostate cancer. Radiat Oncol 2009;4:1–10.
14 Djajaputra D, Li S: Real-time 3D surface-image-guided beam setup in radiotherapy of breast cancer. Med Phys 2005;32:65–75.
15 Bert C, Metheany K, Doppke K, Taghian A, Powell S, Chen G: Clinical experience with a 3D surface patient setup system for alignment of partial-breast irradiation patients. Int J Radiat Oncol Biol Phys 2006;64:1265–1274.
16 Hasan Y, Kim L, Martinez A, Vicini F, Yan D: Image guidance in external beam accelerated partial breast irradiation: comparison of surrogates for the lumpectomy cavity. Int J Radiat Oncol Biol Phys 2008;70:619–625.
17 Riboldi M, Gierga D, Chen G, Baroni G: Accuracy in breast shape alignment with 3D surface fitting algorithms. Med Phys 2009;36:1193–1198.
18 Riboldi M, Book L, Chen G, Taghian A, Baroni G, Gierga D: Quantitative assessment of surface deformation in accelerated partial breast irradiation. Int J Radiat Oncol Biol Phys 2007;69:S669.
19 Gierga D, Riboldi M, Turcotte J, et al: Comparison of target registration errors for multiple image-guided techniques in accelerated partial breast irradiation. Int J Radiat Oncol Biol Phys 2008;70:1239–1246.
20 Cerviño L, Yashar C, Jiang S: Improvement of the stability and reproducibility of deep-inspiration breath hold for left breast irradiation using video-based visual coaching and 3D surface imaging. Med Phys 2008;35:2703.
21 Tarte S, McClelland J, Hughes S, Blackall J, Landau D, Hawkes D: A non-contact method for the acquisition of breathing signals that enable distinction between abdominal and thoracic breathing. Radiother Oncol 2006;81:S209.
22 Xu Q, Hamilton R, Schowengerdt R, Alexander B, Jiang S: Lung tumor tracking in fluoroscopic video based on optical flow. Med Phys 2008;35:5351–5359.
23 Shirato H, Shimizu S, Kitamura K, et al: Four-dimensional treatment planning and fluoroscopic real-time tumor tracking radiotherapy for moving tumor. Int J Radiat Oncol Biol Phys 2000;48:435–442.
24 Shirato H, Shimizu S, Kunieda T, et al: Physical aspects of a real-time tumor-tracking system for gated radiotherapy. Int J Radiat Oncol Biol Phys 2000;48:1187–1195.
25 Shirato H, Suzuki K, Sharp G, et al: Speed and amplitude of lung tumor motion precisely detected in four-dimensional setup and in real-time tumor-tracking radiotherapy. Int J Radiat Oncol Biol Phys 2006;64:1229–1236.
26 Kuo J, Yu C, Petrovich Z, Apuzzo M: The CyberKnife stereotactic radiosurgery system: description, installation, and an initial evaluation of use and functionality. Neurosurgery 2003;53:1235–1239.
27 Gerszten P, Ozhasoglu C, Burton S, et al: CyberKnife frameless stereotactic radiosurgery for spinal lesions: clinical experience in 125 cases. Neurosurgery 2004;55:89–98.
28 Mohan D, Kupelian P, Willoughby T: Short-course intensity-modulated radiotherapy for localized prostate cancer with daily transabdominal ultrasound localization of the prostate gland. Int J Radiat Oncol Biol Phys 2000;46:575–580.
29 Falco T, Shenouda G, Kaufmann C, et al: Ultrasound imaging for external-beam prostate treatment setup and dosimetric verification. Med Dosim 2002;27:3.
30 Peignaux K, Truc G, Barillot I, et al: Clinical assessment of the use of the Sonarray system for daily prostate localization. Radiother Oncol 2006;81:176–178.
31 Fuller C, Thomas C, Schwartz S, et al: Method comparison of ultrasound and kilovoltage X-ray fiducial marker imaging for prostate radiotherapy targeting. Phys Med Biol 2006;51:4981–4993.
32 Morr J, DiPetrillo T, Tsai J, Engler M, Wazer D: Implementation and utility of a daily ultrasound-based localization system with intensity-modulated radiotherapy for prostate cancer. Int J Radiat Oncol Biol Phys 2002;53:1124–1129.
33 Feigenberg, S Paskalev K, McNeeley S, et al: Comparing computed tomography localization with daily ultrasound during image-guided radiation therapy for the treatment of prostate cancer: a prospective evaluation. J Appl Clin Med Phys 2007;8:99–110.

34 Boda-Heggemann J, Kohler F, Kopper B, et al: Accuracy of ultrasound-based (BAT) prostate-repositioning: a three-dimensional on-line fiducial-based assessment with cone-beam computed tomography. Int J Radiat Oncol Biol Phys 2008; 70:1247–1255.

35 McGahan J, Ryu J, Fogata M: Ultrasound probe pressure as a source of error in prostate localization for external beam radiotherapy. Int J Radiat Oncol Biol Phys 2004;60:788–793.

36 Peng C, Kainz K, Lawton C, Li X: A Comparison of daily megavoltage CT and ultrasound image guided radiation therapy for prostate cancer. Med Phys 2008;35:5619–5628.

37 McNair H, Mangar S, Coffey J, et al: A comparison of CT- and ultrasound-based imaging to localize the prostate for external beam radiotherapy. Int J Radiat Oncol Biol Phys 2006;65:678–687.

38 Trichter F, Ennis R: Prostate localization using transabdominal ultrasound imaging. Int J Radiat Oncol Biol Phys 2003;56:1225–1233.

39 Peng C, Kainz K, Lawton C, Li X: CT versus ultrasound image guided prostate cancer radiotherapy: dosimetric impacts. Int J Radiat Oncol Biol Phys 2007;69:S744–S745.

40 Langen K, Pouliot J, Anezinos C, et al: Evaluation of ultrasound-based prostate localization for image-guided radiotherapy. Int J Radiat Oncol Biol Phys 2003;57:635–644.

41 Dobler B, Mai S, Ross C, et al: Evaluation of possible prostate displacement induced by pressure applied during transabdominal ultrasound image acquisition. Strahlenther Onkol 2006;182: 240–246.

42 Serago C, Chungbin S, Buskirk S, Ezzell G, Collie A, Vora S: Initial experience with ultrasound localization for positioning prostate cancer patients for external beam radiotherapy. Int J Radiat Oncol Biol Phys 2002;53:1130–1138.

43 Kupelian P, Willoughby T, Mahadevan A, et al: Multi-institutional clinical experience with the Calypso System in localization and continuous, real-time monitoring of the prostate gland during external radiotherapy. Int J Radiat Oncol Biol Phys 2007;67:1088–1098.

44 Li H, Chetty I, Enke C, et al: Dosimetric consequences of intrafraction prostate motion. Int J Radiat Oncol Biol Phys 2008;71:801–812.

45 Parikh P, Hubenschmidt J, Dimmer S, et al: 4D verification of real-time accuracy of the Calypso system with lung cancer patient trajectory data. Int J Radiat Oncol Biol Phys 2005;63:S26–S27.

46 Mayse M, Parikh P, Lechleiter K, et al: Bronchoscopic implantation of a novel wireless electromagnetic transponder in the canine lung: a feasibility study. Int J Radiat Oncol Biol Phys 2008;72: 93–98.

47 Murphy M, Eidens R, Vertatschitsch E, Wright J: The effect of transponder motion on the accuracy of the Calypso Electromagnetic localization system. Int J Radiat Oncol Biol Phys 2008;72: 295–299.

48 Langen K, Willoughby T, Meeks S, et al: Observations on real-time prostate gland motion using electromagnetic tracking. Int J Radiat Oncol Biol Phys 2008;71:1084–1090.

49 Willoughby T, Kupelian P, Pouliot J, et al: Target localization and real-time tracking using the calypso 4D localization system in patients with localized prostate cancer. Int J Radiat Oncol Biol Phys 2006;65:528–534.

50 Santanam L, Malinowski K, Hubenshmidt J, et al: Fiducial-based translational localization accuracy of electromagnetic tracking system and on-board kilovoltage imaging system. Int J Radiat Oncol Biol Phys 2008;70:892–899.

51 Groh B, Siewerdsen J, Drake D, Wong J, Jaffray D: A performance comparison of flat-panel imager-based MV and kV cone-beam CT. Med Phys 2002;29:967–975.

52 Song W, Chiu B, Bauman G, et al: Prostate contouring uncertainty in megavoltage computed tomography images acquired with a helical tomotherapy unit during image-guided radiation therapy. Int J Radiat Oncol Biol Phys 2006;65: 595–607.

53 Holmes T, Hudes R, Dziuba S, Kazi A, Hall M, Dawson D: Stereotactic image-guided intensity modulated radiotherapy using the HI-ART II helical tomotherapy system. Med Dosim 2008;33: 135–148.

54 Munro P, Bouius D: X-ray quantum limited portal imaging using amorphous silicon flat-panel arrays. Med Phys 1998;25:689–702.

55 Cowen A, Davies A, Sivananthan M: The design and imaging characteristics of dynamic, solid-state, flat-panel X-ray image detectors for digital fluoroscopy and fluorography. Clin Radiol 2008; 63:1073–1085.

56 Yaffe M, Rowlands J: X-ray detectors for digital radiography. Phys Med Biol 1997;42:1–39.

57 Morin, O, Gillis A, Chen J, et al: Megavoltage cone-beam CT: system description and clinical applications. Med Dosim 2006;31:51–61.

58 Sillanpaa J, Chang J, Mageras G, et al: Low-dose megavoltage cone-beam computed tomography for lung tumors using a high-efficiency image receptor. Med Phys 2006;33:3489–3497.

59 Seppi E, Munro P, Johnsen S, et al: Megavoltage cone-beam computed tomography using a high-efficiency image receptor. Int J Radiat Oncol Biol Phys 2003;55:793–803.

60 Sawant A, Antonuk L, El-Mohri Y, et al: Segmented crystalline scintillators: an initial investigation of high quantum efficiency detectors for megavoltage X-ray imaging. Med Phys 2005;32: 3067–3083.

61 Orton E, Robar J: Megavoltage image contrast with low-atomic number target materials and amorphous silicon electronic portal imagers. Phys Med Biol 2009;54:1275–1289.

62 Ploquin N, Rangel A, Dunscombe P: Phantom evaluation of a commercially available three modality image guided radiation therapy system. Med Phys 2008;35:5303–5311.

63 Meeks S, Harmon J, Langen K, Willoughby T, Wagner T, Kupelian P: Performance characterization of megavoltage computed tomography imaging on a helical tomotherapy unit. Med Phys 2005;32:2673–2681.

64 McBain C, Henry A, Sykes J, et al: X-ray volumetric imaging in image-guided radiotherapy: the new standard in on-treatment imaging. Int J Radiat Oncol Biol Phys 2006;64:625–634.

65 Kak AC, Slaney M: Principles of Computed Tomography. New York, IEEE Press, 1988.

66 Hsieh J: Computed Tomography: Principles Design, Artifacts, and Recent Advances. Bellingham, SPIE – The international Society for Optical Engineering, 2003.

67 Steinke M, Bezak E: Technological approaches to in-room CBCT imaging. Australas Phys Eng Sci Med 2008;31:167–179.

68 Thorson T, Prosser T: X-ray volume imaging in image-guided radiotherapy. Med Dosim 2006;31: 126–133.

69 Feldkamp L, Davis L, Kress J: Practical cone-beam algorithm. J Opt Soc Am A Opt Image Sci Vis 1984;1:612–619.

70 Siewerdsen J, Jaffray D: Cone-beam computed tomography with a flat-panel imager: magnitude and effects of X-ray scatter. Med Phys 2001;28: 220–231.

71 Ding G, Duggan D, Coffey C: Characteristics of kilovoltage X-ray beams used for cone-beam computed tomography in radiation therapy. Phys Med Biol 2007;52:1595–1615.

72 Glover G: Compton scatter effects in CT reconstructions. Med Phys 1982;9:860–867.

73 Endo M, Tsunoo T, Nakamori N, Yoshida K: Effect of scattered radiation on image noise in cone beam CT. Med Phys 2001;28:469–474.

74 Mail N, Moseley D, Siewerdsen J, Jaffray D: The influence of bowtie filtration on cone-beam CT image quality. Med Phys 2009;36:22–32.

75 Graham S, Moseley D, Siewerdsen J, Jaffray D: Compensators for dose and scatter management in cone-beam computed tomography. Med Phys 2007;34:2691–2703.

76 Siewerdsen J, Moseley D, Bakhtiar B, Richard S, Jaffray D: The influence of antiscatter grids on soft-tissue detectability in cone-beam computed tomography with flat-panel detectors. Med Phys 2004;31:3506–3520.

77 Love L, Kruger R: Scatter estimation for a digital radiographic system using convolution filtering. Med Phys 1987;14:178–185.

78 Seibert J, Boone J: X-ray scatter removal by deconvolution. Med Phys 1988;15:567–575.

79 Wiegert J, Bertram M, Rose G, Aach T: Model based scatter correction for cone-beam computed tomography; in Flynn MJ (ed): Medical Imaging 2005: Physics of Medical Imaging. San Diego, SPIE, 2005, pp 271–282.

80 Kyriakou Y, Riedel T, Kalender W: Combining deterministic and Monte Carlo calculations for fast estimation of scatter intensities in CT. Phys Med Biol 2006;51:4567–4586.

81 Zbijewski W, Beekman F: Efficient Monte Carlo based scatter artifact reduction in cone-beam micro-CT. IEEE Trans Med Imaging 2006;25:817–827.

82 Maltz J, Gangadharan B, Vidal M, et al: Focused beam-stop array for the measurement of scatter in megavoltage portal and cone beam CT imaging. Med Phys 2008;35:2452–2462.

83 Zhu L, Bennett N, Fahrig R: Scatter correction method for X-ray CT using primary modulation: theory and preliminary results. IEEE Trans Med Imaging 2006;25:1573–1587.

84 Ning R, Tang X, Conover D: X-ray scatter correction algorithm for cone beam CT imaging. Med Phys 2004;31:1195–1202.

85 Virshup G, Suri R, Star-Lack J: Scatter characterization in cone-beam CT systems with offset flat panel imagers. Med Phys 2006;33:2288–2288.

86 Siewerdsen J, Daly M, Bakhtiar B, et al: A simple, direct method for X-ray scatter estimation and correction in digital radiography and cone-beam CT. Med Phys 2006;33:187–197.

87 Ohnesorge B, Flohr T, Klingenbeck-Regn K: Efficient object scatter correction algorithm for third and fourth generation CT scanners. Eur Radiol 1999;9:563–569.

88 Li G, Citrin D, Camphausen K, et al: Advances in 4D medical imaging and 4D radiation therapy. Technol Cancer Res Treat 2008;7:67–81.

89 Maltz J, Gangadharan B, Bose S, et al: Algorithm for X-ray scatter, beam-hardening, and beam profile correction in diagnostic kilovoltage. and treatment megavoltage. Cone beam CT. IEEE Trans Med Imaging 2008;27:1791–1810.

90 Petit S, van Elmpt W, Nijsten S, Lambin P, Dekker A: Calibration of megavoltage cone-beam CT for radiotherapy dose calculations: correction of cupping artifacts and conversion of CT numbers to electron density. Med Phys 2008;35:849–865.

91 Li H, Mohan R, Zhu X: Scatter kernel estimation with an edge-spread function method for cone-beam computed tomography imaging. Phys Med Biol 2008;53:6729–6748.

92 Star-Lack J, Sun M, Kaestner A, et al: Efficient scatter correction using asymmetric kernels; in Ehsan S, Jiang H (eds): Medical Imaging 2009: Physics of Medical Imaging. Lake Buena Vista, SPIE, 2009.

93 Wieczorek H: Effects of Trapping in a-Si:H Diodes. Solid State Phenomena 1995;44–46:957–972.

94 Overdick M, Solf T, Wischmann H: Temporal artifacts in flat dynamic X-ray detectors; in Antonuk LE, Yaffe MJ (eds): Medical Imaging 2001: Physics of Medical Imaging. San Diego, SPIE, 2001, pp 4347–4358.

95 Siewerdsen J, Jaffray D: Cone-beam computed tomography with a flat-panel imager: effects of image lag. Med Phys 1999;26:2635–2647.

96 Starman J, Virshup G, Bandy S, Star-Lack J, Fahrig R: Investigation into the cause of a new artifact in cone beam CT reconstructions on a flat panel imager. Med Phys 2006;33:2288–2288.

97 Mail N, Moseley D, Siewerdsen J, Jaffray D: An empirical method for lag correction in cone-beam CT. Med Phys 2008;35:5187–5196.

98 Mollov I, Toqnina C, Colbeth R: Photodiode forward bias to reduce temporal effects in a-Si based flat panel detectors; in Ehsan S, Jiang H (eds): Medical Imaging 2008: Physics of Medical Imaging. San Diego, SPIE, 2008.

99 Hsieh J, Gurmen OE, King KE: Recursive correction algorithm for detector decay characteristics in CT; in Dobbins JT, Boone JM (eds): Medical Imaging 2000: Physics of Medical Imaging. San Diego, SPIE, 2000, 3298–3305.

100 Xing L, Thorndyke B, Schreibmann E, et al: Overview of image-guided radiation therapy. Med Dosim 2006;31:91–112.

101 Guckenberger M, Meyer J, Wilbert J, et al: Intra-fractional uncertainties in cone-beam CT based image-guided radiotherapy (IGRT) of pulmonary tumors. Radiother Oncol 2007;83:57–64.

102 Thompson B, Hugo G: Quality and accuracy of cone beam computed tomography gated by active breathing control. Med Phys 2008;35:5595–5608.

103 Lu J, Guerrero T, Munro P, et al: Four-dimensional cone beam CT with adaptive gantry rotation and adaptive data sampling. Med Phys 2007;34:3520–3529.

104 Dawson L, Jaffray D: Advances in image-guided radiation therapy. J Clin Oncol 2007;25:938–946.

105 Sonke J, Zijp L, Remeijer P, van Herk M: Respiratory correlated cone beam CT. Med Phys 2005;32:1176–1186.

106 Dietrich L, Jetter S, Tucking T, Nill S, Oelfke U: Linac-integrated 4D cone beam CT: first experimental results. Phys Med Biol 2006;51:2939–2952.

107 Li T, Schreibmann E, Yang Y, Xing L: Motion correction for improved target localization with on-board cone-beam computed tomography. Phys Med Biol 2006;51:253–267.

108 Rit S, Wolthaus J, van Herk M, Sonke J: On-the-fly motion-compensated cone-beam CT using an a priori model of the respiratory motion. Med Phys 2009;36:2283–2296.

109 Zhang Q, Hu YC, Kriminski S, et al: Respiratory motion correction of cone-beam CT in abdomen using a patient-specific motion model. Med Phys 2009;36:2812.

110 Garden K, Robb R: 3-D reconstruction of the heart from few projections – a practical implementation of the McKinnon-Gates algorithm. IEEE Trans Med Imaging 1986;5:233–239.

111 McKinnon G, Bates R: Towards imaging the beating heart usefully with a conventional CT scanner. IEEE Trans Biomed Eng 1981;28:123–127.

112 Leng S, Zambelli J, Tolakanahalli R, et al: Streaking artifacts reduction in four-dimensional cone-beam computed tomography. Med Phys 2008;35:4649–4659.

113 Candes E, Romberg J, Tao T: Robust uncertainty principles: exact signal reconstruction from highly incomplete frequency information. IEEE Trans Inf Theory 2006;52:489–509.

114 Song J, Liu Q, Johnson G, Badea C: Sparseness prior based iterative image reconstruction for retrospectively gated cardiac micro-CT. Med Phys 2007;34:4476–4483.

115 Leng S, Tang J, Zambelli J, Nett B, Tolakanahalli R, Chen G: High temporal resolution and streak-free four-dimensional cone-beam computed tomography. Phys Med Biol 2008;53:5653–5673.

116 Tuy HK: An inversion formula for cone-beam reconstructions. SIAM J Appl Math 1983;43:546–552.

117 Schomberg H, van de Haar P, Baaten W: Cone-beam CT using a C-arm system as front end and a spherical spiral as source trajectory; in Ehsan S, Jiang H (eds): Medical Imaging 2009: Physics of Medical Imaging. Lake Buena Vista, SPIE, 2009;72580F–72512.

118 Zeng K, Chen Z, Zhang L, Wang G: An error-reduction-based algorithm for cone-beam computed tomography. Med Phys 2004;31: 3206–3212.

119 Zhu L, Yoon S, Fahrig R: A short-scan reconstruction for cone-beam CT using shift-invariant FBP and equal weighting. Med Phys 2007; 34:4422–4438.

120 Wang J, Li T, Liang Z, Xing L: Dose reduction for kilovotage cone-beam computed tomography in radiation therapy. Phys Med Biol 2008; 53:2897–2909.

121 Komsta L: A comparative study on several algorithms for denoising of thin layer densitograms. Anal Chim Acta 2009;641:52–58.

122 Buades A, Coll B, Morel JM: A review of image denoising algorithms, with a new one. Multiscale Model Simul 2005;4:490–530.

123 Glover G, Pelc N: An algorithm for the reduction of metal clip artifacts in CT reconstructions. Med Phys 1981;8:799–807.

124 Yin F, Guan H, Lu W: A technique for on-board CT reconstruction using both kilovoltage and megavoltage beam projections for 3D treatment verification. Med Phys 2005;32:2819–2826.

125 Zhu L, Star-Lack J, Green M, Fahrig R: Metal Artifact Correction Using Hybrid kV and MV Imaging. Oral Presentation, 10th International Electronic Portal Imaging and Positioning Devices Workshop (EP12K8). San Francisco, Calif., 2008, pp 114–115.

126 Godfrey D, Yin F, Oldham M, Yoo S, Willett, C: Digital tomosynthesis with an on-board kilovoltage imaging device. Int J Radiat Oncol Biol Phys 2006;65:8–15.

127 Zhang J, Wu QJ, Godfrey D, Fatunase T, Marks LB, Yin FF: Comparing digital tomosynthesis to cone beam CT for positron verification in patients undergoing partial breast irradiation. Int J Radiat Oncol Biol Phys 2009;73:952–957.

128 Kriminski S, Lovelock D, Mageras G, et al: Evaluation of respiration-correlated digital tomosynthesis for soft tissue visualization. Med Phys 2006;33:1988–1989.

129 Perks J, Lehmann J, Chen A, Yang C, Stern R, Purdy J: Comparison of peripheral dose from image-guided radiation therapy IGRT. using kV cone beam CT to intensity-modutated radiation therapy IMRT. Radiother Oncol 2008; 89:304–310.

130 Pouliot J, Bani-Hashemi A, Chen J, et al: Low-dose megavoltage cone-beam CT for radiation therapy. Int J Radiat Oncol Biol Phys 2005;61: 552–560.

131 Sheng K, Chow J, Read P: Decision making in daily or weekly volumetric imaging guidance based on case studies. Med Phys 2008;35:2693–2694.

132 Sontag M, Heron D, Yang Y, Komanduri K, Lalonde R, Huq M: Comparison of 2D-2D versus 3D-3D matching for image guided setup of head and neck cancer patients. Int J Radiat Onco Biol Phys 2007;69:S642.

133 Pouliot J: Megavoltage imaging, megavoltage cone beam CT and dose-guided radiation therapy; in Meyer JL (ed): IMRT, IGRT, SBRT: Advances in the Treatment Planning and Delivery of Radiotherapy. Basel, Karger, 2007, pp 132–142.

134 Tanyi J, Fuss M: Volumetric image-guidance: does routine usage prompt adaptive re-planning? An institutional review. Acta Oncol 2008; 47:1444–1453.

135 Hammoud R, Patel S, Pradhan D, et al: Examining margin reduction and its impact on dose distribution for prostate cancer patients undergoing daily cone-beam computed tomography. Int J Radiat Oncol Biol Phys 2008;71:265–273.

136 Kupelian P, Langen K, Zeidan O, et al: Daily variations in delivered doses in patients treated with radiotherapy for localized prostate cancer. Int J Radiat Oncol Biol Phys 2006;66:876–882.

137 Wang L, Feigenberg S, Chen L, Pasklev K, Ma C: Benefit of three-dimensional image-guided stereotactic localization in the hypofractionated treatment of lung cancer. Int J Radiat Oncol Biol Phys 2006;66:738–747.

138 Sonke J, Lebesque J, van Herk M: Variability of four-dimensional computed tomography patient models. Int J Radiat Oncol Biol Phys 2008; 70:590–598.

139 Korreman S, Juhler-Nottrup T, Boyer A: Respiratory gated beam delivery cannot facilitate margin reduction, unless combined with respiratory correlated image guidance. Radiother Oncol 2008;86:61–68.

140 Spoelstra F, de Koste J, Cuijpers J, Lagerwaard F, Slotman B, Senan S: Analysis of reproducibility of respiration-triggered gated radiotherapy for lung tumors. Radiother Oncol 2008;87:59–64.

141 Keall P, Joshi S, Vedam S, Siebers J, Kini V, Mohan R: Four-dimensional radiotherapy planning for DMLC-based respiratory motion tracking. Med Phys 2005;32:942–951.

142 Cho B, Poulsen P, Sloutsky A, Sawant A, Keall P: First demonstration of combined KV/MV image-guided real-time dynamic multileaf-collimator target tracking. Int J Radiat Oncol Biol Phys 2009;74:859–867.

143 de Crevoisier R, Melancon A, Kuban D, et al: Changes in the pelvic anatomy after an IMRT treatment fraction of prostate cancer. Int J Radiat Oncol Biol Phys 2007;68:1529–1536.

144 Mao W, Riaz N, Lee L, Wiersma R, Xing L: A fiducial detection algorithm for real-time image guided IMRT based on simultaneous MV and kV imaging. Med Phys 2008;35:3554–3564.

145 Eccles C, Brock K, Bissonnette J, Hawkins M, Dawson L: Reproducibility of liver position using active breathing coordinator for liver cancer radiotherapy. Int J Radiat Oncol Biol Phys 2006;64:751–759.

146 Timmerman R: An overview of hypofractionation and introduction to this issue of Seminars in Radiation Oncology. Semin Radiat Oncol 2008;18:215–222.

147 Otto K: Volumetric modulated arc therapy: IMRT in a single gantry arc. Med Phys 2008;35:310–317.

148 Titt U, Vassiliev O, Ponisch F, Dong L, Liu H, Mohan R: A flattening filter free photon treatment concept evaluation with Monte Carlo. Med Phys 2006;33:1595–1602.

149 Jin J, Yin F, Tenn S, Medin P, Solberg T: Use of the BrainLAB ExacTrac X-Ray 6D system in image-guided radiotherapy. Med Dosim 2008;33:124–134.

150 Wagner T, Meeks S, Bova F, et al: Optical tracking technology in stereotactic radiation therapy. Med Dosim 2007;32:111–120.

151 Moiseenko V, Duzenli C, Durand R: In vitro study of cell survival following dynamic MLC intensity-modulated radiation therapy dose delivery. Med Phys 2007;34:1514–1520.

152 Ling C, Gerweck G, Zaider M, Yorke E: Dose-rate effects in external beam radiotherapy redux. Radiother Oncol 2010;95:261–268.

C. Clifton Ling, PhD
Department of Medical Physics
Memorial Sloan-Kettering Cancer Center, 1275 York Avenue
New York, NY 10021-6094 (USA)
Tel. +1 212 639 8301, Fax +1 212 717 3290, E-Mail lingc@mskcc.org

Helical Tomotherapy: Image-Guided and Adaptive Radiotherapy

Patrick Kupelian[a] · Katja Langen[b]

[a]Department of Radiation Oncology, University of California School of Medicine, Los Angeles, Calif., and [b]Department of Radiation Oncology, M.D. Anderson Cancer Center Orlando, Orlando, Fla., USA

Abstract

Helical tomotherapy is a treatment device that is designed to deliver intensity-modulated radiation therapy treatments. Helical tomotherapy systems have been used to treat a wide spectrum of anatomical sites. In addition to its unique delivery technique, the capability to obtain megavoltage-based CT (MVCT) images is highly integrated into the system's image guidance. The introduction of MVCT imaging into clinical practice has prompted a range of technical and clinical investigations. The image quality, image dose and use of MVCT images for dose calculation have been investigated. At the same time, routine clinical use of MVCT imaging has provided a wealth of clinical experience. Both technical and clinical experiences with the MVCT system will be reviewed in this chapter.

Copyright © 2011 S. Karger AG, Basel

Helical tomotherapy has been utilized to treat a wide range of anatomical sites. At the M.D. Anderson Cancer Center in Orlando, patients have been treated with helical tomotherapy since October 2003. A second unit was installed at our center in December 2007. In total, more than 1,200 patients have finished a course of helical tomotherapy treatment at our clinic.

Since megavoltage-based CT (MVCT) scanning capabilities are highly integrated in the treatment device, each treatment is typically preceded by an MVCT image. These MVCT images are used to correct the patient alignment prior to treatment delivery. More than 36,000 MVCT scans have been performed at our center since the inception of the tomotherapy program.

In this chapter, we will provide a brief overview of the system and provide examples of its clinical use. The majority of the chapter is dedicated to the MVCT-based image guidance aspect of helical tomotherapy. Technical aspects, clinical experience and the use of MVCT images for adaptive radiation therapy will be discussed.

System Overview

The system uses a 6-MV linear accelerator that is mounted on a ring gantry. The gantry continuously rotates while the patient is translated though the ring's center. Each gantry rotation is divided into 51 segments (called projections) such that each projection covers about a 7 (360/51) degree angular segment. The radiation field resembles a fan beam with a longitudinal dimension of 1, 2.5 or 5 cm and a lateral dimension of 40 cm. In the lateral direction the beam is divided into 64 segments with a multileaf collimator (MLC). The MLC is a binary collimator, i.e. each leaf is either open or closed. The modulation of the radiation beam is temporal via leaf-specific opening times for each projection. Hence, for each gantry rotation, 51 × 64 (3,264) leaf-opening times are generated by the treatment planning software.

At the time of planning, 1 of 3 commissioned fields (1, 2.5 or 5 cm) is selected for the plan. A smaller field size improves the conformity of the dose distribution in the longitudinal direction, but increases the treatment delivery time. The 2.5-cm fan beam width is most frequently used since it results in acceptable dose conformity and reasonable treatment times. Treatment times are case-specific, but are typically about 5 min or less for prostate cases.

TomoTherapy Inc. has developed a delivery mode that allows the use of static gantry angles. This mode may be preferred for sites such as whole breast treatments where the use of certain selected gantry angles is desirable. The use of variable field collimation and couch speed is also being developed by TomoTherapy. In this delivery mode, the field size and couch speed are dynamic during a given plan. This delivery mode increases dose conformity and reduces treatment times [1]. The static gantry angle mode became commercially available in 2010, and the dynamic field size mode is currently under development.

Due to the unique treatment dynamics, a proprietary treatment planning system is used for helical tomotherapy. Most tomotherapy users use a secondary planning system for contouring, and import the CT image and contour set into the tomotherapy planning system. The planning system and treatment console use a common database. This eliminates the need for a secondary recording and verifying system, though most users have one for scheduling and billing purposes.

The same hardware that is used for treatment delivery is used for MVCT imaging. During MVCT scanning, the nominal energy of the incident electron beam is reduced from 6 to 3.5 MV [2] and the radiation field is collimated in the longitudinal direction to 4 mm. All MLC leaves are open during the MVCT acquisition.

When selecting the MVCT scanning mode on the tomotherapy control console, the user has 2 parameters to select. The scan range needs to be selected first. This is variable and typically a range that encompasses the treatment area is selected. A longer scan region will clearly require more time. A second parameter is the image pitch. The pitch determines the nominal slice thickness. There are 3 selections termed *fine*, *normal* and *coarse*. These correspond to nominal slice thicknesses of 2, 4 and 6 mm. Since the gantry rotates at a constant speed during all imaging procedures, it will take longer to scan the same volume in the fine than normal mode. The scan acquisition time is 5 s per slice. Accordingly, a 10-cm long volume will require 250, 125 and about 83 s to scan in fine, normal and coarse modes, respectively. Due to volume averaging, the quality of the sagittal and coronal images is better for smaller slice thicknesses.

Clinical Applications of Helical Tomotherapy

Helical tomotherapy has been used clinically to treat a wide variety of malignant conditions [3]; prostate cancers [4], head and neck tumors [5], lung cancers [6], complex breast tumors [7], gastrointestinal tumors [8], esophageal cancers [9], and brain tumors [10] have been treated. Helical tomotherapy has also been applied in the stereotactic setting [11, 12].

Several tomotherapy-specific applications have emerged over the last several years. For example, tomotherapy is well suited to treat extended target volumes in a single procedure. Since a fan beam radiation field is rotated around the patient while the patient couch is translated through the center of rotation, the longitudinal extent of the treatment field is limited only by the maximal couch travel distance (160 cm). This allows the treatment of lengthy tumor volumes or multiple tumor volumes in 1 single uninterrupted procedure. This feature has been used for the treatment of craniospinal axis (CSA) volumes. The treatment of the entire CSA in a single radiation procedure eliminates the field junctioning issues and the need to feather the field junction. An additional advantage is that the patient can be in a supine position, which makes the administration of sedation easier. Helical tomotherapy treatments of CSA fields have been described in the literature, and initial clinical results have been reported as well [13, 14].

The use of helical tomotherapy for total body irradiation, total marrow irradiation and total lymphatic irradiation has been pioneered at the City of Hope [15]. In this treatment approach, all target volumes except the lower extremities are treated in a single procedure. The lower extremities are treated on a conventional C-arm linac to reduce treatment time on the helical tomotherapy unit. Reduced acute toxicities and clinical experiences have been reported [16].

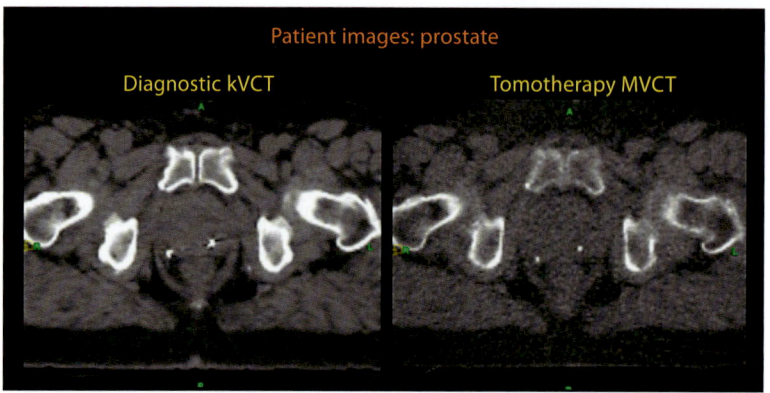

 **Fig. 1.** Diagnostic kVCT and tomotherapy MVCT images. See online supplementary material for additional figure. MV resolution is grainier and may limit evaluation of sagittal images for prostate position.

Daily Megavoltage-Based CT Imaging

Quality of Images

The imaging beam that is used for CT acquisition is in the megavoltage range and the image is hence called a MVCT image. At this higher energy, Compton scattering dominates the photon interactions in all body tissues and the attenuation is proportional to the tissue density. Diagnostic CT scanners use beams of much lower energy, and the bone-to-soft tissue contrast is larger since the bone attenuation for these lower energy beams is partially due to the photoelectric effect. However, bone anatomy and soft tissue structures are visible in MVCT images. Figure 1 shows the diagnostic and MVCT image of a prostate patient. The main difference between diagnostic and MVCT images is that the latter have more extraneous noise. If the noise is expressed as a function of the standard deviation of the Hounsfield units (HU) in a homogeneous phantom, the MVCT images are about twice as noisy as regular CT scans [17].

Daily Image Registration

The primary reason for obtaining MVCT images prior to the treatment delivery is to check and correct the patient position for treatment. This, of course, requires that the daily image is registered with the planning image. A range of image registration tools are available at the tomotherapy operator console. Figure 2 shows an example of MVCT to kVCT image registration. The superior-inferior extent of the MVCT scan is selected by the operator. In general, the scan area encompasses

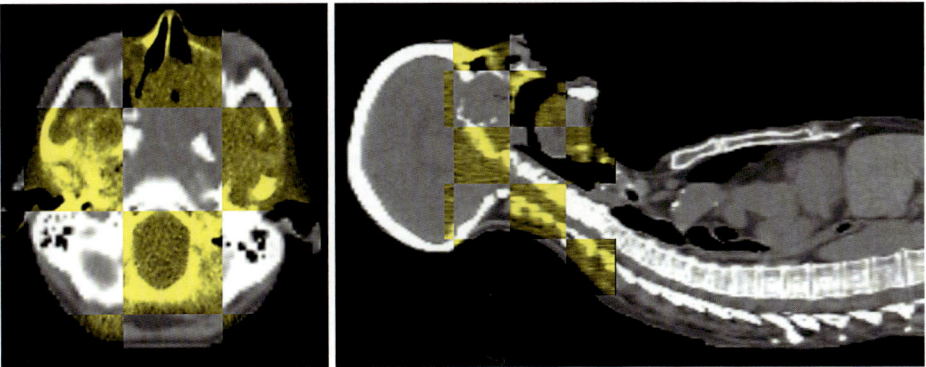

Fig. 2. MVCT to kVCT image registration example for a head and neck patient. The larger kVCT image is shown in gray and the MVCT image is shown in yellow. A checkerboard overlay is displayed. With registration, isodoses can be displayed on acquired images. See online supplementary material.

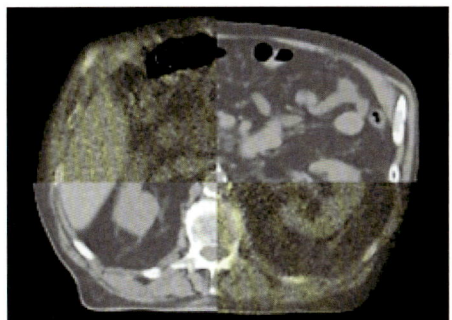

Fig. 3. MVCT to kVCT image registration of scans obtained in the abdominal region.

the treatment area; however, for shortened scan times, a subvolume can be imaged. The registration tool allows checkerboard or partially transparent image overlay. Planning contours and/or dose distributions can be added to help with the registration process. Both manual and automatic registration tools are available. At the user's discretion, the registration can be based on bone or soft tissue anatomy, or implanted tumor markers.

Image registration for lung tumors can be based on the tumor volume if it is surrounded by aerated lung. For less visible tumors, bone anatomy is used to guide the alignment. The MV scans have a slow acquisition time (5 s per slice) and a moving target may appear blurry. For stereotactic applications, this may be an advantage since a range of the tumor positions is averaged during a breathing cycle. Abdominal imaging with tomotherapy is satisfactory, again with grainier images but with a considerable amount of anatomy shown. Figure 3 shows an axial kVCT/MVCT image registration of the abdominal region. In the abdomen, alignment is almost always made to the soft tissues imaged.

The MVCT to kVCT registration for prostate patients has been investigated in our group [18]. The interuser variability of the prostate image registration was reduced for patients who had fiducial markers implanted. Particularly in the superior-inferior direction, the image registration is less subjective if implanted markers are used to guide the registration process. This same problem has been confronted when using kV cone beam platforms.

About 40% of our patients treated using helical tomotherapy have head and neck cancers, and we typically use the upper cervical spine for their alignment. However, the use of implanted markers in head and neck tumors has also been investigated at the M.D. Anderson Cancer Center [19]. It is common for the physicians to provide specific recommendations to the therapists with respect to the MVCT/kVCT image registration process best suited for each patient.

Significance of Daily Setup Errors

Since the MVCT system is highly integrated into the helical tomotherapy hardware and workflow, most helical tomotherapy treatments are preceded by an MVCT scan for image guidance. Most clinics quickly accumulate a vast amount of MVCT and patient setup data. Several groups have analyzed and reported these data. Two examples of such reports are described below.

The group at the Medical College of Wisconsin analyzed 3,800 MVCT scans from 152 patient cases and reported daily alignment shifts [20]. Seven anatomical sites were included in this analysis. In figure 4, the mean daily shifts and their standard deviations are illustrated for each patient. Patients were grouped according to treatment sites, and it was shown that the mean shifts were significantly different depending on the anatomical site. Setup errors in the skull, brain, and head and neck regions were smaller than those observed in sites below the neck.

Chen et al. [9] used MVCT imaging to report setup correction statistics for esophageal patients. They observed *systematic* setup corrections of 1.5, 3.7 and 4.8 mm (standard deviations) in the anterior-posterior, lateral and superior-inferior directions, respectively. Corresponding average *random* setup errors were 2.9, 5.2 and 4.4 mm. They also reported that the setup variations were stable over the course of radiation therapy and that there were no correlations between setup variation results and body habitus.

Significance of Daily Deformation

With the introduction of daily CT-based image guidance, it has become clear that simple couch translations are not always sufficient to realign the patient accurately

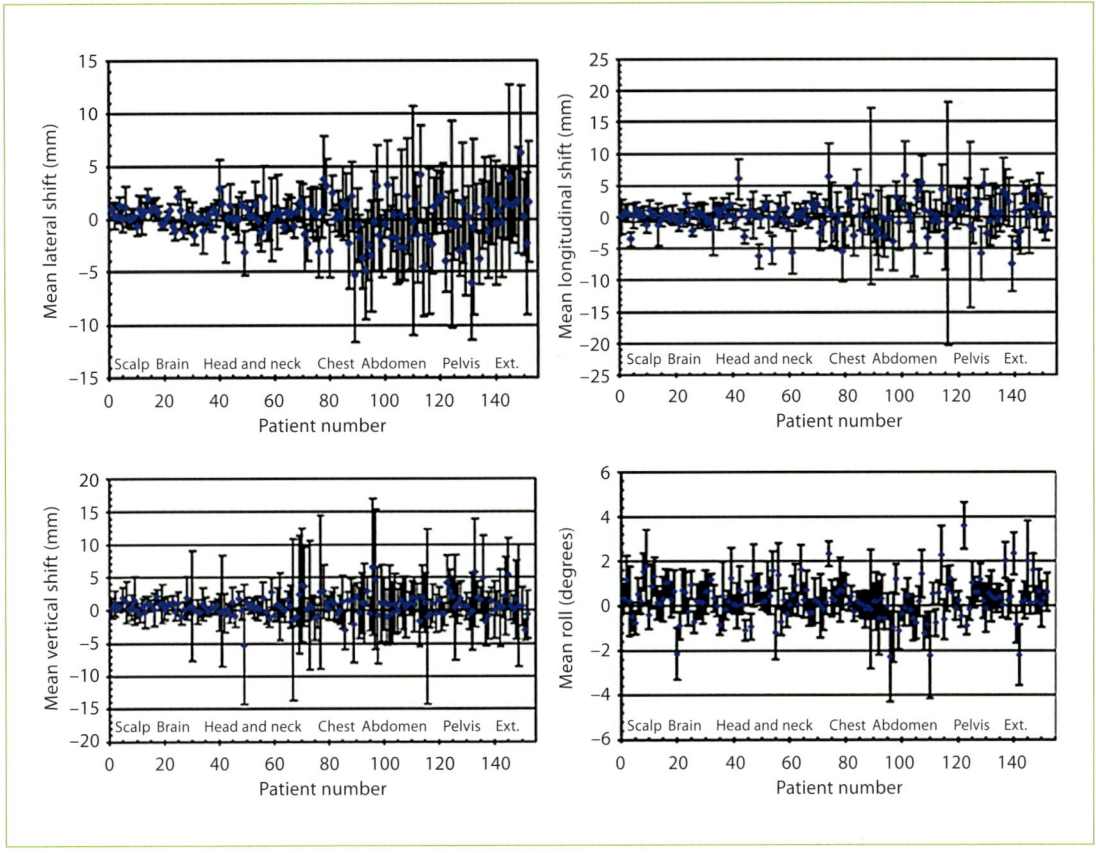

Fig. 4. Patient-specific mean daily alignment shifts and their standard deviations for 152 patients grouped according to the anatomical site (from Li et al. [20]).

prior to treatment. Frequently, we find that the complete MVCT image cannot be accurately registered with the planning CT. One often has to compromise between different parts of the anatomy. This is also the case for alignments based on bone anatomy. Figure 5 shows an example where alignment on bone anatomy at the base of the skull resulted in an alignment compromised in the spinal cord region.

Another example of clinically significant deformation is the shrinkage of lung tumor volumes during the course of treatment, which has been observed using MVCT guidance. We have reported the regressions of lung tumors during helical tomotherapy [21]. Figure 6 shows the degree of tumor regression measured during the course of treatment in 10 patients.

Anatomical tumor changes were documented for esophageal cancers in a report from the City of Hope [9]. Daily MVCT scans have also been used to document shrinkage and migration of parotid volumes during the course of treatment [22].

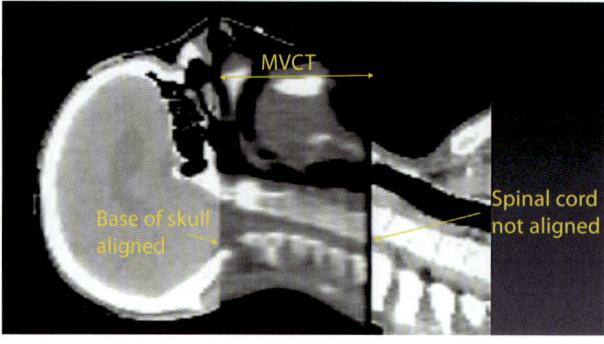

Fig. 5. Example of bone alignment mismatch due to deformation in the neck region.

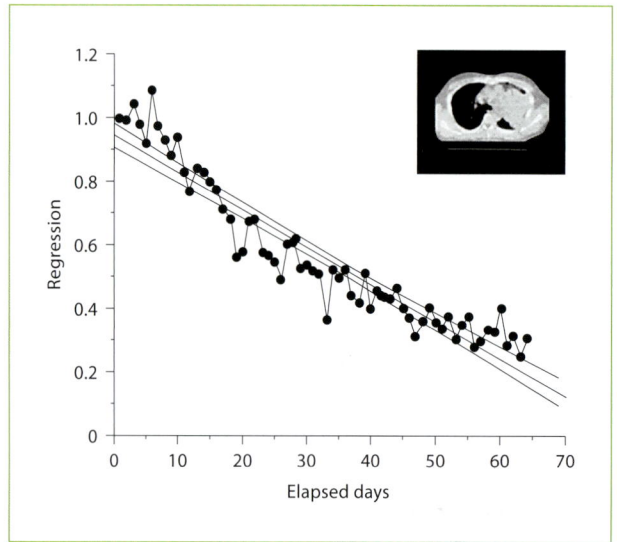

Fig. 6. Regression of non-small cell lung cancers during helical tomotherapy observed on daily MVCT scans [21]. The y-axis is the fraction of regression, and the x-axis is the number of elapsed days from the first treatment day. Ten patients had non-small cell lung cancers, which was treated with helical tomotherapy. There was an average of 27 scans per patient, and an average of 1.2% shrinkage per day (range: 0.6–2.3%).

Imaging Doses

Since MVCT images are typically used prior to each treatment fraction, it is important to understand the additional radiation dose to the patient due to MVCT imaging. While the imaging dose can be measured using phantoms, the ultimate goal is a determination of the imaging doses specific for the complex anatomies of individual patients. To approach this, Shah et al. [23] commissioned an image beam dose calculation method based on CT anatomy, and determined the doses for various MVCT settings. The MVCT scans can be acquired in fine, normal and coarse modes. These settings determine the couch speed and hence the longitudinal slice thickness during MVCT imaging. Compared with the fine mode, the couch travels 2 and 3 times faster during the normal and coarse scans, respectively. The imaging dose scales accordingly, since the dose rate and gantry speed are constant for all 3 imaging modes. It was found that the imaging dose is relatively homogenous throughout the imaged anatomy and that it increases with *decreasing* anatomical thickness. For instance, the image dose is higher in the neck than in the pelvis region. For the 'normal' MVCT scan mode, the MVCT dose ranged from 1 to 1.5 cGy per scan. Table 1 shows the calculated doses for all MVCT settings and for a range of anatomical sites. Overall, the doses delivered by tomotherapy imaging are quite low and do not limit the use of daily image guidance.

The calculated dose distributions due to a single MVCT scan are shown in figure 7. For calculations, it was assumed that the entire CT anatomy was scanned. In reality, only the volume of interest will generally be scanned.

The Benefit of Daily Imaging

In addition to the imaging dose, the daily use of MVCT adds to the patient in-room time. What is the benefit of daily imaging? Can equally acceptable results be achieved with less than daily imaging? The benefit of daily MVCT imaging was evaluated by comparing alternative clinical scenarios in terms of the frequency of imaging and the corrections based on it. A group of patients, all treated with helical tomotherapy and daily image guidance for head and neck or prostate cancers, was reviewed [24, 25]. Since the setup corrections were known for each day based on the daily MVCT, one can deduce the error that would have been made on each day if the image guidance had been omitted. This method has allowed us to evaluate the *remaining setup errors* that would have occurred if different imaging protocols had been used, other than daily. The following scenarios were evaluated:
– (a) No imaging
– (b) Imaging for the initial 1, 3, 5 or 7 fractions only. The remaining fractions are corrected by the mean shifts that were determined in these initial fractions
– (c) Weekly imaging, shifts from the first fraction are applied to all subsequent treatments. If a setup error larger than 3 mm is detected on subsequent weekly imaging, the setup corrections are adjusted accordingly from that point forward

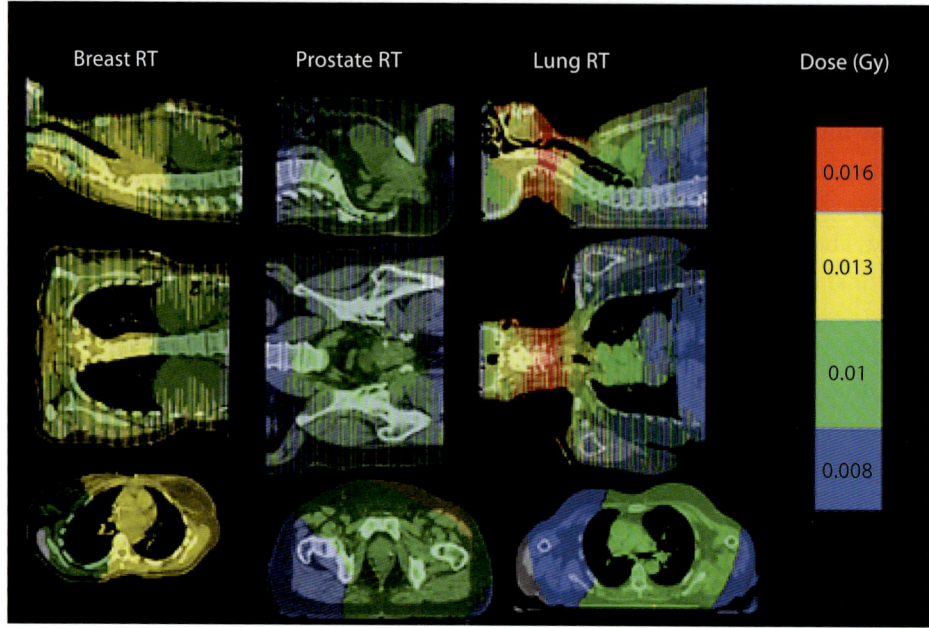

Fig. 7. Imaging dose distributions for helical MVCT (image from [23]).

Table 1. Imaging doses with helical tomotherapy

Imaging setting	Fine cGy	Normal cGy	Coarse cGy
Prostate	2.1	1.0	0.7
Lung	2.1	1.0	0.7
Head	2.7	1.4	0.9
Neck	3.0	1.5	1.0

Data from [23].

- (d) Image first 5 fractions, then weekly. All non-IGRT (image-guided radiation therapy) treatments are corrected by the mean shifts of the first 5 fractions. If weekly imaging reveals a setup correction that is larger than 2 times the standard deviation of the first 5 fractions, the daily setup corrections are adjusted accordingly
- (e) Image every other day and correct non-IGRT fractions by the running mean setup error

The image frequency of these protocols ranges from 0 to 50% of treatment fractions. A zero setup error was assigned to all IGRT sessions. To score the value of each imaging protocol, the fraction of treatments that were subject to a setup error

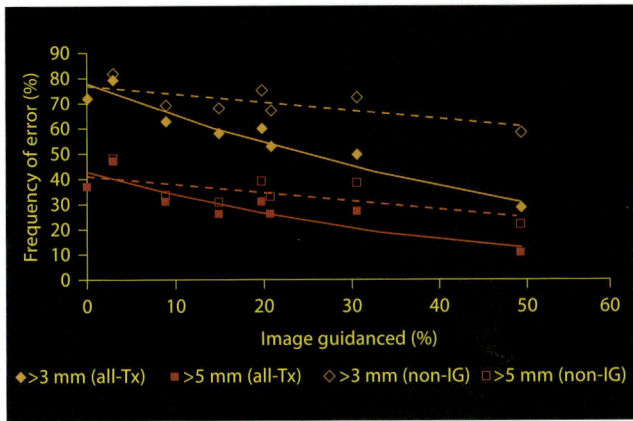

Fig. 8. Reduction of setup errors with increasing use of image guidance [25]. The solid line shows the fraction of all treatments that had a setup error larger than 3 or 5 mm. The dashed line shows the same score for the non-IGRT fractions.

larger than 3 or 5 mm were scored. The score was calculated separately for all treatments and the non-IGRT fractions.

Results for the head and neck treatments are shown in figure 8. The solid line shows the frequency of setup errors for all fractions combined (both IGRT and non-IGRT fractions). The orange and red curves show setup errors with magnitudes of >3 mm and >5 mm, respectively. As the image frequency increased, the fraction of misaligned treatments reduced, even for the non-IGRT fractions. However, with an imaging frequency of 50% (every other day), about 10% of all fractions still had a setup error >5 mm. At the 3 mm level, this fraction increased to 30%. Since all patients were immobilized with thermoplastic masks, the results are notable. Every-other-day imaging still permitted significant residual errors, and the results indicate to us the need for daily volumetric imaging, even in immobilized head and neck cases.

Dose Evaluation and Adaptive Radiotherapy

While the main reason to acquire an MVCT image is image guidance, MVCT images can also be used to evaluate the dosimetric consequences of organ deformation in patients. In earlier sections, it has been pointed out that simple couch translations do not always achieve accurate alignment of the patient for treatment. If the patient's anatomy in the treatment area changes due to weight loss, tumor shrinkage and so forth, simple couch translation will always compromise the dosimetry in part of the treatment anatomy.

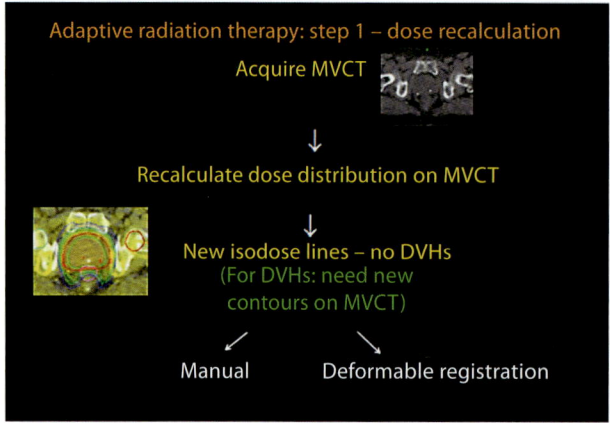

Fig. 9. The workflow of assessing the dosimetric consequence of changes in the patient's anatomy or setup differences in a single MVCT image. DVHs = Dose-volume histograms.

The dosimetric consequence of these deformations in patients is of interest and can be assessed using MVCT imaging. To assess the dosimetric consequence in a single MVCT image is fairly straightforward. The dose needs to be recalculated in the daily CT dataset; contours can be generated manually to generate dose-volume histograms. It is more complicated to assess the cumulative doses since there will be some degree of organ deformation on a daily basis. The dose to each tissue *voxel* needs to be assessed in the daily images and accumulated. Deformable image registration is needed to map the location of a voxel from one image to the next. Once a deformable image registration algorithm is available to map a tissue voxel from one image to the next, it can also be used to transfer the *contour* of a volume of interest from one image to the next, since each voxel is assigned to a contour. A volume of interest is simply a group of voxels that belong to a structure. Hence, deformable image registration can be used to automate the process of evaluating daily images accounting for their changing contours.

Adaptive Radiation Therapy: Step 1 – Dose Recalculation

The use of MVCT images for the assessment of daily dosimetry is illustrated in figure 9. The daily image is acquired. Prior to using the MVCT image for dose calculation, a MVCT HU to the electron density curve must be established. This process is identical to the HU to electron density calibration that is performed for a diagnostic CT scanner. The stability and uniformity of the MVCT HU calibration process has been tested previously [26].

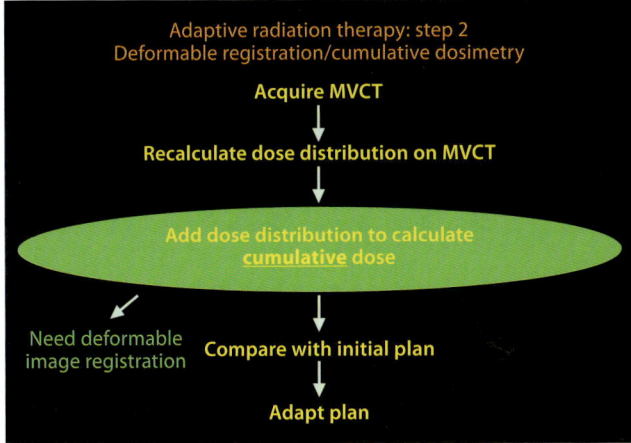

Fig. 10. The process of adaptive radiation therapy.

After the MVCT HU conversion to electron density, the image can be used for dose calculation. At this point, data from the treatment plan is used to calculate the dose distribution in the MVCT image. This process assumes that the treatment plan was delivered as intended. This MVCT dose calculation procedure is integrated into the TomoTherapy system via proprietary software called 'Planned Adaptive'. The current software package allows the user to manually adjust contours such that a second structure set is generated that corresponds to the daily MVCT images. Once the daily structure set is available, daily dose-volume histogram data can be calculated and compared with the dose-volume histogram information for the treatment plan. While this process is labor intensive, it allows the dosimetric evaluation of patient deformation. Using these tools, we have evaluated the variations of prostate, rectum and bladder doses in prostate cancer patients [4]. Large differences in the rectal and bladder dosimetry became evident, while dosimetric variations in the target dose were small in comparison.

The time consuming process of generating the daily contours could be replaced using a deformable image registration dose. A kVCT to MVCT deformable image registration code has been developed by Lu et al. [27].

Adaptive Radiation Therapy: Step 2 – Dose Accumulation

Ultimately, the goal of adaptive planning is to acquire the MVCT on a daily basis, recalculate the dose distribution on each MVCT and evaluate the *cumulative* treatment doses over time. Should differences between the planned and actual de-

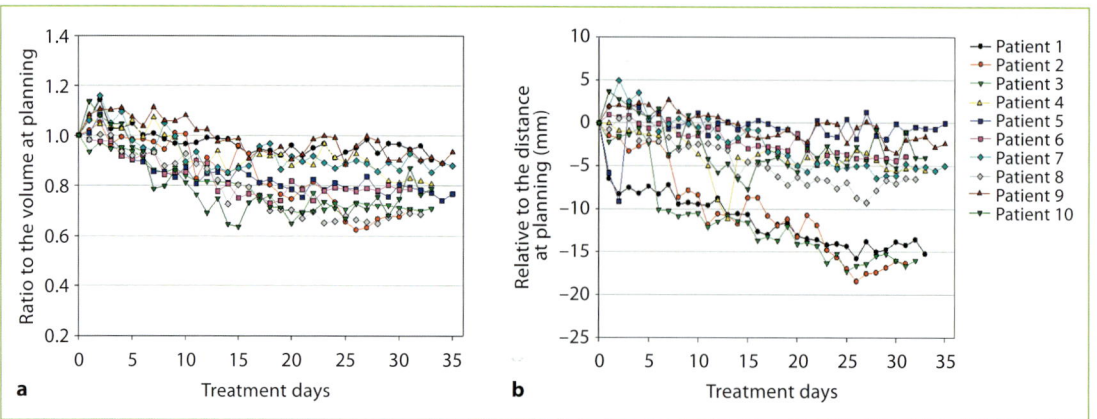

Fig. 11. Parotid volume and position changes during the course of radiation therapy as determined from daily MVCT images [22]. See online supplementary material. **a** Parotid glands volume changes over the fractionated radiation therapy. **b** Distance between the center of mass of the left and right parotid glands over fractionated radiotherapy.

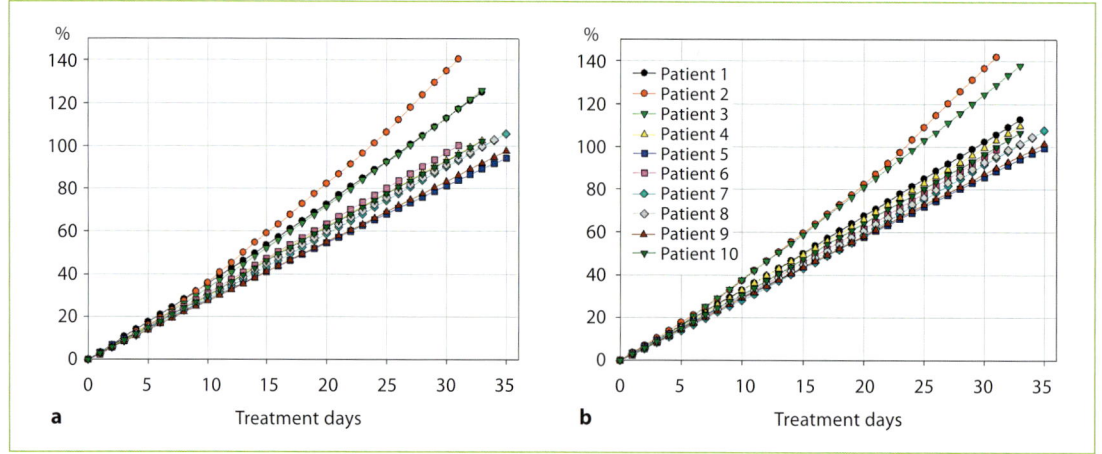

Fig. 12. The running cumulative mean dose to the contralateral (left) and ipsilateral (right) parotid [29]. Ratios of calculated cumulative dose to total planned dose: low-dose side, spared (**a**); high-dose side, spared with compromise (**b**).

livered dose distribution be evident, the user can generate a new plan that adapts to the observed changes in anatomy and/or dosimetry. This adaptive radiation therapy workflow is illustrated in figure 10.

An accurate deformable image registration algorithm is paramount in this process. It is, however, difficult to assess the accuracy of a deformable image registration algorithm. At this point, our group has evaluated a MVCT to kVCT deformable image registration algorithm for the automatic contouring of parotid volumes on MVCT images; the algorithm was developed by Lu et al. [27]. The daily parotid volumes and positions were extracted after the application of the algorithm, and results were compared with data available in the literature [22]. Figure 11 shows the parotid volumes and positions over the course of treatment for each patient.

Trends in parotid volume reductions and migration were comparable with those published by others [28]. Subsequently, the algorithm was used to generate cumulative dose volume histograms for 10 head and neck patients. Significant differences from the planned dose volume histograms were identified in 3 of the 10 patients [29]. Figure 12 shows the running cumulative mean dose to the parotid volumes. In 3 of the 10 patients, a substantial increase in the mean parotid dose was observed. Averaged over all patients, the mean parotid dose increased by about 10%.

Conclusions

Helical tomotherapy is a unique delivery device with in integrated imaging modality. The image quality is sufficient for image guidance applications and images are acquired with a relatively low dose. As a result, the system makes daily image guidance based on volumetric imaging a reality. The daily use of MVCT imaging has provided a wealth of information to users. In addition, regular MVCT imaging opens the avenues for adaptive therapy, an area that shows the greatest promise of improvement in dose delivery accuracy for intensity-modulated radiation therapy.

References

1 Sterzing F, Uhl M, Hauswald H, et al: Dynamic jaws and dynamic couch in helical tomotherapy. Int J Radiat Oncol Biol Phys 2010;76:1266–1273.
2 Jeraj R, Mackie TR, Balog J, et al: Radiation characteristics of helical tomotherapy. Med Phys 2004;31:396–404.
3 Sterzing F, Schubert K, Sroka-Perez G, et al: Helical tomotherapy. Experiences of the first 150 patients in Heidelberg. Strahlenther Onkol 2008; 184:8–14.
4 Kupelian PA, Langen KM, Zeidan OA, et al: Daily variations in delivered doses in patients treated with radiotherapy for localized prostate cancer. Int J Radiat Oncol Biol Phys 2006;66:876–882.
5 Kodaira T, Tomita N, Tachibana H, et al: Aichi cancer center initial experience of intensity modulated radiation therapy for nasopharyngeal cancer using helical tomotherapy. Int J Radiat Oncol Biol Phys 2009;73:1129–1134.
6 Cattaneo GM, Dell'oca I, Broggi S, et al: Treatment planning comparison between conformal radiotherapy and helical tomotherapy in the case of locally advanced-stage NSCLC. Radiother Oncol 2008;88:310–318.
7 Goddu SM, Chaudhari S, Mamalui-Hunter M, et al: Helical tomotherapy planning for left-sided breast cancer patients with positive lymph nodes: comparison to conventional multiport breast technique. Int J Radiat Oncol Biol Phys 2009;73: 1243–1251.

8 De Ridder M, Tournel K, Van Nieuwenhove Y, et al: Phase II study of preoperative helical tomotherapy for rectal cancer. Int J Radiat Oncol Biol Phys 2008;70:728–734.

9 Chen YJ, Han C, Liu A, et al: Setup variations in radiotherapy of esophageal cancer: evaluation by daily megavoltage computed tomographic localization. Int J Radiat Oncol Biol Phys 2007;68:1537–1545.

10 Vaandering A, Lee JA, Renard L, et al: Evaluation of MVCT protocols for brain and head and neck tumor patients treated with helical tomotherapy. Radiother Oncol 2009;93:50–56.

11 Do L, Pezner R, Radany E, et al: Resection followed by stereotactic radiosurgery to resection cavity for intracranial metastases. Int J Radiat Oncol Biol Phys 2009;73:486–491.

12 Sterzing F, Welzel T, Sroka-Perez G, et al: Reirradiation of multiple brain metastases with helical tomotherapy. A multifocal simultaneous integrated boost for eight or more lesions. Strahlenther Onkol 2009;185:89–93.

13 Penagaricano J, Moros E, Corry P, et al: Pediatric craniospinal axis irradiation with helical tomotherapy: patient outcome and lack of acute pulmonary toxicity. Int J Radiat Oncol Biol Phys 2009;75:1155–1161.

14 Penagaricano JA, Papanikolaou N, Yan Y, et al: Feasibility of cranio-spinal axis radiation with the Hi-Art tomotherapy system. Radiother Oncol 2005;76:72–78.

15 Schultheiss TE, Wong J, Liu A, et al: Image-guided total marrow and total lymphatic irradiation using helical tomotherapy. Int J Radiat Oncol Biol Phys 2007;67:1259–1267.

16 Wong JY, Rosenthal J, Liu A, et al: Image-guided total-marrow irradiation using helical tomotherapy in patients with multiple myeloma and acute leukemia undergoing hematopoietic cell transplantation. Int J Radiat Oncol Biol Phys 2009;73:273–279.

17 Meeks SL, Harmon JF Jr, Langen KM, et al: Performance characterization of megavoltage computed tomography imaging on a helical tomotherapy unit. Med Phys 2005;32:2673–2681.

18 Langen KM, Zhang Y, Andrews RD, et al: Initial experience with megavoltage (MV) CT guidance for daily prostate alignments. Int J Radiat Oncol Biol Phys 2005;62:1517–1524.

19 Zeidan OA, Huddleston AJ, Lee C, et al: A comparison of soft-tissue implanted markers and bony anatomy alignments for image-guided treatments of head-and-neck cancers. Int J Radiat Oncol Biol Phys 2010;76:767–774.

20 Li XA, Qi XS, Pitterle M, et al: Interfractional variations in patient setup and anatomic change assessed by daily computed tomography. Int J Radiat Oncol Biol Phys 2007;68:581–591.

21 Kupelian PA, Ramsey C, Meeks SL, et al: Serial megavoltage CT imaging during external beam radiotherapy for non-small-cell lung cancer: observations on tumor regression during treatment. Int J Radiat Oncol Biol Phys 2005;63:1024–1028.

22 Lee C, Langen KM, Lu W, et al: Evaluation of geometric changes of parotid glands during head and neck cancer radiotherapy using daily MVCT and automatic deformable registration. Radiother Oncol 2008;89:81–88.

23 Shah AP, Langen KM, Ruchala KJ, et al: Patient dose from megavoltage computed tomography imaging. Int J Radiat Oncol Biol Phys 2008;70:1579–1587.

24 Kupelian PA, Lee C, Langen KM, et al: Evaluation of image-guidance strategies in the treatment of localized prostate cancer. Int J Radiat Oncol Biol Phys 2008;70:1151–1157.

25 Zeidan OA, Langen KM, Meeks SL, et al: Evaluation of image-guidance protocols in the treatment of head and neck cancers. Int J Radiat Oncol Biol Phys 2007;67:670–677.

26 Langen KM, Meeks SL, Poole DO, et al: The use of megavoltage CT (MVCT) images for dose recomputations. Phys Med Biol 2005;50:4259–4276.

27 Lu W, Olivera GH, Chen Q, et al: Deformable registration of the planning image (kVCT) and the daily images (MVCT) for adaptive radiation therapy. Phys Med Biol 2006;51:4357–4374.

28 Barker JL Jr, Garden AS, Ang KK, et al: Quantification of volumetric and geometric changes occurring during fractionated radiotherapy for head-and-neck cancer using an integrated CT/linear accelerator system. Int J Radiat Oncol Biol Phys 2004;59:960–970.

29 Lee C, Langen KM, Lu W, et al: Assessment of parotid gland dose changes during head and neck cancer radiotherapy using daily megavoltage computed tomography and deformable image registration. Int J Radiat Oncol Biol Phys 2008;70:1563–1571.

Patrick Kupelian, MD
Department of Radiation Oncology, University of California School of Medicine
200 UCLA Medical Plaza, Suite B265
Los Angeles, CA 90095-6951 (USA)
Tel. +1 310 825 9775, Fax +1 310 794 9795, E-Mail pkupelian@mednet.ucla.edu

The CyberKnife in Clinical Use: Current Roles, Future Expectations

Sonja Dieterich · Iris C. Gibbs

Department of Radiation Oncology, Stanford University Medical Center, Stanford, Calif., USA

Abstract

The CyberKnife system deploys a linac mounted on an agile robot and directed under image guidance for stereotactic radiotherapy using nonisocentric beam delivery. A design advantage of the CyberKnife system is its method of active image guidance during treatment delivery. Recent developments in the hardware and software of the system have significantly enhanced its functionality: (a) an optimized path traversal process significantly reduces the robot motion time, resulting in reductions of overall treatment times of at least 5–10 min; (b) to optimize the accuracy of dose calculation in CyberKnife planning/delivery, Monte Carlo algorithms have been introduced; (c) the new IRIS collimator reduces the monitor units required, increases treatment speed and improves conformality and homogeneity of treatment plans; (d) XSight lung tracking, an algorithm for fiducial-less lung tracking, has been developed for peripheral, radio-dense lung tumors with diameters >15 mm; and (e) a sequential optimization planning process incorporates a more flexible approach to optimize the multiple, complex treatment planning criteria used today. The clinical efficacy of CyberKnife radiosurgery for brain/head lesions such as metastases, arteriovenous malformations, acoustic neuromas and meningiomas is well established. Since there is no need for skeletal fixation with the CyberKnife, radiosurgery can be applied to targets beyond the brain, and the technology has been extensively used for stereotactic body radiotherapy, treating targets in many anatomic sites. Currently, clinical studies have been completed or are ongoing for common malignancies including tumors involving the spine, lung, pancreas, liver and prostate.

Copyright © 2011 S. Karger AG, Basel

The CyberKnife System

The CyberKnife was developed in the 1990s at Stanford, where 2 systems are now operational. Last year over 700 patients were treated, and the Stanford clinic now has cumulative experience with over 5,000 patients. The CyberKnife (Accuray Inc., Sunnyvale, Calif., USA) consists of a 6-MV, flattening-filter free linac mounted on an industrial robot (Kuka, Augsburg, Germany) [1]. The patient couch may be conventional or robotic. An X-ray imaging system, consisting of 2 cameras in the ceiling directed obliquely toward amorphous silicon detectors integrated into the floor, is installed in the treatment room. This imaging system is used for 2D-3D image registration [2–5] of the live images to a digitally reconstructed radiograph, which determines the patient position in near real-time and is used to correct for patient motion during treatment. The accuracy of the system is defined by the end-to-end test. This test takes a phantom through the complete treatment planning and delivery process, comparing the location of the delivered 70% isodose with the planned dose. In our experience and in the literature, the observed accuracy ranges from 0.4 to 0.7 mm [3].

A design advantage of the CyberKnife system is its method of active image guidance during treatment delivery. The system is capable of imaging before every treatment beam. However, as several authors have shown [6–9], imaging can be skipped for a small number of beams (e.g. every 3–5 beams) to reduce imaging dose without compromising dose delivery accuracy. Generally, imaging every third to fifth beam or every 20 to 60 s, depending on the compliance of the patient, is sufficient. Typical treatment times are about 20–30 min, and 1–5 fractions are delivered in each treatment course.

It is often thought that the CyberKnife is a stereotactic system and can treat only small tumors; however, the system is able to treat quite large tumors (up to 2,500 cm^3) if they are appropriately planned. The homogeneity of the treatment field is commonly cited in terms of prescribing to the 80% isodose line. Although many physicians prescribe in this manner, it is not a necessity. The treatment can be planned with whatever level of homogeneity requested, as this is a matter of prioritizing the constraints in the plan. This provides operative flexibility in the dosimetry that can be tailored to the clinical intent. For instance, it is possible to create plans mimicking Gamma Knife treatment by using a 50% prescription isodose line resulting in large inhomogeneity. Another example of purposely utilizing inhomogeneous plans is to replicate the dose distribution of an HDR treatment [10]. On the other hand, more homogeneous plans (with only 10–13% inhomogeneity) can be designed for prostate treatments equally well [11].

The CyberKnife represents a major paradigm shift from traditional stereotactic radiosurgery (SRS) technology. The prior SRS delivery systems have employed a fixed gantry with isocentric beam delivery, whereby the radiation dose is given relative to a central point of focus. Dose conformity is achieved by 'sphere packing', a treatment planning technique that creates multiple spherical or ellipsoid dose distributions. The CyberKnife uses an agile robot that is capable of delivering dose at any point within the target. This nonisocentric beam delivery allows optimal dose conformity without compromising dose homogeneity.

Additionally, since there is no need for skeletal fixation with the CyberKnife, radiosurgery can be applied to targets beyond the brain. Currently, clinical studies

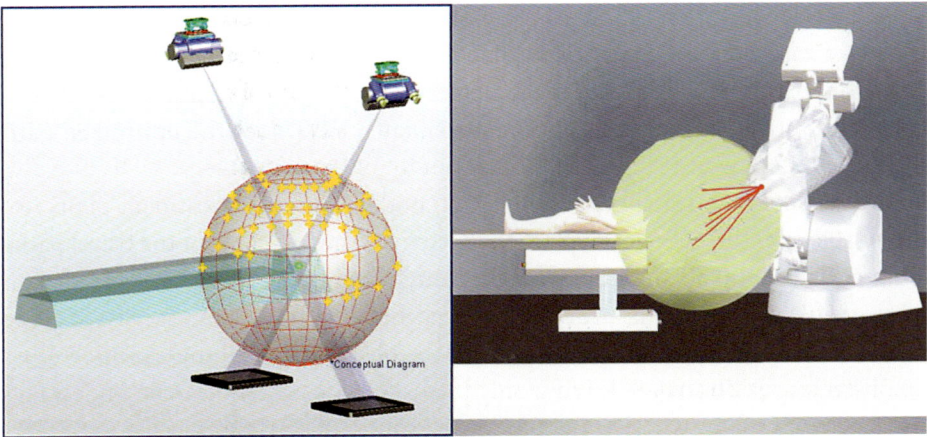

Fig. 1. Left side: treatment nodes located around the patient. Right side: 12 beam directions originating from a node. Figure courtesy of Accuray Inc.

have been completed or are ongoing for common malignancies including tumors involving the spine, lung, pancreas, liver and prostate. In this chapter we review the current roles and future expectations of the CyberKnife system, and provide practical guidelines for its clinical use.

Advances in CyberKnife Treatment Planning and Dose Delivery

Recent developments in hardware and software have significantly enhanced the CyberKnife's functionality from the basic system described above. We will highlight some of these new technical developments and their useful applications for treatment.

Optimized Path Traversal

The CyberKnife was the first robotic device in which a human was permitted to be present within a robot workspace. As safety precautions, the FDA implemented regulations limiting the speed of the robot motion and requiring its travel to occur only along a set of predefined paths, stopping at treatment 'nodes' (fig. 1). At each stopping position, the robot could change the linac beam angle within limits to create 12 beam directions at each node. In the initial design, the robot sequentially travelled to all nodes whether or not treatment monitor units were assigned to them by the treatment plan, which added to the overall treatment time.

After a decade of experience with patient safety and operational reliability of the Kuka robot, federal regulations have been updated. Optimized path traversal is now approved, which reduces the total treatment pathway and improves delivery time while still maintaining safety requirements. How does the optimized path traversal system work?

A complete path set contains about 120 nodes, but most treatment plans use only a subset of them. For each plan, the system creates a matrix of the travel paths from each node to all subsequent nodes being used in the plan. There is also an 'obstacle file' that contains all areas in space which the robot cannot cross, e.g. the couch or the patient safety zone. Based on the obstacle file, all routes which cross an obstacle are eliminated, leaving only the list of safe routes between nodes. Figure 2 shows a robot path designed to move around obstacles projected onto a 2D surface. For instance, if nodes 7 and 8 have no beams with dose, the robot may move directly from node 6 to 9, but only if a safe route exists. Otherwise, a node cannot be skipped, and nodes may be included in the path only for safety reasons. Overall, the robot now traverses the shortest safe distance between nodes with dose. In addition, if one needs to interrupt treatment for any reason, treatment can resume at the point of interruption by skipping nodes already used for treatment and moving to the starting point of the make-up treatment. This optimized path traversal process significantly reduces robot motion time, resulting in reductions of overall treatment times of at least 5–10 min, depending on the level of pre-existing planning efficiency.

Monte Carlo Dose Calculation Algorithm

The ability of algorithms to accurately model dose perturbations in regions of tissue inhomogeneity has been a long-standing problem in medical physics. Simple dose calculation models such as the ray-tracing algorithm, which was initially the only algorithm implemented in the CyberKnife planning system, were known to work very well in homogeneous areas of the body such as the brain. It was also known that such a pencil beam-based algorithm, which uses a simplistic density correction model for inhomogeneities, does not model in vivo dose distributions very well in inhomogeneous areas such as the lung. Further, these inaccuracies become increasingly pronounced in small lesions and treatment fields as used in stereotactic work. Not unique to the CyberKnife, these issues exist for all treatment planning systems for stereotactic delivery that use these older dose calculation algorithms.

To improve the accuracy of dose calculation in CyberKnife planning/delivery, Monte Carlo (MC) algorithms have been available for the fixed collimators since 2008, and for the IRIS collimator since 2009. The MC calculation is based on a

Fig. 2. Concept of optimized path traversal. A robot path designed to travel from node to node around obstacles (shaded areas) is projected onto a 2D space. Safe paths that skip nodes (green arrow) and unsafe paths (red arrow) are mapped. See online supplementary material.

dual-source model of the CyberKnife photon beam using the EGS4/BEAM MC code [12, 13].

1. To create the MC model, 3 beam parameters need to be measured: (a) the in-air output factor for an open collimator, (b) the open field profile and (c) the percent depth dose for the 60-mm collimator.

2. The linac source model is then created from the data. Cone correction factors and energy correction factors are used as parameters to iteratively fit the MC-generated tissue-phantom ratio and off-center ratio to the measured beam data.

3. After those fits are completed, the MC-based output factors are calculated and compared with measured output factors.

4. Clinical commissioning of the MC model is completed by verifying the accuracy of the delivered dose distribution by measurement in an inhomogeneous phantom [14, 15].

In the clinical application, the first plan remains always a pencil beam calculation. The user then can recalculate the pencil beam dose distribution with MC leaving the initial beam directions and monitor units/beam constant. Depending on the desired dose uncertainty, this recalculation will take 5–20 min. If the resulting dose distribution differs significantly from the pencil beam calculation, the user can either change the prescription isodose line or reoptimize the plan using MC. In the current clinical implementation, the first step is to calculate a ray-tracing plan, followed by a recalculation with MC to assess the difference between both plans. The planner then has the option to either use this recalculated plan and adjust the dose prescription accordingly, or reoptimize the plan with MC. Reoptimization is a good option if the beam weighting can be changed so that more beams enter from more homogeneous directions of the patient's body. In the case of a lung lesion in the middle of the lung, where dose differences originate from the change of build-up dose and scatter dose at the interface between the tissues, reoptimization with MC will not result in significant improvement. The physical situation in this case will require a lower prescription isodose line, i.e. larger inhomogeneity to cover the target adequately. Wilcox et al. [16] have shown the ratio of maximum dose in a lung tumor to change between 1.0 (no change) and 1.7 (a 70% change). The exact change is highly dependent on tumor location and size and has to be assessed on a patient-by-patient basis. Current clinical studies such as RTOG 0613 incorporate improved methods for inhomogeneity correction in the dose calculation requirements and will provide the data to adjust dose prescriptions based on older studies such as RTOG 0236 [17].

Other sites often showing inhomogeneity are the thoracic spine (because of proximity to lung tissue), nasopharynx and embolized AVM lesions. By default, these areas and other potential areas of concern should be recalculated with MC to assess how much the inhomogeneities influence the dose distributions. Especially for SRS, one should consider a shift of the isodose line due to the accuracy of the dose calculation on an equivalent level as the spatial dose delivery accuracy.

Before any new dose calculation algorithm such as MC is clinically implemented, the accuracy of the dose calculation must be verified clinically by measurement in a phantom. Since the algorithm requires input parameters that are based on measurements and a model that is adjusted during the commissioning process to match the beam data of a specific machine, each institution must repeat the quality assurance process of dose calculation verification measurements. Published data have shown a high degree of accuracy when using a well-implemented MC dose calculation algorithm [14, 15].

IRIS Collimator

Since the early development of the CyberKnife, the option of adding a mini-multileaf collimator to it has been discussed (though peer-reviewed studies on the advantage of a multileaf have not been published yet). The use of multiple collimators can reduce the monitor units and time needed to deliver a plan [18]. The IRIS collimator [19] was designed to maximize the utilization of multiple field sizes in CyberKnife radiosurgery (see online supplementary video).

The IRIS collimator has 2 collimator banks with 6 leaves each, rotated by 30 degrees relative to each other, thereby creating a near-circular field shape. The apertures are adjustable for each treatment beam from a 60- to 5-mm beam diameter with an aperture reproducibility of ±0.1 mm. In the current implementation, the beam diameters are incremental to match the fixed collimator set. In clinical practice, the IRIS collimator reduces the monitor units required, decreases treatment times and improves conformality and homogeneity of treatment plans.

Sequential Optimization

To take full advantage of the IRIS collimator, a new treatment planning algorithm called 'sequential optimization' (or 'stepwise optimization') was developed [20]. It incorporates a more flexible approach to optimize the multiple, complex treatment planning criteria used today. For example, the priority and tolerance levels on criteria such as conformality, target coverage, organ-at-risk sparing and homogeneity goals can vary considerably between patients based on their disease site, clinical history and other variables. The stepwise optimization process of sequential planning mimics closely the decision making process of the physician when designing a clinical treatment plan.

Contrary to a dose-volume histogram-based planning algorithm, where all criteria are optimized at once and priorities can only be set by relative weights, sequential optimization optimizes 1 criterion at a time in the order of priority, which the user has defined. In addition, the solution of each prior step is maintained, or relaxed by a user-defined value, when the next step is optimized. This method always maintains a feasible solution. The plan quality with respect to one criterion is never impaired in subsequent steps. The idea is to represent every clinical planning criterion as a constraint in an objective function. This method of sequential optimization lends itself to scripting: if a certain subgroup of patients is always planned with the same criteria, dosimetric variability due to differing level of treatment planning skills can be largely eliminated by scripting the plan optimization. It is also possible to quickly tailor the plan to preferences of individual physicians or to specific aspects of patients' histories (e.g. irradiated vs. unirradiated spinal segments, which would have different objectives for the spinal cord).

Figure 3 shows an example of a sequential optimization script for a spine tumor abutting the spinal cord. In step 1, the tumor prescription dose of 24 Gy is set. The relaxation value of 10 Gy is larger than typical, but in this case, where the cord is immediately adjacent to the tumor, some underdosing of the tumor has to be accepted to maintain cord tolerance. Step 2 defines the maximum cord tolerance, 14 Gy, which cannot be compromised (hence the tolerance value of 0 Gy). In step 3, the tumor should be covered as homogeneously as possible. Step 4 uses an (asymmetric) tuning shell to optimize conformality. Right kidney sparing is achieved in step 5. Finally, step 6 optimizes (minimizes) the monitor units; the goal and relaxation values are nominal for this step.

XSight Lung Tracking

Regardless of the technology, direct tracking of lung tumors requires implanted fiducials as surrogates to mark the position of radiographically low-contrast le-

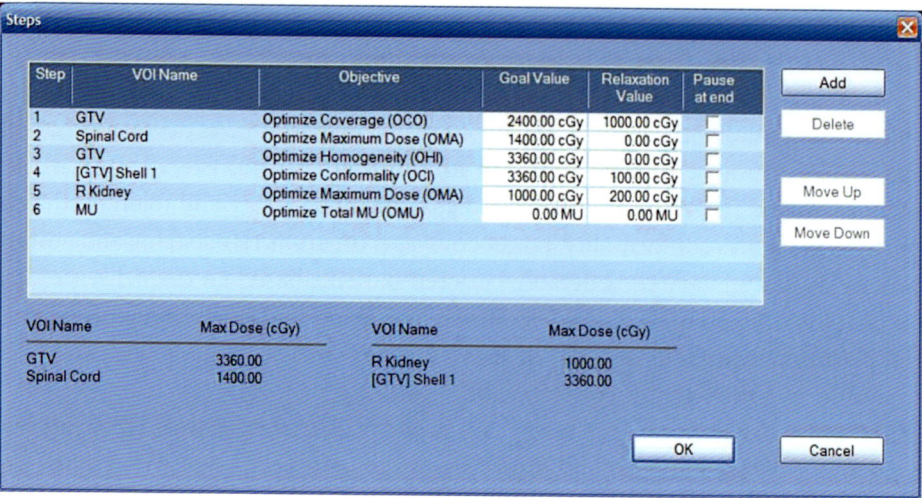

Fig. 3. Sample sequential optimization script.

sions. While the implant procedure is usually safe, with only minor and uncommon complications [21–23], it requires an investment in time, additional risk and added cost. Implanting fiducials is more challenging for more peripheral tumors. Can these markers be eliminated for some cases? XSight lung tracking [24], which is an algorithm for fiducial-less lung tracking, has been developed for peripheral, radio-dense lung tumors with diameters >15 mm.

For XSight lung tracking, the tumor GTV is contoured for treatment planning. This PTV contour is then used by the tracking algorithm to create a matching square in the oblique imaging projection bounding the tumor. The spine is segmented out to improve tracking accuracy. During the treatment, a pattern-similarity matching algorithm [25] is used to identify the tumor in the live X-ray images and compare the tumor position with the digitally reconstructed radiographs to localize it. This localization information is then processed in Synchrony respiratory tracking [26] in the same way as fiducial-based localization.

Current Clinical Applications of CyberKnife

Brain, Base of Skull Tumors

The clinical efficacy of radiosurgery for brain lesions such as metastases, arteriovenous malformations, acoustic neuromas and meningiomas is well established [27–31], and the CyberKnife can be used effectively for these lesions, similarly to other radiosurgical machines. Additional flexibility of the CyberKnife platform has allowed for: (a) efficient treatment strategies employing fractionation with up to 5 fractions, which may benefit lesions adjacent to sensitive critical tissues such as cranial nerves or the spinal cord, and (b) treatment of more generous target sizes such as large meningiomas and brain metastatic resection cavities [32, 33].

In traditional single fraction radiosurgery, tumors that extend to within a few millimeters of the anterior optic nerve structures are generally not considered ideal candidates for therapy. However, using the CyberKnife, our group has treated such perioptic tumors, most commonly with doses of 18–25 Gy in up to 5 fractions. Our studies confirm visual preservation in 94% of subjects at a mean follow-up of 49 months [13].

In another application, we are testing the hypothesis that greater fractionation may increase hearing preservation after acoustic neuroma radiosurgery. The potential role of limited fractionation in achieving this goal is debated, and the ability to generate highly conformal and homogeneous dose distributions using the CyberKnife makes it feasible to explore this hypothesis. Using CyberKnife radiosurgery delivering a total dose of 18–21 Gy in 3 fractions, our treatment series of 61 patients now has at least 3 years of follow-up: 74% of patients have maintained serviceable hearing, no patient developed a new trigeminal nerve injury and local control was 98% with only 1 patient demonstrating progression [34].

In some cases, fractionation allows treatment of somewhat larger volumes than traditional single fraction radiosurgery. As a proof of concept, our group has treated patients following surgical resection of brain metastases with radiosurgery to the resection cavity, despite the larger treatment volumes than conventionally accepted for radiosurgery. Our recent series demonstrated similar rates of local tumor control within the resection site, as compared with whole brain radiotherapy but with more limited brain exposure. Our group is now exploring dose escalation for brain tumors or their resection cavities if 2–5 cm in size, to test the optimal dose for this strategy in a prospective trial; the beginning dose levels are 24–27 Gy in 3 fractions.

Spinal Tumors

Another growing application of radiosurgery is for spinal metastases and primary spine tumors. Many of the early investigations of image-guided spinal radiosurgery were performed using the CyberKnife. This includes one of the largest, a single institution series of 500 patients with spinal metastases reported by Gerszten et al [35]. Overall their rate of long-term pain improvement was 86% and clinical improvement was 84%. As other image-guided technologies begin to report similar rates of efficacy, several prospective trials are currently underway to compare pain control rates achieved by radiosurgery with other palliative approaches.

Lung Cancer

Owing to its frameless image-guidance platform, the CyberKnife has been most extensively used as a technology for stereotactic body radiotherapy (SBRT), treating targets within the spine, lung, abdomen and pelvis. With lung cancer representing one of the most common cancers, and emerging clinical data confirming that SBRT of primary early stage lung tumors is superior to the historical results of conventional radiotherapy, lung cancer is one of the most rapidly growing indications for CyberKnife SBRT.

While surgery remains the standard of care for operable patients, SBRT has emerged as a preferable treatment for medically inoperable patients. RTOG 0236 is a multicenter prospective phase II study of SBRT for peripherally located, non-metastatic stage T1–T3 lung tumors (chest wall invasion only, ≤5 cm in size) in medically inoperable patients. The trial specifies 60 Gy delivered in 3 fractions (equivalent to 54 Gy in 3 fractions when using more accurate dose calculation algorithms). Among 55 evaluable patients enrolled, the 3-year actuarial local control rate was 97.6% at a median follow-up time of 34.4 months. There have been 2 regional relapses, and distant metastasis has been the primary mode of progression in 11 patients [36]. This trial has established the general clinical use of SBRT for such patients. Currently, typical fractionation schedules range from single fraction treatments of approximately 30 Gy for small T1 lesions to multifraction regimens of 54 Gy in 3 fractions (established based on RTOG 0236) for peripheral lesions. Ongoing investigations seek to determine the optimal schedule for centrally located tumors. Given the promising results seen in medically inoperable patients, international randomized trials have been launched to compare surgery with SBRT for medically operable patients including a CyberKnife radiosurgery study (STARS trial) and the ROSEL trial [37].

GI, GU and Other Tumors

Clinical studies using CyberKnife have also demonstrated clinical feasibility and safety for pancreatic and liver tumors [38–41]. Additional trials are required to determine the optimal treatment schedules. Reports of CyberKnife for prostate cancer have begun to emerge that build upon 2 different basic study rationales. The first is the assumption that prostate cancer has a unique biology that favors large doses per fraction, and SBRT regimens have been devised and tested with encouraging early results [42–44]. Another approach simply mimics the dosimetry of high dose rate brachytherapy and tests the feasibility of using CyberKnife noninvasively to deliver similar dosimetry [10, 44]. Two ongoing clinical trials are studying each of these strategies using the CyberKnife radiosurgery system.

The Future of CyberKnife

Image Guidance/Tracking

What are the new directions and opportunities for development of the CyberKnife system? The CyberKnife already has an excellent localization system, which is superior to many other approaches because of its regular imaging and tracking during treatment, not just for patient setup before treatment. Also, the 2D-3D tracking algorithms [2, 4, 5, 24] for correlating orthogonal X-ray pairs to a 3D CT study clearly have advantages: 2D-3D registration is fast, uses a low imaging dose and is excellent for localizing bone structures.

The system does not provide volumetric imaging. However, in hypofractionated treatments, there is less need for adaptive planning approaches because the tumor and other anatomy will not change much during the short courses of therapy. In the CNS and upper head-and-neck regions (e.g. nasopharynx), tumors and organs at risk are fixed relative to skeletal structures and not deforming, changing significantly or moving during the course of a SRS treatment. Why change the system?

Alternative approaches for the localization process could be important contributions to its targeting reliability and range of use. First, volumetric information would be very valuable for localizing soft tissue structures. Second, localization using nonionizing methods, including optical, ultrasound or radiofrequency beacon systems, would be desirable. Third, localization methods permitting prone treatments would expand the usability of the system.

One example in which 3D nonionizing tracking options could improve the system clinically is in respiratory motion tracking (Synchrony). The localization process for Synchrony uses a hybrid system with beacons on the skin monitoring overall respiratory motion and X-rays monitoring tumor targets (either using fiducials as surrogate markers or soft tissue tracking). To achieve accuracy of the treatment delivery, there must be accuracy of the correlation model between skin and tumor throughout all phases of respiration. But in some patients, a good correlation model is difficult to establish; the root cause for this has not been established yet. Irregularity of the respiratory motion itself, although often said to cause tracking inaccuracy, has not been shown to cause significant tracking inaccuracies [26]. Instead, the crux of the problem is the correlation (or lack of it) between the skin and tumor motion, which may change throughout treatment.

This issue is not unique to the CyberKnife system. Of note, we have learned that 4D CT is not necessarily a good predictor of motion at the time of treatment [45], which also indicates that the 4D CT may also not be a good predictor of the correlation model. In some cases, there can be significant differences which could be major concerns, especially for gated forms of treatment delivery.

In the clinical process of localization for Synchrony, an ideal design would involve some form of direct tumor tracking using nonionizing modalities. Possible approaches include: electromagnetic methods (such as the Calypso system using implanted beacons), ultrasound for some tumor sites, optical surface matching in combination with deformable registration or MR. One might even consider an approach using electromyography to use the nerve signal to the diaphragm, instead of the motion of the abdomen, as the surrogate marker for respiration.

New Clinical Applications

Recent publications report active research to expand the uses of noninvasive, frameless SRS systems such as the CyberKnife beyond cancer therapy applications [46]. Many 'minimally invasive' procedures such as radiofrequency ablation, deep-brain stimulation and others are being used for several benign conditions, but these procedures still carry risks of anesthesia complication and infection [47–49]. These are not shared by SRS, which is entirely noninvasive and could be studied as a potential option in these situations, especially for patients who would otherwise not be candidates for treatment.

References

1 Adler JR Jr, Chang SD, Murphy MJ, Doty J, Geis P, Hancock SL: The Cyberknife: a frameless robotic system for radiosurgery. Stereotact Funct Neurosurg 1997;69:124–128.
2 Fu D, Kuduvalli G: A fast, accurate, and automatic 2D-3D image registration for image-guided cranial radiosurgery. Med Phys 2008;35:2180–2194.
3 Ho AK, Fu D, Cotrutz C, et al: A study of the accuracy of cyberknife spinal radiosurgery using skeletal structure tracking. Neurosurgery 2007; 60(2 Suppl 1):ONS147–ONS156, discussion ONS156.
4 Mu Z, Fu D, Kuduvally G: Multiple fiducial identification using the hidden Markov model in image guided radiosurgery. Conference on Computer Vision and Pattern Recognition Workshop, New York, N.Y., 2006.
5 Fu D, Kuduvalli G, Maurer CJ, Allision J, Adler J: 3D target localization using 2D local displacements of skeletal structures in orthogonal X-ray images for image-guided spinal radiosurgery. Int J CARS 2006;1(Suppl 1):198–200.
6 Murphy MJ: Intrafraction geometric uncertainties in frameless image-guided radiosurgery. Int J Radiat Oncol Biol Phys 2009;73:1364–1368.
7 Ma L, Sahgal A, Hossain S, et al: Nonrandom intrafraction target motions and general strategy for correction of spine stereotactic body radiotherapy. Int J Radiat Oncol Biol Phys 2009;75: 1261–1265.
8 Hoogeman MS, Nuyttens JJ, Levendag PC, Heijmen BJ: Time dependence of intrafraction patient motion assessed by repeat stereoscopic imaging. Int J Radiat Oncol Biol Phys 2008;70: 609–618.
9 Murphy MJ, Chang SD, Gibbs IC, et al: Patterns of patient movement during frameless image-guided radiosurgery. Int J Radiat Oncol Biol Phys 2003;55:1400–1408.
10 Fuller DB, Naitoh J, Lee C, Hardy S, Jin H: Virtual HDR CyberKnife treatment for localized prostatic carcinoma: dosimetry comparison with HDR brachytherapy and preliminary clinical observations. Int J Radiat Oncol Biol Phys 2008;70: 1588–1597.

11 King CR, Brooks JD, Gill H, Pawlicki T, Cotrutz C, Presti JC Jr: Stereotactic body radiotherapy for localized prostate cancer: interim results of a prospective phase II clinical trial. Int J Radiat Oncol Biol Phys 2009;73:1043–1048.
12 Deng J, Guerrero T, Ma CM, Nath R: Modelling 6 MV photon beams of a stereotactic radiosurgery system for Monte Carlo treatment planning. Phys Med Biol 2004;49:1689–1704.
13 Deng J, Ma CM, Hai J, Nath R: Commissioning 6 MV photon beams of a stereotactic radiosurgery system for Monte Carlo treatment planning. Med Phys 2003;30:3124–3134.
14 Wilcox EE, Daskalov GM: Accuracy of dose measurements and calculations within and beyond heterogeneous tissues for 6 MV photon fields smaller than 4 cm produced by Cyberknife. Med Phys 2008;35:2259–2266.
15 Sharma SC, Ott JT, Williams JB, Dickow D: Clinical implications of adopting Monte Carlo treatment planning for CyberKnife. J Appl Clin Med Phys 2010;11:3142.
16 Wilcox EE, Daskalov GM, Lincoln H, Shumway RC, Kaplan BM, Colasanto JM: Comparison of planned dose distributions calculated by Monte Carlo and Ray-Trace algorithms for the treatment of lung tumors with CyberKnife: a preliminary study in 33 patients. Int J Radiat Oncol Biol Phys 2010;77:277–284.
17 Xiao Y, Papiez L, Paulus R, et al: Dosimetric evaluation of heterogeneity corrections for RTOG 0236: stereotactic body radiotherapy of inoperable stage I-II non-small-cell lung cancer. Int J Radiat Oncol Biol Phys 2009;73:1235–1242.
18 Poll JJ, Hoogeman MS, Prevost JB, Nuyttens JJ, Levendag PC, Heijmen BJ: Reducing monitor units for robotic radiosurgery by optimized use of multiple collimators. Med Phys 2008;35:2294–2299.
19 Echner GG, Kilby W, Lee M, et al: The design, physical properties and clinical utility of an iris collimator for robotic radiosurgery. Phys Med Biol 2009;54:5359–5380.
20 Schlaefer A, Schweikard A. Stepwise multi-criteria optimization for robotic radiosurgery. Med Phys 2008;35:2094–2103.
21 Kothary N, Heit JJ, Louie JD, et al: Safety and efficacy of percutaneous fiducial marker implantation for image-guided radiation therapy. J Vasc Interv Radiol 2009;20:235–239.
22 Prevost JB, Nuyttens JJ, Hoogeman MS, Poll JJ, van Dijk LC, Pattynama PM: Endovascular coils as lung tumour markers in real-time tumour tracking stereotactic radiotherapy: preliminary results. Eur Radiol 2008;18:1569–1576.
23 Reichner CC, Collins BT, Gagnon GJ, Malik S, Jamis-Dow C, Anderson ED: The placement of gold fiducials for CyberKnife stereotactic radiosurgery using a modified transbronchial needle aspiration technique. J Bronchol 2005;12:193–195.
24 Fu D, Kahn R, Wang B, et al: Xsight Lung Tracking System: A Fiducial-Less Method for Respiratory Motion Tracking; in Urschel HC, Kresl JJ, Luketich JD, Papiez L, Timmerman RD (eds): Robotic Radiosurgery. Treating Tumors That Move with Respiration. Berlin, Springer, 2007, pp 265–282.
25 Penney GP, Weese J, Little JA, Desmedt P, Hill DL, Hawkes DJ: A comparison of similarity measures for use in 2-D-3-D medical image registration. IEEE Trans Med Imaging 1998;17:586–595.
26 Wong KH, Dieterich S, Tang J, Cleary K: Quantitative measurement of CyberKnife robotic arm steering. Technol Cancer Res Treat 2007;6:589–594.
27 Colombo F, Pozza F, Chierego G, Casentini L, De Luca G, Francescon P: Linear accelerator radiosurgery of cerebral arteriovenous malformations: an update. Neurosurgery 1994;34:14–20, discussion 20–21.
28 Kondziolka D, Niranjan A, Lunsford LD, Flickinger JC: Stereotactic radiosurgery for meningiomas. Neurosurg Clin N Am 1999;10:317–325.
29 Lunsford LD, Flickinger J, Lindner G, Maitz A: Stereotactic radiosurgery of the brain using the first United States 201 cobalt-60 source gamma knife. Neurosurgery 1989;24:151–159.
30 Lunsford LD, Kondziolka D, Flickinger JC: Stereotactic radiosurgery for benign intracranial tumors. Clin Neurosurg 1993;40:475–497.
31 Sheehan JP, Sun MH, Kondziolka D, Flickinger J, Lunsford LD: Radiosurgery in patients with renal cell carcinoma metastasis to the brain: long-term outcomes and prognostic factors influencing survival and local tumor control. J Neurosurg 2003;98:342–349.
32 Tuniz F, Soltys SG, Choi CY, et al: Multisession cyberknife stereotactic radiosurgery of large, benign cranial base tumors: preliminary study. Neurosurgery 2009;65:898–907, discussion 907.
33 Soltys SG, Adler JR, Lipani JD, et al: Stereotactic radiosurgery of the postoperative resection cavity for brain metastases. Int J Radiat Oncol Biol Phys 2008;70:187–193.
34 Chang SD, Gibbs IC, Sakamoto GT, Lee E, Oyelese A, Adler JR Jr: Staged stereotactic irradiation for acoustic neuroma. Neurosurgery 2005;56:1254–1261, discussion 1261–1253.
35 Gerszten PC, Burton SA, Ozhasoglu C, Welch WC: Radiosurgery for spinal metastases: clinical experience in 500 cases from a single institution. Spine (Phila Pa 1976) 2007;32:193–199.

36 Timmerman R, Paulus R, Galvin J, et al: Stereotactic body radiation therapy for inoperable early stage lung cancer. JAMA 2010;303:1070–1076.
37 Hurkmans CW, Cuijpers JP, Lagerwaard FJ, et al: Recommendations for implementing stereotactic radiotherapy in peripheral stage IA non-small cell lung cancer: report from the Quality Assurance Working Party of the randomised phase III ROSEL study. Radiat Oncol 2009;4:1.
38 Goodman KA, Wiegner EA, Maturen KE, et al: Dose-escalation study of single-fraction stereotactic body radiotherapy for liver malignancies. Int J Radiat Oncol Biol Phys 2010;78:486–493.
39 Koong AC, Christofferson E, Le QT, et al: Phase II study to assess the efficacy of conventionally fractionated radiotherapy followed by a stereotactic radiosurgery boost in patients with locally advanced pancreatic cancer. Int J Radiat Oncol Biol Phys 2005;63:320–323.
40 Koong AC, Le QT, Ho A, et al: Phase I study of stereotactic radiosurgery in patients with locally advanced pancreatic cancer. Int J Radiat Oncol Biol Phys 2004;58:1017–1021.
41 Schellenberg D, Goodman KA, Lee F, et al: Gemcitabine chemotherapy and single-fraction stereotactic body radiotherapy for locally advanced pancreatic cancer. Int J Radiat Oncol Biol Phys 2008;72:678–686.
42 King CR, Brooks JD, Gill H, Pawlicki T, Cotrutz C, Presti JC Jr: Stereotactic body radiotherapy for localized prostate cancer: interim results of a prospective phase II clinical trial. Int J Radiat Oncol Biol Phys 2009;73:1043–1048.
43 Friedland JL, Freeman DE, Masterson-McGary ME, Spellberg DM: Stereotactic body radiotherapy: an emerging treatment approach for localized prostate cancer. Technol Cancer Res Treat 2009; 8:387–392.
44 Katz AJ, Santoro M, Ashley R, Diblasio F, Witten M: Stereotactic body radiotherapy for organ-confined prostate cancer. BMC Urol 2010;10:1.
45 Minn AY, Schellenberg D, Maxim P, et al: Pancreatic tumor motion on a single planning 4D-CT does not correlate with intrafraction tumor motion during treatment. Am J Clin Oncol 2009;32: 364–368.
46 Lee M, Kalani MY, Cheshier S, Gibbs IC, Adler JR, Chang SD: Radiation therapy and CyberKnife radiosurgery in the management of craniopharyngiomas. Neurosurg Focus 2008;24:E4.
47 Xiaowu H, Xiufeng J, Xiaoping Z, et al: Risks of intracranial hemorrhage in patients with Parkinson's disease receiving deep brain stimulation and ablation. Parkinsonism Relat Disord 2010;16: 96–100.
48 Sixel-Döring F, Trenkwalder C, Kappus C, Hellwig D: Skin complications in deep brain stimulation for Parkinson's disease: frequency, time course, and risk factors. Acta Neurochir (Wien) 2010;152:195–200.
49 Fytagoridis A, Blomstedt P: Complications and side effects of deep brain stimulation in the posterior subthalamic area. Stereotact Funct Neurosurg 2010;88:88–93.

Sonja Dieterich, PhD
Department of Radiation Oncology, Stanford University Medical Center
875 Blake Wilbur Drive, Rm G101F
Stanford, CA 94305-5847 (USA)
Tel. +1 650 736 8380, Fax +1 650 498 5008, E-Mail sonja.dieterich@stanford.edu

II. IMRT and IGRT Clinical Treatment Programs

Image Guidance and the New Practice of Radiotherapy: What to Know and Use from a Decade of Investigation

John Kim[a] · John L. Meyer[b] · Laura A. Dawson[a]

[a]Department of Radiation Oncology, Princess Margaret Hospital, University of Toronto, Toronto, Ont., Canada; [b]Department of Radiation Oncology, Saint Francis Memorial Hospital, San Francisco, Calif., USA

Abstract

Over the past decade, fundamental advances in image-guided radiation therapy (IGRT) have been made that are now being implemented in clinical practice. Imaging technologies to direct and confirm beam accuracy at the time of radiotherapy delivery have been intensively researched and developed. More recently, these imaging data have been used to evaluate and even modify the daily dose delivery of intended treatment plans. The rationale for the use of IGRT, to improve tumor control while limiting normal tissue toxicity, is a universal goal in radiotherapy. Avoidance of unexpected under- or overdosing during treatment is the most important benefit of IGRT, and has led to its integration into the use of advanced radiotherapy planning/delivery technologies for many clinical applications. Evidence-based strategies to effectively use IGRT in the clinic are still emerging. The evolving role of IGRT and some proposed strategies to exploit its potential benefits in the clinic will be presented, emphasizing the perspective of the radiation clinician. Practical strategies will be proposed to exploit the potential benefits of IGRT technologies in the clinic.

Copyright © 2011 S. Karger AG, Basel

The Rationale for Image-Guided Radiation Therapy: Perspectives on the Current Practice of Radiotherapy

Fundamental advances in image-guided radiation therapy (IGRT) have been made that are now being implemented in clinical practice. Yet, many questions remain about how to use IGRT effectively in the clinical setting. Imaging technologies have been intensively researched and developed to direct and confirm beam accuracy at the time of radiotherapy delivery. More recently, these imaging data have been used to evaluate and even modify the daily dose delivery of intended treatment plans. Guidance technologies were rapidly incorporated into the hardware and/or software operations of the delivery units, presenting many opportunities for their daily use, but also resulting in many questions regarding their optimal integration into daily workflow procedures. Clinical outcome evaluations of the current IGRT technologies can determine new directions for basic technical and clinical investigation, and will influence the development of later generations of IGRT technology in very practical ways. In other chapters in this volume, many of the technical limitations and challenges currently facing IGRT usage are reviewed. Additional insights from the clinician's perspective will be presented here.

The Need for Image-Guided Radiation Therapy

From the development of 3-dimensional conformal radiation therapy (3DCRT) to intensity-modulated radiation therapy (IMRT), the consistent rationale for improving radiotherapy delivery has been to increase the therapeutic ratio. More focused dose deposition has allowed increasing the tumor dose and/or decreasing the normal tissue exposure. The technical advances have brought significant reductions in the doses delivered to critical normal tissues and have already resulted in demonstrable improvements in the quality of life of patients with head and neck and certain other cancers [1–6]. The goals of IGRT are the same, since it is used in combination with IMRT or 3DCRT to increase their reliability and accuracy. The intent of IGRT is to reduce uncertainties in the targeting and delivery of treatment, called *residual geometric uncertainties*, so that the differences between the actual delivered doses and the planned doses are negligible (or clinically insignificant) and the benefits of 3DCRT or IMRT are realized.

Imaging during treatment has revealed the extent of tumor motion and defined a central role for IGRT in the therapy delivery process. During fractionated radiotherapy, organ motion (due to breathing, peristalsis and other normal functions) and organ geometric or anatomic changes (due to tumor shrinkage or patient weight loss) may be substantial. IMRT and 3DCRT are *more* sensitive to such changes than 2D radiotherapy, with the potential clinical consequences of unintended lower doses to tumor targets *and* higher doses to adjacent normal tissues. Without image guidance, 3DCRT and IMRT could actually result in deterioration of therapeutic gain compared with less focused approaches. Avoidance of unexpected under- or overdosing during treatment is the most important benefit of effective IGRT strategies, and they have become essential to the use of advanced radiotherapy planning/delivery technologies for many clinical applications.

IGRT opens doors to a new practice of radiotherapy in several possible ways. (a) As discussed, when radiation can be highly conformed to target volumes, radiation doses to normal tissues can be reduced and tumor doses escalated, potentially improving tumor control rates and/or reducing toxicities. (b) The number of treatment fractions may be reduced as larger fractions can be more safely delivered. The most hypofractionated dose-dense form of radiotherapy is stereotactic body radiation therapy (SBRT). It can be implemented only with the most careful image guidance, since small errors in treatment may have significant clinical impact. While SBRT is limited in its applications, the principles of hypofractionation and dose escalation can be broadly applied. Eventually, treatment of all or most tumor sites may change in their dose/fractionation programs based on IGRT and related advanced technologies. (c) Hypofractionation protocols may decrease treatment costs, improve efficiencies and increase patient throughput in working clinics. Some patients who were previously unable because of the long duration of the older treatment courses, such as those living at great distances from the treatment center, may be able to receive radiotherapy. In summary, there are several strong rationales for using IGRT with advanced radiotherapy planning and delivery.

Radiotherapy Practice: Reducing Variability in Dose Delivery

The potential benefit of IGRT is less variation in the radiation doses actually delivered for a given dose prescription. This is an important factor for patient groups on treatment protocols who should be treated with similar doses and techniques. Reducing variability in delivered doses across a population should allow (a) improved interpretation of what doses are actually required for tumor control, (b) better understanding of the dose-volume relationships for toxicity development and (c) clarification of the benefits of radiation modifiers such as sensitizers. More uniform radiation delivery in the control and treatment groups of clinical trials will reduce the heterogeneity of clinical responses within each group, and help define the true differences in outcomes between the groups being studied.

Reducing variability in radiotherapy dose delivery may also improve outcomes. A parallel can be drawn to the experience from a large randomized head and neck cancer trial in which all radiotherapy plans were subjected to expert peer review. Major planning protocol deviations were associated with poorer tumor control outcomes [7]. These results demonstrated the importance of quality control and quality assurance review of treatment planning. But what if these same dose variations were a result of treatment *delivery* variations? IGRT facilitates treatment delivery quality control and quality assurance review. Hence, 'standards' of radiotherapy planning and delivery in clinical practice can now be described, implemented and reviewed for quality control at every step.

A further potential clinical benefit of IGRT, increased awareness of the actual doses delivered to both tumor and normal tissues, should result in a better understanding of normal tissue toxicity relationships and tumor control probabilities. Such relationships may be found to be different from those historically reported, once IGRT is fully integrated into clinical trials and their outcomes mature. With improved understandings of normal tissue dose/volume toxicity risk relationships, the benefits of IMRT will be better exploited by knowing which organs should be spared by what specific volume. Currently, the clinician is faced with making important decisions about normal tissue sparing or avoidance based on limited or incomplete knowledge of normal tissue partial volume tolerances.

Radiotherapy Practice: Using Image-Guided Radiation Therapy Information

IGRT is a dynamic process. Treatment monitoring and IGRT-based decision making require the development of new skill sets while retaining the fundamental principles of radiation oncology practice. Careful treatment requires a clear understanding of the treatment target, typically based on advanced imaging. IGRT involves imaging at planning, imaging at radiation delivery and imaging to monitor tumor response. With imaging data, patients are monitored with assessments that are *on-line* (immediately before the treatment beam is turned on), *real-time* (as the treatment beam is being delivered) or *off-line* (between treatments, often analyzing multiple images acquired in the treatment room after several fractions). Each assessment gives more information about the patient and, at each step, a clinical intervention may be triggered. This intervention may be as simple as repositioning the patient prior to radiation delivery that day. It is also possible that tumor shrinkage or growth, or new metastases may be detected. These are findings which could dramatically change the treatment plan or intent, and bring the overall goals of the treatment plan back into consideration. With the acquisition of this new information in the IGRT-era, what we *visualize* during treatment should influence how we *intervene* in terms of (a) planning target volume (PTV) margins; (b) radiation treatment plans including their replanning, dosimetry modification and dose prescription; and (c) the overall treatment goals. How this is accomplished will be different depending on the individual center and the patient population it treats.

Figure 1 summarizes the role of IGRT in the process of radiotherapy. The circle on the left represents the continuing cycle of a patient's medical evaluation and treatment. First, a patient is seen and a clinical objective is articulated: to control a tumor, reduce recurrence risk or palliate symptoms. A therapy is designed and the intervention initiated. During treatment, patients are monitored, generally with clinical assessments and diagnostic imaging. Sometimes, the focus of the monitoring is on normal tissue toxicity, while at other times it is on tumor response. IGRT

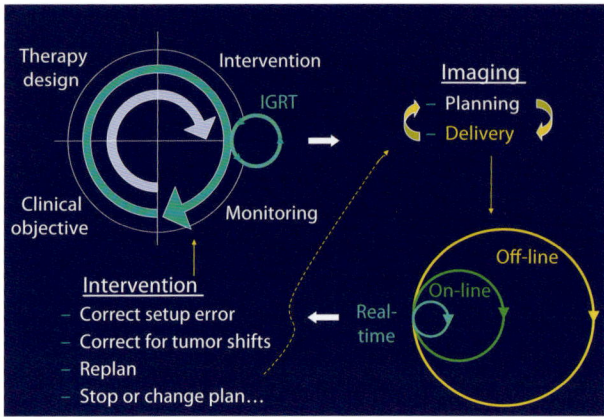

Fig. 1. IGRT in current radiotherapy.

is a central part of that monitoring process. While the IGRT imaging first evaluates accuracy to the plan, it may acquire information on tissue changes (tumor or normal tissue) that prompts redesign of the plan in an interactive way. Such evaluations may occur off-line, on-line or in real-time. Together, they may prompt revision of the intervention, which may be a technical change, plan change or even a basic change in the clinical goals. Effectively used, IGRT carries a central role in the radiotherapy and sometimes the broader cancer management of the patient.

Image-Guided Radiation Therapy Development

Although much has been learned about IGRT in the last decade, the concept of 'image guidance' is not new. Fluoroscopic and plain X-ray imaging were available and sometimes integrated into radiation treatment rooms decades ago. Figure 2 shows a schematic of a cobalt unit with kV imaging portal from Toronto in 1958. Although in-room kV imaging was possible at that time, it was not exploited, perhaps because it was less critical to the treatment programs which were technically simpler, far less conformal and treated to lower doses. As a consequence, they were less sensitive to geometric uncertainties. Also, image acquisition and review processes were not advanced or rapid enough to support the efficiencies and capacities needed in the clinic. Now, many years later, we have a plethora of IGRT technologies, and even more advanced treatment units integrating imaging and treatment in development. Perhaps we are at the stage where the technologies are surpassing our knowledge of how we should best use them to maximally benefit patients.

Table 1 summarizes some lessons learned while implementing and using IGRT. Over the past 10 years, there has been a dramatic change in IGRT technologies,

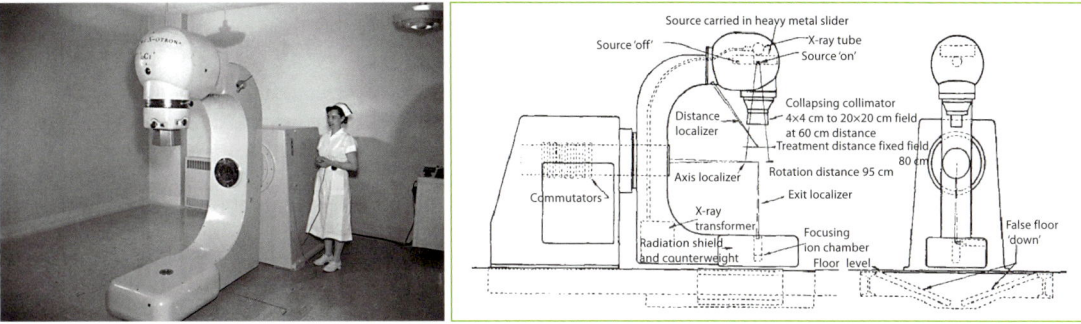

Fig. 2. IGRT at Princess Margaret Hospital/Ontario Cancer Institute in 1958. The cobalt source was housed within the head of the treatment unit. The treatment unit also housed an X-ray transformer and tube that enabled kV portal imaging. Reprinted with permission of the Department of Public Relations photographs and records, 1B.48. University Health Network Archives.

Table 1. General lessons learned from IGRT

---
(1) Increased awareness of setup error
(2) Confirmation of suspected types of setup error
(3) Increased awareness of soft tissue change
    a. shifts in targets and normal tissues relative to vertebral bodies
    b. deformation of internal anatomy
    c. different motion for targets and organs at risk
    d. interfraction,/intrafraction motion
(4) PTV/PRV modifications resulting from IGRT
    a. reduced (e.g. prostate)
    b. increased (e.g. gastric)
    c. variable (e.g. head and neck, gastric)
(5) Dosimetric impacts (still learning)
(6) Impacts on outcomes (still learning)
(7) Implementation of IGRT is more challenging than expected

PTV = Planning target volume; PRV = planning organ at risk volume, defined by the International Commission on Radiation Units and Measurements (ICRU) as a margin to account for motion and setup uncertainties of an organ at risk. PRV principles of uncertainty are similar to those for PTV.

---

and much has been learned about the effects of setup uncertainties for different patient populations. The type and magnitude of soft tissue changes that occur in tumors and normal tissues during a radiotherapy course are only beginning to be realized. Visualizing serial 3D volumetric imaging data on many patients helps increase awareness of the variability in geometric change that can occur, both patient-to-patient and within individual patients.

Challenges in Image-Guided Radiation Therapy

Target and organ motion occur in different ways, each presenting challenging areas for investigation using IGRT.

Interfraction Motion. A common observation from daily IGRT experience is that targeted soft tissues frequently shift in position relative to adjacent bone anatomy (sometimes referred to as a 'baseline shift') [8, 9].

Intrafraction Motion. More recently, there is increased awareness about shifts that can occur during the delivery of a radiation fraction.

Organ Deformation. Most corrective strategies address only linear displacements. However, there are other sources of geometric uncertainty including deformation of organs, rotations of tumors and organs at risk, and variability in motion and positions of normal tissues relative to the targets. These nonlinear uncertainties pose difficult problems for field placement, as current IGRT methodologies are not always able to account for such geometric changes to achieve the planned dosimetry.

Now that we have much more information about how organs move and change, ideal IGRT solutions are perhaps more challenging to clearly define than originally anticipated.

Prioritizing Image Guidance

Anatomic structures must be defined that prioritize the image guidance for individual cases because of the nonlinear uncertainties discussed above. In the initial selection of the IGRT strategy, the physician needs to define the structures that are to be used for image matching (image registration at the time of treatment). Practical questions must be answered, for example 'Should the target volume soft tissues or the vertebral bodies (as a surrogate for the spinal cord) be used for image guidance?' All target volumes and organs at risk should be identified prior to treatment delivery.

Planning Target Volume Expansions Based on Image-Guided Radiation Therapy Strategy

The selection of the IGRT strategy for each case will determine the residual uncertainties likely to occur across the therapy volumes. These need to be accounted for at the time of radiation therapy planning. The larger the therapy field, the larger this concern may be. For instance, comprehensive head and neck irradiation may involve large fields, but primary image matching may be based on one structure (e.g. the vertebral bodies as a surrogate for the spinal cord). In such cases, the target volumes distal from the vertebral bodies may need to have larger PTV margins than the target volumes directly adjacent to the vertebral bodies.

With increased knowledge of geometric uncertainties, a more informed PTV and PRV (planning organ at risk volume) margin design is possible. In general, we have not made comprehensive changes in PTV margins in our radiation oncology department at Princess Margaret Hospital (Toronto, Canada); however, there have been some small but significant changes. For example, prostate cancer PTV margins have been anisotropically reduced in some protocols. For gastric cancer, PTV margins and PRV margins have increased for some individual patients, due to the awareness of the variability in how the organs of the upper abdomen move.

For head and neck cancer, PTV margins are generally 5 mm; however, variable PTVs are now used. One question that has arisen in the IGRT era is whether one uniform PTV margin is relevant for all head and neck radiotherapy. For example, if IGRT is used focusing on the upper cervical spine as the region of interest, then the lower cervical spine and the lower target volumes may move more than the upper target volumes, requiring a larger PTV margin inferiorly (e.g. 7 mm). In contrast, for a small base of skull target, PTV margins may be reduced to 3 mm.

Accumulation of Actually Delivered Doses

Daily volumetric imaging provides the database for determining the actually delivered daily doses. How can this daily dose information be used? The dosimetric benefit of daily IGRT on guiding the remaining therapy course is an important issue that may have far-reaching importance but is poorly understood as yet. Deformable dose accumulation with frequent volumetric imaging can determine the impact of geometric change and even determine the need for continuing daily IGRT in individual cases. Unfortunately, the best way to accumulate doses across the deforming and changing anatomy is unclear and challenging. Deformable dose accumulation methods are investigational, and commercial software is not yet available for clinical use. Other methods to estimate the dosimetric impact of IGRT in populations of treated patients have been explored, and have demonstrated improved homogeneity of delivered doses and better concordance of delivered doses with those prescribed. The accumulation of actually delivered doses in manageable data forms opens many opportunities for improving IGRT outcomes.

Management of Organ Motion

Different organs move differently during respiration, and breathing motion amplitude changes of organs have now been quantified extensively. For instance, dif-

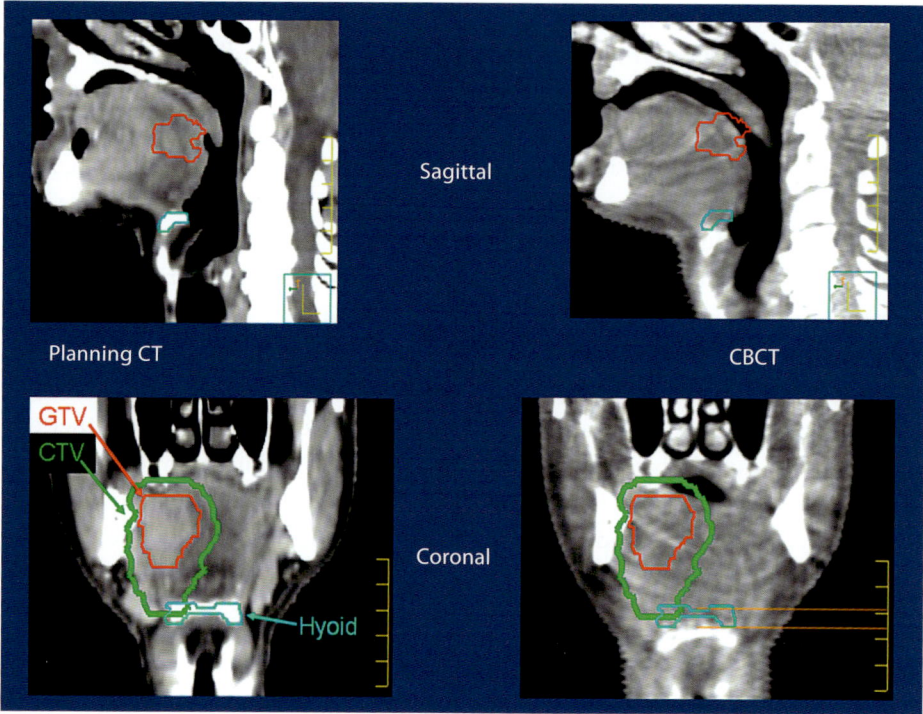

Fig. 3. Baseline shifts in head and neck soft tissues, following alignment of the vertebral body. The red contour is the GTV. The green contour is the CTV, shown from the planning CT overlaid onto the cone-beam CT. The blue contour is the hyoid bone.

ferent structures involved with the planning of adjuvant therapy for gastric cancer, either as avoidance structures or targets, vary in their breathing amplitudes. Typically, there is more motion in the splenic hilum and far less motion in the celiac axis. Some variability in motion has been observed over time, but the organ-to-organ variability has been much larger than the variability of the same organ on repeat measurements. If target volumes and avoidance structures move differently, IGRT solutions are challenging. Similar to amplitude variability, the baseline shifts also vary from organ-to-organ, day-to-day.

Interfraction Motion
Position shifts in soft tissues relative to vertebral bodies have been observed in the head and neck, thorax, upper abdomen and prostate. They tend to occur day-to-day, regardless of what immobilization devices or breathing motion management strategies are used. Figure 3 gives an example of baseline shifts for a patient with a base of tongue squamous cell carcinoma and bilateral lymphadenopathy. He was treated with a PTV margin of 5 mm and image guidance using C1–6 for matching

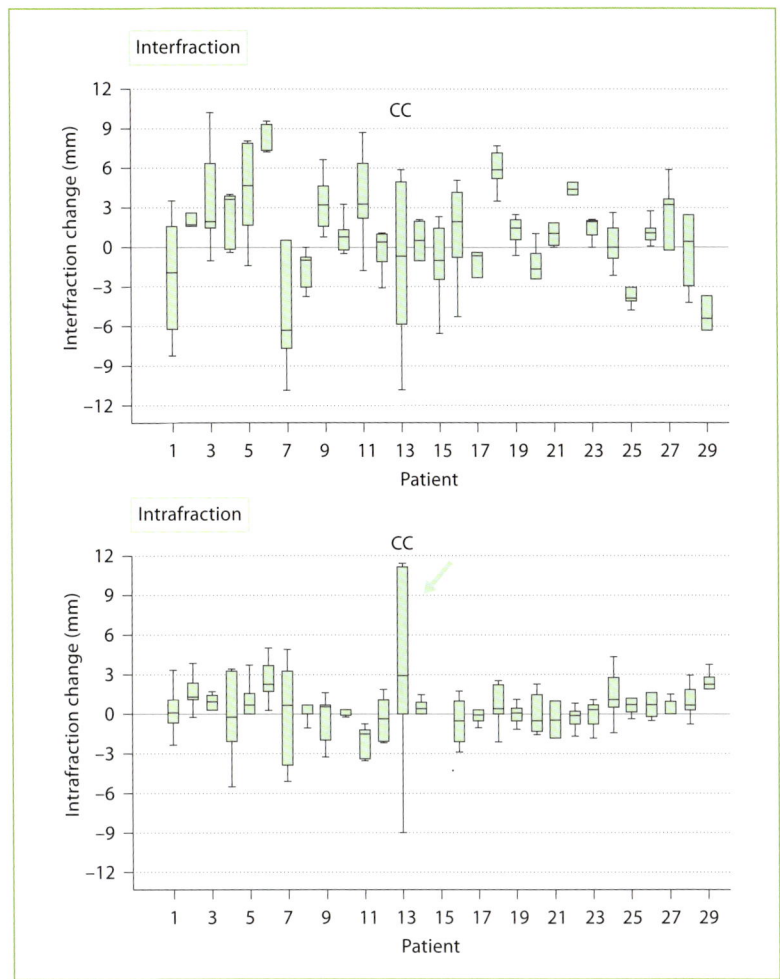

Fig. 4. Box plots of inter- and intrafraction changes in the exhale craniocaudal (CC) liver position, by patient. Interfraction variability in liver position is greater than intrafraction variability. Patients 1–15 were treated during free breathing without abdominal compression, and patients 16–29 were treated with abdominal compression. Lines represent ranges of liver position changes; boxes represent range from 25th to 75th percentile in liver position changes. Arrow identifies Patient 13 who had substantial pain and the largest intrafraction liver position change [9].

so that the spinal cord would be in the same spot every day. Following alignment to the vertebral body, image guidance showed changes in soft tissues of approximately 1 cm, relative to the vertebral bodies, resulting in target position changes over the first 3 fractions. To correct for this, the case was replanned with increased PTV margins around the CTV.

Figure 4 shows data from respiratory correlated kV cone-beam CTs (CBCTs) used to measure shifts in the liver position relative to adjacent vertebral bodies.

Patients were treated with free breathing (patients 1–15) or abdominal compression (patients 16–29). Systematic and random errors were calculated for the group of patients (systematic errors have been defined as displacements occurring at treatment preparation resulting in a displacement of the intended dose distribution relative to the CTV at treatment delivery). The exhale position of the liver can vary substantially from fraction to fraction. Systematic errors up to 5 mm occurred mostly in the cranial-caudal direction [9].

Intrafraction Motion

Organ motion may occur during treatment delivery. The actually delivered doses may differ from the expected to an unknown degree if intrafraction monitoring is not used and targets are only imaged prior to treatment. For example, figure 4 estimates the intrafraction shifts of the liver during abdominal treatments by evaluating the pre- and post-therapy positions of the organ. Fortunately, the intrafraction shifts were generally very small (<3 mm), though this would not have been known without post-treatment evaluation. The uncertainties, especially the systematic ones, were far less than interfraction shifts in liver position relative to vertebral bodies. Of note, one patient in severe pain from his liver metastases had larger intrafraction shifts, indicating the need for patient comfort.

There was not a substantial correlation of intrafraction shifts with treatment time in the figure 4 study, though the mean treatment time was under 12 min – much shorter than in many SBRT series. Interestingly, intrafraction shifts for lung SBRT at our institution were much larger than those observed for liver. The lung SBRT doses were substantially higher, up to 20 Gy per fraction, as were the treatment times. These correlated with intrafraction motion: the vector change in tumor, from the beginning to end of treatment, increased as the treatment time increased, with an apparent threshold of about 30 min.

Strategies to reduce the potential for significant intrafraction shifts include: keeping treatment times short and keeping patients comfortable to avoid voluntary movement. Monitoring intrafraction motion of targets or their surrogates can be performed with optical, electromagnetic and X-ray methods, which are discussed in other chapters in this volume.

Clinical Applications of Image-Guided Radiation Therapy

Prostate Cancer

Clinical applications of IGRT for prostate cancer began more than a decade ago, and many of the IGRT strategies now available were initially evaluated in prostate treatments. These approaches have included fiducial markers, MV imaging, kV

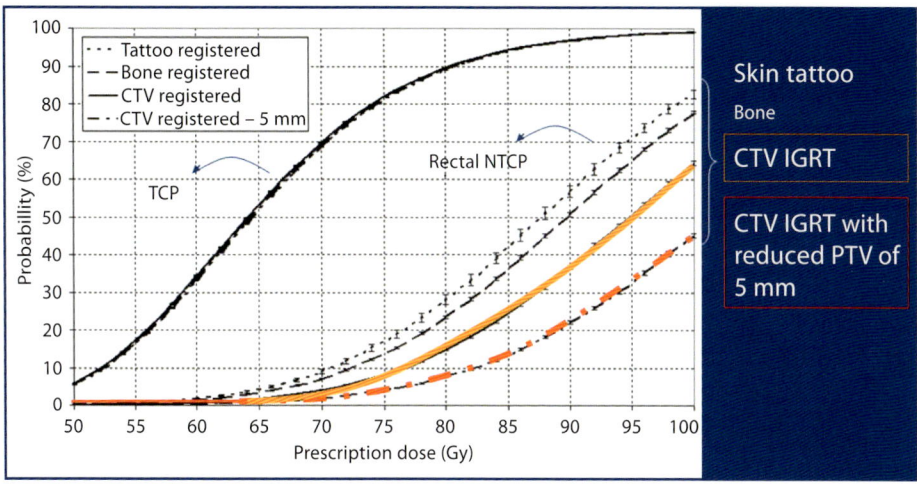

Fig. 5. Reduction of normal tissue toxicity with IGRT and reduced PTV margins in IGRT in prostate cancer radiation therapy. Modified from Song et al. [14].

and MV volumetric imaging, and continuous electromagnetic monitors (e.g. Calypso® System, Seattle, Wash., USA). The ability to perform automated matching of the prostate using kV CBCTs has improved as image quality has improved and clinical experience has developed. For instance, substantial amounts of rectal gas moving during a CBCT acquisition were found to cause artifacts and make an automated match with a planning CT impossible [10–13]. Now, an antiflatulence bowel routine is used during a course of radiation therapy for prostate treatment at our center. Despite the advances in automated matching of CBCTs, reproducibility of matching is generally improved with the use of fiducial markers, and many clinics still use fiducial markers with volumetric imaging for prostate cancer therapy when possible.

Do improved accuracy and precision from IGRT of prostate cancer lead to improved clinical outcomes? Song et al. [14] demonstrated from a modeling experiment in prostate cancer that the maximal reduction of normal tissue toxicity was observed if IGRT was used and if PTVs were reduced (fig. 5).

Several publications have reported that rectal expansion at the time of simulation is an important factor associated with increased risk of PSA relapse after radiotherapy. The first paper, from the M.D. Anderson Cancer Center, showed that rectal diameter was strongly correlated with the biochemical control rate [15]; no IGRT was used in these patients. Dr. Kupelian evaluated the Cleveland Clinic experience, where ultrasound image guidance was used to assist prostate alignment, and found that the adverse effect of rectal size was not as strong. Together, these studies suggest that a systematic error in planning (i.e. a full rectum at planning that is not expanded at treatment), can result in the posterior prostate being missed if there is no image guidance correction used [16].

A different clinical experience reveals the potential *negative* consequence of IGRT when PTV margins are reduced too aggressively [17]. In this work, 238 prostate cancer patients were treated with radiotherapy; image guidance was based on bone anatomy in most patients and on fiducial markers in 25 patients. The prostates were delineated on MRI, and PTV margins were 3 mm. Factors associated with worse outcomes were rectal size, prostate cancer risk group and lower treatment doses. In addition, to their surprise, the use of IGRT was also associated with an *increased* risk of PSA relapse. These were preliminary data, as only 25 patients had fiducial marker-based IGRT, but the poorer outcomes with IGRT were unexpected. The authors hypothesized that perhaps the change in the PTV margin was too much, despite the use of fiducial based IGRT. Also, fiducial based IGRT versus soft tissue IGRT may not account for some residual prostate deformation that may occur. This paper highlights the importance of evaluating clinical outcomes, as well as the dosimetric and geometric changes, when novel technologies are introduced.

Head and Neck Cancers

Substantial changes can occur during head and neck cancer radiotherapy: weight loss, neck flexion, head rotation, tissue deformations and bite block changes. The changes occurring over the duration of a therapy course are addressed below in the section on adaptive therapy. Day-to-day positioning and alignment can also pose significant challenges, despite the use of immobilization tools. These issues especially concern neck flexion relative to the base of skull.

Figure 6 shows a CBCT matched to a planning CT for an advanced nasopharyngeal cancer extending into the cavernous sinus. This is an example of a treatment plan prescribing uncommonly high doses close to critical normal tissues such as the optic chiasm. Although toxicities would be unlikely with accurate plan delivery, small but systematic shifts of even 2 mm could potentially lead to blindness or brainstem necrosis. As a result, the regions chosen for image matching (the clivus and clinoid) were as close as possible to the regions of tumor extension and potential normal tissue toxicity. A reduced PTV margin can be used here (only 3 mm). The limitation of choosing this quite small and superior matching region, compared with a larger or more inferior one, is that neck curvature changes can result in larger displacements of the spinal cord and lymph nodes in the lower neck. To account for this, these lower neck targets and normal tissues require larger expansions for PTV and for PRVs around the cord (5–7 mm).

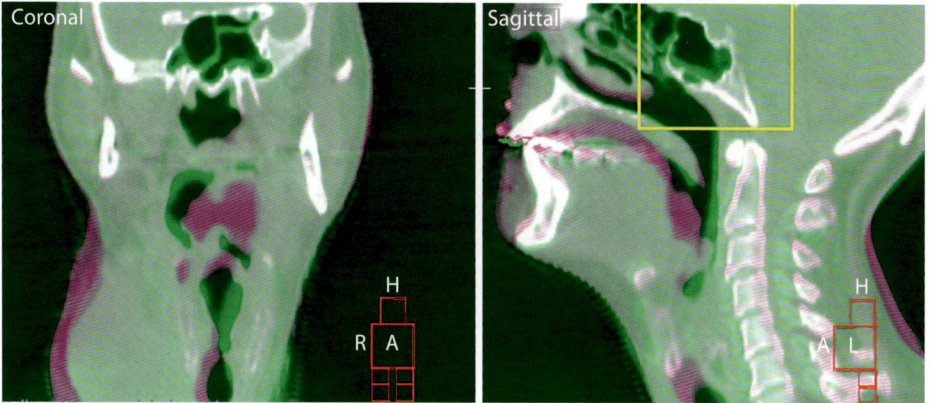

Fig. 6. T4N2M0 nasopharynx cancer. The GTV is in close proximity to the brainstem and chiasm. The cavernous sinus and clivus were chosen as the 'clipbox' (region of interest) for image guidance matching. See online supplementary material.

In IGRT, the acquired volumetric treatment setup image is compared with a reference region of interest from the planning studies and identified by a specified clipbox (fig. 6). Typically, one representative clipbox is chosen to encompass the most clinically important treatment region. For nasopharynx cancer patients treated at Princess Margaret Hospital, the clipbox region is selected to encompass the skull base due to the importance of accurate targeting there and the need to avoid critical organs at risk, such as the brainstem and optic chiasm. The choice of the clipbox region is important. In a recent report from The Netherlands Cancer Institute-Antoni van Leeuwenhoek Hospital, van Kranen et al. [18] demonstrated that the use of one large 'global' clipbox may underestimate and not detect the complexity of setup errors when compared with the use of 8 clipboxes encompassing different head and neck bone regions.

The patient discussed above presents an exception; for patients receiving comprehensive bilateral neck irradiation, we typically use the C1-C6 spine for matching. This ensures that the spinal cord is in the same position every day, though we recognize that the target volume and other normal tissues may move slightly. After bone (vertebral body) matching, the position of the soft tissues are evaluated to ensure that they are not moving more than expected. The PTV margins in this situation are usually 5 mm. As the primary tumor and involved nodes shrink during the therapy course, there will be some degree of change in the delivered doses, which we have verified using repeated treatment CBCTs. Exactly when to replan a patient's therapy remains a matter of judgment since action trigger points have not been formalized.

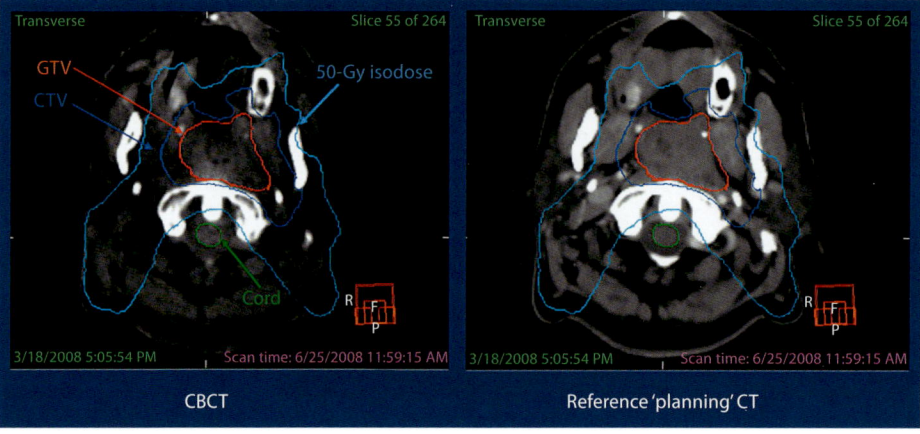

Fig. 7. Isodose display on a CBCT scan may be helpful for IGRT.

Table 2. Lessons learned in head and neck IGRT

(1) Avoid introducing systematic errors at the time of the planning CT, such as imaging the patient when swallowing

(2) Check the positions of targets and critical normal tissues away from the match points of image guidance, regardless of the image guidance strategy used. With rotations or deformations, those tissues may be moving more than anticipated. This is especially true of the spinal cord and nodes in the low neck when matching to high neck structures

(3) For smaller clinical target volumes, where there is no spinal cord or other critical structure near a high dose gradient, using soft tissues near the target for guidance is an option, e.g. early stage larynx cancer may be handled in this way

A helpful step for IGRT in the clinic is to overlay selected isodose lines from the plan onto the daily volumetric CBCT imaging. This improves communication with the therapists regarding what doses and what critical structures are most meaningful for that particular case. For example, in figure 7, the contour in blue is the 50-Gy isodose. Instructions to the therapists may be, 'at IGRT matching, if the isodose moves into the spinal canal, hold treatment and call the physician'. This practice has helped the physicians try to articulate what their thresholds are for change, though it is recognized that the true delivered doses will vary slightly based on the actual anatomy.

Like most centers, we use image guidance based on bone structures as opposed to soft tissues for head and neck sites. IGRT solutions based on soft tissues in this region are not obvious, and avoidance of overtreating the spinal cord, brainstem and other critical structures based on vertebral and base of skull anatomy remains the overriding priority (table 2). Since head and neck targets are almost always soft tissue structures, further investigation may prove useful.

Thoracic Cancers

PTV margins may be potentially reduced with IGRT for lung cancers [19]. Bissonnette et al. [19] demonstrated that geometric uncertainty (PTV) can be reduced with IGRT. Initial displacements were the measured IGRT offsets in relation to the planning (reference) CT scan. Manual or automatic couch correction of initial displacements resulted in residual displacements less than 4 mm, whereas the initial displacements were greater than 4 mm without any correction. Couch corrections could not eliminate all uncertainties and residual errors. Interestingly, residual errors varied depending on whether patients were repositioned using manual couch corrections (shown with hatched lines) or with remote couch corrections (solid lines). With the automated couch corrections, the residual setup error was improved (fig. 8) [19].

Non-SBRT lung cancers are slightly more challenging to treat. Often the edges of the tumor near the heart are not visible, but some simple IGRT strategies can be useful. For example, a carina match can be used for lung cancers tumors near the carina [20, 21].

Gastrointestinal Cancers

With IGRT, the actual tumor cannot always be visualized and surrogates need to be used for image guidance. In liver cancer, the tumor is rarely seen without contrast imaging. However, in patients with liver cancer treated with transarterial chemoembolization using Lipiodol, this agent is well visualized on kV imaging. Other potential surrogates for liver tumors include calcifications, surgical clips, inserted fiducial markers and the whole liver itself. Oral contrast at the time of volumetric image acquisition can be useful to delineate the stomach, duodenum, small bowel or esophagus in situations when these normal tissues specifically require avoidance. These and other issues are discussed in the chapter on IGRT of gastrointestinal cancers.

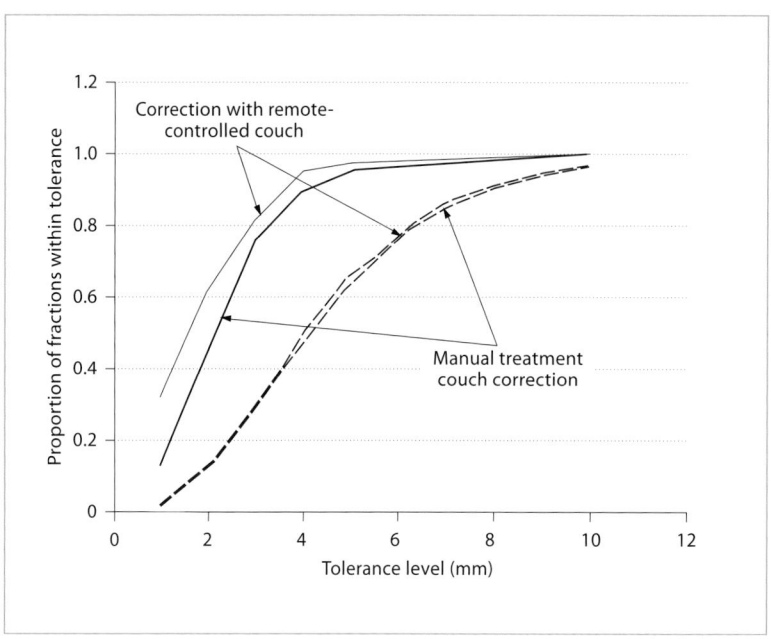

Fig. 8. Efficacy of the IGRT procedure for lung radiation therapy, per treatment fraction. Initial (dashed) and residual (solid) positional errors are shown for both treatment techniques. Using a remote-controlled treatment couch further improved accuracy [19].

Adaptive Radiotherapy

Rationale

Adaptive therapy means assessing patient anatomical changes during a course of therapy and adapting ongoing treatment to achieve the originally prescribed doses accurately. During a fractionated radiotherapy course, large anatomical changes of tumor and normal tissues potentially affecting dosimetry can occur. What is the dosimetric significance of tumor shrinkage and of weight loss in patients during treatment? When is replanning indicated, and how can this be quickly assessed during daily treatment? The answers to these questions are not known, but are subject to ongoing research. In current practice, changes in doses to normal tissues and tumors can be estimated using setup measurements or IGRT data. If the delivered doses differ substantially from those required for tumor control or normal tissue protection, replanning is judged to be indicated. However, given that tumor control and normal tissue toxicity profiles are largely based on clinical data acquired during the pre-IGRT era, these thresholds are themselves associated with a degree of uncertainty. If clinicians record and accumulate the doses actually de-

Table 3. IGRT research opportunities

Smaller/individualized PTV and PRV margins
    Hypofractionation
    Stereotactic body radiation therapy
    New indications for radiotherapy (e.g. liver cancer)
    Reduced organ at risk dose with reduced PTV
Dose accumulation
    Better knowledge of tumor-control probability
    Better knowledge of normal tissue complication probability relationships
    More power to learn about sensitizers, fractionation effects, biologics, etc.
    Better linkage of dose data with functional imaging changes
    Individualized radiotherapy
Adaptive and individualized radiation therapy
    Facilitates novel radiation strategies (chasing a changing tumor)
    Reduced normal tissue toxicity
    Dose painting

---

livered to each point in tumor and normal tissues, then the relationship of delivered doses and outcomes will be better understood. With this motivation to record actual delivered doses if possible, and to change therapy dosimetry if judged necessary, the rationale for adaptive radiation therapy becomes clear.

Research in Adaptive Radiotherapy

Reducing irradiated volumes, as an adaptive strategy in response to a shrinking tumor, should be done with some caution. For example, cancers may respond in an 'erosive' pattern versus 'elastic', meaning that tumor cells may remain throughout the original tumor volume, though reduced in number, rather than shrink toward its center. If this is the case, equally high doses may be required to the entire original tumor volume despite measurable tumor response.

In adaptive research planning studies at Princess Margaret Hospital, our approach has been to treat the initially defined tumor clinical target volume continuously throughout the course of radiation to the prescribed dose [22]. This has often required treatment adaptation based on the changing normal tissue boundaries and volumes. In addition, the same anatomical clinical target volume tissues may have changed in size, shape and position due to tumor response. The goal of our work to date has been to track these geometric changes effectively. However, there are many additional biological questions to be answered based on the rapidity and patterns of tumor response. Can these be used to modify subsequent treatment during a fractionated course? Research is needed to investigate adapting radiotherapy final doses to changing tumor volumes. This should be undertaken using careful clinical trials to better understand the safety of changing the radiation therapy paradigms that have been clinically developed over decades of experience. Table 3 outlines the far-ranging opportunities for clinical trials involving IGRT.

Conclusions

IGRT has provided critical insights into the complexities of geometric (anatomic) changes that can occur during a course of radiation therapy. Initial work showed frequent displacement of soft tissues (tumors and normal tissues) relative to bone anatomy at treatment that could be significant, creating potential for critical normal structures to receive more dose than anticipated and target volumes to receive less. These issues motivated the need for practical solutions to image guidance of treatments so that regions of interest, including PTVs, and avoidance structures could be incorporated – directly or indirectly – into the daily positioning of the patient. In many cases, relatively simple IGRT strategies such as the placement and imaging of fiducial markers can account for the largest systematic geometric uncertainties (e.g. day-to-day baseline shifts in soft tissue targets). Management of organ motion due to periodic respiration, peristalsis and filling of luminal structures, such as the bowel, create the need for more involved IGRT strategies. These will be reviewed in detail in the next chapters of this volume.

IGRT opens a new realm of opportunities to understand far more clearly the dose-volume normal tissue and tumor control relationships that are the basis for radiotherapy treatment. Refining our knowledge of these relationships can broaden the horizons of radiotherapy practice (table 3). Also, we are now able to quantify the tumor and normal tissue volumetric changes occurring during a course of treatment, further improving our ability to assess the effects of therapy on an ongoing basis.

However, there is still much to learn about implementing IGRT in effective, reproducible and efficient treatment programs in the clinic. These issues lead to fundamental questions regarding therapy planning itself. How small can PTV and PRV margins be? What is our ability to image and determine clinical target volumes? IGRT technologies now challenge our abilities to determine the location of biologically active tumor deposits and to understand their potential risks and patterns of spread. For many tumor sites, there are probably diminishing returns for more intricate IGRT solutions since PTV margins cannot be reduced below our current levels of confidence in contouring. In some sites being treated with high precision, as for stereotactic radiotherapy, the needs for ultrasmall margins and for equally precise IGRT solutions are immediate and critical concerns. It is likely that the strategies for IGRT will not be universal, but will need to be adapted to the level of information available in each case and the clinical requirements and goals for each treatment course. The PTV margin and its IGRT implications may become specific to each clinical situation, and may become as much a part of personalized cancer care as in the selection of other cancer treatments.

Finally, objective measures of the clinical benefits of IGRT are largely lacking from the existing literature. Ultimately, clinical outcomes will drive the adoption of IGRT in the clinic; understanding and quantifying the impact of IGRT should be a goal with the highest priority. To this end, IGRT is now being incorporated into clinical trials that should provide meaningful data to guide our clinical strategies.

References

1 Eisbruch A: Reducing radiation-induced xerostomia with highly conformal radiotherapy techniques. J Support Oncol 2005;3:201–202.
2 Eisbruch A: Reducing xerostomia by IMRT: what may, and may not, be achieved (comment). J Clin Oncol 2007;25:4863–4864.
3 Eisbruch A, Lyden T, Bradford CR, et al: Objective assessment of swallowing dysfunction and aspiration after radiation concurrent with chemotherapy for head-and-neck cancer. Int J Radiat Oncol Biol Phys 2002;53:23–28.
4 Eisbruch A, Ship JA, Dawson LA, et al: Salivary gland sparing and improved target irradiation by conformal and intensity modulated irradiation of head and neck cancer. World J Surg 2003;27:832–837.
5 Eisbruch A, Terrell JE: The relationships between xerostomia and dysphagia after chemoradiation of head and neck cancer. Head Neck 2003;25:1082, author reply 1082–1083.
6 Kwong DL, Pow EH, Sham JS, et al: Intensity-modulated radiotherapy for early-stage nasopharyngeal carcinoma: a prospective study on disease control and preservation of salivary function. Cancer 2004;101:1584–1593.
7 Rischin D, Peters L, O'Sullivan B, et al: Phase III study of tirapazamine, cisplatin and radiation versus cisplatin and radiation for advanced squamous cell carcinoma of the head and neck. J Clin Oncol 2008;26:LBA6008.
8 Hugo G, Vargas C, Liang J, Kestin L, Wong JW, Yan D: Changes in the respiratory pattern during radiotherapy for cancer in the lung. Radiother Oncol 2006;78:326–331.
9 Case RB, Sonke JJ, Moseley DJ, Kim J, Brock KK, Dawson LA: Inter- and intrafraction variability in liver position in non-breath-hold stereotactic body radiotherapy. Int J Radiat Oncol Biol Phys 2009;75:302–308.
10 McDermott LN, Wendling M, Sonke JJ, van Herk M, Mijnheer BJ: Anatomy changes in radiotherapy detected using portal imaging. Radiother Oncol 2006;79:211–217.
11 Nijkamp J, Pos FJ, Nuver TT, et al: Adaptive radiotherapy for prostate cancer using kilovoltage cone-beam computed tomography: first clinical results. Int J Radiat Oncol Biol Phys 2008;70:75–82.
12 Rijkhorst EJ, Lakeman A, Nijkamp J, et al: Strategies for online organ motion correction for intensity-modulated radiotherapy of prostate cancer: prostate, rectum, and bladder dose effects. Int J Radiat Oncol Biol Phys 2009;75:1254–1260.
13 Rijkhorst EJ, van Herk M, Lebesque JV, Sonke JJ: Strategy for online correction of rotational organ motion for intensity-modulated radiotherapy of prostate cancer. Int J Radiat Oncol Biol Phys 2007;69:1608–1617.
14 Song W, Schaly B, Bauman G, Battista J, Van Dyk J: Image-guided adaptive radiation therapy (IGART): Radiobiological and dose escalation considerations for localized carcinoma of the prostate. Med Phys 2005;32:2193–2203.
15 de Crevoisier R, Tucker SL, Dong L, et al: Increased risk of biochemical and local failure in patients with distended rectum on the planning CT for prostate cancer radiotherapy. Int J Radiat Oncol Biol Phys 2005;62:965–973.

16 Kupelian PA, Willoughby TR, Reddy CA, Klein EA, Mahadevan A: Impact of image guidance on outcomes after external beam radiotherapy for localized prostate cancer. Int J Radiat Oncol Biol Phys 2008;70:1146–1150.

17 Verellen D, De Ridder M, Storme G: A (short) history of image-guided radiotherapy. Radiother Oncol 2008;86:4–13.

18 van Kranen S, van Beek S, Rasch C, et al: Setup uncertainties of anatomical sub-regions in head-and-neck cancer patients after offline CBCT guidance. Int J Radiat Oncol Biol Phys 2009;73: 1566–1573.

19 Bissonnette JP, Purdie TG, Higgins JA, Li W, Bezjak A: Cone-beam computed tomographic image guidance for lung cancer radiation therapy. Int J Radiat Oncol Biol Phys 2009;73:927–934.

20 Higgins J, Bezjak A, Franks K, et al: Comparison of spine, carina, and tumor as registration landmarks for volumetric image-guided lung radiotherapy. Int J Radiat Oncol Biol Phys 2009;73: 1404–1413.

21 Sonke JJ, Zijp L, Remeijer P, van Herk M: Respiratory correlated cone beam CT. Med Phys 2005;32: 1176–1186.

22 Hwang D, Vakilha M, Breen S, et al: Temporo-spatial changes of enlarged cervical lymph nodes during head and neck cancer IMRT imaged with daily on-line cone-beam CT. Int J Radiat Oncol Biol Phys 2007;69:S429–S421.

Laura A. Dawson, MD, FRCPC
Department of Radiation Oncology, Princess Margaret Hospital, University of Toronto
610 University Avenue
Toronto, ON M5G 2M9 (Canada)
Tel. +1 416 946 2125, Fax +1 416 946 6566, E-Mail laura.dawson@rmp.uhn.on.ca

Intensity-Modulated and Image-Guided Radiation Therapy for Head and Neck Cancers

Karen Pat-Ming Chu · Quynh-Thu Le

Department of Radiation Oncology, Stanford University Medical Center, Stanford, Calif., USA

Abstract

Radiation therapy is a key component of the multidisciplinary treatment of head and neck cancers (HNC), which are ideal tumors for intensity-modulated radiation therapy (IMRT) because of their location and intimate relationship to the surrounding critical structures. Several institutional studies have suggested that IMRT is superior to conventional radiation therapy in salivary preservation and holds promises for improved locoregional control of these tumors. Small randomized studies have supported the role of IMRT in reducing xerostomia and possibly improving quality of life. Target delineation for IMRT in these tumors is complex and requires detailed knowledge of head and neck anatomy and pathways of tumor spread. The advent of image-guided radiation therapy offers a new innovation that can refine IMRT delivery even further. This article focuses on the issues surrounding IMRT target delineation for typical HNC presentations and a discussion on the role of FDG-PET imaging in HNC treatment planning.

Copyright © 2011 S. Karger AG, Basel

Radiation therapy is the single most effective 'drug' against head and neck cancers (HNC) and can be delivered without the typical problems associated with chemotherapy, such as tumor development of drug resistance or inability of the drug to reach the tumor cells due to the abnormal tumor vasculature. In vitro, tumors do not have spontaneous resistance to irradiation. Therefore, improved radiation targeting of tumors will inevitably lead to improved clinical outcomes – an effect already documented with the sequential introduction of 3D conformal radiation therapy, intensity-modulated radiation therapy (IMRT) and image-guided radiation therapy (IGRT) [1–3]. The rationale for using innovative radiotherapy approaches such as IMRT and IGRT for HNC is simply to enhance locoregional control and to reduce normal tissue toxicity. These approaches allow us to deliver higher doses to the tumor while sparing normal tissues, thereby reducing late radiation-related complications such as xerostomia and its dental consequences. In addition, for tumors located near the skull base such as nasopharyngeal carcinoma (NPC), IMRT and IGRT offer means to reduce the risk of long-term neural complications and pituitary dysfunction in treated patients.

However, several uncertainties are associated with IMRT for HNC, including the accuracy of target delineation, intrafraction patient motion, interfraction patient localization, and changes in patient and tumor anatomy during the treatment course. With the steep dose gradients inherent to IMRT, it is imperative that we address these uncertainties to ensure precise radiation delivery in these patients. In this chapter, we will discuss the role of IGRT in addressing these uncertainties for HNC.

Intensity-Modulated Radiation Therapy/Image-Guided Radiation Therapy Planning for Head and Neck Cancers

FDG-PET/CT for Target Delineation

Target delineation is one of the most challenging aspects of IMRT and can be a large source of treatment error if inaccurate. Distortion of the normal anatomy from the tumor itself, prior surgical procedures or variable, unusual patterns of tumor spread can result in significant target volume delineation errors. Therefore, acquiring appropriate imaging studies to guide tumor segmentation is essential.

The three most commonly used imaging studies for HNC staging are CT, MRI and FDG-PET/CT, and each carries specific advantages and disadvantages (we will refer to FDG-PET scanning as PET). Of these, contrast-enhanced CT is the most commonly used for treatment planning. It provides adequate cross-sectional spatial representations of normal anatomy, tumor and involved nodes; allows for accurate radiation dose calculation; and can be used to generate digitally reconstructed radiographs for comparison with portal imaging during radiotherapy. However, a well-known problem with head and neck CT is scatter artifacts from metallic dental fillings, which often obscure normal anatomy and tumor edges, specifically those located in the oral cavity and oropharynx. Further, CT has poor sensitivity and specificity for accurate identification of nodal involvement.

MRI is better than CT for evaluation of soft tissue and skull base involvement. However, it is subject to motion degradation, especially when patients move or swallow during the image acquisitions. It may be difficult to obtain MRI scans in some patients who become claustrophobic in the confined space during the longer procedures. Advances in MRI are being made, and current research has examined the benefit of diffusion-weighted MRI over conventional MRI. In comparison with CT or conventional MRI, diffusion-weighted MRI shows improved sensitivity and specificity of 89–97% in detecting nodal disease, which are promising results [4].

Within the last decade, PET has been applied increasingly to the staging workup of HNC patients. It has the highest specificity and sensitivity for detecting cervical nodal involvement when compared with CT and MRI [5, 6]. However, PET alone has poor anatomic definition, and is more accurate when used in conjunction with CT [7]. Since the introduction of PET, there has been a strong interest in integrating it into radiation treatment planning. Investigations have shown that PET can modify the initial staging of more than 50% of patients at presentation [8]. Similarly, it can modify the target volumes of more than 50% of patients undergoing treatment planning [9, 10], and it can reduce observer differences in tumor delineation. It may allow for up to 25% dose escalation, which can theoretically improve the tumor control probability by 6% for the same normal tissue control probability. These findings suggest that PET/CT is a promising tool to aid with target delineation.

For PET/CT to be accepted as superior to CT alone for target delineation, it must meet 2 different criteria. First, it should reflect the actual *pathologic tumor volume* better than CT. Second, it should reflect the actual *tumor cell burden* better, and hence be more prognostic of treatment outcomes. There are data in the literature that support both of these premises. Regarding the first, Daisne et al. [11] compared preoperative CT, MRI and PET tumor volumes with surgical findings in 9 patients with laryngeal cancers treated with total laryngectomy. While all 3 modalities overestimated the gross tumor volume (GTV) seen in the surgical specimens, the average PET-based GTV correlated best. However, no modality showed the true extent of the primary tumor with complete accuracy, especially in terms of the mucosal and submucosal spread of disease. While PET imaging can aid in tumor volume definitions, no imaging study can replace the information obtained by a thorough physical exam.

The second criterion has also been investigated. At our institution, we have evaluated the prognostic significance of PET-determined tumor volume, known as metabolic tumor volume (MTV), in 85 stage III–IV HNC patients treated with definitive chemoradiotherapy [12]. All patients underwent a PET/CT treatment planning simulation and these images were displayed as maximum intensity projections. First, the maximum intensity uptake (SUV) value within the hypermetabolic tumor was identified. Then, RT Image, a semiautomatic contouring software, was used to delineate the MTV, which was defined as the volume encompassing all the voxels within the tumor that fell within the 50% threshold

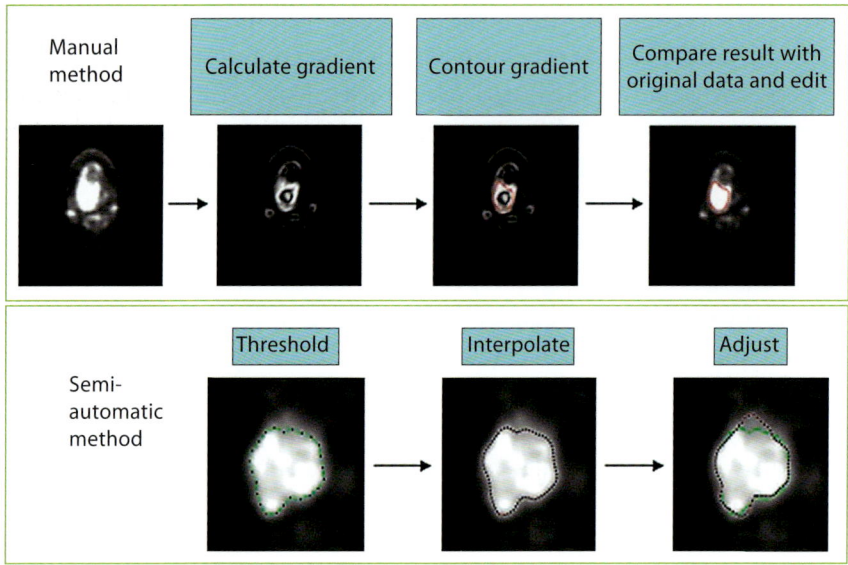

Fig. 1. Semiautomatic segmentation method using RT Image, delineating the MTV based on the 50% threshold of the maximum intensity value [12].

intensity of the maximum intensity value (fig. 1). MTV values were the best predictor for both disease-free (DFS) and overall survival (OS) in these 85 patients when stratified by stage, tumor site, performance status and treatment, whereas there was no relationship between the maximum SUV and outcome. The prognostic significance of MTV found in this study suggests that the PET volume does indeed reflect the viable tumor burden, hence validating its role in target delineation.

Segmentation Approaches for Tumor Definition Using PET

Defining the demarcation between the tumor and normal tissues (segmentation) is often not as clear using PET studies as with CT or MRI, due to the background uptake of FDG in normal tissues, motion during acquisition, partial voluming effects and FDG equilibrium changes. Therefore, several approaches have been proposed for automatic segmentation of PET volumes based on specific criteria:

(a) *SUV-based contouring*: includes all voxels within the tumor that have a SUV above a certain predefined cutoff value such as 2 or 2.5.

(b) *Thresholding approach*: involves outlining the volume as the region that is encompassed by a given fixed percent intensity level relative to the maximum activity in the tumor, e.g. 40–50%.

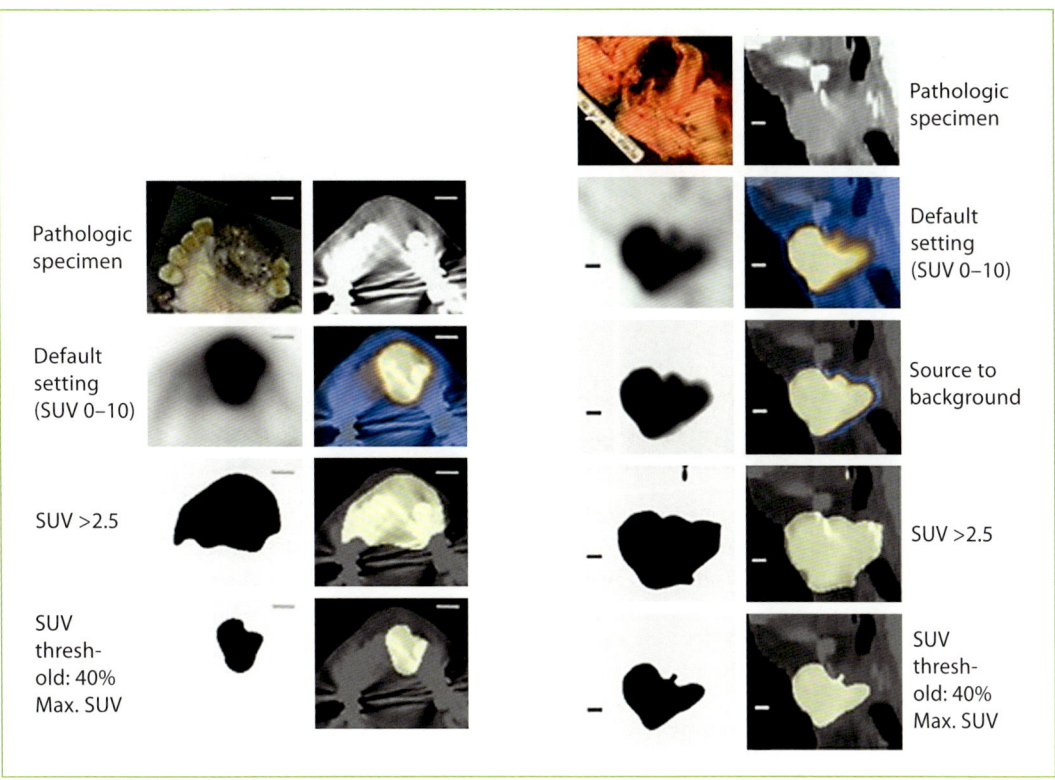

Fig. 2. Comparison of different PET delineation methods to pathology. Maximal SUV or SUV40 is better than using SUV >2.5 or source-to-background ratio to correlate positive lymphadenopathy on PET/CT imaging with surgical pathology [14].

(c) *Background cutoff method*: uses contours of the region showing intensities exceeding a certain predefined standard deviation above the background.

(d) *Source/background algorithm*: involves identifying the optimal threshold based on source/background ratio.

Each approach has its strengths and weaknesses, as reviewed by MacManus et al. [13], but the best one to use for HNC is still unclear. Burri et al. [14] evaluated several of these approaches in 12 tumors in presurgical PET/CT studies, and the final pathological volume was compared with the volumes predicted by each of these approaches (fig. 2). The volume that correlated best with the pathological findings, and was least likely to underestimate the actual tumor size, was the SUV40 threshold method (encompassing all voxels within the primary tumor with an intensity of 40% or greater of the maximal intensity). In a similar study, the authors also found that the GTV50 (the volume encompassing the 50% threshold value of the maximal intensity value within the tumor) had the best correlation

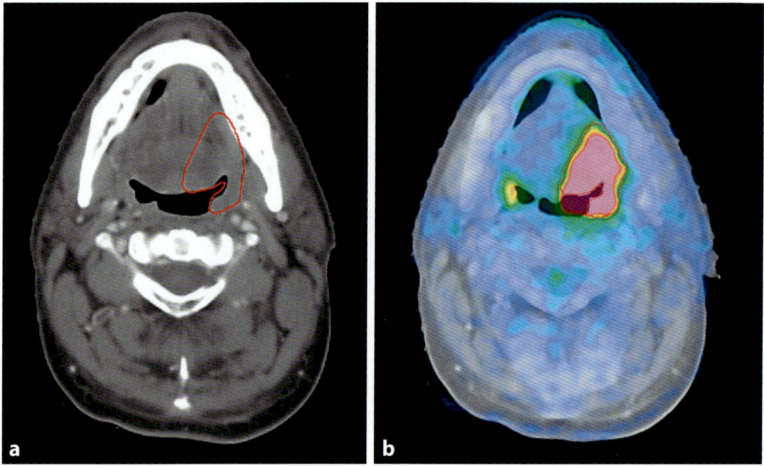

Fig. 3. Example of the halo phenomenon (GTV contoured in red) using our standard approach. **a** The GTV without PET fused to CT. **b** Fused PET and CT demonstrating halo effect using an arbitrarily set window 35,000 and level 30,000 Bq/ml to guide tumor delineation in radiotherapy planning.

with the pathologic findings [15]. These findings and our own experience suggest that a threshold PET volume of 40–50% may be a reasonable start for MTV estimation.

Another interesting but not yet validated approach for PET contouring is known as the halo phenomenon, which has been studied in 3 different tumor sites [16]. Here, the PET scan is arbitrarily set to a window of 35,000 and level of 30,000 Bq/ml. This results in a halo around the FDG avid areas of about 2 mm in thickness (fig. 3). Using this halo edge to guide tumor delineation, Ashamalla et al. [16] showed that this use of the PET/CT improved interobserver variability when compared with CT alone. While it is a convenient approach, studies to validate its accuracy for tumor segmentation are needed.

PET is likely to be a valuable tool for radiation therapy planning in HNC, but its practical incorporation into routine use is technically challenging and requires careful attention to detail. No single methodology for defining the tumor is presently recommended; each technique must be carefully considered and implemented, and other imaging and clinical information should not be ignored.

PET/CT-Based Cervical Nodal Delineation

In any staging PET/CT, some nodes may be strongly FDG-avid and other nodes may not, making the choice of which nodal groups to treat difficult. A study by

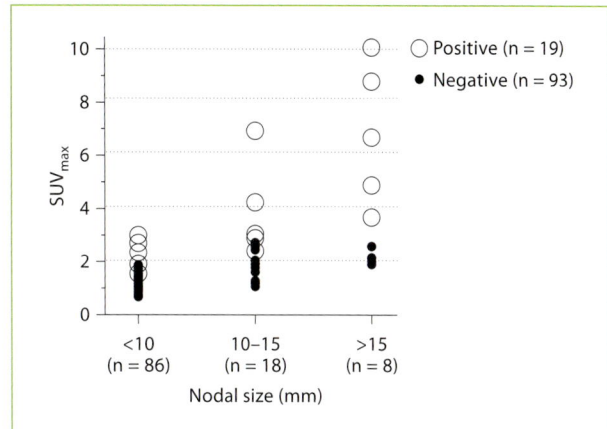

Fig. 4. Correlation between maximal SUV levels on PET imaging and lymph node size with pathologic findings on neck dissection (adapted from Murakami et al. [17]).

Murakami et al. [17] reports the results from 23 patients undergoing neck dissections following PET scanning for HNC. They compared the size and the PET SUV of the nodes with the pathologic findings. In lymph nodes <10 mm in size, the SUV was not helpful. For lymph nodes ≥10 mm in maximum diameter, a SUV threshold of 1.9 captured nearly 100% of positive nodes, though at the expense of specificity, which dropped to 70% (fig. 4). While this approach may lead to overtreatment of some lymph nodes, it ensures coverage of all involved nodes, which otherwise might be missed using CT planning alone.

Situations where PET Is Most Useful for Target Delineation

Induction Chemotherapy
Recently, there has been increased use of induction chemotherapy in patients with locally advanced HNC based on the results of 2 large randomized studies. These studies showed improved survival when the TPF induction regimen (cisplatin, 5-fluorouracil and a taxane – paclitaxol/docetaxol) was compared with cisplatin and 5-fluorouracil alone [18, 19]. Although the benefit of adding induction TPF chemotherapy to the standard cisplatin-based concurrent chemoradiotherapy regimen has not been established in large phase III studies, these early results have led many oncologists to recommend induction chemotherapy for many patients with locally advanced HNC. This poses an added dilemma to the radiation oncologist as to which volume to cover – the pre- or postinduction chemotherapy volume.

While the overall response rates to induction chemotherapy can be high, complete response rates are only 10–17%. A prior study has shown that only 3 logs of

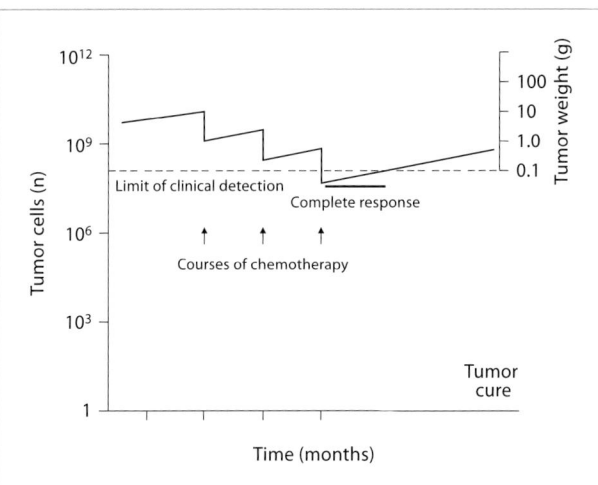

Fig. 5. Induction chemotherapy inducing 3 logs of cell kill will appear as a complete response, but will only decrease number of tumor cells to approximately 10^8 (from Tannock [20]).

cell kill are required to achieve a *clinical* complete response, though several logs of cells remain (fig. 5) [20]. Moreover, tumors do not necessarily shrink concentrically with chemotherapy. These observations argue for the use of prechemotherapy volumes for treatment planning, except for situations when volume shrinkage may be desirable to minimize critical normal tissue complications. These are situations where the tumors overlap critical neural structures such as the spinal cord, brain stem and optic apparatus.

To date, neither PET nor any other imaging study has been found to accurately assess tumor sites following induction therapy for pathologic complete response. Konski et al. [21] evaluated the role of PET/CT in predicting the extent of residual pathologic esophageal cancer following chemoradiotherapy. The postchemoradiotherapy and preoperative PET/CT were not able to predict the extent of residual disease in the surgical specimen. Another study performed a similar evaluation for HNC treated with induction chemotherapy followed by concomitant chemoradiotherapy [22]. Again, there was no correlation between preoperative PET and surgical findings. The situation is likely to be even less favorable following chemotherapy alone, and the results of these studies argue for the use of prechemotherapy tumor volumes for radiotherapy planning.

Due to the increased popularity of induction chemotherapy, a set of guidelines was established by a group of HNC experts to provide clinical practice guidance in this setting. One key recommendation is the necessity of anatomic imaging studies prior to induction chemotherapy to help with tumor delineation [23]. In a study presented at the 2009 ASTRO meeting, a panel of experts was asked to delineate the target volumes for a patient with a base of tongue cancer *after* induction chemotherapy. They were provided CT scans (without PET) before and after che-

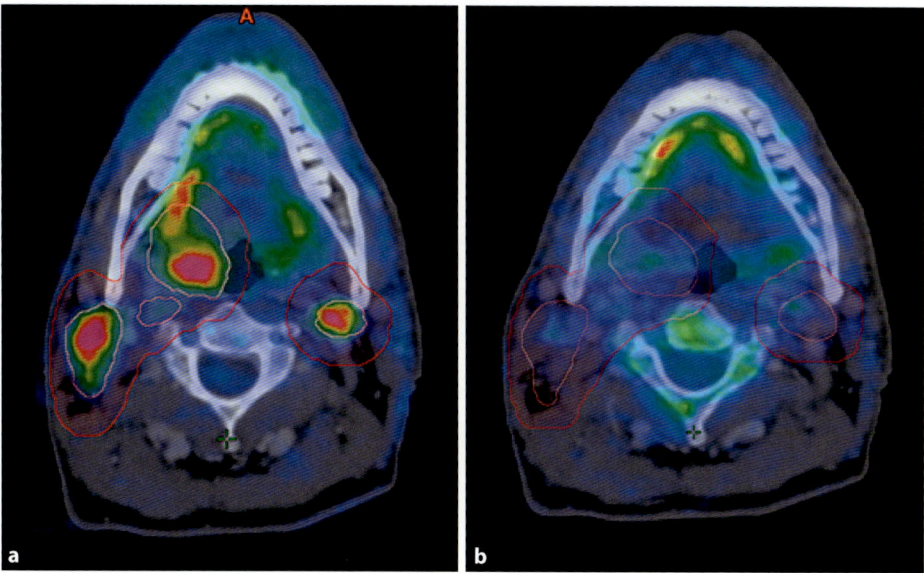

Fig. 6. Example of target volumes in a patient with oropharyngeal cancer (GTV = pink; PTV = red). **a** PET/CT before induction chemotherapy. **b** PET/CT after induction chemotherapy.

motherapy. There was little consensus among these experts when CT imaging alone was provided [24].

At our institution, we perform PET/CT simulations before and after induction chemotherapy. We use the prechemotherapy PET/CT to guide target delineation, with adjustments made for anatomic changes based on the fused postchemotherapy scan. Also, we have found that prechemotherapy PET scans help to identify initially FDG avid lymph nodes that are not seen on later postchemotherapy scans because of treatment response (fig. 6). These nodes would otherwise be missed and would not receive adequate radiation dose.

PET/CT for Adaptive Intensity-Modulated Radiation Therapy Guidance
PET/CT has also been evaluated for adaptive IMRT guidance. In one study, serial CT, PET/CT and MRI were performed in 10 patients approximately every 10 Gy during the course of radiotherapy to evaluate volumetric changes during treatment (fig. 7) [25]. Overall, the pretreatment volume was smaller on PET than CT. Further, both PET and CT demonstrated shrinkage of tumor volume over the treatment course. When adaptive planning was attempted based on these serial imaging studies, the plan using PET volumes yielded the lowest dose to the surrounding normal tissues without compromising the tumor dose. Therefore, the authors recommended using PET when dose escalation beyond 70 Gy is considered.

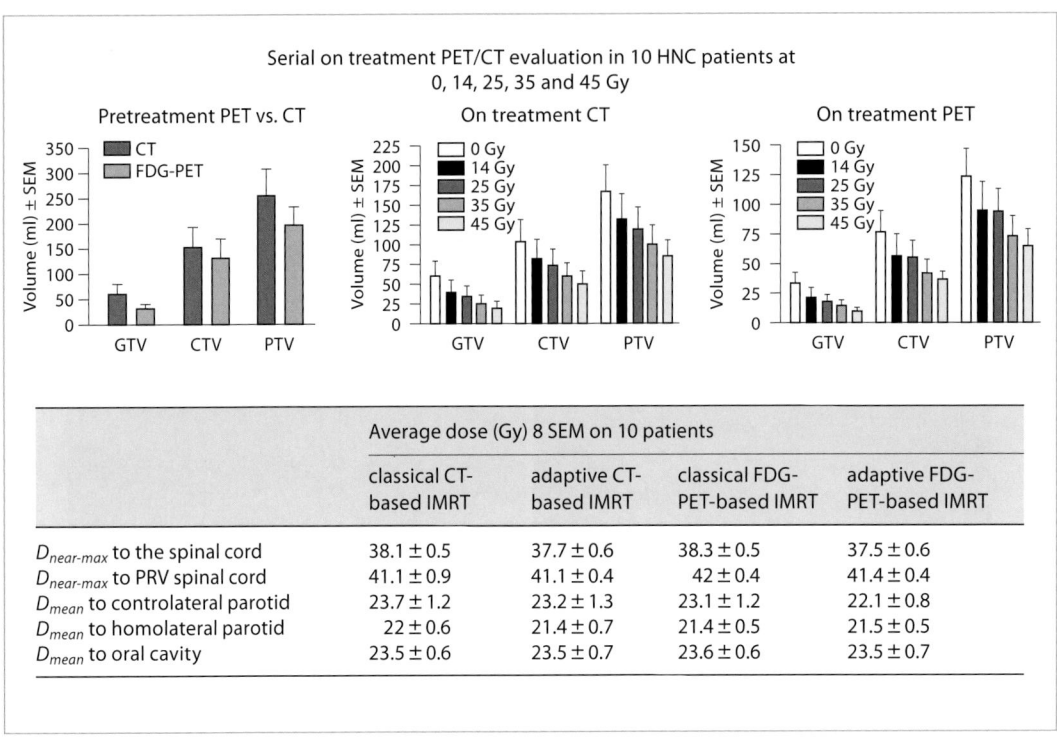

Fig. 7. PET for adaptive IMRT guidance. Volumetric changes during radiotherapy course based on serial PET/CT evaluation (adapted from Geets et al. [25]).

Pitfalls in Using PET Imaging for Radiotherapy Planning

Registration Errors

An important consideration for using PET imaging in radiation treatment planning is the accuracy of image registration between PET and CT. Even when PET is acquired immediately before or after CT in the same treatment position and with adequate immobilization devices, it is still subjected to intrafraction motion due to the amount of time required for PET acquisition (about 30 min). This should be taken into consideration when the tumor is located in a mobile region such as the larynx.

One problem with the use of PET scans is simply error in registration with the CT. Even with an immediate sequencing of 2 CTs for planning (a noncontrast CT for PET and second CT with contrast), the period of about 15 min for these scans may introduce motion changes. Additionally, the best immobilization technique is not necessarily sufficient to prevent registration errors. A recent study from the University of California, San Francisco (UCSF) discusses registration error using

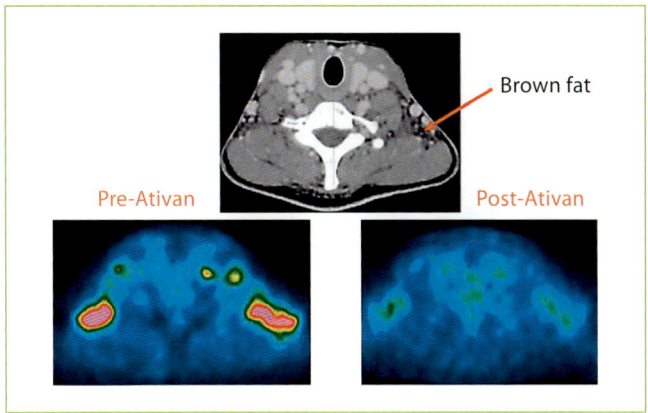

Fig. 8. PET/CT treatment planning: the brown fat factor. False-positive PET findings due to the presence of brown fat in the supraclavicular fossa. Note: CT did not show any obvious node. Pretreatment with Ativan resulted in decreased FDG uptake.

PET and CT if they are not performed on the same unit (that is, the patient is moved from a PET scan couch to a CT scan couch). In this study, they used the skull base to register these scans, so that the further away from this point, the greater the difference. In the lower neck, the registration error was from 10 to 15 mm [26]. In their evaluations, they demonstrated that manual and deformable registration resulted in the least error. Therefore, if a change in couch is required in PET/CT acquisition, a mechanism for manual and deformable registration should be used. Further, regardless of the registration tools used, the extent of registration error must be included in planning target volume (PTV) determination.

Accuracy of PET Imaging – False Positives and Negatives
The discrimination of tumor from benign processes is an important consideration for using PET/CT in HNC staging and treatment planning. False positives in PET imaging occur due to inflammation and increased cellular metabolism. In the head and neck region, this includes the base of tongue and tonsil region where the false positive rate of PET/CT is up to 42% [27, 28]. Incidental findings of pituitary adenomas, thyroid goiters or uptake in laryngeal muscles (even with limited vocalization) may raise concern and delay HNC radiation treatments. Another cause of false FDG activity is the presence of brown fat. Present more commonly in younger patients and females, brown fat is found in the lower neck/supraclavicular region and shows moderate uptake of FDG in PET/CT imaging as shown in figure 8 [29]. This resting brown fat can be minimized with the administration of a benzodiazepine, such as 1 mg Ativan, prior to imaging without compromising tumor

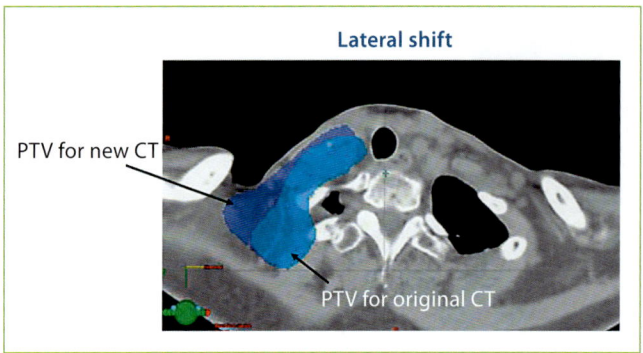

Fig. 9. The effect on nodal PTV from inadequate shoulder immobilization. The light blue area denotes the PTV from the original treatment planning CT and the dark blue area denotes the PTV from the new treatment planning CT.

uptake [30]. However, activity in brown fat can be induced by shivering in cold temperatures often found in the simulator room. This cannot be alleviated with medication and therefore must be appropriately attended to.

In addition to false positives, PET/CT is also susceptible to false negatives. While the physiologic uptake seen in the tonsils and base of tongue could signify a false positive, this same effect can mask the presence of disease. PET/CT is also limited by its difficulty to visualize tumors <1 cm or the presence of necrotic nodal disease [31]. These pitfalls of PET/CT should be taken into consideration when it is used for radiation treatment planning.

Accuracy in the Clinical Application of Intensity-Modulated Radiation Therapy/ Image-Guided Radiation Therapy

Shoulder Immobilization
The standard thermoplastic mask provides adequate immobilization of the head; however, it does not immobilize the shoulders sufficiently – an important IMRT requirement for comprehensive nodal irradiation. There are several home-fabricated and commercially available systems for shoulder immobilization. These include extended head and shoulder thermoplastic masks, various shoulder fixators, arm pullers or pegboards with shoulder straps and 3-point tattoos. Figure 9 illustrates the significant change in the PTV that can occur from poor shoulder immobilization. In this particular case, the patient was unintentionally set up slightly crooked in her body position on the initial treatment planning CT. This required a readjustment in her position and a repeat treatment planning CT. A significant change in the PTV in the lower neck (light blue for the 1st scan and

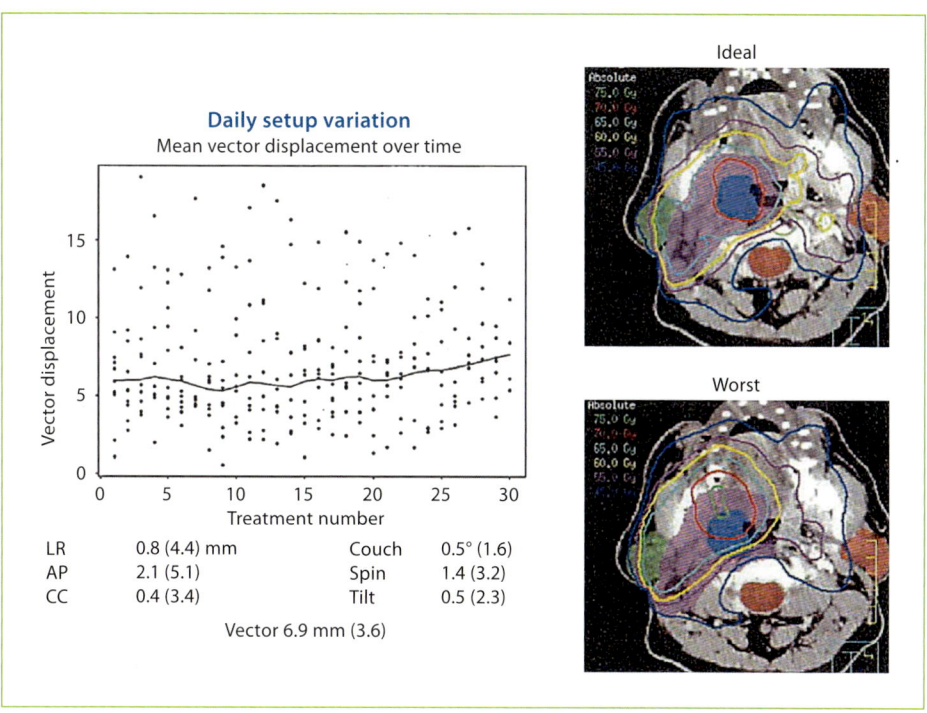

Fig. 10. Quantitation of daily setup variability using an optical system (adapted from Hong et al. [32]). LR = Left-right; AP = anterior-posterior; CC = cranio-caudal.

dark blue for the 2nd scan) was observed. This example illustrates the importance of adequate shoulder immobilization for accurate radiation delivery to the lower neck regions.

Daily Setup Variability
Daily setup variability has been quantified, primarily for the treatment isocenters. Hong et al. [32] from the University of Wisconsin used an optically guided patient localization system to evaluate setup accuracy in 10 patients who were immobilized with thermoplastic masks and a baseplate fixation to the treatment couch. The details of their findings are shown in figure 10. The mean setup error in any single dimension averaged 3.33 mm and the mean composite vector offset was 6.97 mm (standard deviation: 3.63 mm) when all 6 degrees of freedom were accounted for. These setup variabilities translated to a potential decrease in the equivalent uniform dose of up to 21%, which could pose a significant detriment to tumor control.

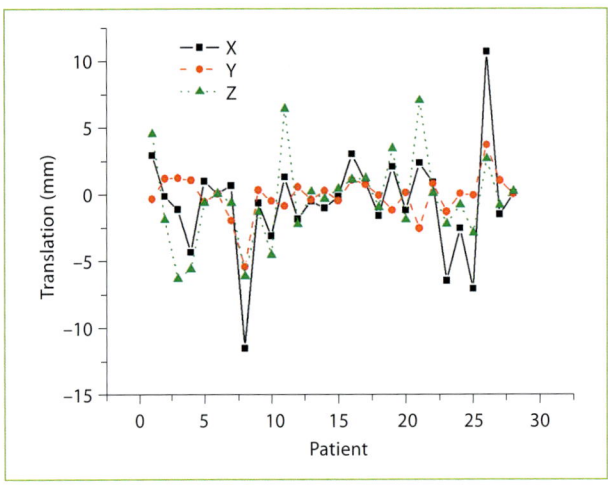

Fig. 11. Plot of translational intrafraction motion in the shoulders in the lateral (X), anterior-posterior (Y) and superior-inferior (Z) direction in 29 patients [33]. Note the amount of motion in some patients.

Intrafraction Variability
Little data exist regarding intrafraction motion in HNC. We have attempted to quantify the intrafraction motion in 29 HNC patients by comparing 2 sequential CT scans from the same day. These were taken approximately 20 min apart, with the patient immobilized in the same treatment position, on the same couch and without being moved (immobilization was accomplished with a customized Accuform head holder, a thermoplastic mask extending from the cranium to below the mandible and a customized peg board) [33]. Potential translational and rotational motions were measured in the lateral, anterior-posterior and superior-inferior directions. While the average motion in the head/neck was <0.5 mm in any direction and 1.02 degrees of rotation, the maximal translation in the shoulders was 11.58 mm in the lateral direction and maximal rotation was 3.27 degrees (fig. 11). Further, 9 patients (31%) revealed a translational motion of >5 mm in the shoulders, emphasizing the uncertainty in the lower neck and shoulders which may affect targeting accuracy in these tumor volumes.

Another important concern is the motion of the larynx upon breathing and swallowing. Investigators at M.D. Anderson Cancer Center (MDACC) used bone anatomy (hyoid bone, thyroid cartilage and the C2 vertebral body) to quantify systematic changes of the larynx with swallowing, and estimated that the difference in the larynx position between simulation and daily treatment could be as great as 1.2 cm [34]. Over the course of treatment, imaging showed that the hyoid spent a greater amount of time in a more superior position, posing a risk of systematic errors depending on its position during simulation. These data emphasize the importance of being aware of certain intrafraction motions during planning, and incorporating them into the PTV margins for adequate tumor targeting for HNC.

Changes in Tumor and Normal Tissue Volumes during Fractionated Intensity-Modulated Radiation Therapy

Changes in the tumor volume and the patient's anatomy over the 6–7 weeks of a fractionated IMRT course present important concerns. Often there is rapid shrinkage of primary and nodal tumor volumes, requiring IMRT replanning. In addition, patients may lose a significant amount of weight, causing poor fit of the initial immobilization system. There are many implications for clinical care.

Weight Loss
Given the complex anatomy of the head and neck and its multiple functions, it is common for patients to lose significant weight due to dysphagia from radiotherapy and nausea/vomiting from chemotherapy. McRackan et al. [35] evaluated 72 patients retrospectively and found that those with a BMI ≤25 were more likely to require a percutaneous endoscopic gastrostomy tube and to have a lower OS compared with those with higher BMI.

To reduce the severity of weight loss during radiotherapy for HNC, a number of strategies have been assessed. Studies have defined several factors that predict for development of late swallowing dysfunction, including patient age, tumor stage and tumor site (larynx/hypopharynx) [36]. Further, Caudell et al. [37] found that swallowing complications such as pharyngoesophageal stricture and percutaneous endoscopic gastrostomy tube dependence were associated with dose to the larynx and pharyngeal constrictors [38]. In particular, the volume receiving ≥60 Gy was an important threshold. These findings highlight the importance of contouring these structures and prioritizing them for avoidance during IMRT planning.

Salivary Gland Changes
Normal glands such as the parotid and submandibular can shrink during radiotherapy. Figure 12 shows the interfractional clinical target volume (CTV) and parotid volumes obtained from a HNC patient treated with IMRT at MDACC [39]. These volumetric changes can result in overdosing of the parotid glands when compared with the intended, planned dose distributions. The literature suggests that there is a parotid dose threshold above which xerostomia will become permanent (approximately 25–30 Gy for the mean parotid dose, or 30 Gy for the dose

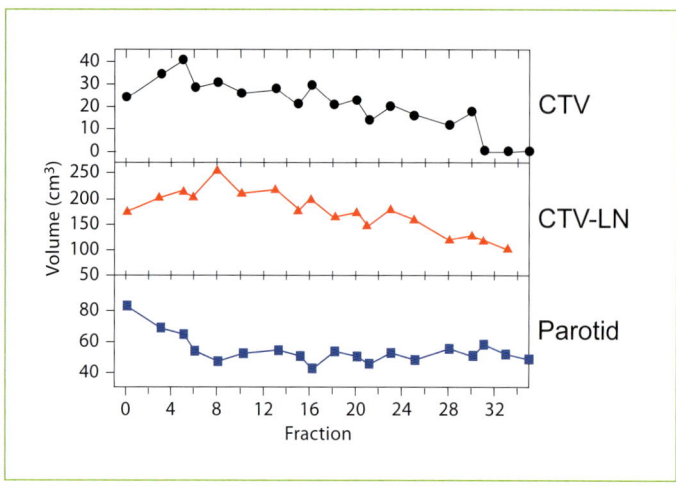

Fig. 12. Quantitation of interfractional anatomic changes using CT-on-rails (adapted from Mohan et al. [39]). LN = Lymph nodes.

received by ≥50% of the parotid volume) [39, 40]. However, these thresholds were established based on a single pretreatment scan; no data take into consideration the shrinkage of the parotid gland during treatment and the actual doses it received, which would be helpful in anticipating and planning for adaptive radiotherapy.

The Role of Image-Guided Radiation Therapy in the Management of Head and Neck Cancer

Addressing Setup Variability with Image-Guided Radiation Therapy
Several IGRT approaches are available to adjust for day-to-day patient setup variability, including several forms of orthogonal kV 2D imaging and in-room CT. While there is great interest in these innovations, there is no consensus as to the best modality or required frequency of use. One study evaluated the interfraction motion using anterior and lateral electronic portal imaging (EPID) in 20 head and neck IMRT cases. While medial-lateral and anterior-posterior systematic errors were not significantly altered based on the frequency of EPID, the craniocaudal error was significantly greater when imaging was obtained less than every second treatment day [41]. Den et al. [42] found that daily CBCT improved treatment accuracy, allowing a 50% reduction in the PTV expansion margin overall. These results show that an active IGRT program can be used effectively to decrease the volume of normal tissue in the high-dose regions.

Image-Guided Radiation Therapy for Adaptive Radiotherapy
When treatment imaging does confirm changes in anatomy, how and when should replanning occur? Once the on-treatment volumetric CT has been obtained, possible adaptive strategies include: (a) on-line (or near real-time) IMRT *complete replanning* or (b) *deforming* radiation intensities of the existing plan based on the anatomic deformations seen on the on-line CT. These two choices could be performed at regular intervals, from daily to weekly, or only after a threshold change occurs, which could provide a third choice, (c) generating a *confidence-limited PTV* based on several initial CT scans for the patient, and correcting the planning only when the actual dose distribution is significantly different from the anticipated one.

Most of these approaches are time consuming, and it is unclear which approach will result in clinically superior treatment outcomes. Several approaches have been proposed for on-line adaptive radiotherapy, a goal that is both important and challenging. At the present time, with lack of clinical data to justify the increased cost work load of adaptive radiotherapy, replanning in the middle of the treatment course is justified mainly for patients who experience significant tumor shrinkage and/or weight loss during the treatment course. We have therefore established the following procedure in our clinics prior to replanning:

(a) If the mask is loose, a new mask is made to match the original patient position as much as possible by comparing the kV radiographs with the initial digital reconstructed radiographs, and a new planning CT is obtained.

(b) Registration of the 2 planning CTs using bone landmarks to detect any potential positioning differences.

(c) Dosimetric evaluation by applying the first IMRT plan to the new CT, while taking into consideration the potential shifts of the isocenter between the 2 CT scans.

(d) Determining whether replanning is necessary based on the adequate dosimetric coverage to the tumor and safe protection to the critical structures such as the brain stem, spinal cord and optic apparatus.

If replanning is deemed necessary, we utilize a deformable image registration tool to transfer the contours from the first CT to the second CT as a starting point. The transferred contours are carefully reviewed and adjusted by the attending physicians. As we mentioned previously, the second IMRT plan uses the same planning dose constraints as the first plan, so creation of the second IMRT plan is usually less labor-intensive. With available deformable image registration and inverse planning tools, replanning becomes practical without too much additional burden on the physicians, physicists and dosimetrists.

> **Target volume definition**
>
> - GTV: Gross tumor on CT/MRI and PET
>
> - CTV: GTV + margin including
>   nasopharynx, retropharyngeal nodes,
>   clivus, skull base,
>   inferior sphenoid sinus,
>   pterygold fossae,
>   parapharyngeal space,
>   posterior nasal cavity and maxillary sinuses
>
> - PTV: CTV + 3–5 mm

Fig. 13. Recommended target volume definition for NPC.

Conclusion

No single automatic contouring method is useful across the wide variety of clinical scenarios representing our patients today, and none of these methods should be recommended to be used alone. It is important to use all modalities: PET, CT, MRI and clinical evaluation (especially for the mucosal extent of the disease). The size and extent of PET/CT volumes are prognostic for treatment outcomes and provide useful information for guiding treatment.

Intensity-Modulated Radiation Therapy for Nasopharyngeal Carcinomas

Target Volume Delineation

Figure 13 shows the recommended definitions for the different target volumes in NPC. Head and neck MRI scans, specifically gadolinium-enhanced, fat-saturated T_1-weighted series with axial and coronal views, are critical for GTV delineation. MRI is superior to CT scanning in defining tumor involvement of the parapharyngeal space, skull base, cavernous sinus and retropharyngeal nodes. For nodal volume delineation, treatment planning PET/CT scans can often be helpful. On rare occasions, one can see parotid nodal involvement in patients with locally advanced NPC (fig. 14); therefore, we do not recommend routine sparing of the ipsilateral parotids. Also shown in figure 14, involvement of level V nodes is very common in NPC, and adequate coverage of level V in the nodal CTV is critical for NPC IMRT planning.

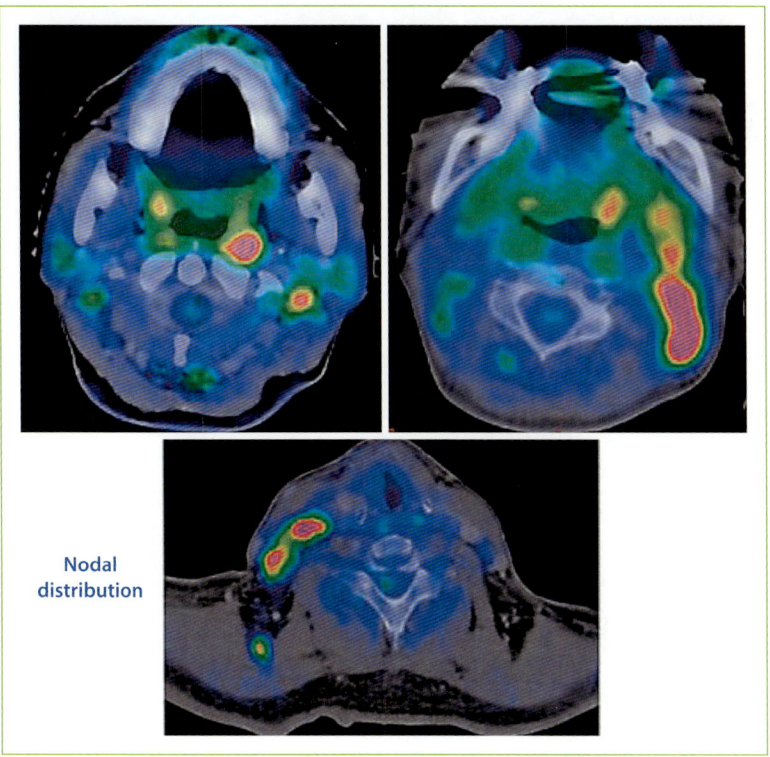

Fig. 14. Treatment planning PET/CT scan of an NPC patient showing different sites of nodal involvement including retropharyngeal nodes, intraparotid node and level 5 neck nodes.

Figure 15 compares treatment volumes for an earlier stage tumor (T1N2) with a more advanced tumor (T4N2). The GTV is determined based on merged images from CT, MRI and PET. For earlier stage disease with no extension outside the nasopharynx as in figure 15a, the CTV includes the posterior aspects of the nasal cavity, maxillary sinuses, the pterygoid fossa with the pterygoid plates and the anterior half of the clivus. In patients with parapharyngeal involvement and/or nasal cavity involvement as in figure 15b, the CTV should include all of the pterygoid fossa and the lateral pterygoid muscle on the side of the parapharyngeal involvement. Further, the nasal cavity should be included when involved with tumor. The CTV70 generated for the primary tumor and involved lymph nodes should include GTV with an added 5 mm margin. To generate the PTV70, we routinely will add approximately 4–5 mm margin to the CTV70 (70 Gy at 2.12 Gy/fraction). Margins for both CTV and PTV are pulled in tighter to spare critical structures such as the brainstem. The PTV56 and PTV52 (1.7 Gy/fraction and 1.58 Gy/fraction, respectively), encompass the clinically negative areas that are at risk of involvement as outlined by the CTV with an additional 5-mm margin listed in

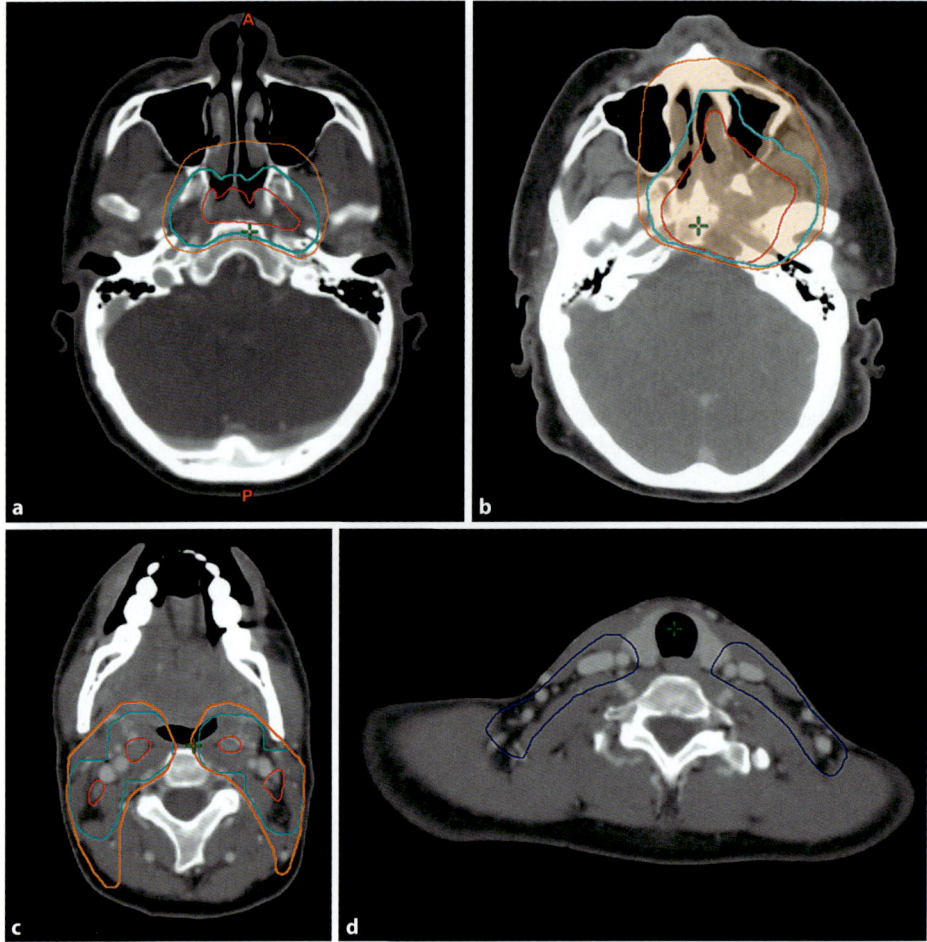

Fig. 15. Target volumes for NPC. **a** T1 tumor limited to the nasopharynx. **b** T4 tumor with parapharyngeal extension as well as cranial nerve involvement and extension into the nasal cavity. **c** Extension of nodal volumes into levels II and V. **d** PTV52 (dark blue) for the lower, uninvolved low-risk areas. GTV = Red; PTV70 = light blue; PTV56 = orange; PTV52 = dark blue.

figure 13 as well as the nodal levels II–V. If level II is involved on the ipsilateral side of the tumor, level Ib is also included.

In patients with tumor involvement of the cavernous sinus and the prepontine cistern, as shown in figure 16a, the GTV needs to extend intracranially to cover all areas of gross disease. The CTV and PTV expansions were small due to the close vicinity of the tumor to the temporal lobes laterally and the brainstem posteriorly. In this particular patient, we performed isocenter portal imaging daily and CBCT weekly before treatment in order to minimize the interfraction setup errors. In a patient with skull base involvement (fig. 16b), the high-risk CTV should include

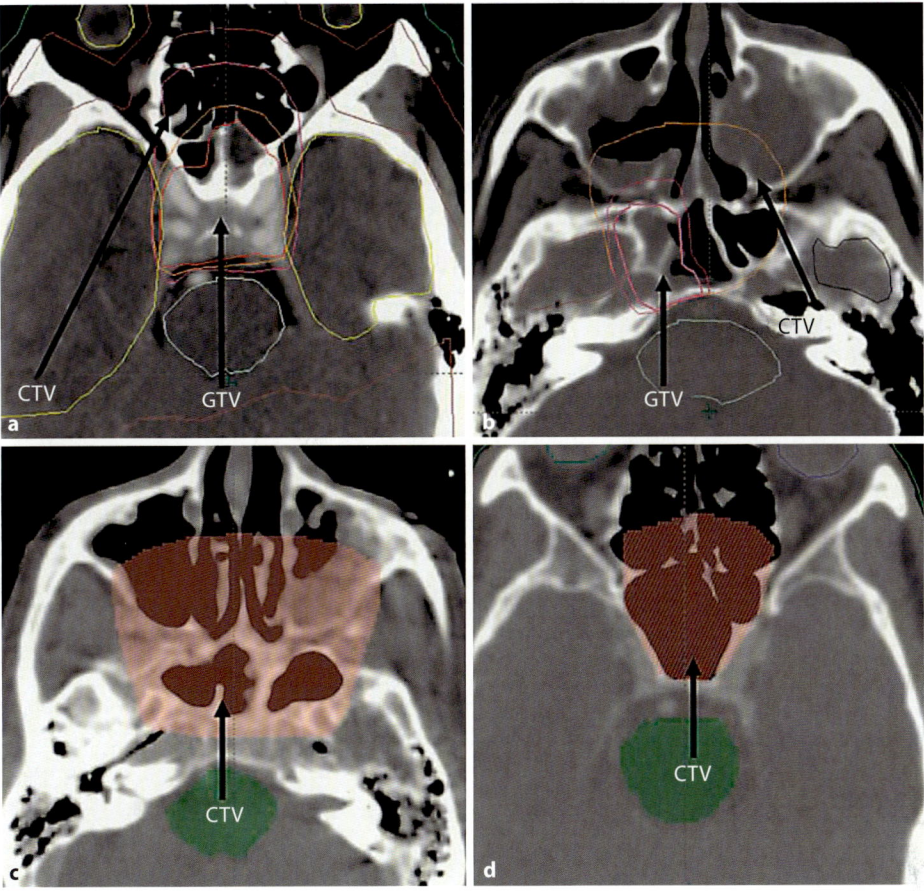

Fig. 16. Examples of GTV and CTV56 delineation (labeled as CTV in the figures). **a** T4 tumor with bilateral cavernous sinus and prepontine cistern involvement. **b** T3 tumor with skull base involvement. **c, d** T2a tumor without either skull base or intracranial extension.

most bone structures of the skull base as well as the entire clivus and any areas of high risk for microscopic disease described earlier. This high-risk volume is treated to 56 Gy in 1.7 Gy/fraction. In patients without imaging evidence of the skull base or intracranial tumor involvement, the CTV56 generally encompasses the pterygopalatine fossae, sphenoid sinuses, part of the ethmoid and maxillary sinuses, and the anterior half of the clivus (fig. 16c, d). We do not routinely cover the cavernous sinus, unless there is cranial nerve or skull base involvement, in order to achieve sparing of the temporal lobes. Approximately 4–5 mm is routinely added to generate each PTV from each CTV volume, to account for daily setup variability. Again, this depends on the proximity of critical structures to the PTV. It is crucial to consult your institutional radiologist for optimal target delineation.

Dose Prescription

- RTOG 0225 and RTOG 0615
  2.12 Gy/fraction/day × 33 to ≥95% of PTV70 (GTV)
  1.80 Gy/fraction/day × 33 to ≥95% of PTV59.4 (high risk)
  1.64 Gy/fraction/day × 33 to ≥95% of PTV54 (low risk)

- MSKCC
  2.34 Gy/fraction/day × 33 to ≥95% of PTV70 (GTV)
  1.80 Gy/fraction/day × 33 to ≥95% of PTV54

- UCSF
  2.12 Gy/fraction/day × 33 to ≥95% of PTV70 (GTV) +/− brachytherapy boost
  (5–7 Gy in 2 fractions)
  1.80 Gy/fraction/day × 33 to ≥95% of PTV60
  1.65 Gy/fraction/day × 33 to ≥95% of PTV54

- Stanford
  2.12 Gy/fraction/day × 33 to ≥95% of PTV70 (GTV) +/− STR boost
  1.70 Gy/fraction/day × 33 to ≥95% of PTV56 (high risk)
  1.64 Gy/fraction/day × 33 to ≥95% of PTV52 (low risk)

Fig. 17. Different dose prescriptions for NPC. STR = Stereotactic radiotherapy. MSKCC = Memorial Sloan Kettering Cancer Center; UCSF = University of California San Francisco.

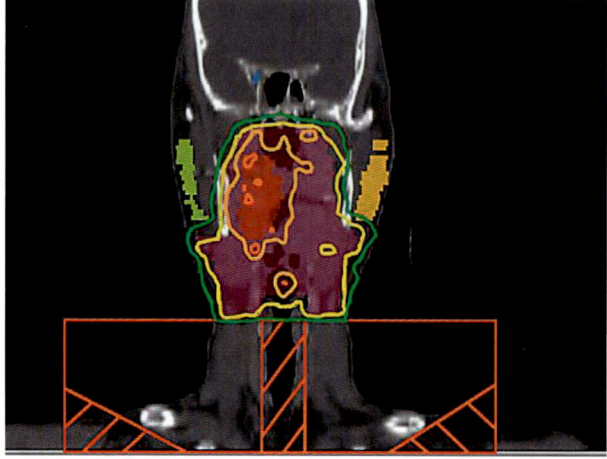

Fig. 18. Illustration of the IMRT technique used for treating NPC at the University of California San Francisco. IMRT is used to treat the primary tumor and upper neck nodes, whereas a conventional supraclavicular field is used to treat the low neck nodes (adapted from Bucci et al., [43]).

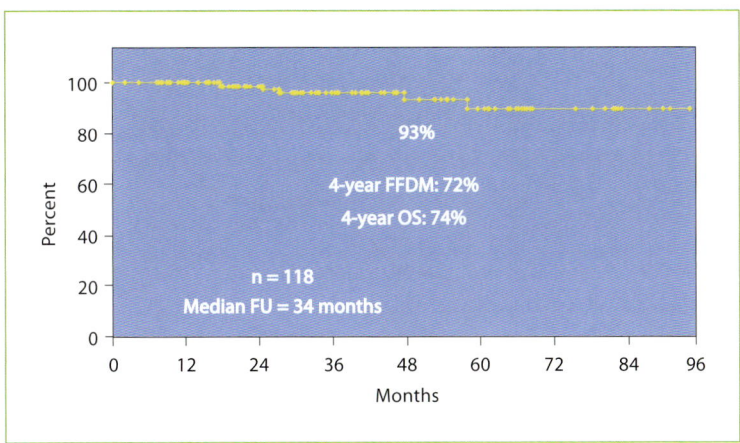

Fig. 19. Freedom from local relapse in 118 patients treated by IMRT for NPC at the University of California San Francisco (adapted from Bucci et al. [43]). FFDM = Freedom from distant metastases; FU = follow-up.

Dose Prescription

Figure 17 shows the different dose prescriptions that have been used for NPC. Most of these prescriptions employ a dose painting approach. For patients with large T2b–T4 tumors, the Stanford group also adds a stereotactic radiotherapy boost to the area of residual disease (see below).

Bucci et al. [43] updated the UCSF IMRT experience at the 2004 ASTRO annual meeting. Figure 18 shows an example of a treatment plan employed at UCSF where IMRT was used to treat the tumor and upper neck nodes, while a conventional anteroposterior supraclavicular field was used to treat the lower neck nodes. The dose prescription is shown in figure 17. One hundred and eighteen patients were treated between 1995 and 2003, of which 74% had stage III/IV tumors, 90% received concomitant chemotherapy and 23% also had a high-dose-rate brachytherapy boost. The mean parotid dose was kept between 28–30 Gy.

Clinical Results for Intensity-Modulated Radiation Therapy of Nasopharyngeal Carcinoma

At a median follow-up period of 34 months, the 4-year local control rate in the UCSF experience was 93% and the OS rate was 74% (fig. 19). The most common site of relapse was distant. Salivary measurement in a subset of these patients revealed that the majority had grade 2 xerostomia at completion of radiotherapy, but recovered to grade 0 or 1 at 2 years of follow-up (fig. 20).

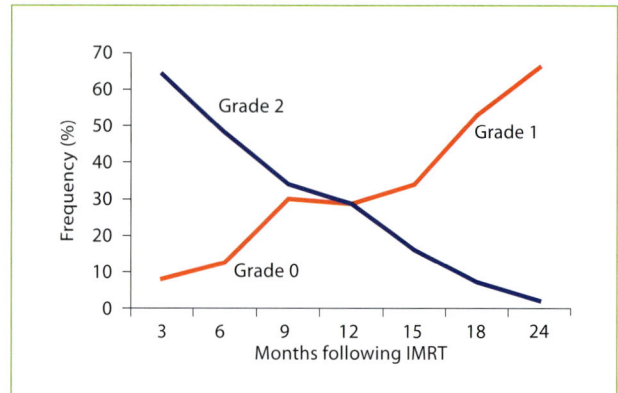

Fig. 20. Frequency of patients with grade 0–2 xerostomia over time after completion of IMRT for NPC (courtesy of Nancy Lee, MD).

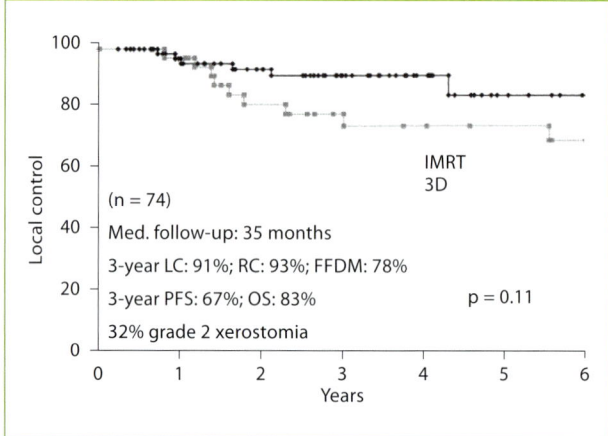

Fig. 21. Memorial Sloan Kettering Cancer Center experience with IMRT for treatment of NPC: IMRT vs. CRT (Wolden, [44]).

Figure 21 and table 1 summarize the IMRT experience for NPC from several institutions. Wolden et al. [44] reported on 50 NPC patients treated with IMRT and concurrent chemotherapy at the Memorial Sloan-Kettering Cancer Center. At a short follow-up period of less than 2 years, the local control rate was 94%. Data from the Prince of Wales Hospital are shown in table 1 [45]. Intracavitary and conformal boosts were used routinely and 30% of the patients also received chemotherapy. At a median follow-up of 29 months, only 4 in-field failures were observed and the predominant site of relapse was distant. These results translated into a local relapse-free survival rate of 92% and an OS rate of 90% at 3 years. Only a quarter of the patients had grade 2 or greater xerostomia. At the 2009 ASTRO Annual Meeting, Zhao et al. [47] (table 1) presented data on 419 NPC patients treated with IMRT at Sun Yat-sen University. The GTV dose ranged from 64 to 70 Gy and 23% of the patients received chemotherapy. At a median follow-up of 48 months, the local control rate was 93% and the OS rate was 83%. Only 4% had grade 2 xerosto-

Table 1. Reported experience by several institutions using IMRT to treat nasopharyngeal cancers

Institution	Patients, n	Treatment technique	Median follow-up months	IMRT disease control and survival (3-year results)	Xerostomia
MSKCC [44]	74 stage III/IV = 77%	DMLC GTV: 70 Gy PTV: 54 Gy chemotherapy (≥stage II): CDDP/5FU	35	LC = 91% RC = 93% FFDM = 78% PFS = 67% OS = 83%	grade 2 = 32%
Prince of Wales Hospital [45]	63 stage III/IV = 57%	DMLC GTV: 66 Gy PTV: 60 Gy N0 neck: 54–60 Gy IC boost (T1–T2a) conformal boost (>T2a)	29	LRFS = 92% DMFS = 79% OS = 90% 13/18 failures = distant	2-year grade 2–3 = 23%
Taiwan [46]	203 stage III/IV = 54.2%	3DCRT (93 patients) or IMRT (110 patients) – DMLC GTV: 64.8–75.6 Gy 80–85% stage III/IV received concurrent chemotherapy	40	LRC = 84.2% DMFS = 82.6% OS = 85.4%	2-year grade 2 = 42% (3DCRT or IMRT)
Sun Yat-sen University [47]	419 stage III/IV = 66%	Corvus 3.0/ MIMiC GTV: 64–70 Gy CTV_1: 55–66 Gy CTV_2: 42–54 Gy chemotherapy = 23% (CDDP/5 FU)	48	(5-year results) LRFS = 93.4% RRFS = 96.7% DMFS = 84.7% DSS = 84.1% OS = 83%	grade 1 = 51% grade 2 = 3.7%
National Cancer Centre, Singapore [48]	195 stage III/IV = 63%	DMLC GTV: 70 Gy PTV_1: 66 Gy PTV_2: 54–60 Gy IC boost (T1–T2): 4.5–5 Gy/ fraction × 1–2 fractions chemotherapy (CDDP/ 5FU)	36.5	LRFS = 89.6% DMFS = 89.2% OS = 94.3%	grade 0–2 (with chemo) = 97% grade 0–2 (no chemo) = 100%
RTOG 0225 [49]	68 stage III/IV = 59%	GTV: 70 Gy PTV: 59.4 Gy chemotherapy (CDDP/5 FU) for ≥T2b and/or N+ disease	31.2	(2-year results) LRFS = 89.3% DMFS = 84.7% OS = 80.2%	grade 2 = 13.5%
Milan [50]	87 stage III/IV = 75%	2D/3DCRT (73%) or IMRT (26%) GTV: 66–70 Gy PTV: 50–54 Gy chemotherapy (CDDP/5 FU or epirubicin: induction +/– concurrent or both)	46	LRC = 93% FFDM = 90% stage III/IV: • concurrent chemo = 56% • induction + concurrent chemo = 92% DFS = 82% OS = 90%	grade 2 = 21% (IMRT) and 80% (2D/3DCRT)

MSKCC = Memorial Sloan Kettering Cancer Center; RTOG = Radiation Therapy Oncology Group; LC = local control; RC = regional control; LRC = locoregional control; FFDM = freedom from distant metastases; PFS = progression-free survival; DSS = disease-specific survival; DMLC = dynamic multileaf collimator; IC = intracavitary brachytherapy; LRFS = local relapse-free survival; DMFS = distant metastases-free survival.

mia and none had grade 3. Similarly, the data reported from the National Cancer Centre in Singapore (using IMRT with a GTV dose 70 Gy and an intracavitary brachytherapy boost) demonstrate promising results with IMRT [48]: the 3-year local control rate was 90% and OS was 94%. Only 3% of the patients had grade 3 xerostomia (with chemotherapy) and none had grade 3 (without chemotherapy).

Presently, there are several prospective studies evaluating the role of IMRT in NPC. At the Memorial Sloan-Kettering Cancer Center, patients with stage II–IV NPC are treated with dose-painting IMRT concurrently with chemotherapy. The dose regimen is described in figure 17. The Radiation Therapy Oncology Group (RTOG) phase II multi-institutional study – RTOG 0225 – has recently published its data which reveal promising results using IMRT with or without chemotherapy ($\geq$T2b or with positive lymph nodes) [49]. The GTV was treated to 70 Gy with an integrated boost. Locoregional-free survival of 89.3% was similar to previous studies listed in table 1. OS was 80.2%. Looking specifically at more advanced stage NPC in this study, patients who had stage IIB–IVB disease demonstrated a 2-year locoregional-free survival of 91.2%, disease metastases-free survival 82.1% and OS of 76.7%. RTOG 0615 has assessed the use of IMRT with chemotherapy and bevacizumab in stage IIB–IVB NPC, and has completed accrual, though toxicity and efficacy results have not yet been reported.

Stereotactic Radiotherapy as a Boost for Nasopharyngeal Carcinoma

The use of stereotactic radiotherapy as a boost for patients with locally advanced NPC at Stanford University was pioneered by D. Goffinet in 1992, prior to the IMRT era. It was first reported by Le et al. [51] in 2003 and recently updated by Hara et al. [52]. The rationale for a using stereotactic boost is to enhance local control, as the local recurrence rates in the prechemotherapy and IMRT eras were as high as 60% depending on tumor stage. In addition, persistent skull base disease can result in quite poor functional outcomes and low salvage rates. Previous experience using 3D conformal radiation therapy and brachytherapy boost suggested that a dose-response relationship exists for NPC. Therefore, we performed this prospective study, evaluating the role of stereotactic radiotherapy in the management of patients with locally advanced NPC.

From September 1992 to September 2004, we treated 82 patients using this approach. The fractionated external beam dose was 66 Gy, delivered using a conventional approach in 46 patients and with IMRT as part of the entire course of treatment in 36 patients. For the stereotaxis, 40% were treated via a frame-based approach and 60% via a frameless approach with CyberKnife. Figure 22a shows an example of a CyberKnife boost plan for a patient with an initial T2b tumor, where a 12 Gy dose was prescribed to the 80% isodose line, covering the entire nasophar-

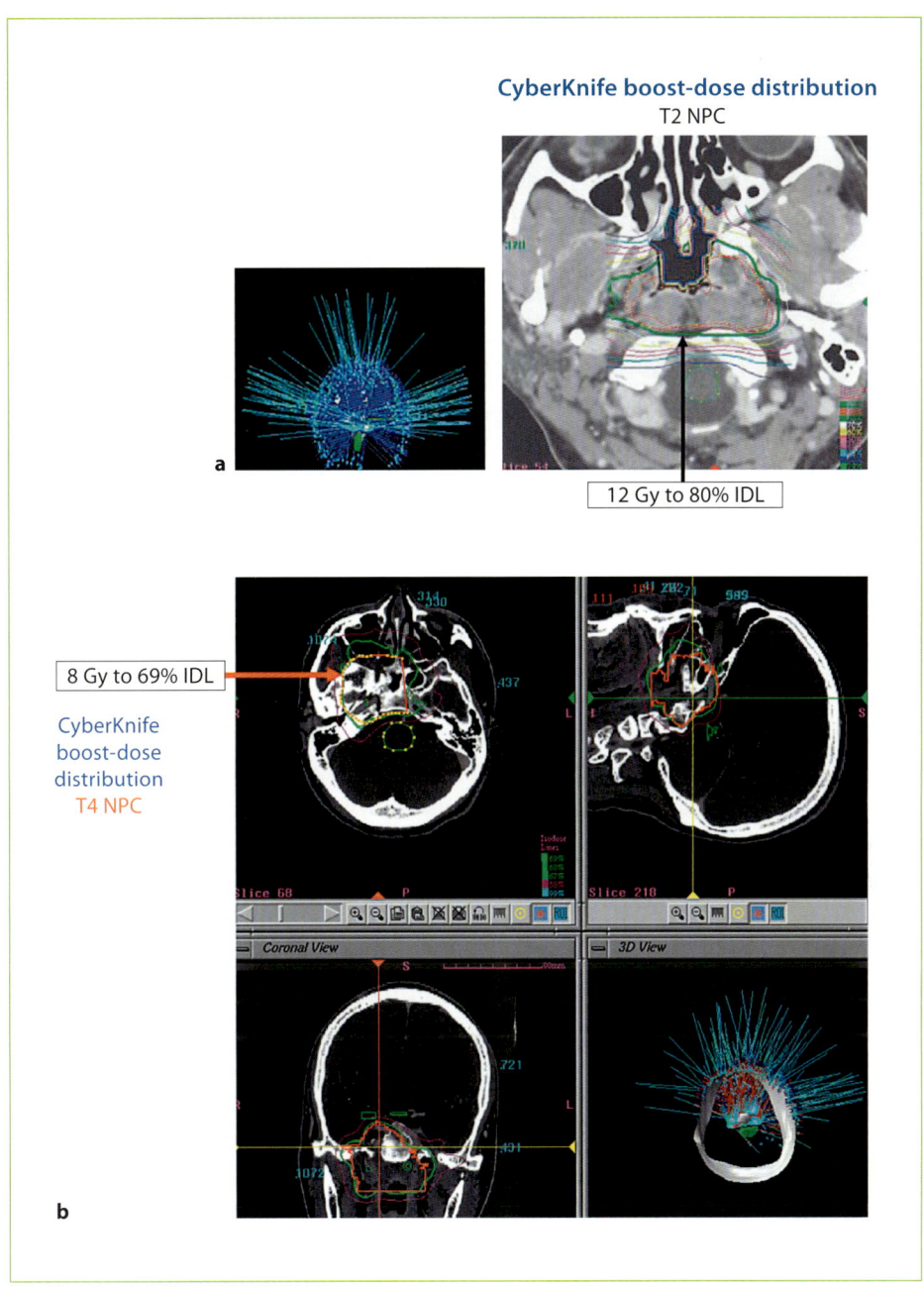

Fig. 22. Examples of stereotactic boost plans in NPC patients. **a** Patient with a T2b tumor receiving 12 Gy prescribed to the 80% isodose line (IDL). **b** Patient with a T4 tumor receiving 8 Gy to the 69% IDL.

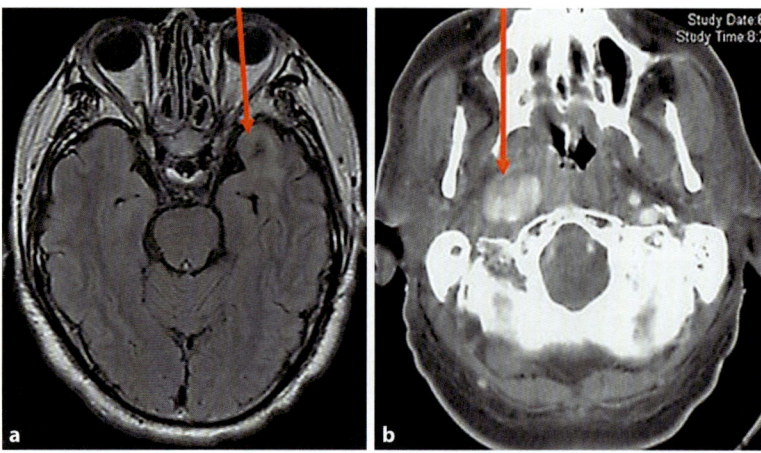

Fig. 23. a Representative MRI axial image showing evidence of temporal lobe necrosis (red arrow) in a NPC patient with a previous T4 tumor. The patient is completely asymptomatic and continues to work full-time as an engineer. **b** CT scan showing a carotid aneurysm (red arrow) requiring coiling in a patient with T2b tumor previously treated with stereotactic boost.

ynx and area of residual tumor. Figure 22b shows another plan for a patient with T4b tumor with skull base involvement and intracranial extension. The patient received 8 Gy to the area of persistent abnormality on MRI; the prescribed dose was lowered to meet the dose constraint to the optic chiasm and the brainstem. Overall, the most commonly prescribed dose was 8–12 Gy (median: 11 Gy, range: 7–15 Gy) and the median prescribed isodose line was 78% (54–95%). Chemotherapy was used in 85% of the patients. Stereotactic therapy was generally delivered approximately 2–3 weeks after completion of fractionated external beam radiotherapy, before the first cycle of adjuvant chemotherapy.

With a median follow-up period of 41 months, only 1 local relapse has been observed and the 5-year Kaplan Meier estimate of local control is 98%. The predominant pattern of relapse was distant despite the frequent use of systemic chemotherapy. The 5-year OS was 69%.

Late toxicity included the following: 1 patient with carotid aneurysm in the face of hypertension (fig. 23), 4 patients with transient cranial nerve paresis, 2 patients with permanent V2 and V3 numbness and 3 patients with retinopathy (1 possibly related to underlying diabetes). Ten patients had radiographic evidence of temporal lobe necrosis; 8 were asymptomatic and 2 experienced seizures. Nine of the 10 patients had initial intracranial extension of their tumors. Given the excellent control rates and incidence of temporal lobe necrosis, we now consider stereotactic boost primarily for large T3 and T4 tumors, and we have adjusted the dose to 10 Gy in 2 fractions.

Intensity-Modulated Radiation Therapy for Oropharyngeal Carcinomas

Target Volume Delineation

For oropharyngeal carcinoma (OPC), treatment planning is routinely performed using contrast-enhanced CT scans. Whenever possible, a recent MRI or PET scan may be merged with the CT to aid target delineation. We have found that the PET scan is primarily useful for localizing the tumor and involved nodes, but not for exact volume delineation as the FDG-avid volume is highly sensitive to the window/level or threshold settings used, as discussed above.

Dose Prescription

The prescription schemes for OPC are similar to those used for NPC. To account for organ motion and patient setup errors, we routinely add a 4- to 5-mm margin to the CTV70 to generate a PTV70 (70 Gy at 2.12 Gy/fraction). In addition, for base of tongue and vallecular tumors, the entire base of tongue and vallecula are contoured in the PTV60 to ensure that these structures receive at least 60 Gy delivered at 1.8 Gy/fraction. Similarly, for tonsil and soft palate tumors, the entire tonsillar fossa (from the pterygoid plate insertion to the pharyngoepiglottic fold) or the entire soft palate is contoured in the PTV60. In certain cases, where there are small, non-FDG-avid nodes (<1 cm in minimal cross-sectional diameter) that are located adjacent to clusters of known involved nodes, we also include them in the PTV60. The PTV52 and PTV56 (52 Gy at 1.58 Gy/fraction; and 56 Gy at 1.7 Gy/fraction) encompass the clinically negative neck, which includes bilateral retropharyngeal nodes and level II–V nodes. In cases where there is level II nodal involvement, level IB nodes are also included in PTV56.

Figure 24 shows a patient with a base of tongue squamous cell carcinoma. Note that bilateral retropharyngeal nodes were included for treatment since this patient had bilateral nodal involvement. We include in the PTV60 nodes that are considered suspicious but do not meet the definition of an involved node (>1 cm in minimal cross sectional diameter on CT, central necrosis, gross extracapsular extension or FDG avidity).

In the IMRT planning process, a useful option is the addition of a hypothetical tissue structure, called a 'tuning structure', to help conform the dose to the target. Figure 25 shows the IMRT dose distributions for a T2N2c tonsil carcinoma without or with the tuning structure placed posteriorly, adjacent to the PTV. Note the improvement in the dose distribution with smaller posterior dose concentrations.

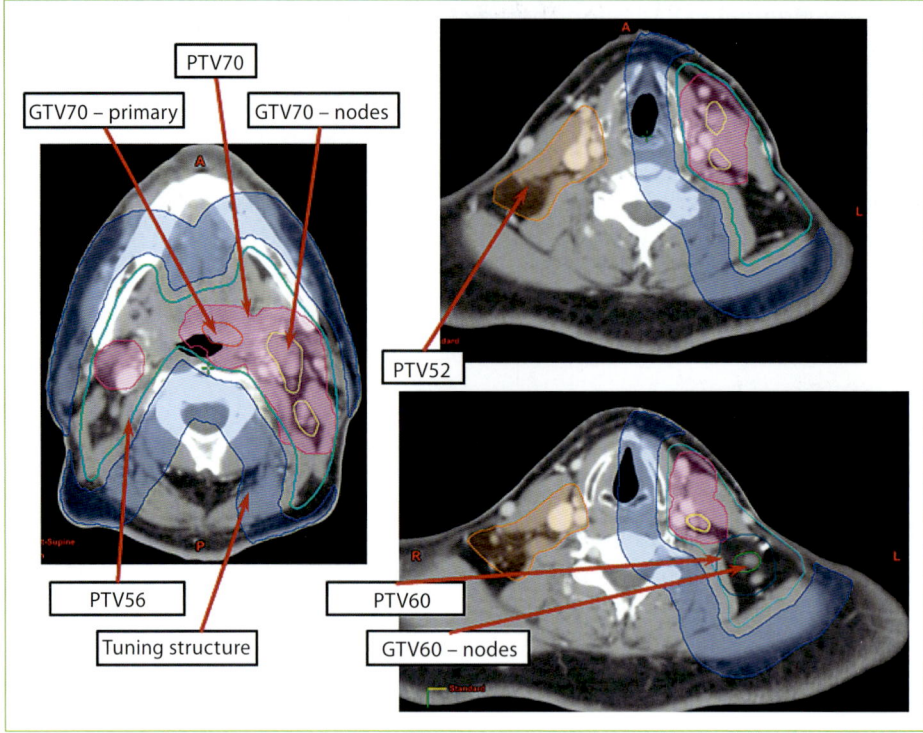

Fig. 24. Axial images showing contour for: GTV70, primary (red) and nodes (yellow); GTV60, nodes (green); PTV70 (magenta); PTV60 (blue); PTV56 (light blue); PTV52 (orange); and tuning structure (blue-shaded region).

Clinical Results for Intensity-Modulated Radiation Therapy of Oropharyngeal Carcinoma

IMRT results for OPC are not as mature as for NPC. De Arruda et al. [2] have published the Memorial Sloan-Kettering Cancer Center experience on IMRT for 50 patients with OPC. Eighty-six percent had concurrent chemotherapy. At a median follow-up point of 18 months, the 2-year estimates of local progression-free, regional progression-free, distant metastasis-free and OS rates were 98, 88, 84 and 98%, respectively. Among the patients with >9 months of follow-up, only 33% had grade 2 xerostomia (fig. 26). Esophageal strictures were noted in 3 patients.

Chao et al. [53] published results on 74 patients treated with IMRT for OPC. At a median follow-up time of 33 months, there were only 10 failures, giving a locoregional control rate of 87% and an OS rate of 87%. Only 9 out of 74 patients had grade 2 xerostomia. At MDACC, 51 patients with oropharyngeal tumors <4 cm in size were treated with IMRT alone and found to have excellent 2-year locoregion-

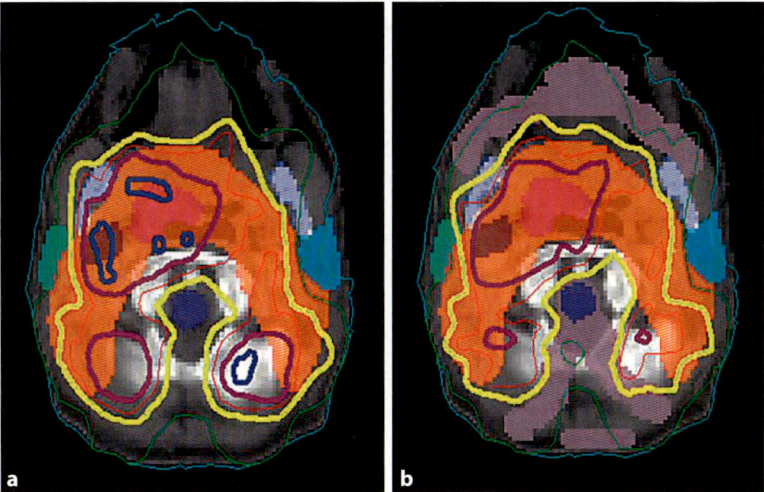

Fig. 25. Tuning structure for improving the dose distribution conformality. **a** The plan was optimized without a tuning structure. **b** The plan was optimized with a tuning structure posteriorly (in light purple). The same dose constraints were used for both cases. Note the reduction of the high-dose region posteriorly. (Dark blue: 73 Gy, purple: 66 Gy, red: 60 Gy, yellow: 54 Gy, green: 40 Gy, and light blue: 20 Gy.)

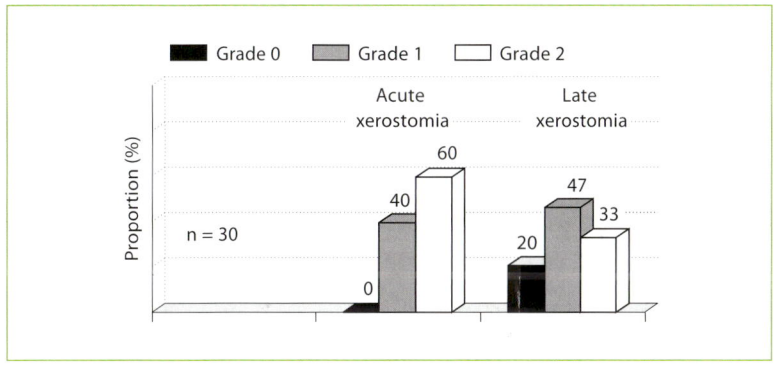

Fig. 26. Frequency of grade 0–2 xerostomia among patients with at least 9 months of follow-up after IMRT (adapted from de Arruda et al. [2]).

al control rates of 93% and OS rates of 93% [54]. While 40% of patients required a gastrostomy tube during radiotherapy, only 4 patients required a feeding tube for more than 6 months. UCSF completed an institutional retrospective study looking at stage III and IV OPC treated with IMRT and chemotherapy [55]. Even with the more advanced stage patients, 3-year local and regional progression-free survival rates were 94% and 94% respectively, and 3-year OS was 83%. We have also evalu-

Table 2. Experience by various institutions with IMRT for OPC

Institution	Patients, n	Treatment technique	Median follow-up months	Disease control and survival	Xerostomia
Mallinckrodt [53]	74	definitive: 31 (70 Gy) post-op: 43 (66 Gy) chemotherapy: 23%	33	4-year results LRC = 87% DFS = 81% OS = 87%	grade 2 xerostomia = 12.2%
MSKCC [2]	50	stage III/IV = 92% GTV: 70 Gy PTV: 59.4 Gy 86% had chemotherapy	18	2-year results LPF = 98% RPF = 88% DMFS = 84% OS = 98%	grade 2 xerostomia = 33%
University Wisconsin [57]	195 IMRT = 52 non-IMRT = 143	GTV: 70 Gy PTV: 54–60 Gy 33% had chemotherapy	30.4	3-year results CSS = 97.7% OS = 88.2% LRC = 96.1%	reported crude incidence of moderate xerostomia = 56% (IMRT group)
MSKCC [58]	293 stage III/IV = 112	IMRT vs. 3DCRT (concomitant boost) GTV: 70 Gy PTV: 54–59.4 Gy chemotherapy: CDDP	46 (3DCRT) 31(IMRT)	3-year results LPF = 95% RPF = 94% DMFS = 86% DFS = 82% OS = 91%	grade ≥2 xerostomia = 12% (IMRT) and 67% (3DCRT)
MDACC [54]	51 tumors <4 cm	GTV: 63–66 Gy PTV: 54 Gy only 5 patients had chemotherapy	45	2-year results LRC = 93% RFS = 87% OS = 93%	xerostomia not reported; feeding tube ≥1 year = 0
UCSF [55]	71 stage III/IV only	GTV: 70 Gy PTV: 54-59.4 Gy chemotherapy: CDDP	33	3-year results LPF = 94% RPF = 94% OS = 83%	grade 2 xerostomia = 33.8%

ated 107 OPC patients treated with definitive IMRT [56]. Concurrent platinum-based chemotherapy was administered to 86 patients (80%) and cetuximab to 8 patients (7%). With a median follow-up of 29 months among surviving patients (range: 4–105 months), the 3-year locoregional control, freedom from distant metastasis, OS and DFS were 92, 92, 83 and 81%, respectively. T stage (T4 vs. T1–3) was predictive of poorer locoregional control ($p = 0.001$), OS ($p = 0.001$) and DFS ($p < 0.001$). Acute toxicity consisted of 58% grade 3 mucosal and 5% grade 3 skin reactions. Six patients (6%) developed grade ≥3 late complications. The most recent RTOG trial published for OPC is a multicenter randomized phase II study looking at IMRT in early-stage disease. The tumor was treated with 66 Gy and chemotherapy was permitted. 2-year survival is promising with an OS 96% and DFS 82% [61]. The data from the literature for OPC patients treated with IMRT are summarized below (table 2).

Table 2 (continued)

Institution	Patients, n	Treatment technique	Median follow-up months	Disease control and survival	Xerostomia
Sanguineti (UT Galveston) [59]	50	DLMC GTV: 66–78 Gy no chemotherapy stage III/IV = 88%	32.6	3-year results LPF = 93.8% RPF = 85.1%	n/a
University of Florida [60]	100 64% oropharynx	GTV: 72 Gy (IMRT with concomitant boost)	3.1 years	3-year results (oropharynx only) FFR = 86% CSS = 92% OS = 84%	xerostomia not reported any grade ≥4 late toxicity = 13%
Stanford [56]	107	DLMC GTV: 66 Gy PTV: 54–60 Gy chemotherapy = 87% (CDDP or cetuximab)	29	3-year results LRC = 92% FFDM = 92% DFS = 83% OS = 81%	xerostomia not reported late grade ≥4 toxicity = 6%
RTOG 0022 [61]	69 T1-2, N0-1 only	GTV: 66 Gy PTV: 54–60 Gy chemotherapy allowed	2.8 years	2-year results LRF = 9% DM = 1/69 patients OS = 95.5% DFS = 82%	grade 2 xerostomia = 16%

MSKCC = Memorial Sloan-Kettering Cancer Center; UCSF = University of California San Francisco; RTOG = Radiation Therapy Oncology Group; LPF = local progression-free survival; RPF = regional progression-free survival; RFS = recurrence-free survival; DMFS = distant metastases-free survival; DM = distant metastases; DFS = disease-free survival; OS = overall survival.

Conclusions

Clinical results to date suggest that IMRT can improve local control rates and reduce xerostomia in patients with NPC and OPC. Long-term follow-up data show that stereotactic radiotherapy used as a boost can provide excellent local control in NPC patients, but at a cost of late toxicity. Future considerations should focus on improving the accuracy of target delineation for these tumors, which can have quite variable presentations in their extent of disease. This will likely include the incorporation of molecular imaging studies in daily radiotherapy treatment planning. Novel systemic therapies are greatly needed to reduce the risk of distant metastases in patients with NPC.

Guidelines for Clinical Practice

Intensity-Modulated Radiation Therapy/Image-Guided Radiation Therapy for Head and Neck Cancer
- It is important to use all modalities: PET, CT, MRI and clinical evaluation (especially for the mucosal extent of the disease). PET/CT for radiotherapy planning can improve consistency in target volume delineation, but is not infallible since false positive and negative results can occur.
- For determination of the GTV of the nodes, all PET-positive nodes should be included. Nodes suspicious on CT should also be included, regardless of the PET scan findings.
- When IMRT is used, adequate immobilization (especially of the shoulders) is needed to minimize both inter- and intrafraction variations.
- IGRT can be used to potentially decrease the PTV margin and reduce the volume of normal tissue receiving high radiation doses; however, its role in day-to-day adaptive planning is still unclear and needs further investigation.

Nasopharynx Cancer Planning
- MRI is superior to CT scanning in defining tumor involvement of the parapharyngeal space, skull base, cavernous sinus and retropharyngeal nodes. For nodal volume delineation, treatment planning PET/CT scans can often be helpful.
- For earlier-stage disease with no extension outside the nasopharynx, the CTV includes the posterior aspects of the nasal cavity, maxillary sinuses, the pterygoid fossa with the pterygoid plates and the anterior half of the clivus (fig. 13).
- In patients with parapharyngeal involvement and/or nasal cavity involvement, the CTV should include all of the pterygoid fossa and the lateral pterygoid muscle on the side of the parapharyngeal involvement. The nasal cavity should be included when involved with tumor.
- The CTV70 generated for the primary tumor and involved lymph nodes should include GTV with an added 5-mm margin. To generate the PTV70, we routinely will add approximately a 4- to 5-mm margin to the CTV70 (70 Gy at 2.12 Gy/fraction).
- The PTV56 and PTV52 (1.7 Gy/fraction and 1.58 Gy/fraction, respectively), encompass the clinically negative areas that are at risk of involvement as outlined by the CTV with an additional 5-mm margin as well as the nodal levels II–V. If level II is involved on the ipsilateral side of the tumor, level Ib is also included.
- See the text for special considerations regarding involvement of cavernous sinus, base of skull and other high-risk areas near critical structures. Consult your institutional radiologist for optimal target delineation, especially in these cases.

Oropharynx Cancer Planning
- Treatment planning is routinely performed using contrast-enhanced CT scans. Whenever possible, a recent MRI or PET scan may be merged with the CT to aid target delineation. PET is primarily useful for localizing the tumor and involved nodes, but not for exact volume delineation.
- The prescription schemes for OPC are similar to those used for NPC. To account for organ motion and patient setup errors, we routinely add a 4- to 5-mm margin to the CTV70 to generate a PTV70 (70 Gy at 2.12 Gy/fraction).

- For base of tongue and vallecular tumors, the entire base of tongue and vallecula are contoured in the PTV60 to ensure that these structures receive at least 60 Gy delivered at 1.8 Gy/fraction. Similarly, for tonsil and soft palate tumors, the entire tonsillar fossa (from the pterygoid plate insertion to the pharyngoepiglottic fold) or the entire soft palate is contoured in the PTV60.
- The PTV52 and PTV56 (52Gy at 1.58 Gy/fraction; and 56Gy at 1.7 Gy/fraction) encompass the clinically negative neck, which includes bilateral retropharyngeal nodes and level II–V nodes. In cases where there is level II nodal involvement, level IB nodes are also included in PTV56.

References

1 Cozzi L, Fogliata A, Bolsi A, Nicolini G, Bernier J: Three-dimensional conformal vs intensity-modulated radiotherapy in head-and-neck cancer patients: comparative analysis of dosimetric and technical parameters. Int J Radiat Oncol Biol Phys 2004;58:617–624.

2 de Arruda FF, Puri DR, Zhung J, Narayana A, Wolden S, Hunt M, Stambuk H, Pfister D, Kraus D, Shaha A, Shah J, Lee NY: Intensity-modulated radiation therapy for the treatment of oropharyngeal carcinoma: the Memorial Sloan-Kettering Cancer Center experience. Int J Radiat Oncol Biol Phys 2006;64:363–373.

3 Webster GJ, Rowbottom CG, Mackay RI: Accuracy and precision of an IGRT solution. Med Dosim 2009;34:99–106.

4 Dirix P, Vandecaveye V, De Keyzer F, Op de Beeck K, Poorten VV, Delaere P, Verbeken E, Hermans R, Nuyts S: Diffusion-weighted MRI for nodal staging of head and neck squamous cell carcinoma: impact on radiotherapy planning. Int J Radiat Oncol Biol Phys 2010;76:761–766.

5 Roh JL, Yeo NK, Kim JS, Lee JH, Cho KJ, Choi SH, Nam SY, Kim SY: Utility of 2-(18F) fluoro-2-deoxy-D-glucose positron emission tomography and positron emission tomography/computed tomography imaging in the preoperative staging of head and neck squamous cell carcinoma. Oral Oncol 2007;43:887–893.

6 Rodrigues RS, Bozza FA, Christian PE, Hoffman JM, Butterfield RI, Christensen CR, Heilbrun M, Wiggins RH 3rd, Hunt JP, Bentz BG, Hitchcock YJ, Morton KA: Comparison of whole-body PET/CT, dedicated high-resolution head and neck PET/CT, and contrast-enhanced CT in preoperative staging of clinically M0 squamous cell carcinoma of the head and neck. J Nucl Med 2009;50:1205–1213.

7 Lardinois D, Weder W, Hany TF, Kamel EM, Korom S, Seifert B, von Schulthess GK, Steinert HC: Staging of non-small-cell lung cancer with integrated positron-emission tomography and computed tomography. N Engl J Med 2003;348:2500–2507.

8 Fleming AJ Jr, Smith SP Jr, Paul CM, Hall NC, Daly BT, Agrawal A, Schuller DE: Impact of (18F)-2-fluorodeoxyglucose-positron emission tomography/computed tomography on previously untreated head and neck cancer patients. Laryngoscope 2007;117:1173–1179.

9 Ciernik IF, Dizendorf E, Baumert BG, Reiner B, Burger C, Davis JB, Lütolf UM, Steinert HC, Von Schulthess GK: Radiation treatment planning with an integrated positron emission and computer tomography (PET/CT): a feasibility study. Int J Radiat Oncol Biol Phys 2003;57:853–863.

10 Paulino AC, Koshy M, Howell R, Schuster D, Davis LW: Comparison of CT- and FDG-PET-defined gross tumor volume in intensity-modulated radiotherapy for head-and-neck cancer. Int J Radiat Oncol Biol Phys 2005;61:1385–1392.

11 Daisne JF, Duprez T, Weynand B, Lonneux M, Hamoir M, Reychler H, Grégoire V: Tumor volume in pharyngolaryngeal squamous cell carcinoma: comparison at CT, MR imaging, and FDG PET and validation with surgical specimen. Radiology 2004;233:93–100.

12 La TH, Filion EJ, Turnbull BB, Chu JN, Lee P, Nguyen K, Maxim P, Quon A, Graves EE, Loo BW Jr, Le QT: Metabolic tumor volume predicts for recurrence and death in head-and-neck cancer. Int J Radiat Oncol Biol Phys 2009;74:1335–1341.

13 MacManus M, Nestle U, Rosenzweig KE, Carrio I, Messa C, Belohlavek O, Danna M, Inoue T, Deniaud-Alexandre E, Schipani S, Watanabe N, Dondi M, Jeremic B: Use of PET and PET/CT for radiation therapy planning: IAEA expert report 2006–2007. Radiother Oncol 2009;91:85–94.

14 Burri RJ, Rangaswamy B, Kostakoglu L, Hoch B, Genden EM, Som PM, Kao J: Correlation of positron emission tomography standard uptake value and pathologic specimen size in cancer of the head and neck. Int J Radiat Oncol Biol Phys 2008; 71:682–688.

15 Wu K, Ung YC, Hornby J, Freeman M, Hwang D, Tsao MS, Dahele M, Darling G, Maziak DE, Tirona R, Mah K, Wong CS: PET CT thresholds for radiotherapy target definition in non-small-cell lung cancer: how close are we to the pathologic findings? Int J Radiat Oncol Biol Phys 2010;77: 699–706.

16 Ashamalla H, Guirgius A, Bieniek E, Rafla S, Evola A, Goswami G, Oldroyd R, Mokhtar B, Parikh K: The impact of positron emission tomography/computed tomography in edge delineation of gross tumor volume for head and neck cancers. Int J Radiat Oncol Biol Phys 2007;68: 388–395.

17 Murakami R, Uozumi H, Hirai T, Nishimura R, Shiraishi S, Ota K, Murakami D, Tomiguchi S, Oya N, Katsuragawa S, Yamashita Y: Impact of FDG-PET/CT imaging on nodal staging for head-and-neck squamous cell carcinoma. Int J Radiat Oncol Biol Phys 2007;68:377–382.

18 Vermorken JB, Remenar E, van Herpen C, Gorlia T, Mesia R, Degardin M, Stewart JS, Jelic S, Betka J, Preiss JH, van den Weyngaert D, Awada A, Cupissol D, Kienzer HR, Rey A, Desaunois I, Bernier J, Lefebvre JL, EORTC 24971/TAX 323 Study Group: Cisplatin, fluorouracil, and docetaxel in unresectable head and neck cancer. N Engl J Med 2007;357:1695–704.

19 Posner MR, Hershock DM, Blajman CR, Mickiewicz E, Winquist E, Gorbounova V, Tjulandin S, Shin DM, Cullen K, Ervin TJ, Murphy BA, Raez LE, Cohen RB, Spaulding M, Tishler RB, Roth B, Viroglio Rdel C, Venkatesan V, Romanov I, Agarwala S, Harter KW, Dugan M, Cmelak A, Markoe AM, Read PW, Steinbrenner L, Colevas AD, Norris CM Jr, Haddad RI, TAX 324 Study Group: Cisplatin and fluorouracil alone or with docetaxel in head and neck cancer. N Engl J Med 2007;357:1705–1715.

20 Tannock IF: Combined modality treatment with radiotherapy and chemotherapy. Radiother Oncol 1989;16:83–101.

21 Konski AA, Cheng JD, Goldberg M, Li T, Maurer A, Yu JQ, Haluszka O, Scott W, Meropol NJ, Cohen SJ, Freedman G, Weiner LM: Correlation of molecular response as measured by 18-FDG positron emission tomography with outcome after chemoradiotherapy in patients with esophageal carcinoma. Int J Radiat Oncol Biol Phys 2007;69: 358–363.

22 McCollum AD, Burrell SC, Haddad RI, Norris CM, Tishler RB, Case MA, Posner MR, Van den Abbeele AD: Positron emission tomography with 18F-fluorodeoxyglucose to predict pathologic response after induction chemotherapy and definitive chemoradiotherapy in head and neck cancer. Head Neck 2004;26:890–896.

23 Salama JK, Haddad RI, Kies MS, Busse PM, Dong L, Brizel DM, Eisbruch A, Tishler RB, Trotti AM, Garden AS: Clinical practice guidance for radiotherapy planning after induction chemotherapy in locoregionally advanced head-and-neck cancer. Int J Radiat Oncol Biol Phys 2009;75:725–733.

24 Rosenthal DI, Fuller CD, Ang KK, Brizel DM, Eisbruch A, Le QT, Lee NY, O'Sullivan B, Duppen J, Rasch CRN: Pre- and postinduction chemotherapy target volume delineation in head and neck cancer: preliminary results from an expert panel. Int J Radiat Oncol Biol Phys 2009;75:S431.

25 Geets X, Tomsej M, Lee JA, Duprez T, Coche E, Cosnard G, Lonneux M, Grégoire V: Adaptive biological image-guided IMRT with anatomic and functional imaging in pharyngo-laryngeal tumors: impact on target volume delineation and dose distribution using helical tomotherapy. Radiother Oncol 2007;85:105–115.

26 Hwang AB, Bacharach SL, Yom SS, Weinberg VK, Quivey JM, Franc BL, Xia P: Can positron emission tomography (PET) or PET/computed tomography (CT) acquired in a nontreatment position be accurately registered to a head-and-neck radiotherapy planning CT? Int J Radiat Oncol Biol Phys 2009;73:578–584.

27 Nabili V, Zaia B, Blackwell KE, Head CS, Grabski K, Sercarz JA: Positron emission tomography: poor sensitivity for occult tonsillar cancer. Am J Otolaryngol 2007;28:153–157.

28 Dong MJ, Zhao K, Lin XT, Zhao J, Ruan LX, Liu ZF: Role of fluorodeoxyglucose-PET versus fluorodeoxyglucose-PET/computed tomography in detection of unknown primary tumor: a meta-analysis of the literature. Nucl Med Commun 2008;29:791–802.

29 Yeung HW, Grewal RK, Gonen M, Schöder H, Larson SM: Patterns of (18)F-FDG uptake in adipose tissue and muscle: a potential source of false-positives for PET. J Nucl Med 2003;44: 1789–1796.

30 El-Haddad G, Alavi A, Mavi A, Bural G, Zhuang H: Normal variants in (18F)-fluorodeoxyglucose PET imaging. Radiol Clin North Am 2004;42: 1063–1081.

31 Pentenero M, Cistaro A, Brusa M, Ferraris MM, Pezzuto C, Carnino R, Colombini E, Valentini MC, Giovanella L, Spriano G, Gandolfo S: Accuracy of 18F-FDG-PET/CT for staging of oral squamous cell carcinoma. Head Neck 2008;30:1488–1496.

32 Hong TS, Tome WA, Chappell RJ, Chinnaiyan P, Mehta MP, Harari PM: The impact of daily set-up variations on head-and-neck intensity-modulated radiation therapy. Int J Radiat Oncol Biol Phys 2005;61:779–788.

33 La TH, Chao M, Xing L, Le Q: Evaluation of intrafraction motion in head and neck cancer during radiotherapy. Int J Radiat Oncol Biol Phys 2007;69:S681–S682.

34 Dong L, Chen YP, Lindberg ME, Garden AS, Rosenthal DI, Sejpal SV, Shah SJ, Morrison WH, Ang KK, Schwartz DL: Interfractional movement of the larynx and oropharynx during radiotherapy. Int J Radiat Oncol Biol Phys 2009;75:S16–S17.

35 McRackan TR, Watkins JM, Herrin AE, Garrett-Mayer EM, Sharma AK, Day TA, Gillespie MB: Effect of body mass index on chemoradiation outcomes in head and neck cancer. Laryngoscope 2008;118:1180–1185.

36 Dirix P, Abbeel S, vanstraelen B, Hermans R, Nuyts S: Dysphagia after chemoradiation for head and neck squamous cell carcinoma: dose-effect relationships for the swallowing structures. Int J Radiat Oncol Biol Phys 2009;75:385–392.

37 Caudell JJ, Schaner PE, Desmond RA, Meredith RF, Spencer SA, Bonner JA: Dosimetric factors associated with long-term dysphagia after definitive radiotherapy for squamous cell carcinoma of the head and neck. Int J Radiat Oncol Biol Phys 2010;76:403–409.

38 Salama JK, Stenson KM, List MA, Mell LK, Maccracken E, Cohen EE, Blair E, Vokes E, Haraf DJ: Characteristics associated with swallowing changes after concurrent chemotherapy and radiotherapy in patients with head and neck cancers. Arch Otolaryngol Head Neck Surg 2008; 134:1060–1065.

39 Mohan R, Zhang X, Wang H, Kang Y, Wang X, Liu H, Ang KK, Kuban D, Dong L: Use of deformed intensity distributions for on-line modification of image-guided IMRT to account for interfractional anatomic changes. Int J Radiat Oncol Biol Phys 2005;61:1258–1266.

40 Dijkema T, Raaijmakers CP, Ten Haken RK, Roesink JM, Braam PM, Houweling AC, Moerland MA, Eisbruch A, Terhaard CH: Parotid gland function after radiotherapy: the combined Michigan and Utrecht experience. Int J Radiat Oncol Biol Phys 2010;78:449–453.

41 Pehlivan B, Pichenot C, Castaing M, Auperin A, Lefkopoulos D, Arriagada R, Bourhis J. Interfractional set-up errors evaluation by daily electronic portal imaging of IMRT in head and neck cancer patients. Acta Oncol 2009;48:440–445.

42 Den RB, Doemer A, Kubicek G, Bednarz G, Galvin JM, Keane WM, Xiao Y, Machtay M: Daily image guidance with cone-beam computed tomography for head-and-neck cancer intensity-modulated radiotherapy: a prospective study. Int J Radiat Oncol Biol Phys 2010;76:1353–1359.

43 Bucci M, Xia P, Lee N, Fischbein N, Kramer A, Weinberg V, Akazawa C, Cabrera A, Fu KK, Quivey JM: Intensity modulated radiation therapy for carcinoma of the nasopharynx: an update of the UCSF experience. Int J Radiat Oncol Biol Phys 2004;60:S317–S318.

44 Wolden SL, Chen WC, Pfister DG, Kraus DH, Berry SL, Zelefsky MJ: Intensity-modulated radiation therapy (IMRT) for nasopharynx cancer: update of the Memorial Sloan-Kettering experience. Int J Radiat Oncol Biol Phys 2006;64:57–62.

45 Kam MK, Teo PM, Chau RM, Cheung KY, Choi PH, Kwan WH, Leung SF, Zee B, Chan AT: Treatment of nasopharyngeal carcinoma with intensity-modulated radiotherapy: the Hong Kong experience. Int J Radiat Oncol Biol Phys 2004;60: 1440–1450.

46 Fang FM, Chien CY, Tsai WL, Chen HC, Hsu HC, Lui CC, Huang TL, Huang HY: Quality of life and survival outcome for patients with nasopharyngeal carcinoma receiving three-dimensional conformal radiotherapy vs. intensity-modulated radiotherapy – a longitudinal study. Int J Radiat Oncol Phys Biol 2008;72:356–364.

47 Zhao C, Xiao W, Han F, Lu L, Huang S, Wu S, Chen C, Chen J, Lin C, Deng X: Long-term results and prognostic factors of primary nasopharyngeal carcinoma patients treated with intensity modulated-radiotherapy. Int J Radiat Oncol Biol Phys 2009;75:S630.

48 Tham IW, Hee SW, Yeo RM, Salleh PB, Lee J, Tan TW, Fong KW, Chua ET, Wee JT: Treatment of nasopharyngeal carcinoma using intensity-modulated radiotherapy – the National Cancer Centre Singapore experience. Int J Radiat Oncol Biol Phys 2009;75:1481–1486.

49 Lee N, Harris J, Garden AS, Straube W, Glisson B, Xia P, Bosch W, Morrison WH, Quivey J, Thorstad W, Jones C, Ang KK: Intensity-modulated radiation therapy with or without chemotherapy for nasopharyngeal carcinoma: radiation therapy oncology group phase II trial 0225. J Clin Oncol 2009;27:3684–3690.

50 Palazzi M, Orlandi E, Bossi P, Pignoli E, Potepan P, Guzzo M, Franceschini M, Scaramellini G, Cantu G, Licitra L, Olmi P, Tomatis S: Further improvement in outcomes of nasopharyngeal carcinoma with optimized radiotherapy and induction plus concomitant chemotherapy: an update of the Milan experience. Int J Radiat Oncol Biol Phys 2009;74:774–780.

51 Le QT, Tate D, Koong A, Gibbs IC, Chang SD, Adler JR, Pinto HA, Terris DJ, Fee WE, Goffinet DR: Improved local control with stereotactic radiosurgical boost in patients with nasopharyngeal carcinoma. Int J Radiat Oncol Biol Phys 2003;56:1046–1054.

52 Hara, W, Loo BW, Goffinet DR, Chang SD, Adler JR, Pinto HA, Fee WE, Kaplan MJ, Fischbein NJ, Le QT: Excellent local control with stereotactic radiotherapy boost after external beam radiotherapy in patients with nasopharyngeal carcinoma. Int J Radiat Oncol Biol Phys 2008;71:393–400.

53 Chao KS, Ozyigit G, Blanco AI, Thorstad WL, Deasy JO, Haughey BH, Spector GJ, Sessions DG: Intensity-modulated radiation therapy for oropharyngeal carcinoma: impact of tumor volume. Int J Radiat Oncol Biol Phys 2004;59:43–50.

54 Garden AS, Morrison WH, Wong PF, Tung SS, Rosenthal DI, Dong L, Mason B, Perkins GH, Ang KK: Disease-control rates following intensity-modulated radiation therapy for small primary oropharyngeal carcinoma. Int J Radiat Oncol Biol Phys 2007;67:438–444.

55 Huang K, Xia P, Chuang C, Weinberg V, Glastonbury CM, Eisele DW, Lee NY, Yom SS, Phillips TL, Quivey JM: Intensity-modulated chemoradiation for treatment of stage III and IV oropharyngeal carcinoma: the University of California San Francisco experience. Cancer 2008;113:497–507.

56 Daly ME, Le QT, Maxim PM, Loo BW Jr, Kaplan MJ, Fischbein NJ, Pinto H, Chang DT: Intensity-modulated radiotherapy in the treatment of oropharyngeal cancer: clinical outcomes and patterns of failure. Int J Radiat Oncol Biol Phys 2010;76:1339–1346.

57 Hodge CW, Bentzen SM, Wong G, Palazzi-Churas KL, Wiederholt PA, Gondi V, Richards GM, Hartig GK, Harari PM: Are we influencing outcome in oropharynx cancer with intensity-modulated radiotherapy? An inter-era comparison. Int J Radiat Oncol Biol Phys 2007;69:1032–1041.

58 Lee NY, de Arruda FF, Puri DR, Wolden SL, Narayana A, Mechalakos J, Venkatraman ES, Kraus D, Shaha A, Shah JP, Pfister DG, Zelefsky MJ: A comparison of intensity-modulated radiation therapy and concomitant boost radiotherapy in the setting of concurrent chemotherapy for locally advanced oropharyngeal carcinoma. Int J Radiat Oncol Biol Phys 2006;66:966–974.

59 Sanguineti G, Gunn GB, Endres EJ, Chaljub G, Cheruvu P, Parker B: Patterns of locoregional failure after exclusive IMRT for oropharyngeal carcinoma. Int J Radiat Oncol Biol Phys 2008;72:737–746.

60 Schoenfeld GO, Amdur RJ, Morris CG, Li JG, Hinerman RW, Mendenhall WM: Patterns of failure and toxicity after intensity-modulated radiotherapy for head and neck cancer. Int J Radiat Oncol Biol Phys 2008;71:377–385.

61 Eisbruch A, Harris J, Garden AS, Chao CK, Straube W, Harari PM, Sanguineti G, Jones CU, Bosch WR, Ang KK: Multi-institutional trial of accelerated hypofractionated intensity-modulated radiation therapy for early-stage oropharyngeal cancer (RTOG 00-22). Int J Radiat Oncol Biol Phys 2010;76:1333–1338.

Quynh-Thu Le, MD
Department of Radiation Oncology, Stanford University Medical Center
875 Blake Wilbur Drive, Rm CC-G228
Stanford, CA 94305-5847 (USA)
Tel. +1 650 498 5032, Fax +1 650 725 8231, E-Mail qle@stanford.edu

Delineating Neck Targets for Intensity-Modulated Radiation Therapy of Head and Neck Cancer

Merav Ben David · Avraham Eisbruch

Department of Radiation Oncology, University of Michigan, Ann Arbor, Mich., USA

Abstract

Experience with intensity-modulated radiation therapy (IMRT) for head and neck cancer is building greater understanding of the requirements for therapy planning. Delineation of the lymphatic targets for IMRT of the head and neck is a crucial step in this planning, and often determines the risks of marginal or out-of-field local/regional tumor recurrence. Definition of the gross tumor volumes needs to take into account both radiological (CT, MRI, PET) and clinical findings. Understanding of the appropriate CTVs is developing based on: (a) established knowledge of the natural history and spread patterns of head and neck cancer, (b) the accruing experience of clinicians using IMRT, and (c) evaluations of patient outcomes following consistent treatment approaches as determined by institution practice patterns and prospective clinical studies. This chapter will outline the important steps in lymphatic target definition for head and neck cancer, and will discuss several special clinical concerns for these patients and their management.

Copyright © 2011 S. Karger AG, Basel

Studies in intensity-modulated radiation therapy (IMRT) of head and neck cancer have mostly concentrated on the sparing of noninvolved tissues from irradiation. Major salivary glands, minor salivary glands dispersed throughout the oral cavity, the mandible and the pharyngeal mucosa and musculature can all be partly spared, improving post-treatment quality of life compared with conventional radiotherapy [1–3]. Because of the close proximity of lymphatic structures to many of these volumes, it is important to understand the precise anatomy of the lymphatic chains at risk.

In cases of nasopharyngeal and paranasal sinus cancers, additional critical normal tissues that may be partly spared using IMRT include the inner and middle ears, temporomandibular joints, temporal brain lobes and optic pathways [4]. In addition to sparing noninvolved tissues, IMRT offers a potential for improving tumor control by reducing the constraints on the total tumor dose imposed by adjacent critical normal tissues, especially the spinal cord and brain stem, which can be problematic with conventional treatment.

The need for accurate selection and definition of the targets for IMRT is especially relevant in head and neck cancer, where a high risk of subclinical local and nodal disease exists and where adequate irradiation of the lymph nodes at risk is crucial for local-regional control and survival. For example, in standard 3-field radiotherapy of oropharyngeal cancer, the first echelon and the retropharyngeal (RP) nodes are inevitably treated when the primary tumor is targeted. In contrast, these nodes will not be adequately irradiated by IMRT if they are not specified as targets on the planning CT. Often the accurate identification of the clinical target volume (CTV) for the regional lymphatics at risk is critical in achieving disease control and normal tissue preservation.

Tumor and Target Volumes in the Neck

Gross Tumor Volume

Understanding the regional lymphatics at risk depends on an accurate evaluation of disease involvement of both the primary site and the regional lymphatics; the extent of the gross tumor volume (GTV) should be known precisely. The physical exam (including mirror and fiber-optic exam), in conjunction with the surgeon's report of direct endoscopy under anesthesia, are important data to define the tumor. In most cases, the simulation contrast-enhanced CT is the only imaging modality required for the delineation of the targets. MRI is limited by its sensitivity to artifacts, difficulty in interpretation, long examination time and cost. However, MRI is an essential adjunct to CT for tumors close to the base of skull, i.e. nasopharyngeal and paranasal sinus cancers, where it provides better detail of the tumor extension and of the parapharyngeal and RP spaces compared with CT [5]. MRI is therefore essential for delineating the targets in these cases.

FDG-PET imaging may provide additional staging information. Clinical evaluations of head and neck cancer patients, where CT, MRI and FDG-PET were all obtained and surgery was then performed to validate the primary tumor extent and lymph node involvement, have reported a small additional benefit of FDG-PET compared with CT and MRI [6, 7]. In cases of recurrent cancer, FDG-PET has demonstrated significantly higher utility compared with CT or MRI [8–10]. We utilize FDG-PET registered with CT as an adjunct to the simulation CT. The GTV may be adjusted by including the composite abnormality of both CT and PET

Table 1. Criteria for including lymph nodes in the GTV

Lymph nodes are included in the GTV if they have any of the radiologic criteria:
(1) diameter >1 cm (or in the case of the jugulodigastric nodes, >1.1–1.5 cm);
(2) smaller than 1 cm with spherical rather than ellipsoidal shape;
(3) contain inhomogeneities suggestive of necrotic centers;
(4) cluster of 3 or more borderline nodes [11];
(5) FDG-PET-positive.

(rather than relying on one modality). Borderline enlarged nodes noted on the CT scan, which might have been included in the subclinical target, become part of the GTV if they are PET avid. Thus, PET can alter therapy by increasing prescribed doses to sites found to be involved, compared with CT alone.

Based on the experience of imaging sciences and radiation oncology practice, this institution has developed some broad policies. Lymph nodes are included in the GTV if they have any of the radiologic criteria shown in table 1.

Clinical Target Volumes in the Neck

The CTV surrounding the primary tumor encompasses tissues judged to be at risk for microscopic, subclinical tumor extension. Neck CTVs consist of the neck levels at risk that do not match the radiologic criteria of node involvement. Factors used for assessing the extent of the CTV margins in each case include tumor site, size, stage, differentiation and morphology (ulcerative or exophytic, infiltrative or pushing front). Rather than simply expanding the GTV uniformly, an approach of outlining the CTV on the planning CT on a slice-by-slice basis is recommended. Knowledge of the anatomical and clinical patterns of tumor extension, clinical judgment, and familiarity with head and neck imaging are necessary for the accurate estimation of the CTV margins around the tumor. Details of CTV outlining around the primary tumor have been discussed elsewhere [12]. In this paper, we will concentrate on the delineation of the neck CTVs.

Lymphatic Drainage
Our knowledge of the pattern and risk of lymphatic drainage from different head and neck sites is based on the classic anatomical work by Rouvière [13, 14], the assessment of the location and prevalence of clinical neck metastases by Lindberg [15], and the large experience with elective neck dissections providing information about microscopic metastases reported by Byers et al. [16], Candela et al. [17] and others. These studies demonstrate that squamous cell carcinomas of the upper aerodigestive tract tend to metastasize to the neck in predictable patterns, gov-

erned by the density and drainage of the lymphatics at each site, and with increasing risk at each level if the adjoining proximal level is involved.

To create an anatomic reference system for neck nodes, surgeons from the Memorial Sloan-Kettering Hospital developed a simple division, revised by Robbins et al. [18] and Robbins [19]. Six levels were defined, each with discrete anatomic boundaries that are apparent during neck dissection. This classification allows standardized reporting of the sites of nodal involvement in the neck, as well as the sites of surgical therapy. A corresponding imaging-based nodal classification, using CT- or MRI-based criteria for these surgical anatomic landmarks, has been developed by head and neck radiologists [20]. In addition, several publications by radiation oncologists have demonstrated how to outline the lymph node neck levels as CTV on planning CT or axial MRI scans [21–23]. The above publications are highly recommended for the reader before pursuing 3D conformal radiation therapy or IMRT to the head and neck for cancer treatment. An extensive review of the literature regarding the risk of metastases to each neck level has recently been published by Gregoire et al. [21]. Another tool is on the Radiation Therapy Oncology Group (RTOG) website, which has a detailed atlas of the head and neck with illustrations of the different neck levels (http://www.rtog.org/hnatlas) [24].

To include a neck region in the CTV, we consider that a 10% or higher risk of metastatic involvement justifies treatment. The factors affecting this risk should be taken into account for each case: tumor stage, size, thickness (3 mm or more is associated with a high metastatic risk for oral cavity tumors), differentiation, keratinization status, lymphatic vessel invasion in the tumor specimen, and whether other neck levels are involved [12].

Defining the Clinical Target Volume in the Neck
In defining the CTVs for head and neck cancer IMRT, the following node regions or levels (fig. 1) should be included. While these general concepts are broadly useful, specific approaches for individual tumor sites are described in detail elsewhere [23].

Contralateral Nodes. These should be included for cases of lateralized cancer where only the ipsilateral neck would ordinarily require therapy (small tonsillar cancers, and retromolar trigone and buccal mucosa cancers). Contralateral neck treatment is always added to the CTV when the ipsilateral neck involvement is greater than N1.

Level II Nodes. Level II is the most frequent nodal metastatic site for tumors in all mucosal sites. This level can be divided into the subdigastric (jugulodigastric) nodes, located below the level at which the posterior belly of the digastric muscle crosses the jugular vein, and more cranially located nodes below the base of skull ('junctional' nodes in Rouvière's [13] terminology). The subdigastric nodes are the main nodes involved when contralateral metastases occur, while the more cepha-

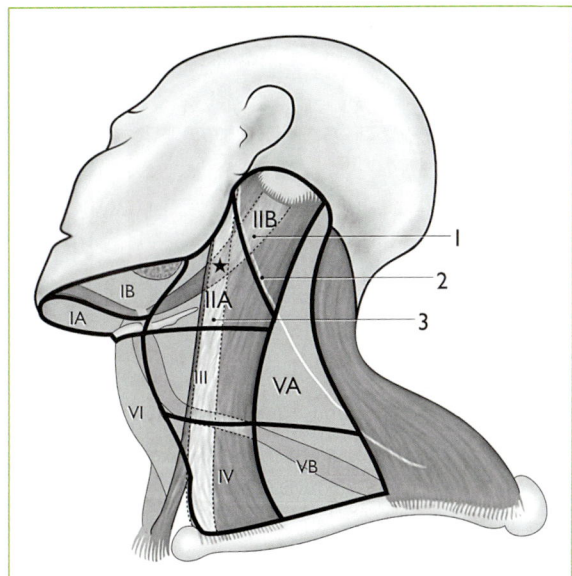

Fig. 1. Neck node levels. 1 = Posterior belly of the digastric muscle; 2 = accessory nerve; 3 = jugular vein; star = jugulodigastric nodes: where jugular vein bisects the posterior belly of the digastric muscle (just below the transverse process of C1).

lad nodes are at risk bilaterally in cases of nasopharyngeal cancer and in the neck side that contains other level II–III metastases [23] (fig. 2, 3).

Level IB and IV Nodes. Levels IB and IV are treated in all cases in the neck side with clinical involvement of levels II or III (fig. 4).

Level V Nodes. Level V is treated in the neck side with significant (>N1) involvement of levels II–IV, in all cases.

RP Nodes. The RP nodes are treated bilaterally in all cases of oropharyngeal and hypopharyngeal cancer with clinical involvement of levels II–IV (in cases of early lateralized oropharyngeal tumors with small N1 disease, treat ipsilaterally).

Level VI Nodes. Level VI nodes are treated in all cases with clinical involvement of level IV nodes.

Special Considerations for Target Delineation

To better understand the validity of the IMRT target delineations used at this center and others, a series of 133 patients treated for head and neck cancer were studied. If local-regional failure occurred after treatment (21 cases), the sites of failure were carefully examined (especially near the base of skull) in relation to the IMRT targets that were used for each case [25, 26]. All patients had non-nasopharyngeal head and neck squamous cell carcinomas and underwent courses of curative, parotid-sparing IMRT.

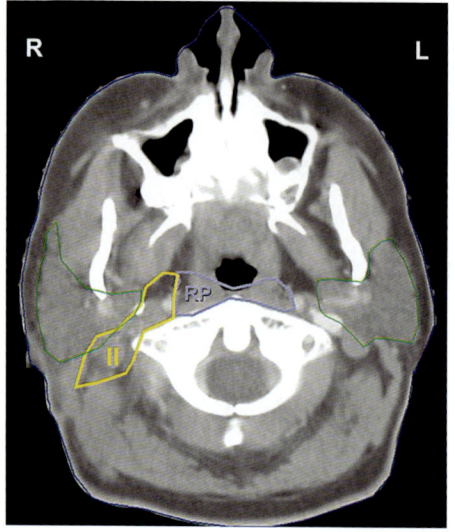 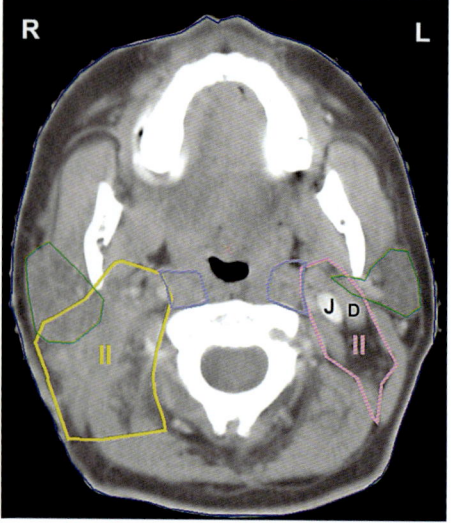

Fig. 2. Case 1 (described below), showing an axial CT image *superior* to the transverse process of C1. The patient has a right oropharyngeal cancer with grossly involved right level II nodes. On the right, level II nodes are outlined for treatment as well as the bilateral RP nodes to the base of the skull. On the left, level II is not outlined at this superior level.

Fig. 3. Case 1, showing an axial CT image *inferior* to the transverse process of C1. The nodal targets are now outlined in level II on the left to ensure coverage of the jugulodigastric nodes.

In this study, the *contralateral* neck was clinically node-negative, but was judged to be at a high risk of subclinical disease and therefore treated in all patients. The uppermost level II nodal target was the subdigastric node group. To assure its coverage, the uppermost CTV was delineated at the axial CT image in which the posterior belly of the digastric muscle crossed the jugular vein.

In the *ipsilateral* neck, which was node-positive in most patients, the uppermost level II CTV was delineated through the base of the skull (fig. 2, 3). The uppermost RP nodal target was delineated at the level of the top of the C1 vertebral body, accommodating Rouvière's [13] description of the location of the lateral RP nodes.

At a median follow-up period of 32 months, 21 of the 133 patients (16%) had local-regional recurrences (17 were in-field and 4 were marginal). Findings from this study and from similar work at other institutions highlight the following special concerns with the use of IMRT for head and neck cancer.

Retropharyngeal Nodes

Two of the marginal recurrences in our study occurred near the base of skull, corresponding to the lateral RP nodes. Both patients had oropharyngeal cancer (one

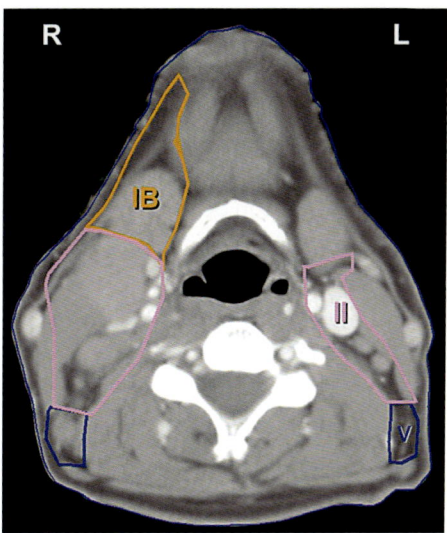

Fig. 4. Case 1, showing an axial CT image in the mid-neck. Level IB is outlined as a target in the ipsilateral (right) neck, due to the gross involvement of adjacent level II nodes, not in the contralateral, clinically uninvolved neck.

tonsillar and one base of tongue), and presented with N0 necks. In each case, the RP nodes had been defined as targets, and the cranial most extent of the CTV was delineated at the top of C1. However, the epicenter of the recurrence volume of the two marginal RP recurrences lay cranial to the top of C1. One of these recurrences was in the ipsilateral lateral RP nodes, and the other was at the contralateral base of skull and was extensive.

The findings of RP nodal recurrences whose epicenters were cranial to the top of C1 were unexpected. Details in the literature about the pattern of metastases to the RP nodes in non-nasopharyngeal cancer are scant [27, 28]. In IMRT of nasopharyngeal cancer, these nodes are usually included in the GTV or the CTV of the primary tumor that extends cranially to the base of the skull. In IMRT of non-nasopharyngeal cancer, there is a need for a conscientious effort to determine first, whether these nodes should be included as targets, and second, how to delineate them. Until more comprehensive information is available, and given the recurrences we have observed, we recommend the inclusion of the lateral RP nodes as targets in all cases of locally advanced oropharyngeal cancer, whether or not there is evidence of involvement of other neck levels. Our results suggest that in cases that are at risk of RP nodal metastases, these targets should extend more cranially than the top of C1, from the level of the upper nasopharynx to the base of skull. Both the ipsilateral and the contralateral nodes should be included in the RP nodal target. This is consistent with the recent consensus guidelines [29] (fig. 2, 3) and case reports in the literature outlining concerns similar to ours [30–32].

Extension from Disease Involvement

The third recurrence in our series was in a patient with tongue cancer who had significant involvement of the low neck. After irradiation, the patient achieved complete response and later recurred in level VI. This recurrence highlights the risk of recurrence at any neck level when adjoining levels are involved. If lymphatic flow obstruction occurs due to metastases, retrograde lymphatic flow may result [33]. Connections from the pretracheal/prelaryngeal nodes (level VI) to the jugular nodes (levels III–IV) have been described by Rouvière [13]. These connections, in tandem with flow obstruction at level IV, explain this marginal recurrence at level VI. Level VI nodes should be treated in all cases with clinical involvement of level IV nodes.

Prior Surgery

The fourth marginal recurrence was observed in a patient who had had a neck dissection more than a year prior to recurrence of an oral cavity cancer, then treated with local excision and IMRT. He recurred again in an unpredicted anatomic site: the subcutaneous tissues of contralateral level I. This marginal recurrence demonstrates the effect of collateral lymphatics, which have been demonstrated to fully develop 1 or more years after surgery and to extend to unpredictable neck sites [13]. These collaterals develop in the submental and submandibular areas, including subdermal tissues. We currently exclude from IMRT patients who have had major neck surgery more than 1 year previously, due to the uncertainty regarding their targets. Similar findings demonstrating aberrant lymphatic drainage after neck dissection have been reported by other investigators [34, 35].

Supraclavicular Treatment

Chao et al. [36] reported the outcomes of 126 head and neck cancer patients treated with IMRT, which was used only in the upper neck to spare the salivary function. The lower neck was treated with a conventional anterior low neck port abutted to the inferior IMRT dose distribution border. Five patients had recurrence in the supraclavicular area (28% of recurrences). This demonstrates the importance of contouring the regional lymph nodes in the low neck in cases where they are at high risk, such as cases with significant nodal involvement of the high neck, and delivering radiation to that area with appropriate dose and energy. Preferably, the low neck targets should be included in the IMRT treatment plan when they are at high risk of metastases.

Preservation of Swallowing Structures

Radiotherapy effects on the swallowing structures of the head and neck may cause permanent toxicities and their avoidance – especially the pharyngeal constrictor muscles (PC) that are part of the posterior pharyngeal wall – should be considered

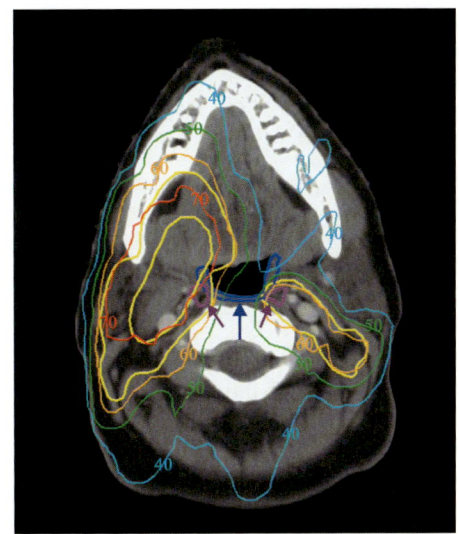

Fig. 5. Example of an IMRT plan. Dose distributions (Gy) are labeled. Blue (long arrow), PC; purple (short arrows), lateral RP nodal sites; yellow, planning target volumes. The medial RP nodes are not included in the targets, and the medial parts of the PC are outside the high doses [37].

when possible. However, sparing of the swallowing structures must take into account the RP nodes, which are important sites of metastases in advanced head and neck cancer. The RP nodes consist of lateral and medial groups, and the medial ones are close to the center of the PCs. However, the medial RP nodes have rarely been involved with metastasis in head and neck cancer. Can they and other structures involved in swallowing be excluded from IMRT treatment?

At the University of Michigan, we performed a prospective study on 73 patients to assess the clinical and functional results of chemoradiotherapy for oropharyngeal cancer IMRT to spare the important swallowing structures [37]. IMRT planning was performed with the intent of sparing noninvolved parts of the swallowing structures: PCs, glottic and supraglottic larynx, and esophagus, as well as the oral cavity and major salivary glands. At a median follow-up of 36 months, 3-year disease-free and locoregional recurrence-free survivals were 88 and 96%, respectively. At 1 year after therapy, observer-rated dysphagia was absent or minimal (scores 0–1) in all patients except 4: 1 who was feeding-tube-dependent and 3 who required a soft diet. These results indicate that IMRT aiming to reduce dysphagia can be performed safely for oropharyngeal cancer and has high locoregional tumor control rates. In this study, IMRT planning objectives included the dosimetric sparing of the parts of the swallowing structures (i.e. PCs, esophagus, and glottic and supraglottic larynx, as well as the major salivary glands and oral cavity, which were outside the targets. Targets included the lateral, but not medial, RP nodes in all patients. The GTV and CTV were each expanded uniformly by 3 mm to yield the planning target volume. For IMRT, 70 and 59–63 Gy were delivered to the GTV and CTV, respectively, in 35 daily fractions. An example of an IMRT plan is provided in figure 5.

Table 2. Additional recommendations for target delineation based on recent studies

- Lateral RP nodes should be included as targets in all cases of locally advanced oropharyngeal cancer, whether or not there is evidence of involvement of other neck levels. The medial RP nodes are not included in the targets, and the medial parts of the PCs are outside the high doses.
- In cases that are at risk of RP nodal metastases, these targets should extend more cranially than the top of C1, from the level of the upper nasopharynx to the base of skull.
- In patients with lateral primary tumors that are at high risk of RP nodal involvement, both the ipsilateral and the contralateral nodes should be included in the RP nodal target.
- Patients who have had major neck surgery more than 1 year previously should be excluded from IMRT due to the uncertainty regarding their targets.
- It is important to contour the regional lymph nodes in the low neck when they are at risk, such as cases with significant nodal involvement of the high neck, and to deliver irradiation to the low neck with appropriate dose and energy. Preferably, the low neck targets should be included in the IMRT treatment plan in these cases.

Conclusion

In summary, these analyses suggest the additional recommendations shown in table 2 for target volume delineation in IMRT, and augment those already understood.

Target Doses and Techniques

Dose

The current policy at this institute is to deliver 70 Gy to the GTV in 35 fractions (2 Gy per fraction). The high-risk CTV, which is around the GTV and the first echelon nodes, receives 63 Gy (in 1.8-Gy fractions, biologically equivalent to 60 Gy at 2.0 Gy per fraction). The lower-risk areas receive 60 Gy (1.7 Gy per fraction, biologically equivalent to 54 Gy at 2.0 Gy per fraction). These specifications are very similar to the policies at UCSF [38], Washington University [36], University of Iowa [35] and the RTOG in the nasopharyngeal protocol [39].

IMRT Planning

The greater the number of beam angles, the greater the conformality that can be achieved, but each field takes additional time for planning and execution, thereby reducing the efficiency of the clinic. The question is what degree of planning provides clinically significant benefit for the patient? At our institution, we have found that about 7 equidistant fields are sufficient for most head and neck IMRT courses. In a complex case, such as a nasopharynx tumor with involved lymph nodes, 8 or 9 equidistant fields may give a slightly better result and may be used. Beyond this, a greater number of delivery angles, and the technologies that produce them, seem unlikely to provide additional benefit and their value would need to be demonstrated in clinical evaluations.

Adaptive Replanning

If a patient's contour changes and the mask no longer fits, because of weight loss or tumor shrinkage, then it may be necessary to bring the patient back to simulation, try to reproduce the original position, and to reconstruct the mask. If the original treatment position cannot be reproduced, then occasionally the planning must be entirely repeated. Large external contour changes may also require that replanning be performed to dosimetrically maintain the original targets. However, this is not expected or intended based on tumor regression alone. Currently, our conviction is that the extent of gross tumor before irradiation continues to be the target for therapy to full dose, even as the gross disease shrinks during treatment. Investigation may guide us on this issue, which does need to be studied further, but this is our current policy.

Chemotherapy and Radiotherapy

Whenever one uses a higher radiation dose per fraction than is standard, patients may not tolerate the addition of concurrent chemotherapy acceptably. Concern about this has been raised over the past few years. Certain RTOG radiation protocols are now using 2.2 Gy per fraction or more to some areas of tumor treatment. If one decides to give a higher dose per fraction than usual to the gross disease, such as 2.2 Gy or more, one should take into account that chemotherapy probably should probably not be given. The exception to this may be for the treatment of nasopharynx tumors, where higher doses per fraction may be better tolerated. Conversely, if one thinks that chemotherapy is indicated for a patient because of the extent of disease, then current evidence suggests that a standard dose per fraction to the gross disease is preferable.

  Should the planning target volume or CTV be reduced when concurrent or induction chemotherapy is the treatment program? While this is worthy of investigation, it is not recommended at present. The same planning target volume

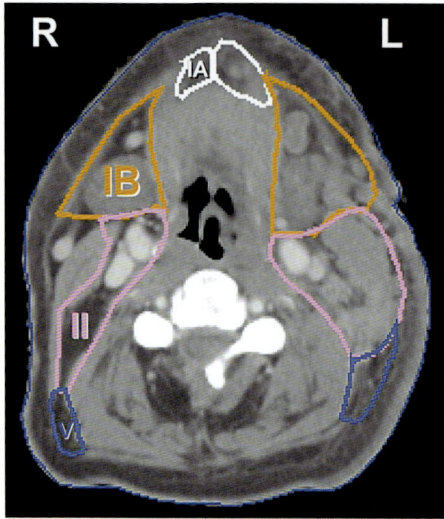

Fig. 6. Case 2. Note the gross involvement of level IA nodes on the left, and the resultant outlining of level IA nodes on the right on the CTV.

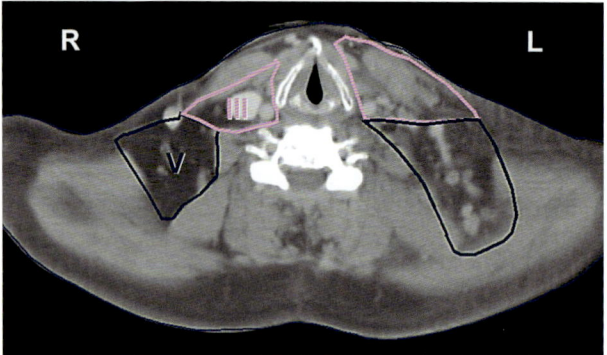

Fig. 7. Case 2. Level V is involved bilaterally. Note the posterior extent of the disease on the left. An anterior low neck field prescribed to 3 cm depth would underdose the deepest nodes of level V. In cases with a high risk of low neck involvement, we prefer to include the low neck in the IMRT plan.

and CTV are indicated as would be used without chemotherapy, and both the fields and doses should be based only on the clinical indications for that particular tumor. It is true that chemotherapy adds to the overall rates of local/regional control for many tumors, but one cannot reliably reduce the extent of the target volumes, or the dose to those volumes, in any individual case at this time.

Case Studies

Case 1
A 57-year-old male with stage T2 N2B M0 carcinoma of the right tonsil (fig. 2–4).

Treatment
IMRT concurrent with chemotherapy. Doses were 70 Gy to CTV1, 64 Gy to CTV2 and 60 Gy to CTV3. In the upper neck, CTV2 extended to the base of skull ipsilaterally and included level II and RP nodes (fig. 2). On the contralateral side of the neck (L), level II extended only to the level in which the jugular vein bisected the posterior belly of the digastric muscle (just below the transverse process of C1), to ensure coverage of the jugulodigastric nodes (fig. 3). At low level II on the right, an enlarged lymph node was observed with gross extracapsular invasion into the sternocleidomastoid muscle (fig. 4). Therefore, level II was outlined with generous margins into the surrounding tissues and the muscle. In contrast, the contralateral level was outlined to include just the fatty tissue harboring level II nodes. Also, level IB was included in the CTV on the right due to level II gross involvement and was not included in the left CTV, the side that was node-negative.

Result
This patient is currently free of disease.

Case 2
A 37-year-old female with stage T2 N2c M0 squamous cell carcinoma of the nasopharynx (fig. 6, 7).

Treatment
IMRT concurrent with chemotherapy. Doses were similar to case 1. Figure 6 shows grossly involved nodes in levels IA, IB, and II on the left. Note the difference in the outlining of level II on the left compared with the clinically noninvolved level II on the right. Levels IA–V were outlined as targets bilaterally. Figure 7 shows grossly involved levels III–IV and low level V (level VB) on the left. Note the depth in tissue in which the involved nodes are observed.

Result
This patient achieved local-regional disease control but failed with lung metastases.

Guidelines for Clinical Practice

Gross Tumor Volume Delineation/Treatment
Lymph nodes are included in the GTV if they have the following radiologic criteria:
- diameter 11 cm (or in the case of the jugulodigastric nodes, 11.1–1.5 cm);
- smaller than 1 cm with spherical rather than ellipsoidal shape;
- contain inhomogeneities suggestive of necrotic centers;
- cluster of 3 or more borderline nodes [11]; and
- FDG-PET-positive

Clinical Target Volume Delineation/Treatment
- Contralateral nodes: for cases of lateralized cancer where only the ipsilateral neck would ordinarily require therapy (small tonsillar cancers, and retromolar trigone and buccal mucosa cancers); the contralateral neck treatment is always added to the CTV when the ipsilateral neck involvement is greater than N1.
- Level II nodes: level II (see fig. 1) neck nodes are the most frequent metastatic site for tumors in all mucosal sites. These nodes can be divided into the subdigastric (jugulodigastric) nodes, located below the level at which the posterior belly of the digastric muscle crosses the jugular vein, and more cranially located nodes below the base of skull ('junctional' nodes in Rouviere's terminology). The subdigastric nodes are the main nodes involved when contralateral metastasis occur, while the more cephalad nodes are at risk bilaterally in cases of nasopharyngeal cancer and in the neck side that contains other level II–III metastasis [23] (fig. 2, 3).
- Level IB and IV nodes: levels IB and IV are treated in all cases in the neck side with clinical involvement of levels II or III (fig. 4).
- Level V nodes: level V is treated in the neck side with significant (1N1) involvement of levels II–IV, in all cases.
- Retropharyngeal nodes: the RP nodes are treated bilaterally in all cases of oropharyngeal and hypopharyngeal cancer with clinical involvement of levels II–IV (in cases of early lateralized oropharyngeal tumors with small N1 disease, treat ipsilaterally).
- Level VI nodes: level VI nodes are treated in all cases with clinical involvement of level IV nodes.

Dose Prescription (currently in use at the University of Michigan)
- GTV: the current policy at this institute is to deliver 70 Gy to the GTV in 35 fractions (2 Gy per fraction).
- High-risk CTV: the high-risk CTV, which is around the GTV and the first echelon nodes, receives 63 Gy (in 1.8 Gy fractions, biologically equivalent to 60 Gy at 2.0 Gy per fraction).
- Low-risk CTV: the lower risk areas receive 60 Gy (in 1.7 Gy per fraction, biologically equivalent to 54 Gy at 2.0 Gy per fraction).

Additional Considerations
- Lateral RP nodes should be included as targets in all cases of locally advanced oropharyngeal cancer, whether or not there is evidence of involvement of other neck levels. The medial RP nodes are not included in the targets, and the medial parts of the PCs are outside the high doses.
- In cases that are at risk of RP nodal metastases, these targets should extend more cranially than the top of C1, from the level of the upper nasopharynx to the base of skull.
- In patients with lateral primary tumors that are at high risk of RP nodal involvement, both the ipsilateral and the contralateral nodes should be included in the RP nodal target.
- Patients who have had major neck surgery more than 1 year previously should be excluded from IMRT due to the uncertainty regarding their targets.
- It is important to contour the regional lymph nodes in the low neck when they are at risk, such as cases with significant nodal involvement of the high neck, and to deliver irradiation to the low neck with appropriate dose and energy. Preferably, the low neck targets should be included in the IMRT treatment plan in these cases.

References

1 Eisbruch A, Kim HM, Terrell JE, et al: Xerostomia and its predictors following parotid-sparing irradiation of head-and-neck cancer. Int J Radiat Oncol Biol Phys 2001;50:695–704.
2 Lin A, Kim HM, Terrell JE, et al: Quality of life after parotid-sparing IMRT for head-and-neck cancer: a prospective longitudinal study. Int J Radiat Oncol Biol Phys 2003;57:61–70.
3 Eisbruch A, Schwartz M, Rasch C, et al: Dysphagia and aspiration after chemoradiotherapy for head-and-neck cancer: which anatomic structures are affected and can they be spared by IMRT? Int J Radiat Oncol Biol Phys 2004;60:1425–1439.
4 Fung K, Lyden TH, Lee J, et al: Voice and swallowing outcomes of an organ-preservation trial for advanced laryngeal cancer. Int J Radiat Oncol Biol Phys 2005;63:1395–1399.
5 Som PM: The present controversy over the imaging method of choice for evaluating the soft tissues of the neck. AJNR Am J Neuroradiol 1997;18:1869–1872.
6 Schechter NR, Gillenwater AM, Byers RM, et al: Can positron emission tomography improve the quality of care for head-and-neck cancer patients? Int J Radiat Oncol Biol Phys 2001;51:4–9.
7 Daisne JF, Duprez T, Weynand B, et al: Tumor volume in pharyngolaryngeal squamous cell carcinoma: comparison at CT, MR imaging, and FDG PET and validation with surgical specimen. Radiology 2004;233:93–100.
8 Kubota K, Yokoyama J, Yamaguchi K, et al: FDG-PET delayed imaging for the detection of head and neck cancer recurrence after radio-chemotherapy: comparison with MRI/CT. Eur J Nucl Med Mol Imaging 2004;31:590–595.
9 Anzai Y, Minoshima S, Wolf GT, Wahl RL: Head and neck cancer: detection of recurrence with three-dimensional principal components analysis at dynamic FDG PET. Radiology 1999;212:285–290.
10 Lapela M, Grenman R, Kurki T, et al: Head and neck cancer: detection of recurrence with PET and 2-(F-18)fluoro-2-deoxy-D-glucose. Radiology 1995;197:205–211.
11 van den Brekel MW, Stel HV, Castelijns JA, et al: Cervical lymph node metastasis: assessment of radiologic criteria. Radiology 1990;177:379–384.
12 Eisbruch A, Foote RL, O'Sullivan B, et al: Intensity-modulated radiation therapy for head and neck cancer: emphasis on the selection and delineation of the targets. Semin Radiat Oncol 2002;12:238–249.
13 Rouvière H: Lymphatic System of the Head and Neck (translated by Tobias M). Ann Arbor, Edward Brothers, 1938.
14 Mukherji SK, Armao D, Joshi VM: Cervical nodal metastases in squamous cell carcinoma of the head and neck: what to expect. Head Neck 2001;23:995–1005.
15 Lindberg R: Distribution of cervical lymph node metastases from squamous cell carcinoma of the upper respiratory and digestive tracts. Cancer 1972;29:1446–1449.
16 Byers RM, Wolf PF, Ballantyne AJ: Rationale for elective modified neck dissection. Head Neck Surg 1988;10:160–167.
17 Candela FC, Shah J, Jaques DP, Shah JP: Patterns of cervical node metastases from squamous carcinoma of the larynx. Arch Otolaryngol Head Neck Surg 1990;116:432–435.
18 Robbins KT, Medina JE, Wolfe GT, et al: Standardizing neck dissection terminology. Official report of the Academy's Committee for Head and Neck Surgery and Oncology. Arch Otolaryngol Head Neck Surg 1991;117:601–605.
19 Robbins KT: Integrating radiological criteria into the classification of cervical lymph node disease. Arch Otolaryngol Head Neck Surg 1999;125:385–387.
20 Som PM, Curtin HD, Mancuso AA: An imaging-based classification for the cervical nodes designed as an adjunct to recent clinically based nodal classifications. Arch Otolaryngol Head Neck Surg 1999;125:388–396.
21 Gregoire V, Coche E, Cosnard G, et al: Selection and delineation of lymph node target volumes in head and neck conformal radiotherapy. Proposal for standardizing terminology and procedure based on the surgical experience. Radiother Oncol 2000;56:135–150.
22 Nowak PJ, Wijers OB, Lagerwaard FJ, Levendag PC: A three-dimensional CT-based target definition for elective irradiation of the neck. Int J Radiat Oncol Biol Phys 1999;45:33–39.
23 Wijers OB, Levendag PC, Tan T, et al: A simplified CT-based definition of the lymph node levels in the node negative neck. Radiother Oncol 1999;52:35–42.
24 Million RR, Cassisi NJ: Management of Head and Neck Cancer: A Multidisciplinary Approach, ed 2. Philadelphia, Lippincott, 1994.
25 Eisbruch A, Marsh LH, Dawson LA, et al: Recurrences near base of skull after IMRT for head-and-neck cancer: implications for target delineation in high neck and for parotid gland sparing. Int J Radiat Oncol Biol Phys 2004;59:28–42.

26 Dawson LA, Anzai Y, Marsh L, et al: Patterns of local-regional recurrence following parotid-sparing conformal and segmental intensity-modulated radiotherapy for head and neck cancer. Int J Radiat Oncol Biol Phys 2000;46:1117–1126.

27 King AD, Ahuja AT, Leung SF, et al: Neck node metastases from nasopharyngeal carcinoma: MR imaging of patterns of disease. Head Neck 2000; 22:275–281.

28 McLaughlin MP, Mendenhall WM, Mancuso AA, et al: Retropharyngeal adenopathy as a predictor of outcome in squamous cell carcinoma of the head and neck. Head Neck 1995;17:190–198.

29 Gregoire V, Levendag R, et al: Anatomic boundaries for delineation of nodal levels in the N0 neck. 2004. http://www.rtog.org/hnatlas/main.html.

30 Hasegawa Y, Matsuura H: Retropharyngeal node dissection in cancer of the oropharynx and hypopharynx. Head Neck 1994;16:173–180.

31 Ballantyne AJ: Significance of retropharyngeal nodes in cancer of the head and neck. Am J Surg 1964;108:500–504.

32 Amatsu M, Mohri M, Kinishi M: Significance of retropharyngeal node dissection at radical surgery for carcinoma of the hypopharynx and cervical esophagus. Laryngoscope 2001;111:1099–1103.

33 Fisch U: Lymphography of the Cervical Lymphatic System. Philadelphia, Saunders, 1968.

34 de Arruda FF, Puri DR, Zhung J, et al: Intensity-modulated radiation therapy for the treatment of oropharyngeal carcinoma: the Memorial Sloan-Kettering Cancer Center experience. Int J Radiat Oncol Biol Phys 2006;64:363–373.

35 Yao M, Dornfeld KJ, Buatti JM, et al: Intensity-modulated radiation treatment for head-and-neck squamous cell carcinoma – the University of Iowa experience. Int J Radiat Oncol Biol Phys 2005;63:410–421.

36 Chao KS, Ozyigit G, Tran BN, et al: Patterns of failure in patients receiving definitive and postoperative IMRT for head-and-neck cancer. Int J Radiat Oncol Biol Phys 2003;55:312–321.

37 Feng FY, Kim HM, Lyden TH, et al: Intensity-modulated chemoradiotherapy aiming to reduce dysphagia in patients with oropharyngeal cancer: clinical and functional results. J Clin Oncol 2010; 28:2732–2738.

38 Lee N, Xia P, Fischbein NJ, et al: Intensity-modulated radiation therapy for head-and-neck cancer: the UCSF experience focusing on target volume delineation. Int J Radiat Oncol Biol Phys 2003;57:49–60.

39 Lee N, Harris J, Garden AS, et al: Intensity-modulated radiation therapy with or without chemotherapy for nasopharyngeal carcinoma: radiation therapy oncology group phase II trial 0225. J Clin Oncol 2009;27:3684–3690.

Avraham Eisbruch, MD
Department of Radiation Oncology, University of Michigan
1500 E. Medical Center Drive
Ann Arbor, MI 48109 (USA)
Tel. +1 734 936 9337, Fax +1 734 763 7370, E-Mail eisbruch@umich.edu

Motion Management and Image Guidance for Thoracic Tumor Radiotherapy: Clinical Treatment Programs

Billy W. Loo, Jr.[a] · Brian D. Kavanagh[b] · John L. Meyer[c]

[a]Department of Radiation Oncology, Stanford University Medical Center, Stanford, Calif.,
[b]Department of Radiation Oncology, University of Colorado School of Medicine, Anschutz Medical Campus, Aurora, Colo., and [c]Department of Radiation Oncology, Saint Francis Memorial Hospital, San Francisco, Calif., USA

Abstract

Managing target motion first requires understanding the nature of the motion characteristic of the tumor in the individual patient. It is important to have effective immobilization and patient training strategies to help reduce motion, and then to design appropriate margins and compensation for the residual motion that is quantified. Especially when considering complex, technically demanding treatments that require a degree of patient cooperation, careful patient selection is needed to ensure that the potential benefits of the treatment design are actually realized. Finally, accurate treatment hinges critically on verification – of overall positioning, of target and organ motion at the time of treatment, and of the performance of the selected treatment strategy. Properly selected imaging methods are central to this verification process. This discussion will present practical solutions for motion management and image guidance of radiotherapy for thoracic tumors, and most of these concepts are widely applicable to treatment of other tumor sites as well.

Copyright © 2011 S. Karger AG, Basel

There are many sources of tumor and organ motion, and respiratory motion has been the subject of considerable study for the thoracic tumors. What is the magnitude of this problem for radiotherapy targeting? This depends on a number of factors. Respiratory motion has different characteristics from site to site in the thorax: from the lung apices superiorly to the diaphragms, from the mediastinum centrally to the peripheral lung zones, and so forth. Also, while there are trends based on anatomic location, the degree of motion at each site differs substantially from patient to patient [1]. Some patients will have extreme amounts of lung mo-

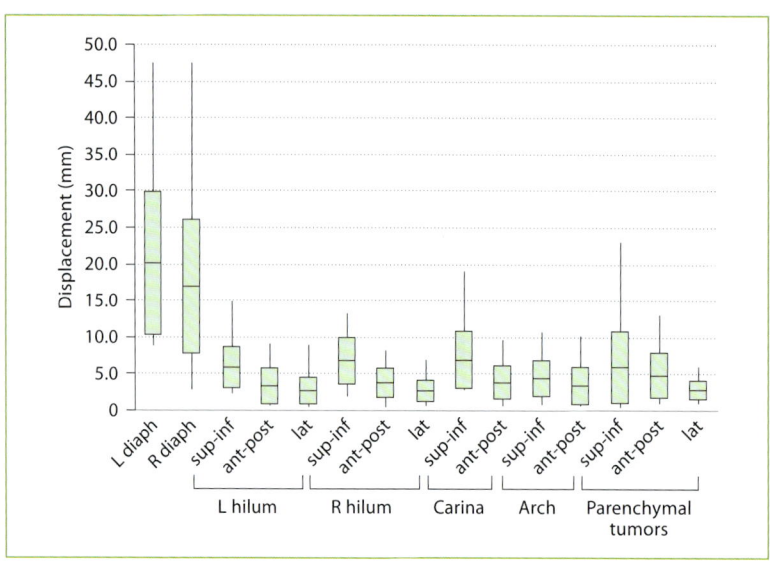

Fig. 1. 4D CT-based measurements of organ and tumor motion in different locations in the thorax. Although there is a trend toward a larger magnitude of motion closer to the diaphragms and in the superior-inferior direction, large error bars indicate that individual variability between patients is substantial.

tion near the diaphragm while others will have very little, and the amount of motion at each site relative to other sites is also quite individual (fig. 1).

The motion itself consists of several components, including aspects of translation, rotation and deformation that may combine in complex ways. During each breath, there may be rotation, stretching and compression of the tumor, and often its motion is not a simple up-and-down pattern (fig. 2). Finally, it may not be the same motion over time, whether considering a single day's treatment delivery (especially if prolonged), day to day between treatments, or a several week therapy course as tumor-related changes gradually occur. As a result, simply applying a uniform PTV expansion to account for motion (or even an expansion that is larger in the superior-inferior direction) based on population-derived motion estimates rather than knowledge of an individual's tumor motion will likely result in suboptimal targeting for that patient.

Practical Strategies for Managing Respiratory Motion

Although the subject of respiratory motion management receives much attention, it is worth noting that its importance in patient management is surpassed by more fundamental clinical decisions addressing case selection and evaluation for accu-

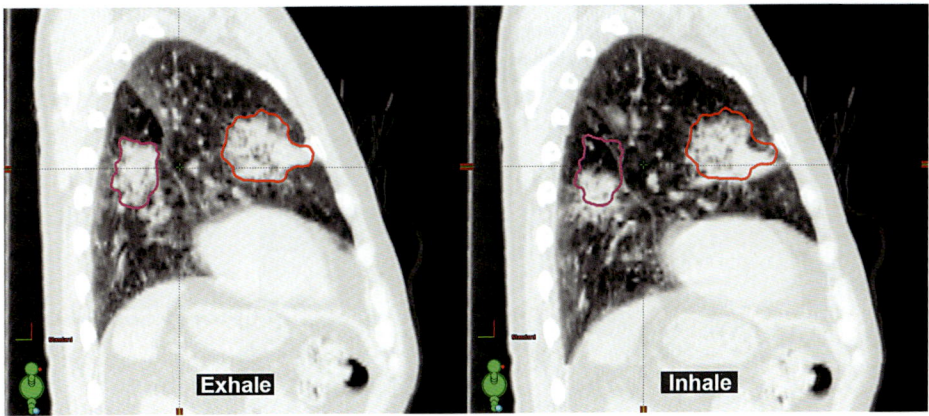

Fig. 2. Illustration of different components of tumor motion in a patient with 2 partially solid tumors in separate lobes. In addition to differing degrees of translation, each tumor demonstrates different rotational motion as well as deformation (compression/stretching) between exhale and inhale portions of the breathing cycle.

Table 1. Clinical priorities for accurate thoracic radiotherapy

Clinical priority	Needed intervention
(A) Accurate target definition	High-quality imaging at simulation (PET-CT, CT contrast)
(B) Accurate delivery of 3D radiation therapy	Identification of an effective image guidance approach before planning
(C) Adequate dose to target while respecting normal tissue tolerances	Planning of highly conformal radiotherapy including IMRT in appropriate cases
(D) Accurate assessment of motion	4D CT and related active imaging
(E) Sophisticated motion compensation strategies	Respiratory gating, tracking, etc.

rate thoracic radiotherapy. Only when these are first in place does it make sense to tackle the more subtle and technically complex issues of motion management. When selected, accurate thoracic radiotherapy involves several priorities that must be clinically solved, as shown in table 1. These priorities are sequential, and a solution for each must be met before the next can be reasonably considered. For instance, the first priority is to obtain accurate target definition. If the tumor cannot be seen adequately, then there is little reason to embark on the steps toward high precision therapy planning and delivery.

Assuming that the appropriate infrastructure is in place, the steps in addressing tumor motion specifically are: (1) characterizing the tumor motion, (2) selecting and implementing a motion management strategy and (3) verifying accurate radiotherapy delivery using image guidance at treatment. Different solutions exist

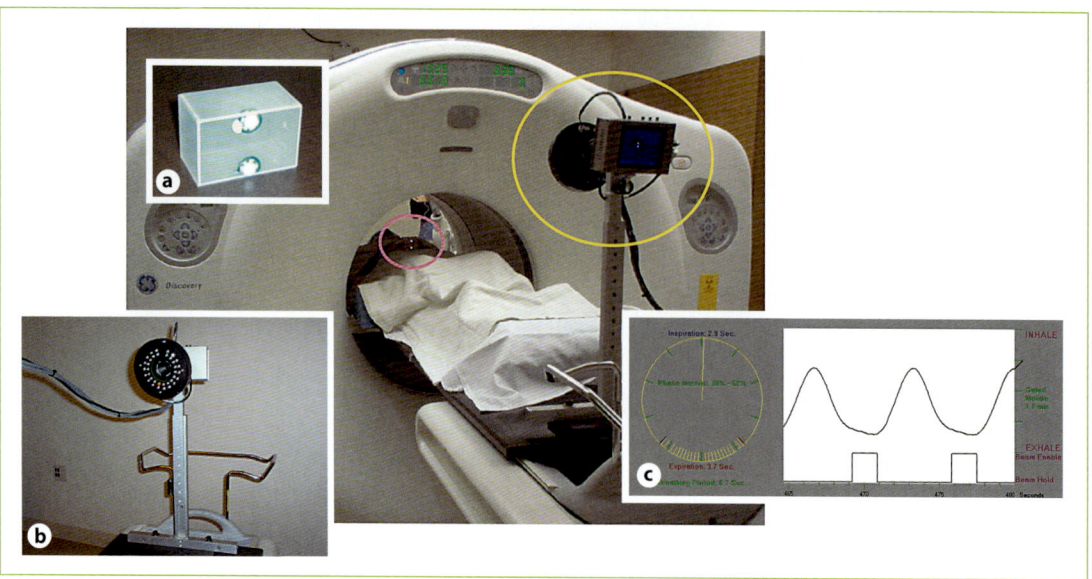

Fig. 3. Implementation of 4D simulation. A PET-CT scanner is configured for radiotherapy simulation with a flat couch-top and alignment lasers in the room. **a** Marker for tracing respiratory motion, which is placed on the external body surface, generally the upper abdomen. **b** An infrared light source and camera record images of the marker device as the patient breathes. **c** Software extracts the trajectory of reflectors on the marker from the images, indicating the respiratory phase, and allows stamping the dynamically acquired CT and/or PET data with the phase of the breathing cycle during which they are acquired.

for each of these steps, and there are useful alternatives for each. While there are no exclusively correct solutions, there are unifying principles and thought processes that define all of them. The implementations that we describe here, while far from exhaustive, illustrate each of the operational steps needed for the effective management of respiratory motion.

1. Characterizing Tumor Motion

How do we capture tumor motion in a data format that is useful for therapy planning and delivery? There are several tools available for this. Many centers are introducing 4D imaging capabilities and these will be discussed here, though there are other strategies that give similar information.

During 4D CT scanning, as shown in figure 3, a reflector is placed on the patient and an infrared camera monitors its position. It detects the patient's body surface movement through the breathing cycle, which is recorded as the CT images are continuously acquired. This permits each image to be tagged according

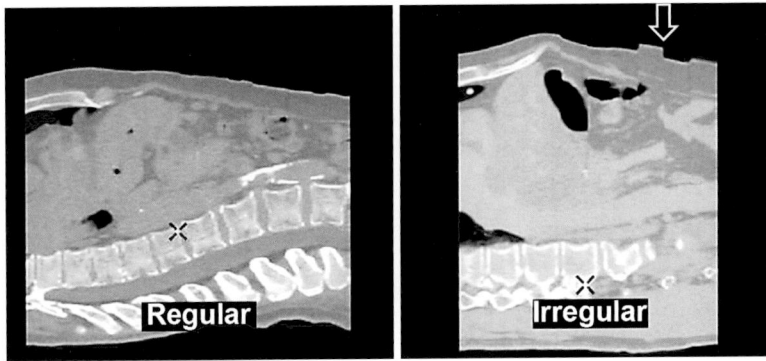

Fig. 4. Artifacts in 4D CT may occur when breathing is irregular, as cycle-to-cycle variability causes discontinuities (arrow) between the images from adjacent scanner bed positions. These become manifest when images are assembled together to form a virtual large volume data set. Such artifacts can potentially compromise accurate target delineation.

to its point in the breathing cycle. A limitation of current technologies is that the field of view of CT scanners in the craniocaudal direction is much less than the length of the thorax, and only a limited segment of the body is scanned at a time. Thus, continuous CT images are acquired at one bed position for at least 1 full respiratory cycle, the table is advanced, additional dynamic images are acquired, and so on, until the desired field of view is entirely imaged. Images from different bed positions corresponding to the same portion of the respiratory cycle are sorted together, creating a moving CT scan of the full region of interest for 1 virtual breathing cycle, although in reality the individual segments were acquired during different breathing cycles [2]. The virtual breathing cycle is divided into time segments, or phases, each of which is associated with a 'frame' of the CT 'movie'. A common but arbitrary convention is to divide the breathing cycle into 10 equally spaced phases, starting with maximum inhale as the initial phase.

The same types of strategies can be applied to PET imaging, and have been implemented in some commercial systems. This overcomes a limitation of static PET, in which motion leads to blurring of the tracer uptake. This can obscure the true tumor extent, or the distinction between tumor and nonmalignant soft tissue such as atelectasis in the lung (see Case Example below).

It is important to recognize some of the characteristics and limitations of 4D CT. In patients who are regular breathers, the quality of the 4D scan can be very good. In those with irregular breathing, there may be artifacts apparent at the junctions between adjacent table positions. This is because images from different bed positions, acquired during different breathing cycles, may not be reproduced consistently (fig. 4). If these occur in areas of tumor, they may introduce potential errors in the contouring; one must be able to take this into account.

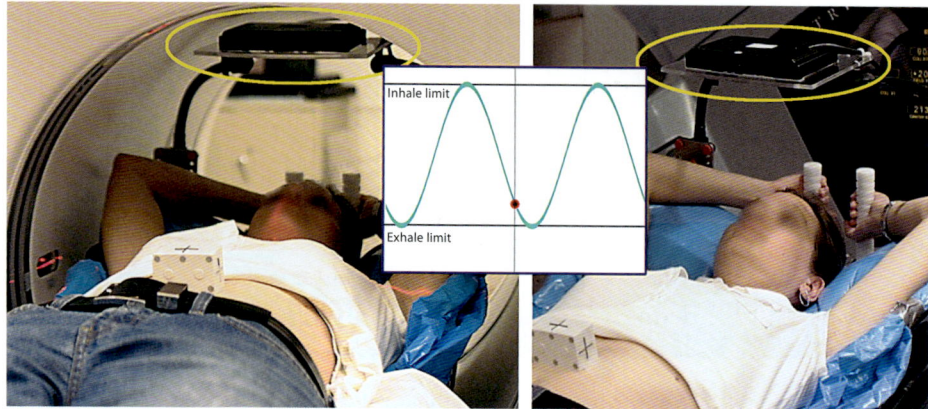

Fig. 5. Prototype audio-visual biofeedback system for helping patients regularize their breathing, to optimize 4D CT acquisition for planning and respiratory motion management during treatment. A screen with speakers (circled in yellow), designed to be compatible with the patient positioning device and fit in both the simulator and treatment machine, provides an image for the patient to follow (**inset**). The vertical position of the red ball reflects the patient's current body surface excursion, and the patient is instructed to make the red ball follow the blue trace, which is a regular pattern derived from the patient's own natural average breathing period and amplitude. Courtesy of Dr. Paul Keall.

It is also important to recognize that the 4D CT captures the motion only during the breathing cycles that it has sampled, although it is often displayed as a time loop giving the semblance of continuous breathing. This captured range of motion may or may not be the same as that during treatment. We do use this information to characterize the motion and guide what margins to use to encompass it. However, even a 4D CT does not give complete information on a patient's breathing pattern over time. Therefore, having a confirmation method at the time of treatment is critically important.

There are strategies to help improve the quality of the 4D scans. In patients with lung disease, a common problem is in achieving a regular breathing pattern. A voice prompt can be played at the time of the CT scan, which gives the patient audio feedback and coaching toward regular breathing. This is available in commercial systems (e.g. the Varian RPM system). At Stanford, we have found that this works for most patients with this problem. Further improvement may be possible through audio-visual biofeedback, in which the patient is shown his own breathing pattern and given audio and visual cues to make it more regular (fig. 5); this is an area of ongoing research [3]. The same coaching strategies can then be used at the time of treatment if necessary.

2. Strategies to Manage Respiratory Motion

Compensating for Tumor Motion in the Treatment Plan
Motion-Inclusive Targeting
Once the motion has been characterized, the next task is to compensate for this motion. Allowance for tumor motion has long been recognized as basic to the target definitions adopted by the International Commission on Radiation Units and Measurements (ICRU) [4, 5]. In their model, the gross tumor volume (GTV) represents the extent of tumor definable by imaging and examination and the clinical target volume (CTV) encompasses sites of potential microscopic extension of tumor, but the planning target volume (PTV) specifies an expansion to ensure coverage that includes certain geometric uncertainties. In ICRU terminology, this expansion consists of an *internal margin* to account for organ motion, and a *setup margin* to account for uncertainties in patient and treatment machine positioning and reproducibility. In practice, this is accomplished by creating an intermediate structure, the *internal target volume* (ITV) consisting of the CTV with the addition of the internal margin. The final PTV adds the setup margin.

In the past, the amount of internal margin expansion for lung targets has often been generalized from motion observed in populations of patients, rather than measured in the individual patient being treated, and has generally been assumed to be greatest in the superior-inferior direction. But the high degree of individual variability in respiratory tumor motion means that such population-based margins are likely to be suboptimal for any given patient and to result in unnecessary lung irradiation, inadequate tumor coverage, or both.

4D CT can be used to guide the crafting of appropriately individualized ITVs. One method for doing so is to define the ITV such that it covers the entire extent of motion identified on 4D CT, which may be referred to as a 'motion envelope' targeting approach. The ITV would be constructed, for example, by starting with a copy of the CTV contoured on one phase of the 4D CT image, then expanding it to cover the CTV as visualized on each phase of the 4D CT scan. When the magnitude of the motion is small compared with the tumor size, it is generally sufficient to contour only on the maximum inhale and exhale images to define the range of motion. Another useful tool for defining the ITV is a maximum intensity projection reconstruction of the 4D CT data, which is a projection of the entire 4D CT series onto a single 3D CT image set. Technically, the value of each 4D voxel is set to its maximum value at that location over *all* the phases, creating the maximum intensity projection (fig. 6). In the case of a small tumor surrounded by lung parenchyma, this provides a tracing of its motion envelope. It is important to recognize that when the tumor abuts other soft tissue anatomy, such as the chest wall or mediastinum, the path of the tumor can be obscured by the adjacent soft tissues in a maximum intensity projection reconstruction.

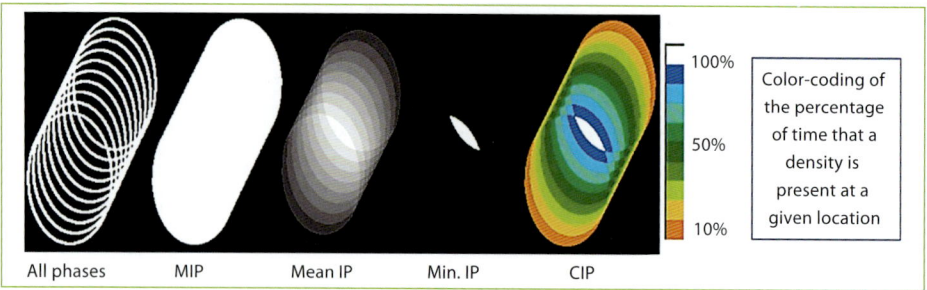

Fig. 6. Illustration of the potential use of pixel-based intensity projection protocols from 4D CT scans. All positions of the tumor identified on the 4D CT (all phases) can be represented on a single 3D CT scan as a maximum intensity projection (MIP). Alternatively, they can be represented as a mean or minimum intensity projection, or even color-coded for the percent of time present at a given location (color intensity projection; CIP). MIP is commonly used to develop a motion-inclusive PTV. Courtesy of F. Lagerwaard and S. Senan, VU University Medical Center, Amsterdam, The Netherlands.

There has been debate about the correctness of the motion envelope target definition. On the one hand, it perhaps *overestimates* the presence of tumor at the extreme positions, since it does not take into account that tumor spends unequal durations at different portions of its trajectory. It might be better to weight the radiation dose to the portion in which the tumor spends the most time [6, 7] as might be determined from the color intensity projection in figure 6. On the other hand, because the 4D CT images represent samples in time rather than a truly continuous measurement, and because they reflect motion estimated from just a few breathing cycles that are again just one sample of all possible breathing cycles, a motion envelope based on a single 4D CT scan might *underestimate* the full extent of potential motion during treatment. Nevertheless, we regard the motion envelope technique as producing a reasonable estimate of the appropriate ITV definition for most patients, recognizing that other suitable approaches may be used.

Respiratory Gating

A more complex strategy is to use respiratory gating so that the beam is turned on only during a portion of the breathing cycle. The main purpose is to allow the reduction of the lung volume irradiated, since the tumor is treated only over a limited part of its path. For tumors that move a large amount, the reduction in treatment volume can be quite significant with gating.

There is also a potential theoretical advantage of gating in the context of IMRT, where we might be concerned about the interplay between leaf motion and organ motion [8]. The dose distribution in IMRT is achieved by the delivery of radiation through many small field segments delivered sequentially in time, and the dose sums properly in space assuming a static target. This assumption is violated by motion, since individual beamlets targeting the same point in the tumor volume may in fact be delivered to different points as it moves. If implemented properly, gating could help reduce the impact of the interplay effect by virtually freezing motion, the way a well-timed strobe light appears to freeze an object moving in a cyclical fashion.

When using gating, the effective motion is reduced to that which is residual during the beam-on period, or gating 'window.' As such, a 'gated ITV' can be defined based on 4D CT estimation of the amount of motion occurring within the gating window (see Case Example below). It is important to understand that most commercial systems use a surrogate of tumor motion, such as a marker on the external body surface, as the signal for triggering the beam on and off. When using such a system, it is critical to verify by imaging at the time of treatment that the internal anatomy is reproduced as planned when the external marker signals the beam to be on.

Tumor Tracking

The next level of complexity is to follow the tumor with the beam, with the goal of further reducing the margins and the lung volume treated. This also improves the duty cycle, as the beam is on during the entire treatment time. The target volume definition in this method requires a detailed understanding of the tracking strategy: while the 'tracked ITV' can now eliminate expansion for the respiratory excursion being tracked, the ITV must still account for motions not followed by the tracking system [9]. For example, the CyberKnife system (the only commercially available system to implement tumor tracking at present) follows the tumor by translations of the beam, but does not correct for tumor rotation or deformation during respiration.

In the CyberKnife system, the linear accelerator is moved by a robotic manipulator to follow the tumor trajectory as determined by an automated image guidance system, as explained in detail below. It is also possible in principle to move a computer-controlled multileaf collimator to track a target during breathing, and this is an area of active research [10]. Because of the high complexity and dynamic nature of this treatment strategy, image guidance with a substantial degree of automated analysis is required.

Reducing Tumor Motion
Breath Hold

One way of minimizing the motion of a tumor is for a patient to hold his or her breath at a consistent depth of inspiration during imaging and treatment. A deep inspiration breath hold, for example, has the theoretical advantages of both stabilizing the target and reducing the percent of normal lung parenchyma receiving a

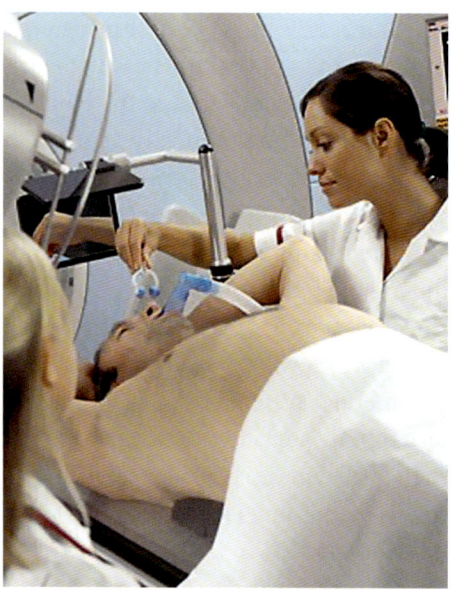

Fig. 7. Image of the patient position for treatment using the Elekta Active Breathing Coordinator (www.Elekta.com). Here the mouthpiece is in place, and the nose clip is being positioned. See online supplementary material for step-by-step demonstration.

high dose [11]. This method can be accomplished with directed patient coaching, with or without the aid of commercially available systems such as the Elekta Active Breathing Coordinator device (fig. 7).

The patient must be cooperative in using this valve-like mechanism, which monitors and intermittently controls the patient's breathing. Patient education and rehearsal are needed at the time of simulation. The patient is asked to develop a practiced level of inspiration and then to hold his breath at that same depth. The system monitors his respiratory cycle, and displays it to him as a tracing on a screen. When the system is deployed, the patient's breathing is effectively frozen at that depth of inspiration by a valve mechanism attached to the mouthpiece breathing device. Nasal breathing is prevented with a nose clip. Most patients can hold their breath for 8–12 s in this position. Throughout the entire breath hold-relaxation process, the patient has a control button that will immediately release the valve and allow normal breathing to resume in the event of any distress.

Voluntary or device-aided breath holding must be used during both the CT scan acquisition for planning and the treatment delivery. The depth of breathing must be consistent in both settings in order for the treatment delivery process to match the planned dose distribution accurately.

Abdominal Compression

With abdominal compression as a form of motion management, the goal is to modify the anatomic pattern of breathing to be less dependent on superior-inferior diaphragmatic excursion and more reliant upon chest wall intercostal expansion to modulate airflow in and out of the lungs. Diaphragm motion causes greater lung and tumor motion than chest wall expansion.

The University of Texas Southwestern group evaluated how much pressure is required to stabilize the diaphragm [12]. They performed a simple but elegant set of experiments using a modified body frame system with an abdominal compression plate; a pressure gauge measured the amount of compression. This allowed them to correlate the amount of compression pressure with the change in diaphragm motion (the superior excursion limit of the diaphragm). The use of no compression at all was compared with medium or high compression force (average 48 or 91 N, respectively). The mean overall motion was 14 mm with no compression versus 8 or 7 mm for medium or high compression. A significant difference in the control of both superior-inferior and overall motion of tumors was seen with the application of medium or high compression force. The clinical significance of this magnitude of reduction in excursion will vary directly with the volume of the tumor to be targeted and inversely with the volume of normal lung. The larger the tumor, the more value to reducing its motion in terms of the volume of lung exposed.

3. Image Guidance of the Motion Management Strategy

After characterizing the tumor motion and planning for it, one must be certain that the delivered treatment accomplishes its goal of covering the target. There are several forms of image guidance used to confirm that the intended treatment is the one that is actually delivered. No matter what overall strategy is used, it will be improved by minimizing the time spent in performing its procedures. Prolonged sessions on the treatment table allow drift in the therapy accuracy. Reducing this time is perhaps the simplest and most general method of improving accuracy, and is an important aspect of quality assurance.

There are at least 3 steps in daily image verification: first, to bring the body into the correct position; second, to confirm that the target itself is correctly positioned; and third, to verify that the motion management is correct for that day. It is important to first confirm the overall positioning of the patient. The body is a nonrigid structure. In larger fields, it is unlikely that all points will precisely match up initially, or that simple x-y-z translational shifts will achieve such perfect matches. Other types of corrections may be required: the patient may need to reposition his arms, shoulders, or twist or roll to one side or the other. Once patient positioning is confirmed, target verification including motion management can proceed.

What image verification method is best? For larger fields, where one needs to line up multiple areas of the body, kV orthogonal pair images have the advantage

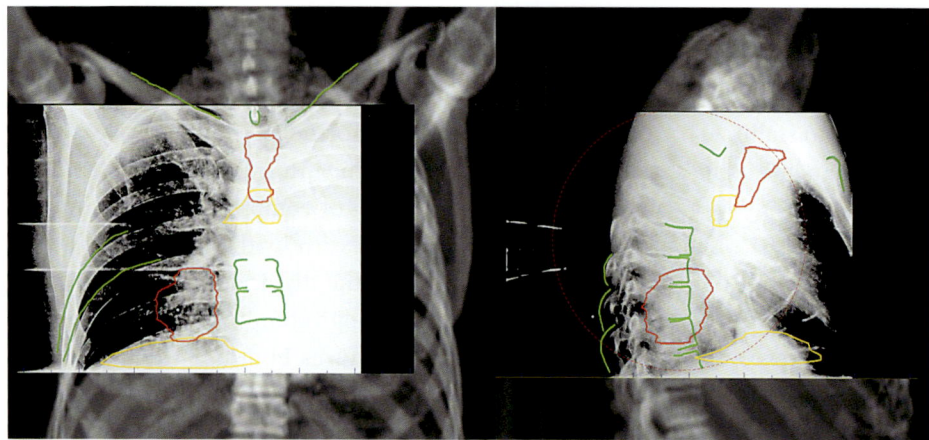

Fig. 8. Verification of body alignment on kV imaging by comparison with digitally reconstructed radiographs and anatomy overlays. The GTVs (red contours) are projected, and nearby bony structures (green contours) and the carina and diaphragm (yellow contours) are drawn on the digitally reconstructed radiographs, then overlaid on the daily kV orthogonal pair images to verify correct positioning of all anatomic landmarks. Proper positioning often requires manipulation of specific body parts or body rotations, rather than simply couch shifts. Emphasis is placed on matching anatomy most closely related to the targets.

of permitting a rapid assessment of a larger field of view. Imaging in fluoroscopic mode gives dynamic images in addition. Tomographic imaging [e.g. cone-beam CT (CBCT) or another in-room CT system] has the advantage of visualizing soft tissue structures in 3 dimensions, and is valuable when the tumor is not easily visualized on planar imaging as is commonly the case. However, it generally suffers from poor time resolution, resulting in motion blurring or artifacts. Respiratory-correlated CBCT has been described, and commercial implementations are anticipated soon [13].

Orthogonal Pair Planar Imaging
Body alignment is first verified by comparing orthogonal pair planar images with corresponding digitally reconstructed radiographs. For visual landmarks on the DRR, the Stanford group asks for a projection of the GTV, and some of the normal anatomy close to where the most careful matching is needed. Other structures that are easily visible on the kV verification films (usually adjacent bony structures, the dome of the diaphragm and the carina) are added to the digitally reconstructed radiograph. These outlines can then be projected on the kV images. On the daily verification images, we attempt to line up the bone and soft tissue structures by a combination of couch shifts and manipulations of the position of the body and individual body parts (fig. 8).

Fluoroscopy for Gating Verification

If the tumor itself is visible on planar projections, a gated portal image can be helpful. It is acquired when triggered by the same signal as engages the treatment beam itself. While this is the most direct way to confirm appropriate gating, in cases of discrepancy there is no straightforward way to apply a correction based on this information. Fluoroscopy at the treatment machine, such as with the addition of an integrated kV imaging system, is helpful in this regard. In the Varian implementation, for example, a user-defined outline can be projected on the fluoroscopic image at the treatment machine before delivering treatment (fig. 9). This outline can be designed to conform to the shape of the PTV (generally on an AP projection), and will change color based on the external signal indicating when the beam would be on or off. In this particular system, the aperture outline is created by setting up a 'dummy' treatment field that conforms to the desired shape, which the system then converts to an overlay on the fluoroscopic view. One can then verify visually that the tumor is within the appropriate position range when the beam would be on, and the gating thresholds can be adjusted if necessary to eliminate a discrepancy. If the depth of breathing is too different from that at simulation, adjusting the gating thresholds may be insufficient compensation, and additional strategies such as coached breathing or, ultimately, resimulation may be required. If the tumor is not readily visible on a fluoroscopic projection, another internal anatomic surrogate may be used, for example the dome of the diaphragm for tumors in the lower lobe, or the carina for tumors/nodes in the hilum or mid-mediastinum, both of which are generally visible on fluoroscopy.

In-Room CT

While planar images primarily provide visualization of high contrast structures such as bone or air-filled organs, cone-beam or other in-room CT devices allow 3D visualization of soft tissues. This is particularly valuable in the context of stereotactic ablative radiotherapy (SABR) of lung tumors, also known as stereotactic body radiation therapy, that are not easily visualized on planar images, especially if there are reasons to avoid fiducial marker implantation. Because of less error tolerance when treating with very few fractions, even proper alignment of bones and other anatomic surrogates may not be adequate confirmation of accurate positioning. The acquisition time of standard CBCT is long compared with the respiratory period resulting in blurring and other artifacts. As such, CBCT is most suitable when tumor motion is small. Even with greater motion, it can be useful since the extent of blurring may roughly approximate the motion envelope, at least when the tumor is surrounded by air and is visible (fig. 10).

Figure 10 shows the significance of image guidance and of minimizing delivery time. At treatment, when bone anatomy was properly aligned, the tumor extended

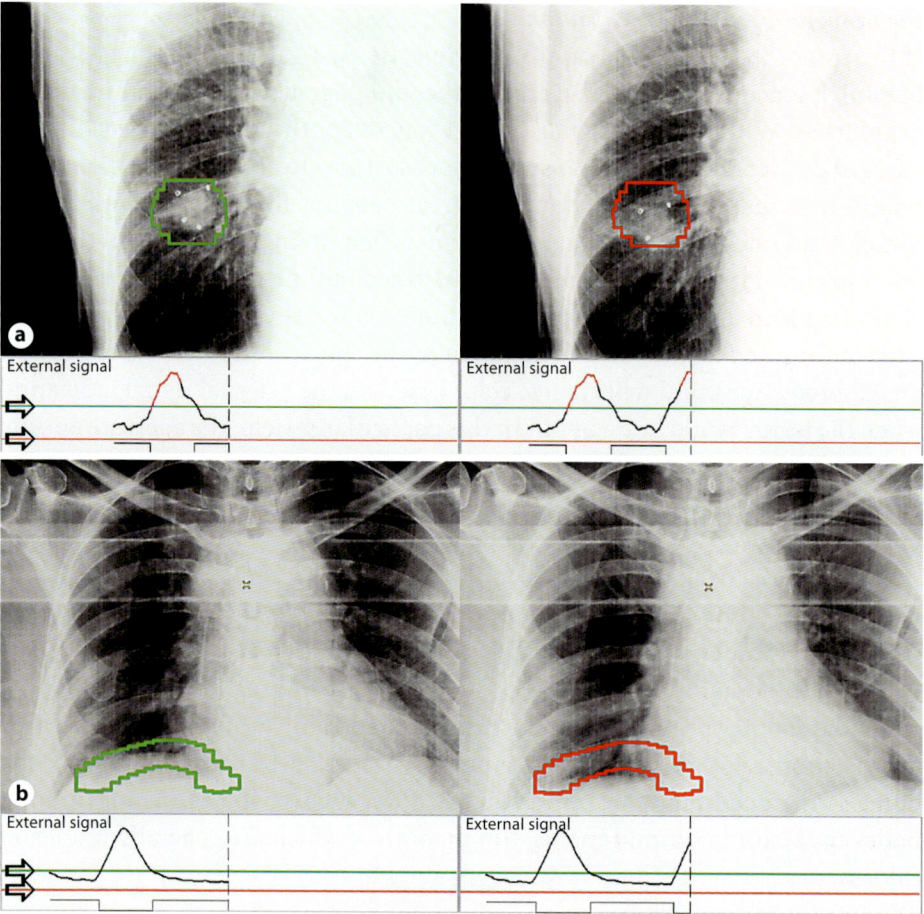

Fig. 9. Fluoroscopic verification of respiratory gating. At treatment, the radiation beam is triggered based on the signal representing the height of the external surface marker on the body (trace under each image). Prior to each treatment, fluoroscopy with an overlay of the PTV outline is performed as a simulation of the gated treatment. The outline turns green (indicating that the beam would be on) when the trace is between the 2 threshold markers (green and red horizontal lines, indicated by the arrows), the positions of which are set by the operator. The thresholds are adjusted each day to ensure that the outline turns green when the tumor is within it, and red when it is not. **a** An example of a tumor readily visible on kV fluoroscopy. **b** An example of using the dome of the right hemidiaphragm as an internal anatomy surrogate for a tumor in the lower lobe that is not readily visible on fluoroscopy. In this case, the outline conforms to the measured excursion of the diaphragm during the chosen gating interval.

outside the PTV. In fact, as can be seen from the motion blurring on CBCT, the motion envelope was shifted, indicating a different baseline depth of breathing compared with simulation. Furthermore, throughout the course of a single fraction, the baseline shifted additionally. These mismatches were corrected by appro-

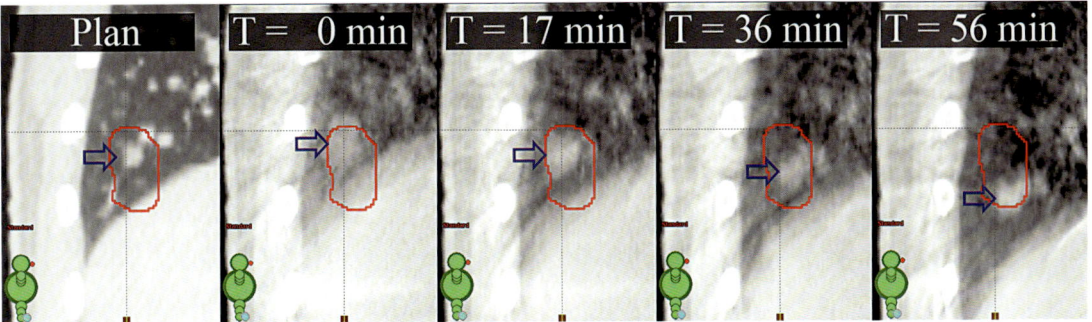

Fig. 10. CBCT verification in a case of single fraction stereotactic ablative radiotherapy for a small lung metastasis (arrow). Based on the planning 4D CT (left panel, exhale phase shown), a motion inclusive PTV was designed. In this case, the plan used multiple treatment fields, and CBCT was acquired prior to treatment delivery for each field (2nd to 5th panels). The visualized tumor extended outside the PTV, requiring shift of the patient (see text).

priate couch translations, but otherwise would not have been caught without repeated imaging. This also illustrates the advantage of minimizing the treatment time. At Stanford, we currently use either dynamic conformal or intensity-modulated arc therapy with a high dose rate (1,000–2,400 MU/min) mode for SABR, so that delivery of the entire fraction can be completed within 10 min from the end of the pretreatment CBCT acquisition, minimizing the opportunity for drifts in patient position or physiologic state.

Tumor Tracking

The dynamic tumor tracking process of the CyberKnife system deserves special mention, as it is the only tracking implementation that is commercially available currently. As discussed, dynamic tracking based on the multileaf collimator is actively under development and would be more generally applicable in linac treatment systems. The CyberKnife (fig. 11) is specifically designed for highly hypofractionated treatments, as in SABR, and the tracking pertains primarily to the treat-ment of lung and upper abdominal tumors with their periodic respiratory movement. Metallic fiducial markers must be implanted in or around the tumor, using CT-guided percutaneous or endoscopic procedures. An optional markerless tracking capability is also available for treatment of a selected minority of lung tumors, though its clinical performance needs to be validated more widely. At Stanford, platinum coils are preferred as the fiducial markers [14].

Two separate imaging systems are employed in this tumor tracking schema [15]. The first (a) is the pair of X-ray imagers that acquire orthogonal images of the fiducial markers. From these images, automatic image analysis software extracts the 3D location of the markers. Of note, the in-room imagers produce intermittent

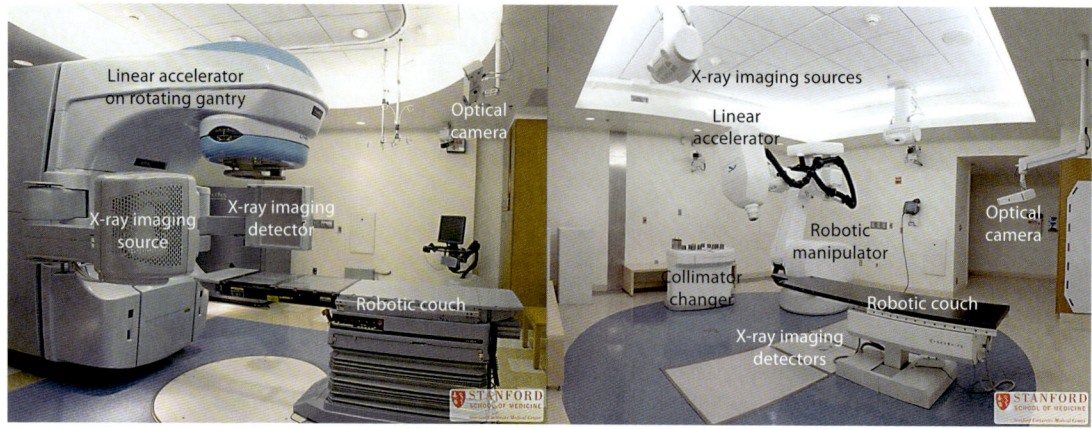

Fig. 11. Two image-guided respiratory motion management radiotherapy systems currently in use at Stanford Cancer Center. Left: Varian Trilogy; right: CyberKnife.

static images rather than continuous images. The second (b) is an optical camera that continuously monitors the position of 3 infrared light emitting diodes, placed on the external surface of the patient's chest at the beginning of the treatment session. As the patient breathes, the trajectories of the external markers follow the breathing cycle, including any variations in the breathing pattern. The key to dynamic tracking with this arrangement is the correlation of the intermittent X-ray localization of the internal markers (a) with the continuous signal of the external markers (b).

Prior to the initiation of treatment, a series of X-ray image pairs is acquired, timed such that the internal fiducial coordinates from the various portions of the breathing cycle are all sampled (the signal from the optical camera reports the full respiratory cycle). A correlation model is then built, which generates a continuously calculated position of the center of mass of the internal fiducials from the external marker coordinates. It is this calculated/predicted position that the beam follows dynamically, using the robotic manipulator. Intermittent X-ray images continue to be acquired throughout the delivery. Each new measurement of internal marker position is compared with the corresponding predicted position, and the correlation error is calculated. Treatment continues as long as the correlation error is below a user-defined threshold, generally 3 mm. If it is exceeded, treatment stops, additional images are acquired, and if necessary, the correlation model is rebuilt from the new images. Otherwise, the correlation model is simply updated with the newest data point. Gradual changes to the breathing pattern throughout treatment are accommodated in this way. Importantly, while the image analysis and corrections happen automatically, it is incumbent on the user to provide vigilant supervision, including manual review of the automatically extracted fiducial contours as they are acquired during treatment, to ensure accurate performance of the system.

Case Example: Clinical Procedures for Respiratory Gating

To illustrate and expand on the details of the respiratory gating process, we present a case treated in the Department of Radiation Oncology at Stanford and discuss the procedures used step by step.

A 58-year-old man with a prior history of a squamous cell carcinoma of the left lung treated with pneumonectomy, presented 5 years later with a stage IIIA squamous cell carcinoma of the opposite lung, involving the right lower lobe and mediastinal (paratracheal) lymph nodes. Definitive treatment to 70 Gy with concurrent chemoradiation was planned. The need was for small margins and highly conformal radiotherapy to spare his remaining lung tissue.

Patient Positioning and Initial Fluoroscopic Simulation

Before obtaining a volumetric data set, we positioned the patient on a conventional simulator with fluoroscopy. It is important to create a comfortable and reproducible position for the patient. For the 'arms up position,' which maximizes beam access to the thorax, adequate arm support is essential. Here, we used a wing board (Med-Tec/Civco) combined with a custom-formed cushion, either a binary foam mold or a vacuum set styrofoam bead bag. It is important to confirm with patients that they are using no effort of their own to maintain their arm position. Otherwise, given the length of the imaging and treatment procedures, discomfort – sometimes severe – may compromise the setup stability and reproducibility.

An optical marker is placed on the patient's upper abdomen to check the regularity of breathing, in this case using the RPM system (Varian). This helps to decide if the patient will need coaching or other strategies to facilitate the 4D CT scan and subsequent treatment.

Fluoroscopy then gives a first assessment of the patient's positioning, the tumor's visibility, and the related tumor and organ motion. We confirmed that the patient had a regular breathing pattern and was able to cooperate fully with the complex simulation procedures. Using the conventional simulator for these steps saves time on the PET/CT equipment later, which improves efficiency and reduces radiation exposure to personnel from the injected radiopharmaceutical. We then completed the simulation by acquiring a 4D PET-CT.

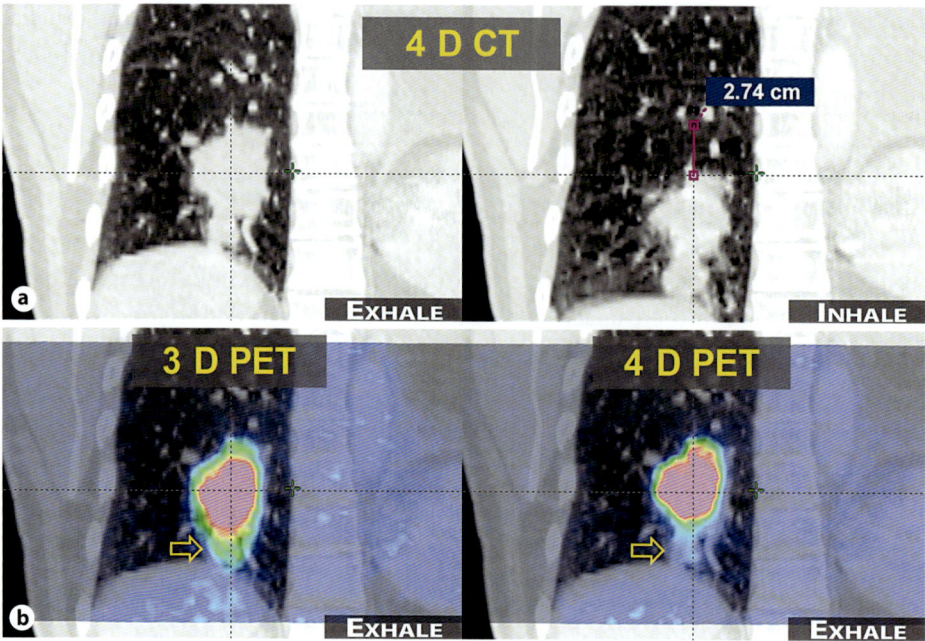

Fig. 12. a 4D CT demonstrates a large degree of tumor motion, nearly 3 cm in the superior-inferior direction. **b** 4D PET reconstruction eliminates the blurring artifact caused by this motion (arrow), revealing that the lower portion of the CT abnormality is atelectasis rather than tumor, minimizing unnecessary irradiation to the lung.

Target Delineation and Selection of Motion Management Strategy

Reconstruction of the 4D CT revealed a great deal of motion of the primary tumor, almost 3 cm in the superior-inferior direction, likely resulting from the lower lobe location and the fact that the patient has only 1 lung serving for all of his ventilation. This degree of motion led to significant blurring of the PET signal, obscuring the true extent of the hypermetabolic lesion. However, applying 4D reconstruction techniques to the PET eliminated the blurring artifact and showed that the inferior-most part of the CT abnormality was not hypermetabolic and likely represented atelectasis. The GTV was delineated based on this multimodality imaging information (fig. 12).

Should gating be done? There are a few criteria that can be used to decide. Obviously if there is large tumor motion, one would think about gating. Patient selection is also important, as the patient must be able to cooperate. How much motion is enough to warrant gating? This has not been definitively answered. Ideally, the answer should be dosimetric, and determined by the amount of normal tissue spared by a gated versus nongated plan. In practice, it is not possible to perform 2

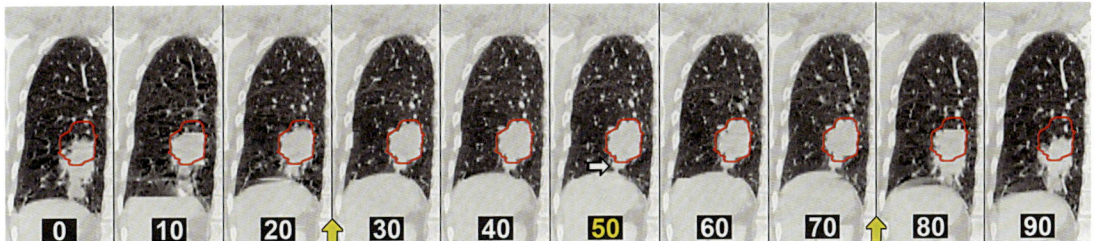

Fig. 13. Construction of the gated ITV. Maximum exhale is at 50% in this case. More abrupt displacements in tumor position are found between the 20 and 30% phases, and between the 70 and 80% phases (yellow arrows), whereas the tumor position is quite stable between the 30 and 70% phases. The gated ITV is thus constructed as the motion envelope within the 30–70% phases (red contour). The area of atelectasis identified by 4D PET-CT is excluded (white arrow).

or more plans on every patient for such comparisons. If the tumor motion seen on 4D CT is greater than 1 cm, gating is often considered. Only a minority of patients with thoracic cancer, about 10–20%, undergo respiratory gating at Stanford. In this case, because of the large degree of motion, the importance of maximally sparing the one remaining lung, and the ability of the patient to cooperate well, we selected a gating approach.

Determination of Margins

After contouring the GTVs, we must decide what margins to use to encompass them in the PTV. As described above, if gating is used a full motion-inclusive margin is not necessary. But some margin is required, since the tumor is not entirely frozen and there will be some residual motion even with gating. How much margin is required? The residual motion during gating can be estimated from the 4D CT to help form the gated ITV (fig. 13). While there is some controversy regarding optimal portion of the cycle to use for gating, we routinely gate during the exhale phase because the tumor position tends to be the most stable and reproducible at that time. By convention, we divide the breathing cycle into 10 equally spaced segments, or phases, numbered from 0 to 90%, where 0% is maximum inhale. Maximum exhale tends to be reached at the midpoint between the maximum inhales or slightly later, usually at the 50–60% phase. In this case it was determined to be at 50%, where the heights of the tumor and diaphragm peak. We then look for the phases around maximum exhale over which the tumor position remains relatively stable. In this case, the tumor displacement accelerates between the 20 and 30% phases, and between the 70 and 80% phases. Thus, we form the gated ITV from the motion envelope of the tumor within the 30–70% phases, comprising half of

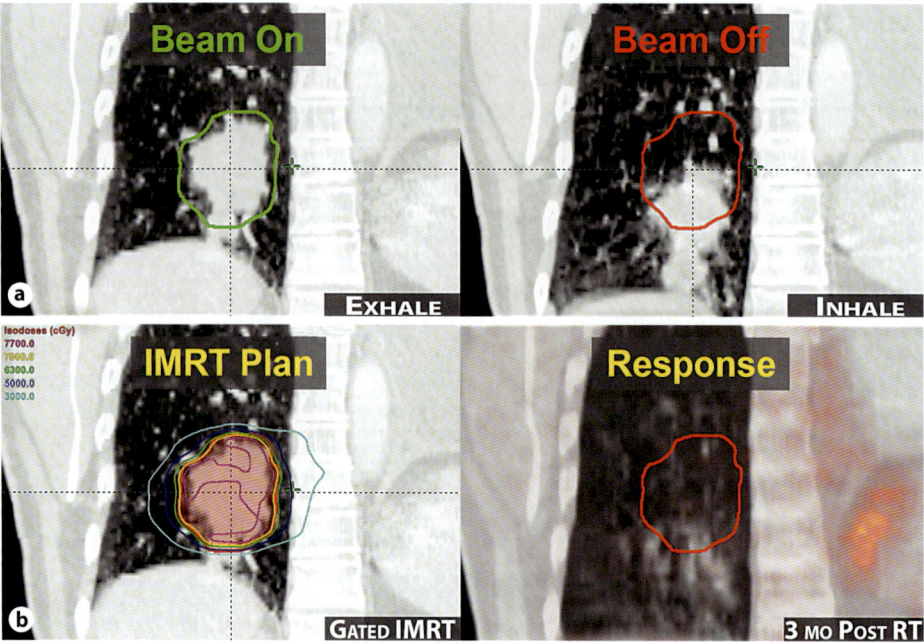

Fig. 14. a Respiratory gated treatment was prescribed. A highly conformal IMRT plan was generated (**b**, left), resulting in a complete radiographic response at 3 months post-treatment (**b**, right), and no evidence of disease at 3 years.

the breathing cycle. This would correspond to a gating duty cycle of 50%, which is quite reasonable from a practical standpoint.

The final PTV is constructed by adding a setup margin to the ITV that is institution-specific, based on the available technology and systematic assessment of positioning reproducibility for the institution. We used a 0.5-cm setup margin. This produced a PTV 30% smaller than if a motion-inclusive approach were used.

Of note, the actual gating interval used during treatment is not determined by this procedure. As described above, it is determined daily on treatment by fluoroscopic verification to set the gating thresholds such that the beam triggers on when the tumor is within the designated PTV. This procedure is used to estimate what residual motion would be realistically achievable for a given patient using a gating approach.

Treatment and Outcome

A 6-field IMRT plan was generated, producing a highly conformal dose distribution respecting conservative normal tissue constraints given the limited residual lung volume. Treatment was delivered using daily fluoroscopic image verification as described above.

Treatment was well tolerated with no pneumonitis. Three months after treatment, PET-CT revealed a complete radiographic response (fig. 14). There was no evidence of disease 3 years later.

References

1 Maxim PG, Loo BW, Shirazi H, et al: Quantification of motion of different thoracic locations using four-dimensional computed tomography: implications for radiotherapy planning. J Radiat Oncol Biol Phys 2007;69:1395–1401.
2 Pan T, Lee TY, Rietzel E, Chen GTY: 4D-CT imaging of a volume influenced by respiratory motion on multi-slice CT. Med Phys 2004;31:333–340.
3 Venkat RB, Sawant A, Suh Y, George R, Keall P: Development and preliminary evaluation of a prototype audiovisual biofeedback device incorporating a patient-specific guiding waveform. Phys Med Biol 2008;53:N197–N208.
4 Prescribing, recording and reporting photon beam therapy. ICRU Report 50. Bethesda, International Commission on Radiation Units and Measurements, 1993.
5 Prescribing, recording and reporting photon beam therapy. ICRU Report 62 (Supplement to ICRU Report 50). Bethesda, International Commission on Radiation Units and Measurements, 1999.
6 Wolthaus JWH, Schneider C, Sonke JJ, et al: Mid-ventilation CT scan construction from four-dimensional respiration-correlated CT scans for radiotherapy planning of lung cancer patients. Int J Radiat Oncol Biol Phys 2006;65:1560–1571.
7 Li X, Zhang P, Mah D, Gewanter R, Kutcher G: Novel lung IMRT planning algorithms with non-uniform dose delivery. Med Phys 2006;33:3390–3398.
8 Berbeco RI, Pope CJ, Jiang SB: Measurement of the interplay effect in lung IMRT treatment. J Applied Clin Med Phys 2006;7:33–42.
9 Loo BW, Thorndyke BR, Maxim PG, et al: Determining margin for target deformation and rotation in respiratory motion-tracked stereotactic radiosurgery of pancreatic cancer. Int J Radiat Oncol Biol Phys 2005;63:S31.
10 Keall PJ, Gattell H, Pokhrel D, et al: Geometric accuracy of a real-time target tracking system with dynamic multileaf collimator tracking system. Int J Radiat Oncol Biol Phys 2006;65:1579–1584.
11 Hanley J, Debois MM, Mah D, et al: Deep inspiration breath-hold technique for lung tumors: the potential value of target immobilization and reduced lung density in dose escalation. Int J Radiat Oncol Biol Phys 1999;45:603–611.
12 Heinzerling JH, Anderson JF, Papiez L, et al: Four-dimensional computed tomography scan analysis of tumor and organ motion at varying levels of abdominal compression during stereotactic treatment of lung and liver. Int J Radiat Oncol Biol Phys 2008;70:1571–1578.
13 Sonke JJ, Zijp L, Remeijer P, van Herk M: Respiratory correlated cone beam CT. Med Phys 2005;32:1176–1186.
14 Hong JC, Yu Y, Rao AK, Dieterich S, et al: High retention and safety of percutaneously implanted endovascular embolization coils as fiducial markers for image-guided stereotactic ablative radiotherapy of pulmonary tumors. Int J Radiat Oncolo Biol Phys 2010, E-pub ahead of print.
15 Schweikard A, Glosser G, Bodduluri M, Murphy MJ, Adler J: Robotic motion compensation for respiratory movement during radiosurgery. Comput Aided Surg 2000;5:263–277.

Billy W. Loo, Jr., MD, PhD, DABR
Department of Radiation Oncology, Stanford University Medical Center
875 Blake Wilbur Drive, Rm CC-G227
Stanford, CA 94305-5847 (USA)
Tel. +1 650 736 7143, Fax +1 650 725 8231, E-Mail BWLoo@Stanford.edu

Breast Cancer

Intensity-Modulated Radiotherapy for Breast Cancer: Advances in Whole and Partial Breast Treatment

Julia R. White[a] · John L. Meyer[b]

[a]Department of Radiation Oncology, Medical College of Wisconsin, Milwaukee, Wisc., and
[b]Department of Radiation Oncology, Saint Francis Memorial Hospital, San Francisco, Calif., USA

Abstract

Intensity-modulated radiotherapy (IMRT) can improve dose distributions through the treated breast and also reduce radiation doses to adjacent normal tissues including the contralateral breast, heart and lung with appropriate planning. Analyses demonstrate that the quality of radiation dose distribution does affect clinical results, and that outcomes are enhanced through improved planning and dose delivery methods. To achieve these results, it is essential to carefully define tissue volumes for treatment or avoidance, select technologies that can potentially conform fields to those volumes, use comprehensive planning methods, and assess their results in terms of objective dose constraints. IMRT can also be used to boost the region of tumor excision concurrently with whole breast treatment, an approach now being evaluated in on-going clinical studies. Partial breast irradiation (PBI) has been proposed as an alternative to irradiation of the entire breast for early-stage breast cancer patients undergoing breast conservation treatment. Numerous single institution phase II studies have demonstrated promising results, and the American Society of Radiation Oncology (ASTRO) has defined a suitable group of low-risk patients for PBI treatment off protocol at this time. IMRT has been proposed as an alternative to 3D conformal radiotherapy (3DCRT) for external beam PBI to improve the dose conformality to target volumes and the sparing of normal tissues. There are an increasing number of institutions evaluating and using IMRT instead of 3DCRT for PBI because of the potential treatment advantages for the breast cancer patient.

Copyright © 2011 S. Karger AG, Basel

Intensity-Modulated Radiotherapy and the Goals of Breast Radiotherapy

There is an important scientific basis for using advanced planning technologies to achieve the goals of postlumpectomy radiotherapy, both for whole and partial breast treatment. These well-established objectives are to: (a) minimize local or in-breast cancer recurrence, (b) preserve the cosmetic appearance of the breasts, (c) minimize adjacent normal tissue radiation toxicity, and (d) maintain disease-free, mastectomy-free and overall survival rates. Advanced planning and delivery technologies including intensity-modulated radiotherapy (IMRT) have already shown significant contributions toward achieving these goals and enhancing patient outcomes, and their further development, testing and integration into clinical programs are important and immediate objectives for breast cancer therapy.

Prospective randomized trials have compared breast conservation therapy with mastectomy and demonstrated clear findings. First, when optimal local therapy is delivered, it is equivalent to mastectomy in achieving breast local-regional control and overall patient survival [1–6]. Also, overview data show that local control achieved with radiotherapy after breast conserving surgery has an effect on ultimate mortality rates, including an improvement in patient survival of approximately 5% for node-negative patients and about 7% for node-positive patients [7]. Although these outcome data are based on 2D radiotherapy methods, they indicate substantial benefit of treatment. Increasing the accuracy of radiotherapy can only add to these outcome results for cancer control and normal tissue preservation.

While these prospective trials have been pivotal in guiding current radiotherapy practice, their treatment methods reflect the practice patterns of the time the trials were designed, in some cases more than 30 years ago. In many or most cases, target and normal tissue volumes were poorly or minimally defined. Dose distributions were examined in only a small portion of the breast using single-plane 2D planning methods, yet even these limited dosimetric samplings demonstrated significant dose heterogeneities – both over- and underdosing of intended breast volumes. This level of planning remained the predominant approach for intact breast treatment for many years. In the decade 1990–1999, fewer than 20% of breast radiotherapy cases underwent CT scanning for planning and fewer than 10% had multiplane dosimetry [8–10]. However, practice patterns for breast treatment are now changing rapidly and affirm a common recognition that such advancements are greatly needed.

Areas of Needed Improvement

Earlier radiotherapy planning methods yielded outcomes that showed room for improvement. Using 2D approaches, about 88% of patients had good cosmetic outcome, though only 41% were considered excellent. Of concern, 12–15% of cases were considered cosmetic failures. Current efforts should attempt to eliminate such results.

Reports from several major centers indicate that poor outcomes have often been related to how the radiation dose was distributed through the breast, as measured by: (a) dose heterogeneity, (b) total whole breast and boost doses, (c) the number of radiation fields used, and (d) the use of compensation and other technical approaches to improve dose homogeneity (table 1) [11–14]. Based on these analyses, the quality of radiation dose distribution does affect clinical results, and outcomes are enhanced through improved planning and dose delivery methods.

Strategies to Improve Outcomes

To optimize the therapeutic gain achieved by breast radiotherapy, one must be able to improve the rates of in-breast cancer control and/or reduce the rates of toxicity for the patient (which should include the burden of the treatment process itself). These goals can be approached through advanced radiotherapy methods including IMRT. Simply stated, technologies can potentially: (a) improve dose homogeneity within the target and/or (b) improve the avoidance of important normal tissues. These fundamental goals will be the topics of the next 2 sections. In addition, clinical work on breast cancer control with more focal treatment programs, such as partial breast irradiation (PBI), may be major advances. From a practical standpoint, hypofractionated partial breast approaches may be significant in reducing the number of treatment visits and possibly improving cosmetic and therapeutic outcomes; approaches to PBI will be discussed later in this chapter.

Improving Dose Homogeneity

Improving Dosimetry

Recent work has been directed toward improving 3D dose homogeneity in treatments, and there are emerging data indicating that these efforts are yielding worthwhile results. Planning methods for breast cancer have gradually evolved from 2D to CT-based 3D methods, later adding multiple fields, forward-planned compensation techniques and, finally, complex inverse-planned beam modulation. For the purposes of this discussion, we will consider methods beyond 2D to comprise advanced planning, though some published experiences did not utilize the full benefits of the IMRT planning and delivery technologies now available. Yet they provide insightful results.

Several studies have compared 2D and advanced planning methods for breast cancer treatment [15–17]. At William Beaumont Hospital, simple IMRT methods with 6–8 segments reduced the volume of breast receiving more than 110% of the dose. Recent work in Australia has evaluated several advanced IMRT approaches (forward planning with electronic compensation, inverse planning, multiple fields

Table 1. Radiation factors associated with late fibrosis and cosmetic failure: physician-assessed cosmetic results

Institution	n	Observation period years	Results (%)				Radiation factors associated with poorer cosmesis
			excellent	good	fair	poor	
Tufts University [11]	234	4.2	41	47	9	3	heterogeneous dose boost use of >2 fields
Harvard/JCRT [12, 13]							
Before 1981	504	8.9	58	28	10	4	breast dose >50 Gy use of >2 fields
1982–85	655	5.6	73	23	3.5	0.5	boost dose >18 Gy implant boost
Washington University [14]	458	4.4	38	44	15	4	use of >2 fields breast dose >50 Gy no compensator filters

Table 2. Cambridge breast trial: IMRT vs. standard planning

Dosimetry	IMRT plan, cm³ (%)	2D plan, cm³ (%)	p
Breast volume	1,349	1,308	
Volume breast receiving >107%*	10.5 (0.6)	44.5 (2.9)	<0.0005
Volume breast receiving <95%*	133 (9.9)	181 (13.8)	<0.0005

* Percent of the prescribed dose. IMRT plans show improvement in breast dose homogeneity.

in both coplanar and non-coplanar approaches). All produced significant improvements in dose distribution compared with 2D methods.

The Cambridge Breast IMRT Trial, recently reported by Barnett et al. [18], is a randomized study involving 1,145 patients. CT simulations were performed and target volumes delineated for each patient, then simple 2D plans were generated using paired tangent fields and 6- to15-MV photons. The plan dose-volume histograms were reviewed and considered unacceptable if a volume greater than 2 cm³ exceeded 107% of the prescription dose, which occurred in fully 71% (815) of plans! These cases were then replanned using IMRT approaches and the 2 plans compared. The results, shown in table 2, demonstrate significant improvement in dose homogeneity with IMRT. Of note, their uncomplicated IMRT method used forward planning involving 4 segments and 2 large open fields, plus 2 additional fields typically weighted to 10% of the original treatment beams. Wedges in the inferior/superior direction were used in about 75% of cases.

Table 3. Reduction in acute skin toxicity associated with the use of advanced treatment planning in the randomized multi-institution Canadian trial [19]

Endpoint	IMRT, %	2DRT, %	p
Skin toxicity grades 3–4	27.1	36.7	0.06
Moist desquamation, all	31.2	47.8	0.002
Moist desquamation, IMF	26.5	43.5	0.001
Pain grades 2–4	23.5	25.5	0.68

2DRT = 2D radiotherapy; IMF = inframammary fold.

Table 4. Reduction in late breast toxicity associated with the use of advanced treatment planning in the Royal Marsden randomized trial [20]

Site	Palpable induration (percent of cases)				p
	year 2		year 5		
	IMRT/3DC	2D	IMRT/3DC	2D	
Center of breast	16	27	21	32	0.02
Pectoral fold	12	27	22	29	0.006
Inframammary fold	16	29	17	24	0.009
Boost site	37	54	37	61	<0.001

Improving Outcomes

The question remains, does the correction of radiotherapy dose inhomogeneity using advanced planning and delivery technologies improve the cosmetic outcomes? The randomized Cambridge Trial, described above, should ultimately show this, and 2 other trials have already demonstrated such results. A Canadian study evaluated acute radiation toxicities by randomizing 331 patients, following breast conserving surgery, to standard 2D wedge planning with single transverse contours or to IMRT with field-in-field forward planning and 4–7 segments [19]. Table 3 shows that the use of IMRT resulted in a reduction in skin toxicity, especially moist desquamation ($p < 0.001$). On multivariate analysis, reduction in moist desquamation was associated with only 2 variables: smaller breast size ($p < 0.001$) and use of breast IMRT ($p = 0.003$)

A study from the Royal Marsden Hospital evaluated the benefit of IMRT in reducing late toxicities of treatment [20]. 306 patients were randomized to 2D planning or IMRT using forward planning and 3–4 segments. Breast appearance was evaluated at 1, 2 and 5 years, and was found to be improved at each of these time points. There was a reduction in breast induration at both 2 and 5 years (table 4). Of interest, there was no significant difference in patient-reported outcomes including breast discomfort or breast hardness, or in measures of quality of life.

There are several large retrospective studies that add weight to these prospective studies. Work at Fox Chase Cancer Center, reported by Freedman et al. [21] in 2009, showed that IMRT plans resulted in reduced acute toxicity when compared with 2D plans. Investigations at Beaumont Hospital [22, 23] showed that plans excluding skin and lung by using a median of 6 segments (range: 3–12) reduced acute toxicities of dermatitis, edema and hyperpigmentation, and reduced late effects of edema and hyperpigmentation. Work at Emory University had a longer follow-up of 6–7 years, and again used a simplified type of IMRT planning using MLC and dynamic wedges [24]. This again showed reduction in grades 1 and 2 dermatitis.

Conclusions: Improving Breast Dosimetry

Overall, the findings indicate that one can improve objective clinical outcomes through attention to dose homogeneity, and that IMRT technology can improve the homogeneity of breast radiotherapy. One randomized trial and several retrospective studies have convincingly shown that this will achieve a reduction in acute dermatitis. In the Royal Marsden trial, IMRT was associated with reduced fibrosis and improved breast appearance, as determined by photos and physician assessments.

It should be recognized that many of these trials used CT-based planning methods that were not technically complicated, and many employed forward planning strategies with field-in-field approaches and dynamic wedges to improve dose homogeneity. These are planning methods beyond 2D, and represent early steps toward improving dose distribution within the breast. They have already shown measurable clinical improvement in acute toxicities, and suggest that additional steps to improve dose distributions may further enhance outcome results.

Avoidance of Organs at Risk

Dose to Heart

More developed approaches to IMRT may further reduce unwanted dose to nearby organs at risk. Returning to the study from Australia [17], 4 different IMRT approaches were used in their work: simpler field-in-field, inverse planning through the same tangent field, and more complex coplanar multifield IMRT and noncoplanar multifield multisegment IMRT. When they compared each of these ap-

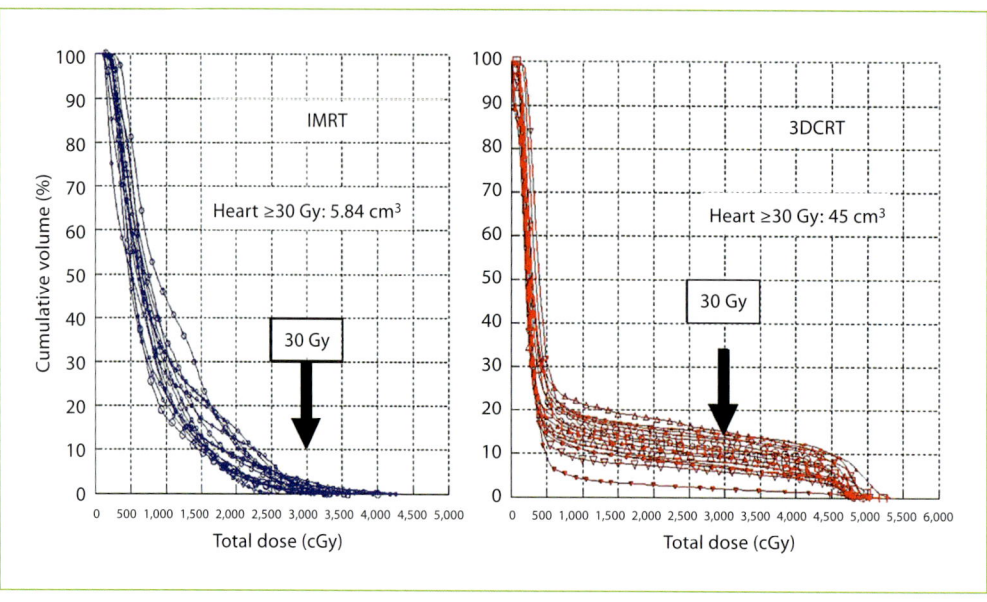

Fig. 1. Heart dose in plans using IMRT or 3DCRT. 14 left-sided cases with unfavorable anatomy were planned with these 2 approaches. Volume of heart receiving ≥30 Gy was greatly reduced by IMRT planning [26].

proaches with 2D approaches for avoidance of the heart, simpler IMRT approaches did not reduce the heart V_{30}, but the more complex ones did show benefit. A study from Alabama [25] used simple forward-planned field-in-field IMRT, inverse-planned IMRT and helical tomotherapy, and achieved greater heart sparing in terms of heart V_5 and V_{30} with the field-in-field planning. A study reported by Lohr et al. [26] compared IMRT with CT-based 3D approaches in 14 patients. Dose-volume histograms demonstrated that IMRT methods reduced the volume of heart receiving ≥30 Gy (fig. 1). The maximum dose to the heart was also reduced, from about 49 to about 35 Gy. In sum, the planning method (including the complexity of the IMRT planning) can determine the degree of heart sparing. Although it is expected that some patients will benefit from this planning, it remains unclear what objective clinical measurements should be used to best correlate patient outcome with these reduced heart doses.

Dose to Lung

The planning studies discussed above have shown mixed results for the use of advanced technologies to reduce lung exposure. The study from Memorial Sloan-

Table 5. Lung exposure using different IMRT approaches

Author	n	Assessment measure	IMRT type	Result IMRT	2D	p
Chiu [16]	15	D_{05}	intensity	97.9	95.3	0.07
		mean dose %	map	25.6	24.9	0.01
Fong [17]	20	mean dose (Gy)	E-IMRT	9.6	9.9	0.23
			T-IMRT	7.8		0.00003
			CP-IMRT	12.9		0.00006
			NCP-IMRT	8.2		0.0001
Caudell [25]	10	mean/V_5/V_{20}	FIF	2.3/7.3/3.9	–	–
			IP-DLC	2.5/8.4/4.3	–	–
			TOMO	5.9/38/4.9	–	–

Note that the particular approach may result in substantially different rates of lung exposure. D_{05} = Dose to hottest 5% of planning target volume; V_5/V_{20} = volume receiving 5, 20 Gy or greater; E = electronic compensators; T = tangential beam; CP = coplanar multifield; NCP = non-coplanar multifield; FIF = field-in-field; IP-DLC = inverse-planned dynamic multileaf collimation IMRT; TOMO = tomotherapy.

Kettering Cancer Center [16] that used an intensity mapping approach showed only modest improvement in lung doses compared with 2D methods. The Australian study [17] using increasingly complex IMRT approaches, discussed above, showed variable results with respect to the lung, and the multifield coplanar IMRT approach actually produced higher lung doses than simpler field-in-field approaches. Caudell et al. [25] showed that tomotherapy approaches could result in higher lung exposures, at least for the mean and low-dose V_5 values. It is clear that one must be careful in using each planning method: some IMRT approaches may potentially increase lung exposure especially at lower dose levels without attention to this concern (table 5).

Dose to the Contralateral Breast

Planning approaches using IMRT convincingly show that dose to the contralateral breast can be diminished [16, 17, 25, 27]. Work reported by Bhatnagar [27] used actual TLD measurements obtained 4 cm from the edge of the field toward the contralateral breast, and showed a highly significant reduction in exposure using simple field-in-field planning compared with physical wedges.

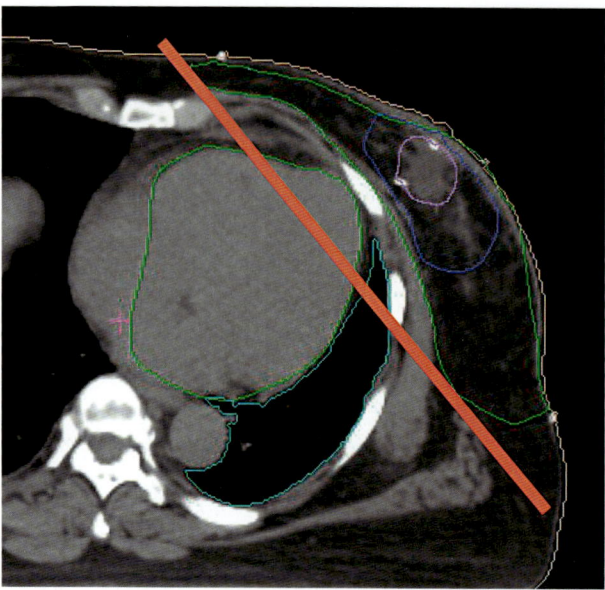

Fig. 2. The use of tangent fields (like that described by the colored straight line) often does not permit adequate coverage of breast target volumes and also avoidance of heart and lung volumes. Unless these volumes are defined, such deficiencies remain unrecognized.

Conclusions: Avoidance of Dose to Organs at Risk

Interpretation of these studies is problematic given the variable methods and measures used, but some conclusions can be drawn. IMRT can reduce radiation dose to the heart, though this is somewhat method-dependent. This is also true of using IMRT to reduce lung dose, since the particular method can result in quite different exposure values. Advanced planning consistently can achieve reduction in radiation dose to the contralateral breast, though the amount is modest. For all of these organs at risk, more complex IMRT approaches often have advantages. As yet, the clinical benefits of these dose reductions to normal tissues are undocumented, but may be quite significant in the long-term.

Clinical Treatment Planning

Defining Organs at Risk

It is important to define the volumes for treatment, and to use treatment methods that adequately conform fields to those targets. Unless target volumes are

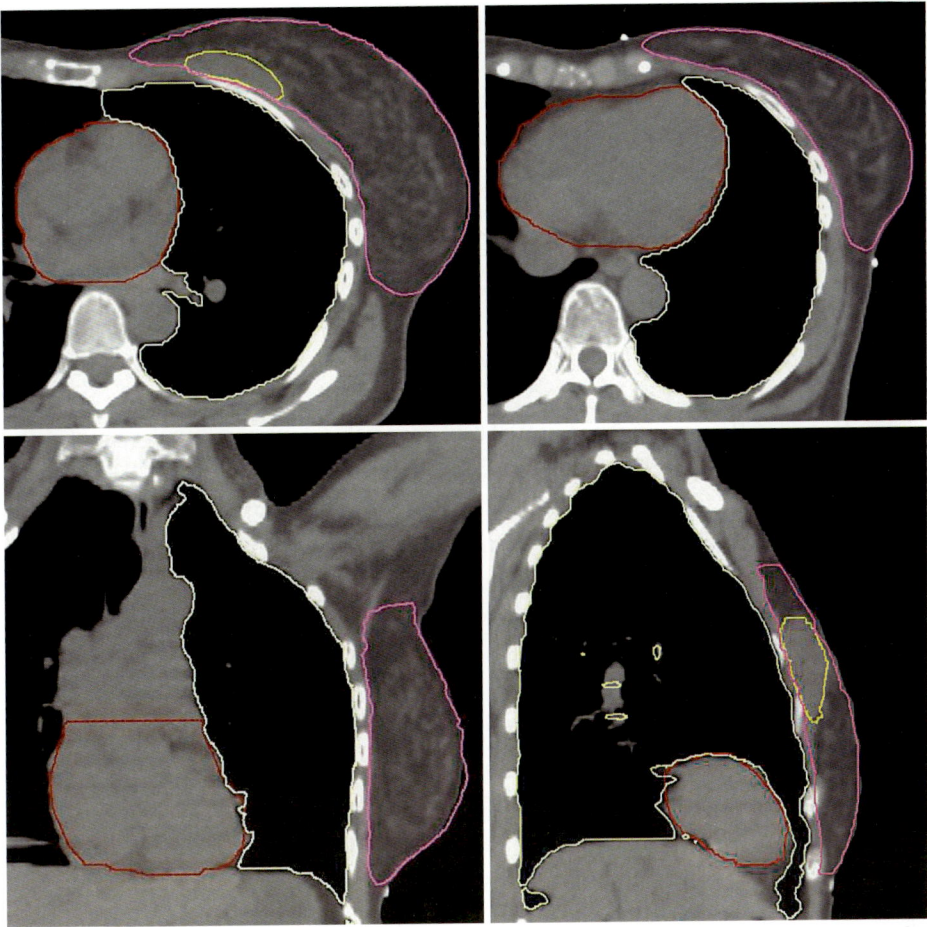

Fig. 3. RTOG example of segmentation for breast radiotherapy, showing breast (purple), lumpectomy GTV (yellow), heart (red) and lung (white). Additional information for planning is found at the RTOG website (www.rtog.org).

defined, it is impossible to know whether adequate targeting and avoidance have been achieved (fig. 2). Throughout this volume, discussions emphasize the importance of understanding and defining the exact targets for treatment. The RTOG has brought together an expert panel to define the anatomical boundaries of the regional node volumes for breast cancer; their recommendations are available at the RTOG website (www.rtog.org) and a useful example is shown in figure 3.

It can be challenging to adequately exclude skin, lung and heart volumes and at the same time provide acceptable margins for target volumes unless advanced technologies are used. This was demonstrated in a study reported by Van der Laan [28]. Twenty-five left-sided breast cancers were used to compare 2D and 3D conformal radiotherapy (CRT) plans for whole breast irradiation (WBI; 50-Gy whole breast dose plus 16-Gy boost). When using 3DCRT approaches, dose-volume histogram coverage of the target volumes (whole breast and boost volumes) were greatly improved, though the sparing of the lung (determined by the maximum V_{20}) and heart (V_{30}) were somewhat compromised. Therefore, 3DCRT approaches may add benefit but may not be sufficient, and some form of IMRT may be required for adequate organ sparing.

Motion Management and Image Guidance

How often is gating or another method of respiratory management required for breast cancer irradiation? This need reflects the dose gradients and margins being used for the target volumes, as well as the patient's anatomy and breathing physiology. Concern for motion management also increases with the size and extent of the field, and is more important in cases receiving postmastectomy regional radiotherapy. Often this relates primarily to avoidance of heart and lung volumes, and a series of objective constraints should be used for these and other important structures. If these constraints cannot be met, then motion control may become necessary. These concerns are not different than those used for other tumor sites. In reality, only a small proportion of patients require gating or deep breath hold approaches since diaphragmatic (and not chest wall/intercostal) breathing predominates in most patients when they are at rest. For patients receiving hypofractionated partial breast approaches, daily image guidance seems highly appropriate for most cases.

Homogeneity Guidelines

Standards for acceptable homogeneity are not universally defined. At the Medical College of Wisconsin (MCW), we expect to achieve ≥95% coverage of the target regions with ≥95% of the prescribed dose. The maximum doses must be less than 108%, as this value has been associated with toxicity. Breast size should not be a factor in limiting the maximum, since our group and others have shown that homogeneous doses can be obtained by using appropriate treatment planning.

Postmastectomy Radiotherapy

Work at the MD Anderson Cancer Center [29] compared 3 CT-based methods for treatment of supraclavicular and axillary nodes following mastectomy: posterior

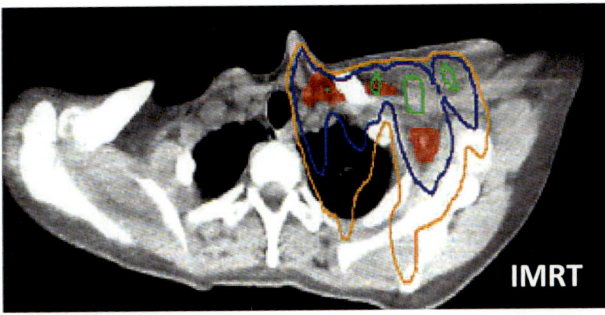

	PAB	AAB	IMRT
Conformality index	0.067	0.085	0.163
V_{105} treatment volume (cm³)	70	55.9	0.163
Mean lung dose (Gy)	36.6	35.6	29.3

Fig. 4. Improvement in 3 measures of dose distribution by IMRT planning for postmastectomy regional irradiation. PAB = Posterior axillary boost; AAB = anterior axillary boost; V_{105} = volume receiving 105% or more of the prescribed dose [29].

axillary boost, anterior axillary boost and IMRT. The conformality index, V_{105} treatment volume and mean lung dose were all improved by IMRT applications (fig. 4). Because of the larger treatment volumes, postmastectomy cases may be among those that benefit most from IMRT.

Concomitant Boost Radiotherapy

IMRT can be used to boost the region of tumor excision concurrently with whole breast treatment. In 2009, Chadha et al. [30] reported the work at Beth Israel Medical Center in New York: 40.5 Gy in 2.7-Gy fractions was delivered to the breast volume while the boost volume received 45 Gy in 3.0-Gy fractions. An example is shown in figure 5. With a median follow-up of 2.3 years, there have been no in-breast recurrences and the degree of acute dermatitis observed has been very mild. Similar work has been reported by others (table 6) and indicates that this approach is generally feasible [31–33]. It may be an attractive use of IMRT to accelerate WBI and give the possible advantages of hypofractionation to the highest risk tissues near the resection cavity.

Several important clinical questions have been developed into a proposal by the RTOG for a phase III randomized trial (RTOG 1005). This will compare (a) hypofractionation using 15 fractions and concomitant boost with (b) WBI of 50 Gy in standard fractionation and sequential boost. There is a target accrual goal of 2,300

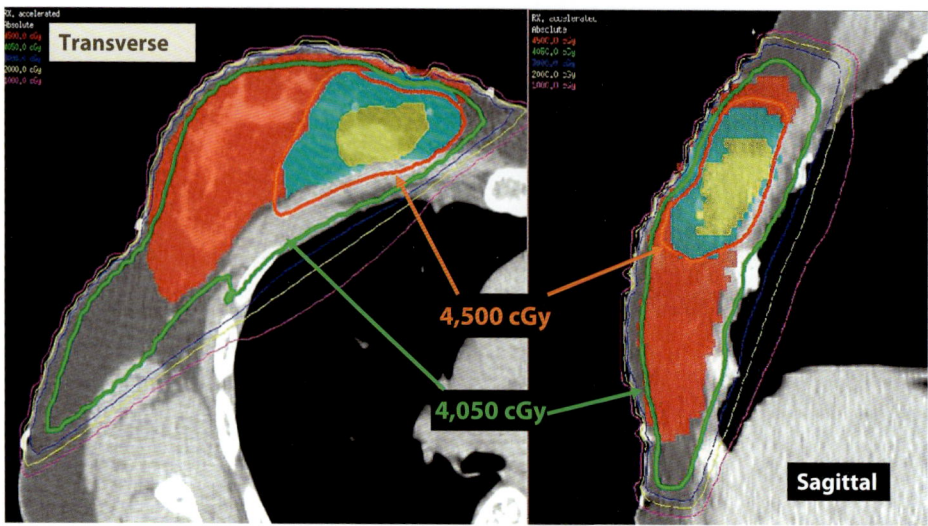

Fig. 5. Forward-plan IMRT using concomitant photon boost [30].

patients. Additional goals of the study include further definition of the dose standards appropriate for target volumes and organs at risk, standardization of IMRT and 3DCRT terminologies, and volume definitions for treatment.

Partial Breast Radiotherapy

Selection of Cases for Partial Breast Irradiation

Accelerated PBI targets conformal irradiation to the breast tissue immediately adjacent to the lumpectomy cavity, and has been proposed as an alternative to irradiation of the entire breast for early-stage breast cancer patients undergoing conservation treatment. Numerous single institution phase II studies have demonstrated promising results, including rates of in-breast cancer recurrence of only 3–5% at 2–8 years of follow-up. It should be recognized that the outcomes from these early PBI studies have been influenced by the selection of fairly low-risk breast cancers: patients had primary tumors that averaged 1 cm in size, were node-negative and ER/PR positive, and most patients were over 65 years of age.

Three phase III clinical trials are actively accruing and randomizing breast cancer patients to receive either PBI or standard whole breast irradiation (WBI) following lumpectomy. The eligibility criteria for enrollment in current trials are now broader, as can be seen in table 7, and will better test the extent that PBI can achieve in-breast tumor control rates equivalent to standard WBI in all patients subsets

Table 6. Reported programs with simultaneous integrated boost using IMRT

Institution	n	Total dose/fraction dose, Gy	Acute toxicity grade, %		
			1	2	3
Fox Chase Cancer Center [31]	75	WB-IMRT: 45/2.25, L-PTV: 56/2.8	65	2	0
University of Groningen [32]	90	WB-3DCRT: 50.4/1.8, L-PTV: 64.4/2.3	60	31	1.1
New York University [33]	91	WB-IMRT: 40.5/2.7, L-PTV: 48/3.2	58.5	8.1	0.9

WB = Whole breast; L-PTV = lumpectomy PTV.

Table 7. Phase III randomized trials comparing PBI to WBI following lumpectomy in early-stage breast cancer

Phase III randomized trial	Date opened	Targeted accrual, n	PBI method	Eligibility
NSABP B-39/RTOG 0413	3/2005	4,300	3DCRT, MST, MCT	stage 0–II, <3 cm, N-0 – N-1 (<3 LN+), age >18 years
GEC-ESTRO	11/2004	1,170	MCT	stage 0–II, <3 cm, N-0 – N_{mi}, negative margins (2 mm), age >40 years
RAPID OCOG	1/2006	2,128	3DCRT	stage 0–II, <3 cm, N-0, age >40 years, excludes infiltrating lobular disease

MCT = Multicatheter brachytherapy.

undergoing lumpectomy for breast cancer. While practitioners await the outcome of these trials, the American Society of Radiation Oncology (ASTRO) convened a task force to review the existing data on PBI and develop a consensus statement about patient selection criteria and best practices for PBI outside the context of a clinical trial (table 8). The *suitable* group is defined by the typical low-risk breast cancer features that constitute the basis of most existing PBI outcomes data; these patients are endorsed for treatment off of a clinical trial. Patients with breast cancer features in the higher-risk categories are viewed as *cautionary* or *unsuitable* for treatment off of a clinical trial.

Selection of the Partial Breast Irradiation Method

PBI was initially developed with brachytherapy methods, and the most mature outcome data are from interstitial experiences [34–36]. The subsequent develop-

Table 8. ASTRO Consensus Statement: patient selection for treatment with PBI outside the context of a clinical trial

Factors	'Suitable' group	'Cautionary' group	'Unsuitable' group
Patient factors			
Age, years	≥60	50–59	<50
BRCA1/2 mutation	not present	NA	present
Pathologic factors			
Tumor size, cm	≤2	2.1–3.0	>3
T stage	T1	T0 or T2	T3 or T4
Margins	negative by at least 2 mm	close (<2 mm)	positive
Grade	any	NA	NA
LVSI	no	limited/focal	extensive
ER status	positive	negative	NA
Multicentricity	unicentric only	NA	if present
Multifocality	clinically unifocal with total size ≤2 cm	clinically unifocal with total size 2.1–3.0 cm	if microscopically multifocal >3 cm in total size or if clinically multifocal
Histology	invasive ductal or other favorable subtypes	invasive lobular	NA
Pure DCIS	not allowed	≤3 cm in size	if >3 cm in size
EIC	not allowed	≤3 cm in size	if >3 cm in size
Associated LCIS	allowed	NA	NA
Nodal factors			
N stage	pN0 (i−, i+)	NA	pN1, pN2, pN3
Nodal surgery	SN Bx or ALND	NA	none performed
Treatment factors			
Neoadjuvant therapy	not allowed	NA	if used

BRCA1/2 = BRCA1/2 germline; LVSI = lymphatic vascular invasion; DCIS = ductal carcinoma in situ; EIC = extensive intraductal component; LCIS = lobular carcinoma in situ; SN Bx = sentinel node biopsy; ALND = axillary lymph node dissection; NA = not available.

ment of the MammoSite (MST) balloon catheter, a simplified and more user-friendly brachytherapy method, expanded the access of PBI to patients [37]. The major advantage of brachytherapy methods is the elimination of inter- and intrafraction setup errors inherent in external beam treatment delivery. The major disadvantages are the need for additional surgical intervention for placement of the device, additional scarring, patient discomfort, infection risk and disruption of bathing for the several days that the device is in place. External beam methods for PBI, primarily with 3DCRT, were developed more recently and have more limited follow-up data [38–41]. However, external beam 3DCRT appeals to many practitioners and is the most frequently used method (approx. 70% of cases) for

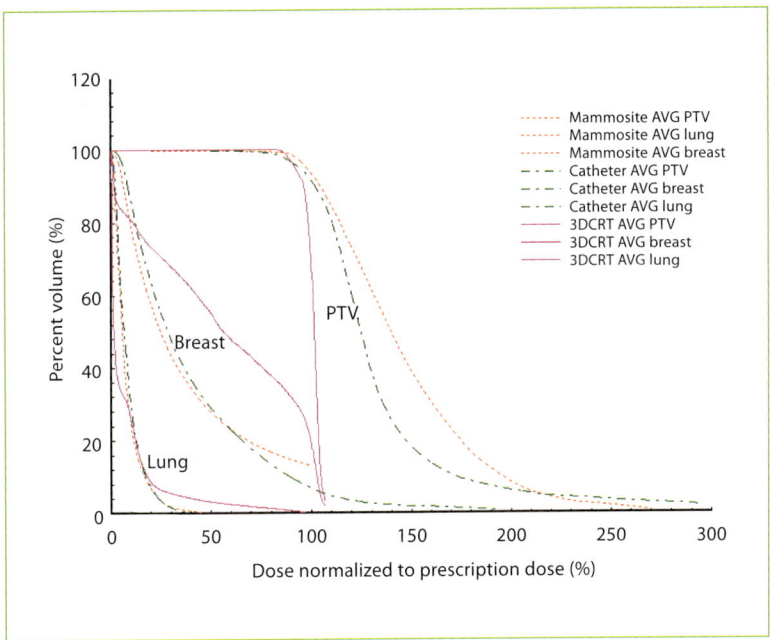

Fig 6. Dose-volume histogram averages (AVG) for PTV, lung and breast from 5 cases each of multicatheter, MammoSite and 3DCRT PBI. The breast dose outside the PTV is higher with 3DCRT PBI.

delivery of PBI on the NSABP B39/RTOG 0413 randomized clinical trial. For this NSABP trial, and also the RAPID trial, the specified clinical target volume (CTV) is a 1.5-cm expansion around the lumpectomy cavity and the planning target volume (PTV) is a 1-cm expansion around the CTV.

There have been numerous dosimetric studies comparing and contrasting PBI delivery using multicatheter or balloon brachytherapy, or 3DCRT [42, 43]. For instance, an analysis from MCW using equivalent uniform dose, tumor control probability and normal tissue complication probability metrics evaluated dose inhomogeneity from the 3 PBI methods. For the multicatheter (cath), MST and 3DCRT PBI methods, the average equivalent uniform dose (normalized to 2-Gy daily fractions) was found to be 42.2 Gy (cath), 46.4 Gy (MST) and 46.9 Gy (3DCRT), respectively; the average tumor control probability was estimated to be 94.8% (cath), 99.14% (MST) and 99.17% (3DCRT), respectively. The normal tissue complication probability was found to be very small for normal breast and lung tissues for all 3 methods, but these tissues did receive higher doses in the 3DCRT PBI plans (fig. 6).

Table 9. Comparison of MammoSite brachytherapy (BRT), 3DCRT and IMRT for PBI at Rush University: summary of dosimetric comparisons

	BRT	3DCRT	IMRT
V_{90}	99.3 (1)	99.9 (0.1)	100.0 (0)
V_{95}	97.9 (1.6)	99.4 (0.4)	99.9 (0.1)
V_{100}	94.9 (2.4)	92.3 (2.2)	93.7 (2.0)
V_{110}	84.2 (4)	0.99 (2.2)	0.8 (1.4)
Ipsilateral breast V_{50}	29.2 (10.1)	55.8 (12.4)	46.2 (12.9)
Maximum skin dose, cGy	3,543 (1,045)	3,569 (122)	3,677 (139)
Ipsilateral lung V_{30}	5.4 (3.5)	6.7 (4.6)	1.9 (2.4)
Contralateral breast V_3	0	0	0
Heart V_5	11.6 (8.7)	4.1 (2.7)	1.2 (2.0)
PTV volume, cm^3	94.3 (18.5)	184.3 (54.6)	184.3 (54.6)

All values are means; values in parentheses are standard deviations. Values without units are percentages [44].

Intensity-Modulated Radiotherapy and Partial Breast Irradiation

Advantages of Intensity-Modulated Radiotherapy for Partial Breast Irradiation

IMRT has been proposed as an alternative to 3DCRT for external beam PBI and potentially offers further improvement in normal tissue sparing. A dosimetric comparison of 3DCRT, IMRT and MST methods for 15 patients receiving PBI from Rush University [44] demonstrated that all 3 methods were highly successful in achieving target coverage, including comparable PTV coverage assessed for V_{90}, V_{95} and V_{100} (table 9). The ipsilateral breast V_{50} was lowest for MST, but IMRT spared more breast outside the PTV than the 3DCRT method. Additionally, IMRT spared heart and ipsilateral lung better than either the MST or 3DCRT techniques.

The dose distribution and normal tissue sparing from PBI delivered with IMRT versus 3DCRT was reported for 56 cases from the University of Colorado [45]. In this dosimetric comparison, the 3DCRT PBI was delivered with an average of 4 fields (range: 3–7), and inverse-planned IMRT was delivered with an average of 6 fields (range: 4–9). For PTV coverage, the V_{90} was excellent for both 3DCRT and IMRT (though the V_{95} showed better coverage with 3DCRT: 96% for 3DCRT and 88% for IMRT, $p < 0.010$). For breast sparing outside the PTV, the ipsilateral breast V_{50} was less with IMRT (47% for 3DCRT and 42.1% for IMRT, $p < 0.01$). In patients with a PTV/breast ratio >25%, there was even more sparing of the ipsilateral breast with the IMRT plans. IMRT also reduced the doses delivered to lung and heart: the lung V_{20} with 3DCRT was 2.3 and 1.2% with IMRT ($p < 0.01$); the heart V_5 with 3DCRT was 1.7 and 0.6% with IMRT ($p = 0.040$). These comparisons show small but significant advantages for IMRT in sparing adjacent normal tissues during PBI.

Intensity-Modulated Radiotherapy/Partial Breast Irradiation Clinical Results

There are limited clinical outcome data from the delivery of PBI with IMRT. Recently, investigators in Colorado reported early results for 55 patients treated prospectively with IMRT PBI using 38.5 Gy total dose (3.85 Gy delivered twice daily for 10 fractions) [45]. The CTV was a 1-cm uniform expansion around the lumpectomy cavity, and the PTV added a 1-cm margin around the CTV. The treated patients all had stage I (node-negative) disease and their median age was 61 years; tumors had a median size of 9 mm and 94% were estrogen receptor-positive. The median follow-up time is still less than 1 year, but no patient has experienced more than grade 1 toxicity.

The University of Florence recently reported acute skin toxicity and dosimetry results on the first 259 patients (of a targeted 520 accrual) in a phase III clinical trial randomizing patients after lumpectomy to conventional WBI versus IMRT PBI (30 Gy delivered in 6 Gy per fraction) [46]. For planning, the CTV gave a 1-cm uniform expansion around the lumpectomy cavity and the PTV added 1 cm to the CTV; 4 coplanar fields were adopted and the PTV V_{95} was 96% on average. Despite the accelerated fractionation, the acute skin toxicity rates were reduced with IMRT: 22% grade 1 and 19% grade 2 with standard WBI compared with only 5% grade 1 and 0.8% grade 2 with IMRT PBI.

The volume of treatment may be important for cosmesis. Another prospective trial, from the University of Michigan, delivered IMRT PBI only during deep inspiration breath hold to 34 women after lumpectomy [47]. The dose delivered was 38.5 Gy over 10 fractions given b.i.d. The CTV was a 1-cm uniform expansion around the lumpectomy cavity, and the PTV expanded the CTV by 1 cm additionally; the beamlet IMRT was inverse-planned. After a 2.5-year median follow-up, there were no in-breast recurrences, but 7 patients (22%) developed unacceptable cosmetic outcomes and led to a premature closure of this prospective trial. This toxicity was related to the treatment volume, assessed by both V_{50} and V_{100}. For instance, the mean V_{50} (the ipsilateral breast volume receiving >19.25 Gy) was significantly less in the patients who had acceptable cosmesis (34.6%) than in those with unacceptable results (46.1%).

At MCW, IMRT PBI has been delivered in the prone position to gain more motion control of the anterior thoracic wall. Initially, tomotherapy was investigated for PBI delivery [48]. A dosimetric study of 10 prone-positioned patients, previously treated using conventional WBI, confirmed that compliance with the NSABP B39/RTOG 0413 guidelines for PBI could be achieved [49]: the mean V_{95} was 99.7% and the V_{50} for the ipsilateral breast was limited to 39.2%. The only exception was the dose to the contralateral breast, which was a maximum of 1.8 Gy (range: 1.2–2.0) and slightly greater than the protocol's limit of 1.2 Gy. Four patients were then treated with tomotherapy for IMRT PBI with plans that replicated the previous

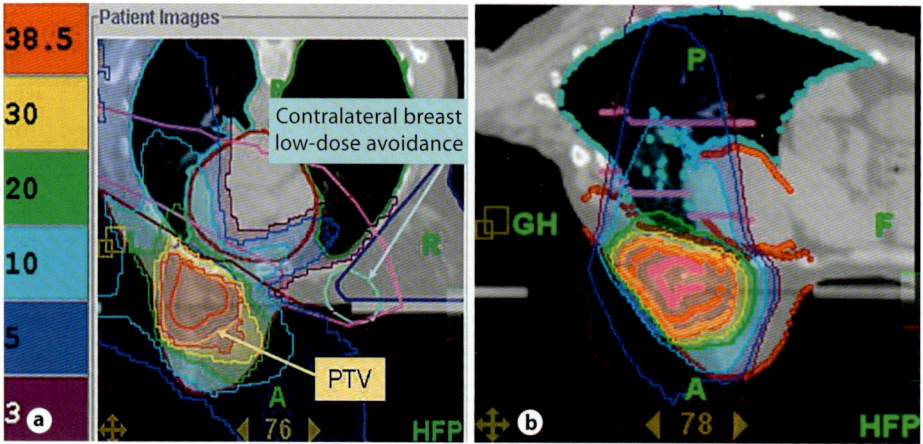

Fig. 7. Isodose distributions on axial and sagittal kV CT slices using helical tomotherapy. Both views are through the midpoint of the PTV. **a** Axial view. For contralateral breast avoidance, IMRT planning kept the 3-Gy isodose lines away from the contralateral breast. Note also the 38.5-Gy coverage of the PTV. **b** Sagittal view.

dosimetry study (fig. 7). Given the contralateral breast dose, and since the lumpectomy seroma and surgical clips sometimes were only faintly visible on the MVCT images, IMRT PBI was further explored using a fixed gantry accelerator and kV CT-on-rails for image-guided therapy. Quality plans could be delivered for another 16 patients on this prospective trial. The average PTV/breast volume was 16% (range: 9–34). After a median follow-up of 12 months, there were no recurrences, and good cosmesis was reported in 95–100% of cases in both physician and patient assessments.

Conclusions: Intensity-Modulated Radiotherapy for Partial Breast Irradiation

The use of IMRT for PBI delivery is promising and offers advantages compared with 3DCRT in terms of sparing normal tissues including ipsilateral breast outside the PTV, heart and lung. While centers are actively pursuing this approach, additional study is warranted before it should be accepted as standard practice since the follow-up periods of existing studies are short and certain issues remain unresolved. In particular, definitions for delineating CTV and PTV margins now vary, and consistent definitions are needed. Also, smaller target volumes and higher dose gradients will require tighter management of motion during therapy delivery, and strategies to accomplish this need further refinement and clinical experience for partial breast treatment.

Guidelines for Clinical Practice

Whole Breast Irradiation
- Clinical outcomes for WBI can be improved through attention to radiation dose homogeneity within the PTV, and IMRT can improve this homogeneity. Already, IMRT methods have been shown to improve breast-related acute toxicity of treatment. At MCW, we expect to achieve ≥ 95% coverage of the target regions with ≥ 95% of the prescribed dose; the maximum doses must be less than 108%.
- Advanced treatment planning including IMRT can reduce radiation doses to normal tissues (ipsilateral breast outside of the PTV, contralateral breast, heart and lung). These dosimetric benefits are method-dependent and require careful attention to the details of planning. As yet, the clinical benefits of these dose reductions are largely undocumented, but may be significant in the long-term.
- It is important to define CTV, PTV and normal tissue volumes for treatment and avoidance, and to use treatment methods that adequately conform fields to those regions. To achieve this, 3DCRT approaches may add benefit over 2D but may not be sufficient, and some form of IMRT may be useful for adequate organ sparing in at least some patients.
- Concern for motion management increases with the size and extent of the field, and may be more important in cases receiving postmastectomy regional radiotherapy.
- IMRT can be used to boost the region of tumor excision concurrently with WBI, an approach that is being explored at several major centers. It may be an attractive use of IMRT to accelerate treatment and give the possible advantages of hypofractionation to the highest risk tissues near the resection cavity.

Partial Breast Irradiation
- Numerous single institution phase II studies have demonstrated promising results with PBI, including rates of in-breast cancer recurrence of only 3–5% at 2–8 years of follow-up. ASTRO guidelines define a group suitable for PBI off of a clinical trial. These patients have low-risk breast cancer features that constitute the basis of most existing PBI outcomes data.
- For the NSABP B39/ RTOG 0413 and RAPID trials, the specified CTV is a 1.5-cm expansion around the lumpectomy cavity and the PTV is a 1-cm expansion around the CTV. Other investigations use smaller margins, and there is need for consistent terminology, treatment volume definitions and dose constraints for PBI.
- IMRT for PBI is a promising approach and offers advantages compared with 3DCRT in terms of sparing normal tissues including ipsilateral breast outside the PTV, heart and lung. To date, there are limited clinical outcome data from IMRT delivery for PBI; its relative benefit compared with 3DCRT PBI is yet to be determined, but is being actively explored.

References

1 Arriagada R, Le MG, Rochard F, Contesso G: Conservative treatment versus mastectomy in early breast cancer: patterns of failure with 15 years of follow-up data. Institut Gustave-Roussy Breast Cancer Group. J Clin Oncol 1996;14:1558–1564.
2 Blichert-Toft M, Rose C, Andersen JA, et al. Danish randomized trial comparing breast conservation therapy with mastectomy: six years of life-table analysis. Danish Breast Cancer Cooperative Group. J Natl Cancer Inst Monogr 1992;11:19–25.
3 Fisher B, Anderson S, Bryant J, et al: Twenty-year follow-up of a randomized trial comparing total mastectomy, lumpectomy, and lumpectomy plus irradiation for the treatment of invasive breast cancer. N Engl J Med 2002;347:1233–1241.
4 Jacobson JA, Danforth DN, Cowan KH, et al: Ten-year results of a comparison of conservation with mastectomy in the treatment of stage I and II breast cancer. N Engl J Med 1995;332:907–911.
5 van Dongen JA, Voogd AC, Fentiman IS, et al: Long-term results of a randomized trial comparing breast-conserving therapy with mastectomy: European Organization for Research and Treatment of Cancer 10801 trial. J Natl Cancer Inst 2000;92:1143–1150.
6 Veronesi U, Cascinelli N, Mariani L, et al. Twenty-year follow-up of a randomized study comparing breast-conserving surgery with radical mastectomy for early breast cancer. N Engl J Med 2002;347:1227–1232.
7 Clarke M, Collins R, Darby S, et al: Effects of radiotherapy and of differences in the extent of surgery for early breast cancer on local recurrence and 15-year survival: an overview of the randomized trials. Lancet 2005;366:2087–2106.
8 Solin L, Fowble B, Martz K, et al: Results of the 1983 Patterns of Care process survey for definitive breast irradiation. Int J Radiat Oncol Biol Phys 1991;20:105–111.
9 Shank B, Moughan J, Owen J, et al: The 1993–94 patterns of care process for irradiation after breast conserving surgery – comparison with the 1992 standard for breast conservation treatment. Int J Radiat Oncol Biol Phys 2000;48:1291–1299.
10 Pierce LJ, Moughan J, White J, Winchester DP, Owen J, Wilson JF: 1998–1999 patterns of care study process survey of national practice patterns using breast-conserving surgery and radiotherapy in the management of stage I-II breast cancer. Int J Radiat Oncol Biol Phys 2005;62:183–192.
11 Wazer D, Dipetrillo T, Schmidt-Ullirch R, et al: Factors influencing cosmetic outcome and complication risk after conservative surgery and radiotherapy for early-stage breast carcinoma. J Clin Oncol 1992;10:356–363.
12 De La Rochefordiere A, Abner A, Silver B, et al: Are cosmetic results following conservative surgery and radiation therapy for early breast cancer dependent on technique? Int J Radiat Oncol Biol Phys 1992;23:925–931.
13 Olivotto IA, Rose MA, Osteen RT, et al: Late cosmetic outcome after conservative sugery and radiotherapy: analysis of causes of cosmetic failure. Int J Radiat Oncol Biol Phys 1989;17:747–753.
14 Taylor M, Perez C, Halverson K, et al: Factors influencing cosmetic results after conservation therapy for breast cancer. Int J Radiat Oncol Biol Phys 1995;31:753–764.
15 Kestin LL, Sharpe MD, Frazier RC, et al: Intensity modulation to improve dose uniformity with tangential breast radiotherapy: initial clinical experience. Int J Radiat Oncol Biol Phys 2000;48:1559–1568.
16 Chui CS, Hong L, McCormick B: Intensity-modulated radiotherapy technique for three-field breast treatment. Int J Radiat Oncol Biol Phys 2005;62:1217–1223.
17 Fong A, Bromley R, Beat M, et al: Dosimetric comparison of intensity modulated radiotherapy techniques and standard wedged tangents for whole breast radiotherapy. J Med Imaging Radiat Oncol 2009;53:92–99.
18 Barnett GC, Wilkinson J, Moody AM, et al: A randomized controlled trial of forward-planned radiotherapy (IMRT) for early breast cancer: Baseline characteristics and dosimetry results. Radiother Oncol 2009;92:34–41.
19 Pignol JP, Olivotto I, Rakovittch, et al: A multi-center randomized trial of breast intensity-modulated radiation therapy to reduce acute radiation dermatitis. J Clin Oncol 2008;26:2085–2092.
20 Donovan E, Bleakley N, Denholm E, et al: Randomized trial of standard 2D radiotherapy (RT) versus intensity modulated radiotherapy (IMRT) in patients prescribed breast radiotherapy. Radiother Oncol 2007;82:254–264.
21 Freedman GM, Tianyu L, Nicolaou N, et al: Breast intensity-modulated radiation therapy reduces time spent with acute dermatitis for women of all breast sized during radiation. Int J Radiat Oncol Biol Phys 2009;74:689–694.

22 Haroslia A, Kestin L, Grillis I, et al: Intensity-modulated radiotherapy results in significant decrease in clinical toxicities compared with conventional wedge-based breast radiotherapy. Int J Radiat Oncol Biol Phys 2007;68:1375–1380.

23 Vicini FA, Sharpe M, Kestin L, et al: Optimizing breast cancer treatment efficacy with intensity-modulated radiotherapy. Int J Radiat Oncol Biol Phys 2002;54:1336–1344.

24 McDonald MW, Godette KD, Butker EK, et al: Long-term outcomes of IMRT for breast cancer: a single-institution cohort analysis. Int J Radiat Oncol Biol Phys 2008;72:1031–1040.

25 Caudell JJ, De Los Santos JF, Keene KS, et al: A dosimetric comparison of electronic compensation, conventional intensity modulated radiotherapy, and tomotherapy in patients with early-stage carcinoma of the left breast. Int J Radiat Oncol Biol Phys 2007;68:1505–1511.

26 Lohr F, El-Haddad M, Dobler B, et al: Potential effect of robust and simple IMRT approach for left sided breast cancer on cardiac mortality. Int J Radiat Oncol Biol Phys 2009;74:73–80.

27 Bhatnagar AK, Brandner E, Sonnik D, et al: Intensity modulated radiation therapy (IMRT) reduces the dose to the contralateral breast when compared to conventional tangential fields for primary breast irradiation. Breast Cancer Res Treat 2006;96:41–46.

28 Van der Laan HP, Dolsma WV, Maduro JH, et al: Dosimetric consequences of the shift towards computed tomography guided target definition and planning for breast conserving radiotherapy. Radiat Oncol 2008;3:6.

29 Wang X, Yu TK, Salehpur M, et al: Breast cancer regional radiation fields for supraclavicular and axillary lymph node treatment: is the posterior axillary boost field technique optimal? Int J Radiat Oncol Biol Phys 2009;74:86–91.

30 Chadha M, Woode R, Silanpaa J, et al: Three-week accelerated whole breast radiation therapy with concomitant boost for early stage breast cancer (abstract). Int J Radiat Oncol Biol Phys 2009;75(suppl 1).

31 Freedman GM, Anderson PR, Goldstein LJ, et al: Four-week course of radiation for breast cancer using hypofractionated intensity modulated radiation therapy with an incorporated boost. Int J Radiat Oncol Biol Phys 2007;68:347–353.

32 Van der Laan HP, Dolsma WV, Maduro JH, et al: Three dimensional conformal simultaneously integrated boost technique for breast conserving therapy. Int J Radiat Oncol Biol Phys 2007;68:1018–1023.

33 Formenti SC, Gidea-Addeo D, Godlberg JD, et al: Phase I-II trial of prone accelerated intensity modulated radiation therapy to the breast to optimally spare normal tissue. J Clin Oncol 2007;25:2236–2242.

34 Chen PY, Vicini FA, Benitez P, et al: Long-term cosmetic results and toxicity after accelerated partial-breast irradiation: a method of radiation delivery by interstitial brachytherapy for the treatment of early-stage breast carcinoma. Cancer 2006;106:991–999.

35 King TA, Bolton JS, Kuske RR, et al: Long-term results of wide-field brachytherapy as the sole method of radiation therapy after segmental mastectomy for T(is,1,2) breast cancer. Am J Surg 2000;180:299–304.

36 Arthur DW, Winter K, Kuske RR, et al: A phase II trial of brachytherapy alone after lumpectomy for select breast cancer: tumor control and survival outcomes of RTOG 95-17. Int J Radiation Oncol Biol Phys 2008;72:467–473.

37 Vicini F, Beitsch PD, Quiet CA, et al: Three-year analysis of treatment efficacy, cosmesis, and toxicity by the American Society of Breast Surgeons MammoSite Breast Brachytherapy Registry Trial in patients treated with accelerated partial breast irradiation (APBI). Cancer 2008;112:758–766.

38 Formenti SC, Truong MT, Goldberg JD, et al: Prone accelerated partial breast irradiation after breast-conserving surgery: preliminary clinical results and dose-volume histogram analysis. Int J Radiat Oncol Biol Phys 2004;60:493–504.

39 Taghian A, Kozak K, Doppke K, et al: Initial dosimetric experience using simple three-dimensional conformal external-beam accelerated partial-breast irradiation. Int J Radiat Oncol Biol Phys 2006;64:1092–1099.

40 Vicini FA, Chen P, Wallace M, et al: Interim cosmetic results and toxicity suing 3D conformal external beam radiotherapy to deliver accelerated partial breast irradiation in patients with early-stage breast cancer treated with breast conserving therapy. Int J Radiat Oncol Biol Phys 2007;69:1124–1130.

41 Vicini FA, Winter K, Wong J, et al: Initial efficacy of RTOG 0319: three dimensional conformal radiation therapy (3D-CRT) confined to the region of the lumpectomy cavity for stage I/II breast carcinoma. Int J Radiat Oncol Biol Phys 2010;77:1120–1127.

42 Bovi J, Qi SQ, White J, et al: Comparison of three accelerated partial breast irradiation techniques: treatment effectiveness based on biological models. Radiother Oncol 2007;84:226–232.

43 Weed DW, Edmundson GK, Vicini FA, et al: Accelerated partial breast irradiation: a dosimetric comparison of three different techniques. Brachytherapy 2005;4:121–129.

44 Kahn AJ, Kirk MC, Mehta PS, et al: A dosimetric comparison of three-dimensional conformal intensity – modulated radiation therapy, and MammoSite partial – breast irradiation. Brachytherapy 2006;5:183–188.

45 Rusthoven KE, Carter DL, Howell K, et al: Accelerated partial breast intensity-modulated radiotherapy resulted in improved dose distribution when compared with three dimensional treatment planning techniques. Int J Radiat Oncol Biol Phys 2008;70:296–302.

46 Livi L, Buonomici FB, Simontacchi G, et al: Accelerated partial breast irradiation with IMRT: new technical approach and interim analysis of acute toxicity in a phase III randomized trial. Int J Radiat Oncol Biol Phys 2010;77:509–515.

47 Jagsi R, Ben-David M, Moran J, et al: Unacceptable cosmesis in a protocol investigating intensity modulated radiotherapy with active breathing control for accelerated partial breast irradiation. Int J Radiat Oncol Biol Phys 2010;76:71–78.

48 Morrow N, Stepaniak C, White J, et al: Intra and interfractional variation for prone breast irradiation: an indication for image-guided radiotherapy. Int J Radiat Oncol Biol Phys 2007;69:910–917.

49 Kainz K, White J, Herman J, et al: Investigation of helical tomotherapy for partial breast irradiation of prone positioned patients. Int J Radiat Oncol Biol Phys 2009;74:275–282.

Julia R. White, MD
Department of Radiation Oncology, Medical College of Wisconsin
8701 Watertown Plank Road
Milwaukee, WI 53226 (USA)
Tel. +1 414 805 4485, Fax +1 414 805 4369, E-Mail jwhite @ mcw.edu

Image-Guided Radiotherapy Strategies in Upper Gastrointestinal Malignancies

Anand Swaminath · Laura A. Dawson

Department of Radiation Oncology, Princess Margaret Hospital, University of Toronto, Toronto, Ont., Canada

Abstract

Organ motion due to breathing, peristalsis and deformation presents challenging problems for the delivery of highly conformal radiotherapy to upper abdominal targets, despite the many advancements in the technology of radiation planning and delivery. It is important to understand and account for this motion to avoid treatment inaccuracies, especially systematic errors that could potentially impact the probability of tumor control or increase the risk of normal tissue toxicity. Various image guidance tools can be utilized from the outset of radiation planning through treatment to minimize introducing such errors. These strategies include: assessment of breathing motion (with or without breath hold) prior to simulation, 4D CT simulation and cine MRI to evaluate tumor/organ motion, and image guidance on the treatment unit using kV fluoroscopy and 4D cone-beam CT. Together, image guidance methods can provide greater assurance that concordance exists between planned and delivered doses during a course of radiotherapy.

Copyright © 2011 S. Karger AG, Basel

Advances in technology for the planning and delivery of radiotherapy have resulted in the ability to administer highly precise treatment to patients with upper gastrointestinal malignancies. However, organ motion and deformation can be substantial, and are important factors to consider in this group of patients. Understanding these issues clearly can directly contribute to improvement in target delineation and reduction of systematic errors throughout a course of radiotherapy. The goal is a potential increase in the likelihood of tumor control and reduction in normal tissue toxicity. Throughout treatment planning and delivery, imaging and image-guided radiotherapy (IGRT) are important tools to predict safe margins for tumor targets, and to assess daily residual geometric uncertainties that may impact the likelihood of tumor control and normal tissue toxicity [1]. Ultimately, IGRT increases the concordance between the planned and delivered doses during a course of radiotherapy.

Immobilization and Motion Management in Upper Abdominal Malignancies

Motion management is challenging in radiotherapy for upper abdominal targets, since different organs at risk may move in different ways during respiration. Many regions within the upper abdomen need to be avoided, yet they may have different breathing motion patterns and amplitudes, and different baseline shifts relative to the vertebral bodies from day to day [2]. Additionally, treatment target volumes may be irregular in shape, and may extend across several organs moving differently. In addition to breathing motion, changes caused by peristalsis and variable filling of luminal structures such as the stomach and duodenum must be considered. All of these geometric changes in the targets may occur between treatment fractions and/or within each fraction of therapy. Finally, there may be changes occurring in the volumes of the tumor and the adjacent normal tissues during a fractionated treatment course.

Of all of the potential errors resulting from these changes, from planning through treatment, the most important are the systematic ones. These are the errors that repeat with each therapy fraction, causing the greatest potential impact on tumor control and normal tissue toxicity [3]. Systematic errors can occur during radiation planning through inaccurate target volume delineation. They may also occur during a treatment course due to persistent organ/target shifts or volume changes. Cone-beam CT (CBCT) or other images obtained in treatment position can often reveal these variations from the planned volumes. Soft tissue changes or anatomic deformations may require replanning to avoid clinically significant geographic misses. Discovering and correcting these systematic errors during a course of radiotherapy improves the chance of achieving the treatment efficacy that was hoped for during planning [4].

For all patients treated, whether for definitive or adjuvant therapy (with or without chemotherapy), there are several tools available to assess motion. As part of the simulation process, motion of the tumor or the organ at risk can be assessed for each patient. To measure the amplitude of breathing motion, available tools include kV fluoroscopy, 4D CT [5] and cine MRI [6]. At the time of treatment, further assessments of motion can be made with MV fluoroscopy in real-time, AP and lateral kV fluoroscopy, and kV CBCT [7]. With respect to CBCT, respiratory correlated (or 4D) CBCT is now available to assess the impact of breathing motion using volumetric image guidance [8].

Another aspect of IGRT treatment verification that is not often discussed, but may be useful for some patients, is visualization through the MV treatment beam aperture itself. For non-IMRT, or for simple IMRT, occasionally the aperture may permit identification of one or more structures, such as an air-diaphragm interface or rib interface, which show that the beams are in the appropriate position. When possible, the radiation therapists can use aperture views to verify the therapy beams relative to the reference beams, assuring that large shifts are not present.

In summary, there are multiple ways to ensure more precise and safer delivery of radiotherapy to upper abdominal malignancies. From planning to delivery, their practical application will be discussed below in the context of specific cases, which illustrate the IGRT procedures used at the Princess Margaret Hospital in Toronto for these tumors.

Image-Guided Radiotherapy for Pancreatic Cancer

Case 1

A 50-year-old woman with borderline resectable pancreatic cancer was treatment planned as part of her participation in a neoadjuvant radiochemotherapy trial [9]. The protocol included full-dose gemcitabine and oxaliplatin chemotherapy given concurrently with lower-dose radiotherapy, 30 Gy in 15 fractions, followed by a second cycle of chemotherapy, subsequent surgery and postoperative chemotherapy.

Initial Assessment and Selection of Motion Management
In this case, given the use of full-dose gemcitabine concurrently with irradiation, the impact of a geographical miss is increased since it would increase the risk of toxicity to the luminal structures, including the small bowel and duodenum. Therefore, vigilance in assuring the proper identification of targets at simulation and treatment is essential.

Table 1 summarizes the program for respiratory management of upper abdominal malignancies at Princess Margaret Hospital. This begins with an initial patient assessment and education session prior to simulation, which is intended (1) to evaluate their gross respiratory function, and (2) to screen for their breath hold capability using an active breathing coordinator [10, 11]. This requires a degree of patient cooperation and learning of the procedures. If patients have severe respiratory disease (COPD) documented, or are dyspneic on minimal effort, no attempt is made to use breath hold procedures. If they do not have significant COPD, their ability to perform this is briefly assessed. In this session, priority is placed on developing a communication system to instruct patients on proper breath hold tech-

Table 1. PMH program for respiratory management of upper abdominal malignancies

Patient education session
Initial assessment of gross respiratory function
Screen for breath hold (Active Breathing Coordinator)
AP kV fluoroscopy, using AP kV CBCT
Free breathing motion measured
Breathing motion measured using abdominal compression device
Repeat exhale breath hold reproducibility (using diaphragm or other fiducial as a surrogate for upper abdominal position)
3- to 5-second breath holds (with assessment of drifts)
1 max. breath hold (typically 20–30 s)
Cine MR – with or without abdominal compression
4D CT – in simulation/treatment position with or without abdominal compression or breath hold planning CT (if breath holds are reproducible)

niques. From this initial assessment, a decision is made regarding whether they are appropriate candidates for a more comprehensive breath hold screening and instruction session to be given later.

The remainder of the patient assessment recreates the actual treatment conditions on the therapy unit. With the patient in an appropriate immobilization device (usually a chest board with arms abducted over the head), kV fluoroscopy is used to assess free-breathing motion on the treatment unit using the kV CBCT system. If motion is more than a certain threshold (5 mm) then abdominal compression is introduced and its benefit in reducing motion assessed [12]. Patient comfort must be considered as well, especially for these upper abdominal cancers since malignant pain can be exacerbated by the compression plate. Another strategy to reduce motion is to coach patients to perform shallow breathing voluntarily, though they must be able to continue this throughout the course of treatment similarly. This requires their full understanding, cooperation and ability.

In practice, most patients are at least screened for suitability for breath hold procedures [10, 13]. Repeated breath holds, typically in the exhale phase of respiration, are carried out. At least 3 breath holds are repeated for 5 s, and the variability in position of the diaphragm or stent is recorded using kV fluoroscopy. One must ensure that the position remains stable during the breath hold; large single drifts or multiple small drifts are generally undesirable. Subsequently, the length of breath hold is recorded. Breath holds of at least 20 s are recommended.

The method of motion management for each patient is then selected, and is used for the subsequent planning CTs and ultimately for treatment. A cine MRI at the time of simulation (see below) can also be used to assess 3D motion, with or without abdominal compression, since it can show soft tissues and tumors direct-

Table 2. CT simulation protocol for pancreatic/liver malignancies

Helical exhale voluntary breath hold CT
Primary dataset for radiotherapy planning
Oral contrast
IV contrast (2 ml/kg at 5 ml/s; scanning for 50–60 s)
Venous phase imaging (imaging taken 50–60 s after injection) for pancreas/metastases,
Arterial (taken at 15–25 s after injection) and venous phase imaging for hepatocellular carcinoma
4D CT scan to follow immediately
Evaluate for artifacts (i.e. quality of 4D reconstruction)
Registration to the helical exhale breath hold CT
Exhale to exhale pancreas/liver registration

ly – as opposed to fluoroscopy, which evaluates organ positions based on surrogate structures [6]. The MRI study can reconfirm the selection of the motion management strategy.

This patient had no respiratory disease and was comfortable learning breath hold at the initial session. At imaging, she had less motion on fluoroscopy than is often seen: 8 mm of diaphragm motion and 5 mm of stent motion (the stent was placed for obstructive jaundice). With compression, there was only a 2-mm further reduction of the amplitude of these motions. Breath hold was not reproducible with respect to stent position: ±2 mm motion relative to the vertebral bodies was documented with repeat breath holds, though no drift was seen. Cine MR confirmed similar amplitude of motion, with and without abdominal compression. The decision was made to treat the patient under free breathing, without abdominal compression and without breath holds, since she had a very small amplitude of breathing motion.

Simulation
Following breathing assessment, the particular strategy (breath hold, compression, free breathing) best suited for immobilization is used for subsequent simulation (table 2). At CT simulation, a CT with oral and IV contrast is obtained to best delineate the target. The IV contrast is similar to that used in diagnostic imaging, and our imaging includes the venous phase for better visualization pancreatic cancer.

(1) Contouring is based on a helical CT scan obtained during voluntary exhale breath hold. This becomes the primary data set, since it has the best diagnostic quality imaging to delineate the tumor and also the organs at risk.

(2) This helical CT scan is immediately followed by a 4D CT to quantify motion for planning, in addition to the kV fluoroscopy and cine MR previously men-

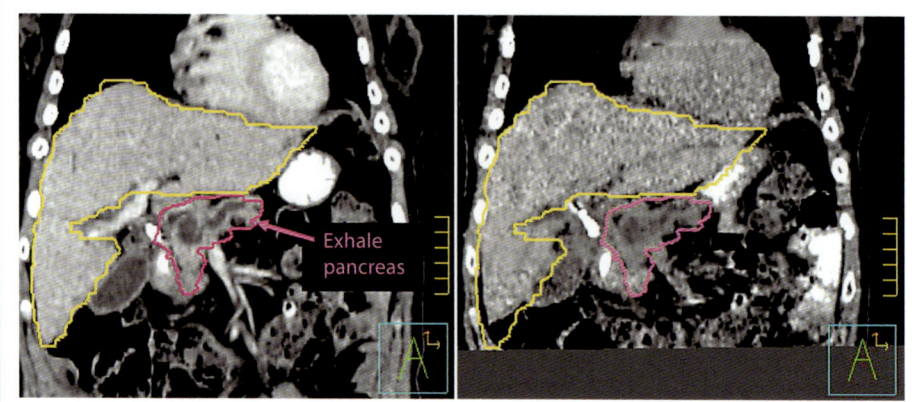

a Exhale helical breath hold contrast CT   4D CT exhale phase

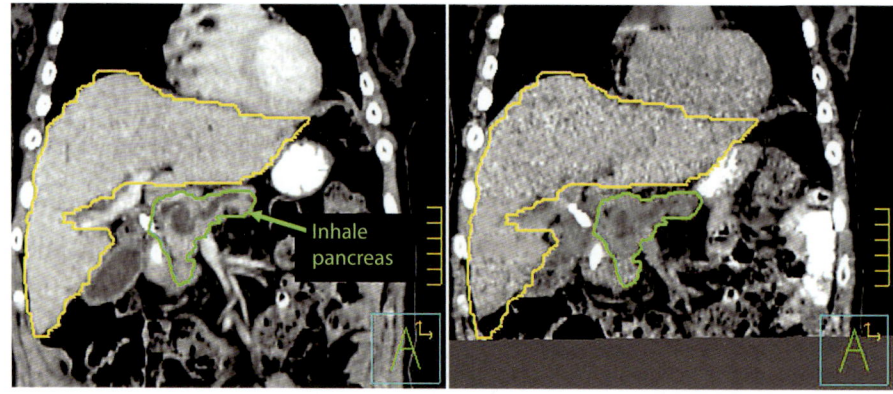

b Exhale helical breath hold contrast CT   4D CT inhale phase

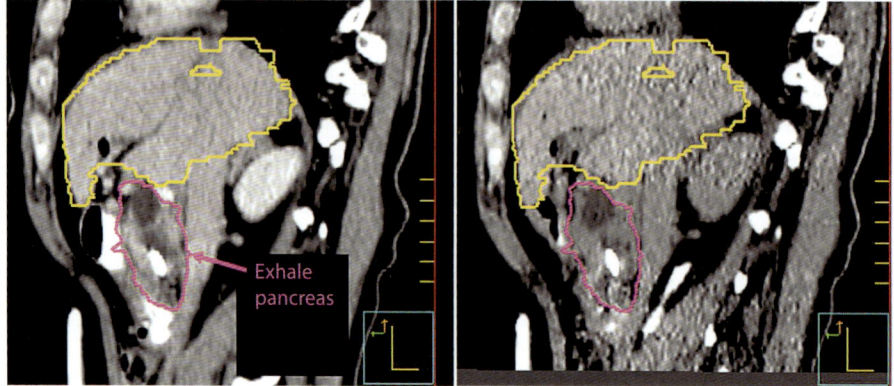

c Exhale helical breath hold contrast CT   4D CT exhale phase

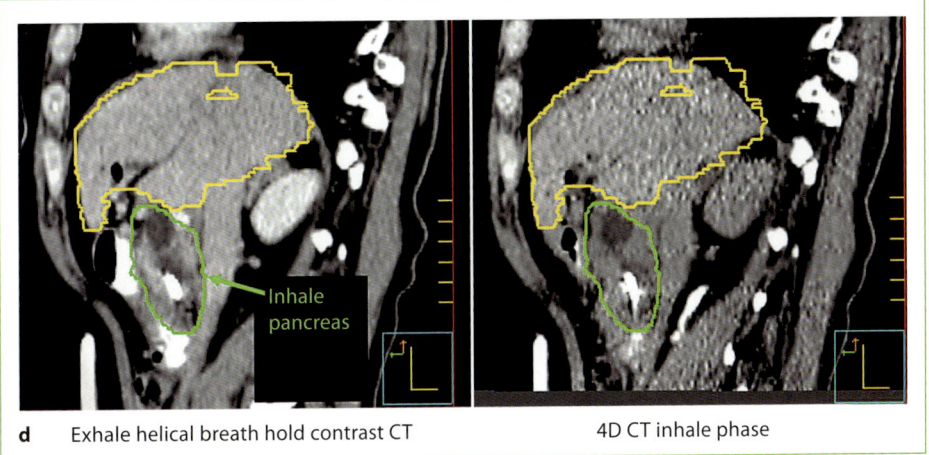

| d | Exhale helical breath hold contrast CT | 4D CT inhale phase |

Fig. 1. CT planning images for pancreatic radiotherapy in a patient with minimal organ motion. **a, b** Coronal images of exhale breath hold helical CT registered to exhale (**a**) and inhale (**b**) 4D CT images. **c, d** Sagittal images at similar reference point on helical CT fused to exhale (**c**) and inhale (**d**) 4D CT images. Note the minimal differences in position of liver and pancreas from exhale to inhale, suggesting a small amount of motion in this example. See online supplementary material.

tioned. Soft tissues on the 4D CT (2) are registered to the primary dataset (1), so that an estimate of the amplitude of breathing motion in the soft tissues can be determined in the data set being used to delineate the target. The information derived from the 4D CT along with the cine MRI and kV fluoroscopy are used in order to best determine the final PTV margins on the targets [14].

Planning

At the time of radiation planning, several structures are contoured (in both the inhale and exhale phases of the 4D CT) that may help with image guidance: the GTV itself, the stents (if they exist) and the duodenum. As shown in figure 1a, the helical contrast breath hold scan for this patient is shown on the left. Notice the exhale positions of the pancreas and liver on this scan. On the right, the 4D CT exhale phase is shown, and is registered to the exhale helical breath hold CT so that the soft tissues align. A comparison can be made between the exhale breath hold helical CT and the inhale 4D CT, as seen in figure 1b. The inhale contour of the pancreas in this case does not change appreciably compared with the exhale helical scan contour. Figures 1c and d demonstrate sagittal views at a similar reference point. Only slight movement of the upper abdominal organs is present, particularly with respect to the liver, but again minimal motion of the tumor volumes is seen. When comparing the helical scans with the 4D scans, it is clear that the

Table 3. General IGRT approach for free breathing/compression for pancreas patients

kV CBCT
Half gantry rotation (180 degrees)
Bone matching (for anteroposterior, medial-lateral and rotational alignment)
AP and lateral fluoroscopy
Stent motion measured
Patient aligned based on stent (for cranial-caudal alignment)
kV CBCT
60-second single gantry rotation (360 degrees)
Oral contrast given to visualize luminal structures
Free breathing CBCT
soft tissue match of pancreas, duodenum and stents
shifts ± verification scans for offsets larger than 3–5 mm

helical CT provides better tumor definition for contouring. However, the 4D CT allows determination of the impact of tumor motion, and is critical in determining PTV margins for these patients [15, 16].

In the treatment protocol for this patient, the following target volumes used are described below. There was no elective nodal irradiation in this particular case. PTV margins are anisotropic, and individualized based on motion, as well as the best estimate of the residual offsets of the stent. For this patient, final margins were based on free breathing motion on 4D CT, cine MR and fluoroscopy.

The following target volumes were created (and their expansions):

Gross tumor volume (GTV) represented the primary tumor and nodes (if any were involved, though there were none for this patient).

Clinical target volume (CTV): 5- to 10-mm expansion to the GTV.

Planning target volume (PTV): expansions to the CTV of 6 mm right and left, 5 mm cranial and posterior, 10 mm inferior (as this was the greatest direction of breathing amplitude) and 8 mm anterior (as the tumor also moved anteriorly as seen on the cine MR). *The main direction of breathing motion is generally in an inferior and anterior direction with reference to the exhale planning CT scan.*

Image Guidance at Treatment
Since the 4D CT in exhale and inhale phases showed little motion in this patient, orthogonal fluoroscopy and free breathing CBCT were selected for image guidance. Minimal motion artifact (such as blurring of organ edges due to respiration) would be expected at the time of treatment, due to the lack of motion seen at the time of initial fluoroscopy, cine MR and 4D CT. Both kV fluoroscopy and CBCT are always performed at the first treatment fraction (table 3). On the fluoroscopy, stent motion is measured and the patient is aligned based on the stent position. Additionally, the amplitude of motion is measured, and is compared with what

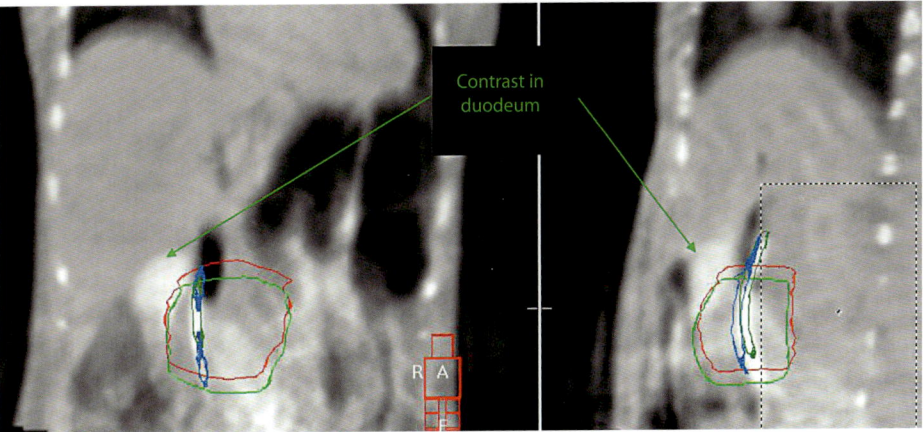

Fig. 2. CBCT images of pancreas patient in treatment position prior to radiation delivery, showing position of stent relative to contours delineated on 4D CT, and contrast in the duodenum.

was recorded at the time of planning. Patients are asked to perform shallow breathing if the amplitude is too large. For the CBCT, a 60-second single gantry CBCT rotation is obtained, with oral contrast prior to scanning, and with the patient in free breathing (as determined during the planning process).

On day 1, the physician is present to evaluate the imaging, assess the motion of the stent and look at the quality of the CBCTs. Data related to the degree of motion and different body habitus which may impact soft tissue alignment are still being confirmed. For the remainder of therapy, if the stent amplitude of motion is more than 7 mm on fluoroscopy, CBCT would be assessed, the isodoses overlaid and a decision made to proceed or hold therapy. On the kV CBCT, if there are soft tissue changes in the pancreas or duodenum, or changes in stent position, the physician would be called to make a decision. These decisions could range from holding therapy, continuing therapy or replanning if there is a substantial change (e.g. >1 cm consistent offset in any direction). It is rare to replan or adapt radiation for these patients. Occasionally, a simple re-CT simulation can determine if there has been substantial organ deformation. If there are no significant issues with position after the first 1–2 fractions of treatment, it is generally not necessary to continue with CBCT imaging daily as long as the stent match is acceptable on kV fluoroscopic images.

Figure 2 shows a free breathing CBCT for image guidance, where the inhale (blue) and exhale (green) stent are aligned. Contrast in the duodenum can assist in identification of soft tissue planes to improve the daily soft tissue match and help ensure against systematic errors. As both exhale and inhale stent contours have been drawn based on the 4D planning CT, matching them to the free breathing CBCT scan in a patient with minimal motion should provide a good overall match most of the time.

Case 2

A 63-year-old woman with resectable pancreatic cancer was treated on the same neoadjuvant chemoradiotherapy trial as the first patient, with 30 Gy in 15 fractions of radiation delivered concurrently with full-dose gemcitabine and oxaliplatin, followed by chemotherapy alone, surgery and subsequent chemotherapy.

Motion Management/Image-Guided Radiotherapy Strategy
This case contrasts with the first because imaging with kV fluoroscopy showed large amplitude of motion, 35 mm for the diaphragm and 25 mm for the stent during free breathing. Abdominal compression reduced this to 30 and 15 mm, respectively. The patient was able to hold her breath for more than 1 min, but had substantial drift (4 mm) and lack of reproducibility/stability (±12 mm) of these structures with repeat breath holds. Cine MRI showed a similar degree of motion of the pancreas, but the addition of compression reduced the motion by 3–5 mm. It was decided to treat this patient using abdominal compression, with its advantage in reducing the amplitude of motion, compared with breath hold where reproducibility was quite poor.

Simulation and Planning
The 4D CT scans, registered to the helical CT scan, are shown in figure 3. The helical CT was used to define the GTV and the normal structures, as in the previous case. The overlay of the helical CT onto the exhale phase of the 4D CT shows the degree of motion. The stents, contoured on the 4D CT, can help give guidance on the motion. What is striking, in comparison with the previous case, is the obvious difference in the magnitude of motion when comparing the helical exhale breath hold scan to the 4D exhale and inhale datasets.

Planning targets including GTV and normal tissues were contoured on the helical CT. GTV, stent and duodenum were also contoured on the exhale and inhale phases of 4D CT, to aid in IGRT. The case has a larger PTV expansion than in Case 1, because of the irregular breathing and greater amplitude of motion. For this patient, the following target volumes were used:

GTV represented the evident tumor, contoured on the helical CT.
CTV: 5- to 10-mm expansion to the GTV.
PTV: expansions to the CTV of 6 mm right and left, 5 mm cranial and posterior, 20 mm inferior and 10 mm anterior.

Image Guidance at Treatment
A first day trial setup is useful, to (1) evaluate motion and position at the treatment unit prior to therapy, (2) confirm breathing amplitude, (3) determine if free breathing kV CBCT is suitable for real-time soft tissue IGRT and (4) evaluate respiratory-sorted kV CBCT off-line. Respiratory correlated kV CBCT is performed using tools that

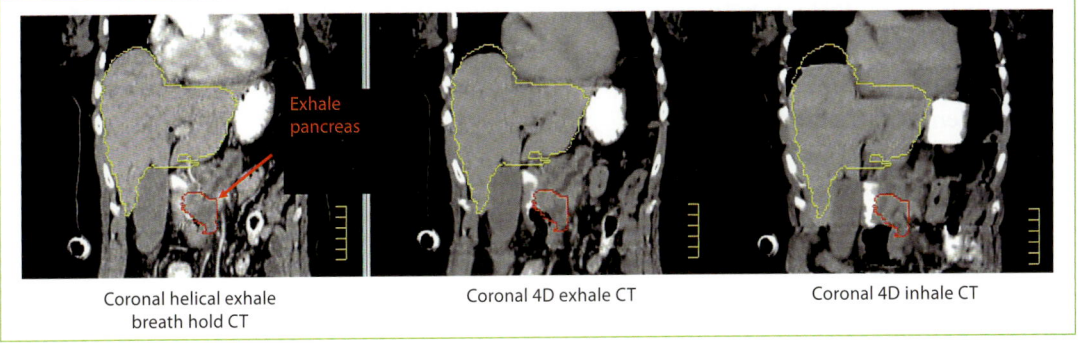

Fig. 3. CT planning images for pancreatic radiotherapy in a patient with a large amplitude of breathing motion. After registering the exhale breath hold CT to the 4D CT datasets, a large difference in organ position is seen between exhale and inhale.

sort the free breathing CBCT images into bins that correspond to various phases of the breathing cycle. This is performed off-line; the tools are not clinically available while the patient is in the treatment position, and must be performed later in the work day prior to the first fraction of treatment. By doing this, a direct comparison can be made between soft tissues of the exhale planning CT and the exhale image of the sorted CBCT. This *off-line* comparison is used to determine if any residual systematic errors exist after the *on-line* daily image guidance based on the free breathing kV CBCT [17]. Analysis and adjustment of these offsets can therefore be recognized, and recommendations regarding optimal image matching to the free breathing kV CBCT can be made prior to the patient commencing treatment. Daily respiratory correlated CBCTs can be obtained to determine reproducibility of image matching, or, in challenging cases, to help resolve problems with matching to free breathing CBCTs.

Image-Guided Radiotherapy for Liver Cancer

Case 3

A 71-year-old man presented with hepatitis B and hepatocellular carcinoma involving the inferior right liver lobe and abutting the large bowel. It was not suitable for other therapies. This patient was enrolled on an in-house stereotactic body radiotherapy 6-fraction protocol for hepatocellular carcinoma [18].

Motion Management/Image-Guided Radiotherapy Strategy
The pre-CT simulation for this patient was similar to that described above (table 1). The diaphragm was selected as a surrogate for liver position. The patient

was uncomfortable with breath hold, and the diaphragm did not show very much motion on free breathing (7 mm). Abdominal compression had little effect on this motion (only a 1-mm reduction). For these reasons, free breathing during treatment was selected.

Simulation and Planning

The CT simulation procedures were similar to those shown in table 2 for pancreas cancer, except that imaging included (1) arterial phase imaging (at 15–25 s after IV injection) to show the primary tumor better and (2) venous phase imaging (at 50–60 s after IV injection) to evaluate for portal venous thrombosis and to capture washout of tumors, which can occur in this disease. Additionally, a contrast enhanced MRI was used to aid volume definition, which can be especially useful in patients unable to receive IV contrast for CT [19]. For this patient, there was a small GTV within the liver. Volumes were determined on an exhale breath hold helical scan, and a larger PTV expansion was used inferiorly and anteriorly for any predicted motion during treatment. Certain dose contours representing the limits of normal tissue tolerances (duodenum, stomach, large/small bowel, kidney, spinal cord) can also be exported to the treatment unit for image guidance. In this patient, GTV, normal structures (particularly the duodenum, which was dose limiting) and isodose volumes for 30, 33 and 36 Gy were exported, to be included in image registration to the CBCT at the time of treatment. The following target volumes were delineated for this case:

GTV: disease visualized on the arterial enhancing phase, and corresponding washout on venous phase

CTV: 5-mm expansion to the GTV within the liver parenchyma.

PTV: expansions to the CTV of 10 mm inferior, 6 mm anterior and 5 mm in other directions.

Image Guidance at Treatment

Figure 4 shows the CBCT for this patient. One can see contrast in the luminal gastrointestinal structures (duodenum, large bowel). A free breathing CBCT was used for soft tissue image guidance in this situation. The contours from the exhale phase of the planning CT are overlaid, with the 36 Gy isodose shown. This is the dose limit for certain luminal structures on this protocol, which uses hypofractionation (just 6 fractions) in delivering the radiation [18]. Assessment of these isodoses allows for appropriate communication with the therapists and instructions to not proceed with radiotherapy if the isodose line in question falls within the luminal structure visualized on the CBCT. The physicians are always called to the treatment unit on the first treatment day to review the CBCT and make recommendations for managing future fractions.

For patients with liver cancer, the entire liver volume is initially used to achieve treatment positioning based on soft tissue registration. Then, other tissues outside

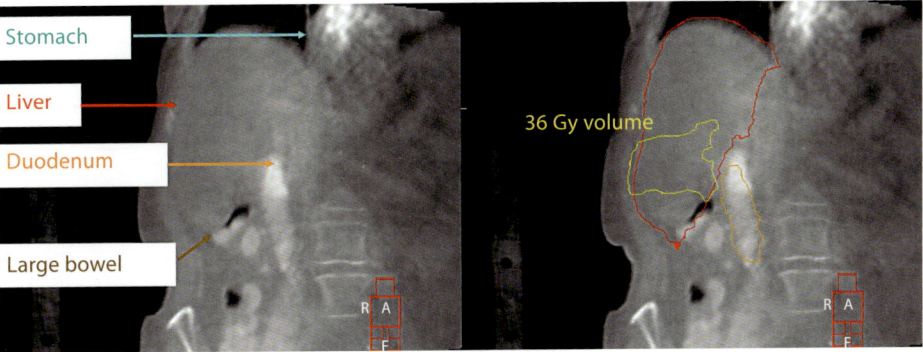

Fig. 4. CBCT images of a liver cancer patient in treatment position. Contours generated from planning CT, as well as critical isodose lines can be evaluated to ensure no overdosing of organs at risk takes place.

of this registration volume are also compared. An acceptable liver-to-liver match, planning to treatment, is possible when there is limited amplitude of motion (i.e. 5–7 mm or less), but is more challenging in cases with larger amplitudes or substantial organ deformation. Under these circumstances, the soft tissue matching is intentionally biased toward areas near the target volumes. Use of respiratory correlated CBCT is also a useful tool in those situations for off-line analysis. Also, off-line deformable image registration can be used to give a better estimate of residual errors [20]. There are situations where radio-opaque surrogates within the liver (surgical clips, previous Lipiodol uptake from embolization procedures) can be used to better define the soft tissue match. On occasional instances, the tumor itself is evident on the CBCT and can be used for matching (fig. 5). IV contrast-enhanced CBCT imaging is another attractive option that is under investigation in order to better delineate tumor during treatment. In patients who are treated in controlled breath holds, breath hold CBCTs can be obtained. Like respiratory correlated CBCT, they are evaluated off-line either prior to treatment (after a trial setup) or between fractions [21]. Similar recommendations can also be made to the therapists following breath hold CBCTs to minimize the potential for systematic errors [22].

Conclusions

Several strategies using image guidance can be employed to ensure more precise, accurate delivery of radiotherapy to upper abdominal malignancies. Each IGRT approach involves important considerations that must be integrated into and

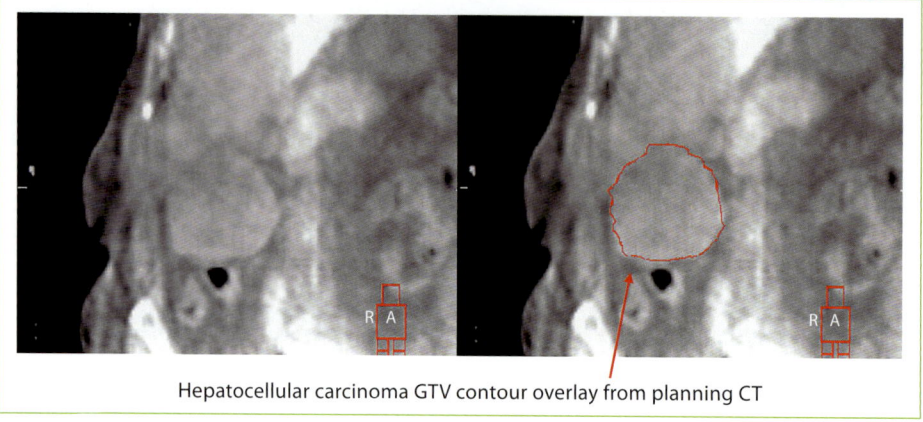

Hepatocellular carcinoma GTV contour overlay from planning CT

Fig. 5. CBCT images of a liver cancer patient with tumor visible within inferior aspect of liver.

consistently applied throughout radiation planning and treatment delivery. Appropriate immobilization and assessment of breathing strategies provide the first steps to achieving the goal of minimizing risk of systematic errors. High-quality diagnostic planning CTs and 4D CTs are important for target delineation. Evaluation of motion using kV fluoroscopy, 4D CT and cine MRI provide information on appropriate PTV expansions. At the treatment unit, 2D fluoroscopy and 3D kV CBCT are complimentary tools for image-guided delivery. Oral contrast administration is useful to detect soft tissue planes. Use of radio-opaque materials (stent, clips) can aid in improving the accuracy of the image registrations. Other methods such as respiratory-correlated CBCT, deformable image registration, and IV contrast-enhanced CBCT are tools that can aid in challenging situations. Finally, as IGRT continues to evolve, there will be increasing understanding, experience and ability to use these tools to provide more clinically meaningful patient outcomes maximizing tumor control and reducing normal tissue toxicity.

Guidelines for Clinical Practice

- Motion management is challenging in radiotherapy for upper abdominal targets since different organs at risk may move in different ways during respiration. There are several tools available to assess motion.
- During simulation, available tools to assess motion include kV fluoroscopy, 4D CT and cine MRI. kV fluoroscopy may be used to assess free-breathing motion on the treatment unit using the kV CBCT system.

- If motion is more than a certain threshold (5 mm) then abdominal compression may be considered and its benefit in reducing motion assessed. Patients also may be screened for the suitability of breath hold procedures. A cine MRI at the time of simulation can also be used to assess 3D motion. Following breathing assessment, the particular strategy (breath hold, compression, free breathing) best suited for immobilization is used for subsequent simulation.
- At CT simulation, CT with oral and IV contrast is obtained to best delineate the target. A helical CT scan is used as the primary data set, as it has the best diagnostic quality imaging to delineate the tumor and also the organs at risk. This can be supplemented with a 4D CT to quantify motion for planning.
- At the time of treatment, further assessments of motion can be made with MV fluoroscopy in real time, AP and lateral kV fluoroscopy, and kV CBCT. With respect to CBCT, respiratory-correlated (or 4D) CBCT is now available to assess the impact of breathing motion using volumetric image guidance.

Acknowledgements

The authors would like to acknowledge financial support from the Canadian Cancer Society, National Cancer Institute of Canada (NCIC grant 018207) and Elekta Oncology Systems for funding some of the research presented in this review.

References

1. Dawson LA, Jaffray DA: Advances in image-guided radiation therapy. J Clin Oncol 2007;25:938–946.
2. Langen KM, Jones DT: Organ motion and its management. Int J Radiat Oncol Biol Phys 2001;50:265–278.
3. van Herk M: Errors and margins in radiotherapy. Semin Radiat Oncol 2004;14:52–64.
4. Balter JM, Brock KK, Litzenberg DW, et al: Daily targeting of intrahepatic tumors for radiotherapy. Int J Radiat Oncol Biol Phys 2002;52:266–271.
5. Ford EC, Mageras GS, Yorke E, et al: Respiration-correlated spiral CT: a method of measuring respiratory-induced anatomic motion for radiation treatment planning. Med Phys 2003;30:88–97.
6. Kirilova A, Lockwood G, Choi P, et al: Three-dimensional motion of liver tumors using cine-magnetic resonance imaging. Int J Radiat Oncol Biol Phys 2008;71:1189–1195.
7. Jaffray DA, Siewerdsen JH, Wong JW, et al: Flat-panel cone-beam computed tomography for image-guided radiation therapy. Int J Radiat Oncol Biol Phys 2002;53:1337–1349.
8. Sonke JJ, Zijp L, Remeijer P, et al: Respiratory correlated cone beam CT. Med Phys 2005;32:1176–1186.
9. Study of gemcitabine and oxaliplatin with radiation therapy in patients with pancreatic cancer. http://clinicaltrials.gov/ct2/show/NCT00456599.
10. Dawson LA, Eccles C, Bissonnette JP, et al: Accuracy of daily image guidance for hypofractionated liver radiotherapy with active breathing control. Int J Radiat Oncol Biol Phys 2005;62:1247–1252.
11. Dawson LA, Brock KK, Kazanjian S, et al: The reproducibility of organ position using active breathing control (ABC) during liver radiotherapy. Int J Radiat Oncol Biol Phys 2001;51:1410–1421.
12. Herfarth KK, Debus J, Lohr F, et al: Extracranial stereotactic radiation therapy: set-up accuracy of patients treated for liver metastases. Int J Radiat Oncol Biol Phys 2000;46:329–335.
13. Dawson LA, Eccles C, Craig T: Individualized image guided iso-NTCP based liver cancer SBRT. Acta Oncol 2006;45:856–864.

14 Brock KK: Image registration in intensity-modulated, image-guided and stereotactic body radiation therapy. Front Radiat Ther Oncol 2007;40:94–115.
15 Chen GT, Kung JH, Rietzel E: Four-dimensional imaging and treatment planning of moving targets. Front Radiat Ther Oncol 2007;40:59–71.
16 Rietzel E, Chen GT, Choi NC, et al: Four-dimensional image-based treatment planning: target volume segmentation and dose calculation in the presence of respiratory motion. Int J Radiat Oncol Biol Phys 2005;61:1535–1550.
17 Case RB, Sonke JJ, Moseley DJ, et al: Inter- and intrafraction variability in liver position in non-breath-hold stereotactic body radiotherapy. Int J Radiat Oncol Biol Phys 2009.
18 Tse RV, Hawkins M, Lockwood G, et al: Phase I study of individualized stereotactic body radiotherapy for hepatocellular carcinoma and intrahepatic cholangiocarcinoma. J Clin Oncol 2008;26:657–664.
19 Hussain SM, Semelka RC: Hepatic imaging: comparison of modalities. Radiol Clin North Am 2005;43:929–947, ix.
20 Brock KK, Dawson LA, Sharpe MB, et al: Application of a novel deformable image registration technique to facilitate classification, tracking and targeting of tumor and normal tissue. Int J Radiat Oncol Biol Phys 2004;60:S226–S227.
21 Hawkins MA, Brock KK, Eccles C, et al: Assessment of residual error in liver position using kV cone-beam computed tomography for liver cancer high-precision radiation therapy. Int J Radiat Oncol Biol Phys 2006;66:610–619.
22 Eccles C, Brock KK, Bissonnette JP, et al: Reproducibility of liver position using active breathing coordinator for liver cancer radiotherapy. Int J Radiat Oncol Biol Phys 2006;64:751–759.

Laura A. Dawson, MD, FRCPC
Department of Radiation Oncology, Princess Margaret Hospital, University of Toronto
610 University Avenue
Toronto, ON M5G 2M9 (Canada)
Tel. +1 416 946 2125, Fax +1 416 946 6566, E-Mail laura.dawson@rmp.uhn.on.ca

Radiotherapy Planning for the Lymphomas: Expanding Roles for Biologic Imaging

Richard Hoppe

Department of Radiation Oncology, Stanford University Medical Center, Stanford, Calif., USA

Abstract

Radiotherapy planning now uses advanced technologies to accurately image and assess the extent of disease for treatment. PET scanning has become established as perhaps the most important imaging study for patients with Hodgkin's disease. With respect to initial staging, FDG-PET is more sensitive overall than CT scanning. PET can detect disease at sites that do not meet size criteria by CT. Also, PET is more specific than CT alone because of the functional information that it provides. However, some disease may still escape PET imaging, and false negative results can occur. With respect to treatment response, PET has now become accepted as the most important response measure for the lymphomas. Current protocols are investigating the benefit of this information for radiotherapy planning, and even the possible elimination of radiotherapy in patients completely responding to chemotherapy. For radiotherapy planning, PET/CT should be obtained prior to and after chemotherapy; both scans give important information for the design of the radiation treatment. This chapter will review specific guidelines for planning radiotherapy based on these new imaging capabilities.

Copyright © 2011 S. Karger AG, Basel

Over the past few years, significant advances in medical technologies have proven useful for the planning and delivery of radiotherapy for the lymphomas. Biologic imaging with FDG-PET has been especially important, as demonstrated in recent clinical studies, and has influenced the clinical approach to the lymphoma patient.

In the 1970s, in addition to a chest X-ray, the most important examination that could be obtained on a patient with lymphoma was the lymphogram. While it was very helpful in defining the extent of disease, it did not allow one to image some very important parts of the anatomy, such as the spleen which is so often involved in the lymphomas. When CT imaging became available, it was an enormous ad-

Table 1. Influence of PET scanning on initial staging of Hodgkin's disease

Authors	Patients n	Stage[1]		Treatment[1]		
		lower %	higher %	less intense %	more intense, %	Total %
Bangerter et al. [1]	44	3	13	2	11	14
Partridge et al. [2]	44	7	41	2	23	25
Munker et al. [3]	73	3	29			
Naumann et al. [4]	88	8	13	8	10	18
Rigacci et al. [5]	186	1	14	1	5	6
Stanford	44	0	5	0	11[2]	11

[1] Based on PET. [2] Slight extension of radiotherapy fields = 4; more chemotherapy = 1.

vance. Early PET scans were interesting, but often difficult to interpret and fraught with many false positives, such as the brown fat phenomenon. But the development of merged PET/CT imaging enabled clinicians to differentiate and, in most cases, to interpret lymph nodes involved with disease versus false positive areas seen on PET alone.

This discussion will address issues involved with the use of PET imaging for lymphomas: initial staging, response evaluation and radiation therapy planning. The roles of other new planning and delivery technologies, generally discussed in this volume, will also be considered with regard to the lymphomas. The term PET will be used with the understanding that this implies modern FDG-PET/CT-fused imaging. Some of the clinical trials using PET as a decision point for determining overall therapy and evaluating its specific roles for radiotherapy planning will also be addressed.

Initial Staging

Hodgkin's Disease

PET scanning has become established as perhaps the most important imaging study for patients with Hodgkin's disease. Table 1 shows early studies by several different groups looking at the impact of PET scanning on initial stage determination [1–5]. Occasionally patients received a lower stage based upon the PET scan results (about 5%) but, much more commonly, they advanced to a higher stage of disease, observed in 25–30% of cases.

How often did that stage determination affect therapy? Identification of new sites of disease does not necessarily affect the overall treatment plan, and this was

Table 2. Role of PET in follicular lymphomas at the Peter McCallum Cancer Centre, Melbourne [6]

Stage before PET	Patients n	Stage after PET		
		stage unchanged	change to stage II	change to stage III–IV
Stage I	26	14 (54%)	4 (15%)	8 (31%)
Stage II	16	9 (56%)	n.a.	5 (31%)

shown in these series. PET affected the treatment plan much less often than it affected the stage. For the patients whose stage was lowered, intensity of treatment was lowered only 2–3% of the time. Even in patients whose stage advanced, the treatment became more intense only 10–20% of the time. This is because patients with all stages of disease were included in these studies. Chemotherapy played an important role in many of these cases, not only in advanced disease but also in early stage disease. In many cases, the combined modality programs were the same, regardless of the PET scanning results.

In general, with respect to initial staging, the FDG-PET scan is more sensitive than the CT scan. PET can detect disease at sites that do not meet size criteria by CT, which is a common observation in clinical practice. Also, PET is more specific than CT alone because of the functional information that it provides. However, some disease may still escape PET imaging, and false negative results occasionally occur. Overall, PET is not perfect, but is an excellent tool in staging Hodgkin's disease.

The false positive rate for PET is uncertain since most of the studies shown above did not have histopathological correlation. Still, common sense may be applied to the interpretation of the data. For example, if a patient had obviously enlarged supraclavicular nodes on the left and there was an adjacent axillary lymph node on PET scan, one could reasonably assume that the axillary node represented disease, based on the known patterns of spread of Hodgkin's disease.

What is the impact of PET scanning on therapy and survival? It is so important that it is now considered essential in the workup of patients with Hodgkin's disease, and is included as such in the NCCN Practice Guidelines along with chest X-ray and diagnostic CT scans of the chest, abdomen and pelvis.

Non-Hodgkin's Lymphomas

How useful is PET for staging the lymphomas? Wirth et al. [6] from the Peter McCallum Cancer Centre in Melbourne reported their experience with patients who had follicular lymphomas (table 2). They evaluated patients referred to Radiation Oncology with stage I or II disease, where radiation therapy alone is the treatment of choice. In 46% of the 26 stage I patients, PET imaging resulted in a change of the stage to II, III or IV. Similarly, 31% of the initially 16 stage II patients had an advance of their stage to III or IV. These stage changes had a significant impact on the treatment of these patients. In their experience, 17 of 42 patients (41%) would have been undertreated in the absence of the PET scan results. Therefore, if one ignores or does not have access to PET imaging, many such patients will not be treated adequately.

The reported experiences in the pre-PET era for patients with stage I or II follicular lymphoma treated with radiation therapy alone usually involved field treatment. In the studies from the Princess Margaret Hospital (Toronto) [7], British National Lymphoma Investigation [8] and Stanford University [9], there was a remarkable consistency among the results: the freedom from relapse or disease-free survival was 40–50% in each of these large series of patients. The remaining 50–60% of patients had progression of disease, a proportion only somewhat larger than the number that would have had a change of their stage with modern PET scanning. Perhaps many of the patients who relapsed did so in sites that were not identified by conventional imaging, but would have been identified by PET imaging. Perhaps current staging with PET will further improve outcome results through more accurate case selection.

Response Evaluation

In lymphoma patients who present with large mediastinal masses, especially those with Hodgkin's disease, conventional CT has been notoriously inaccurate in assessing whether there is residual disease at the completion of therapy (fig. 1). The usual clinical scenario was this: a residual mass was seen on CT, requiring serial scanning to make sure that there was no progression of disease in follow-up. Only after numerous scans were completed could one conclude that the patient was without disease.

There have been several series looking at the differential accuracy of CT versus PET for defining treatment response. A series by De Wit et al. [10] evaluated after therapy CT scan results. When CT (without PET) showed apparent success in achieving complete response, 27% of the patients still relapsed; the CT scans were falsely negative. When CT showed apparent failure to achieve complete response, only 21% of the patients relapsed. So, there was little difference in the rate of relapse between those responding completely by CT and those not completely responding by CT. But if there was a complete response by PET, only 4% of the patients relapsed, whereas 45% did so if there was a failure to achieve complete response by PET.

At Stanford, we have evaluated patients treated on the Stanford V protocol who had locally advanced or stage III-IV disease. In this trial, patients received 12

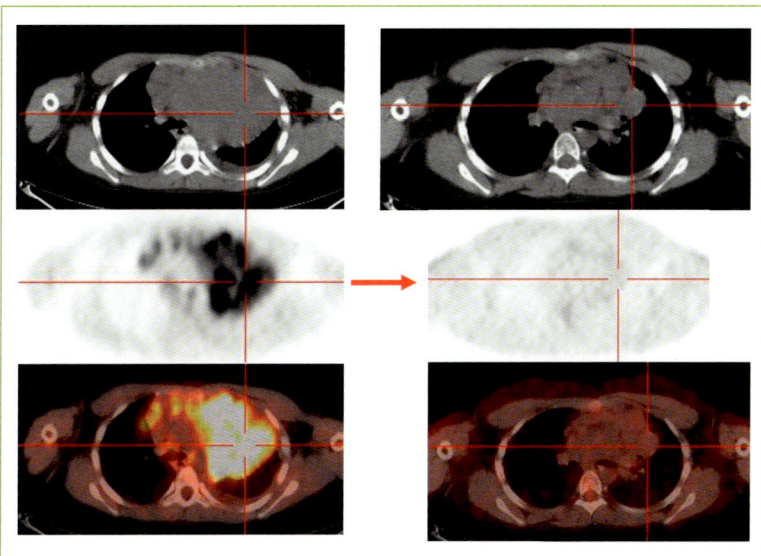

Fig. 1. CT vs. PET for evaluation of treatment response. Left: pretreatment fused PET/CT; right: post-treatment. Looking at the CT images, one can see remarkable abnormality remaining after treatment. However, the PET uptake has normalized.

weeks of chemotherapy followed by radiotherapy, but were rescanned with PET after the completion of chemotherapy and prior to irradiation. The outcomes of patients with positive scans versus negative ones are shown in figure 2. Patients with negative scans had excellent outcomes. Those with positive scans had poor outcomes, despite receiving subsequent radiotherapy.

FDG-PET has now become accepted as the most important measure of response for the lymphomas. In the NCCN response criteria, complete response can be declared regardless of the size of the residual CT abnormality, as long as the PET scan is negative.

One of the more interesting analyses of PET response as an early response indicator was published by Gallamini et al. [12] in 2007. They evaluated their patients by PET after only 2 cycles of chemotherapy. In addition, they divided the patients according to conventional clinical factors using the International Prognostic Score (IPS) for Hodgkin's disease, each factor being independently significant. Figure 3 shows the PET response for patients, divided into groups: those with 0–2 IPS factors present versus 3 or more factors. In predicting outcome, the PET results trumped even the established IPS. For patients who were PET-negative after 2 cycles of ABVD, the cure rates were extremely high, regardless of the IPS, whereas patients who had a positive PET scan had very poor outcomes regardless of their IPS.

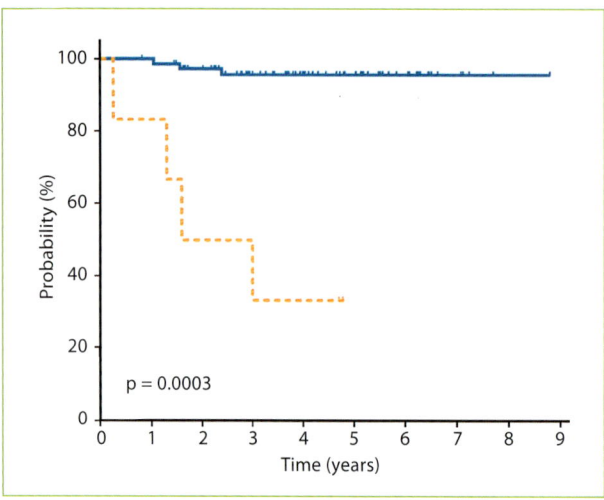

Fig. 2. PET scanning in Hodgkin's disease following chemotherapy, prior to radiotherapy for patients on the Stanford V protocol. Blue: negative PET scans, 75 patients. Orange: positive PET scans, 6 patients. Probability = Freedom from disease progression [11].

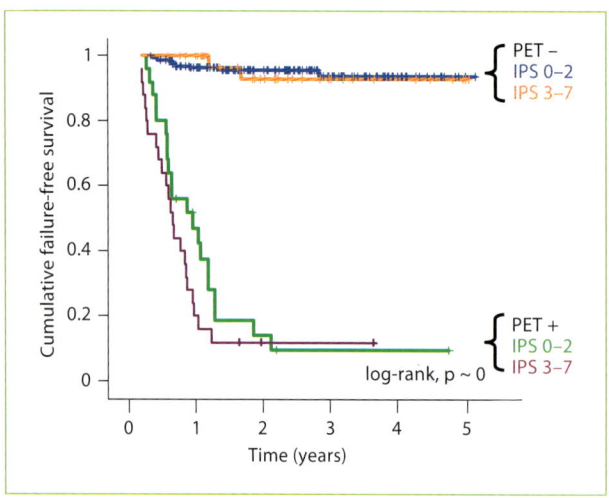

Fig. 3. PET as an early response indicator after ABVD for 2 cycles in patients with Hodgkin's disease [12]. Blue, orange lines: PET-negative. Green, purple lines: PET-positive. IPS is the number of factors for the patient group.

In summary, with respect to response evaluation, PET is superior to conventional imaging such as CT, especially for evaluating the mediastinum. It is strongly predictive of residual disease, based on subsequent relapse of those patients. A negative PET scan is predictive of long-term disease-free survival, but it is not perfect. Overall, PET is the most effective means for treatment evaluation, and is incorporated into the revised response criteria for lymphoma. PET may be useful as an early response indicator, to provide for more adapted therapy. Currently, there are clinical trials testing this concept.

Radiation Therapy Planning

Field Size

How is PET best used for radiation therapy planning, and what are the important variables to consider in designing treatment for lymphoma cases? How are we using the information from PET to maximize outcomes/minimize toxicities for these patients – many of whom have 40- to 50-year life expectancies if they are successfully treated? The important issues, not surprisingly, are fundamentally those of field size and dose.

Historically, until the 1990s, quite large, wide fields were used to treat Hodgkin's disease, especially when the treatment was radiation therapy alone – subtotal or total lymphoid irradiation. This changed dramatically with the introduction of combined modality therapy. In favorable and unfavorable presentations of stage I Hodgkin's disease, there were clinical trials comparing outcomes for patients treated with more extensive fields versus those with more limited field treatment, defined as involved fields (that is, the involved nodal regions). Those randomized clinical trials failed to show any difference in outcomes. As a result, combined modality programs have reduced the size of the radiation fields to involved field therapy.

The question now is, can fields be reduced even further, from treating the involved lymphoid regions to just treating the involved lymph nodes themselves? To address this issue, protocols have been initiated by European clinical trials groups. We have evolved from subtotal or total lymphoid irradiation to involved field therapy to perhaps just treating the individual involved nodes.

Dose

Previously, doses in the range of 36–44 Gy were prescribed. Again, this has changed dramatically with the introduction of combined modality therapy for these diseases. In the favorable presentations of stage I or II, 30 Gy is now standard, but many protocols have incorporated even lower doses. Both in Europe and in the United States, protocols using 20 Gy for favorable presentations seem to be as effective as 30 Gy. Perhaps this will apply even for unfavorable presentations of stage I or II, although this has not yet been shown to be the case for patients with large mediastinal masses.

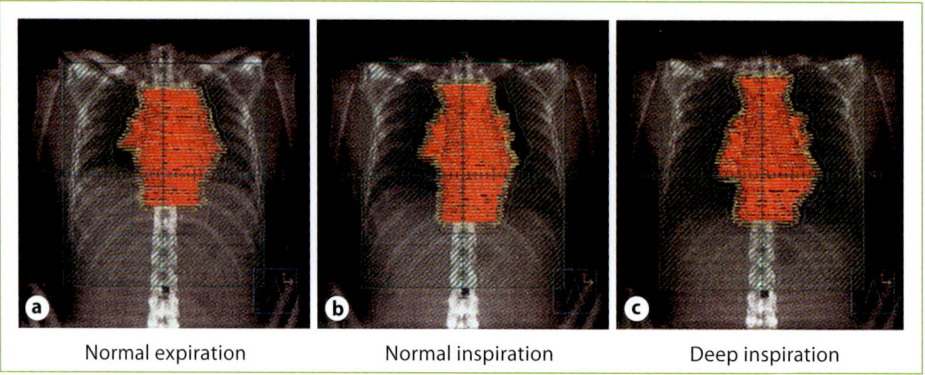

Fig. 4. Respiratory gating/breathing control for radiation therapy planning/treatment. PTV (colored area) and block definitions (hatched area) for normal expiration (**a**), normal inspiration (**b**) and deep inspiration (**c**). Note the considerable change in mediastinal volumes with even normal breathing. Respiratory management strategies have the potential to considerably reduce lung exposure [14].

New questions now present themselves. Can we further refine how we use dose based on early PET response criteria? Until the 1990s, therapy involved anterior/posterior treatment planning or opposed lateral fields with limited attention to organ motion. Now, conformal therapy techniques (with an option for IMRT) are used, and planning can even involve respiratory gating with PET/CT simulation. The bottom line is, will these efforts result in improved patient outcomes?

Advanced Treatment Planning

3D conformal therapy contributes to the treatment of lymphomas in several ways. It allows one to confirm the adequacy of treatment field coverage anatomically much more effectively than one could with 2D imaging. The ability to develop dose-volume histograms (DVHs) facilitates better prediction of complications and reduction of risk for these patients, many with long life expectancies. Also, for selected presentations, inverse treatment planning and intensity-modulated radiotherapy (IMRT) can be useful.

One of the important issues is the accuracy and agreement among physicians in defining the target volume for treatment. Michael Barton evaluated conventional 2D planning for Hodgkin's disease using radiation therapy alone. He sent simulator films to a number of radiation oncologists in Australia and had them draw the lung blocks, and found that there was wide variation in their designs [13]. Better results probably can be obtained by outlining tumor volumes on CT, although there have been no equivalent studies until recently. The EORTC has or-

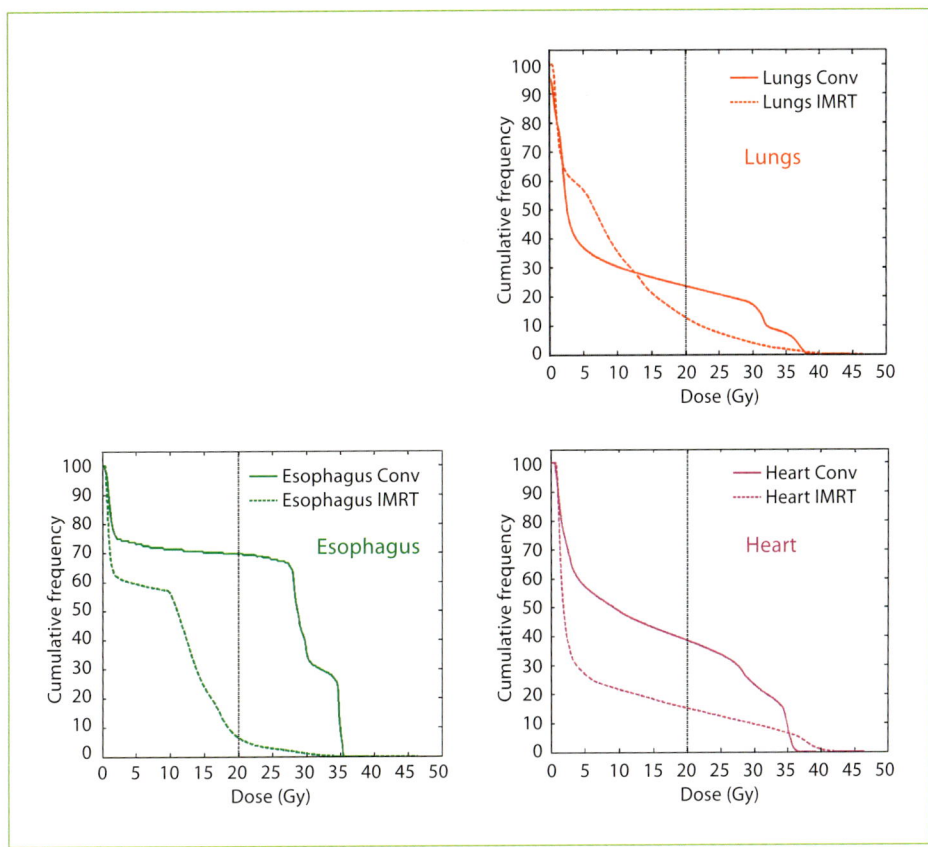

Fig. 5. DVHs for IMRT vs. conventional radiotherapy: the contribution of advanced planning techniques on reduction in exposure of lung, esophagus and heart volumes. Solid lines: conventional planning; dashed lines: IMRT.

ganized a group of oncologists to evaluate the consistency of gross tumor volume (GTV) identification, using PET/CT imaging for 3D conformal therapy.

3D conformal therapy for the lymphomas (a) allows computerized confirmation of the adequacy of treatment fields, (b) provides DVHs that give precise calculation of doses to organs at risk and (c) enables inverse treatment planning and IMRT when needed. Evaluating DVHs improves one's ability to put constraints on treatment fields and to limit heart and lung toxicities in patients undergoing radiation therapy to the mediastinum (fig. 4). Algorithms to predict, for example, the risk of pneumonitis based on the V-20 or the mean lung dose can be developed and can be essential in planning lymphoma therapy. Lung DVHs allow selection of cases for active breathing control or gating, and can facilitate more conformal fields. Respiratory gating can be used to significantly reduce lung exposure for mediastinal radiotherapy of Hodgkin's disease [14].

Occasionally IMRT is used in the treatment of lymphomas, although not nearly as often as cancers at other sites. Figure 5 compares treatment plans using conventional conformal therapy with AP/PA fields versus IMRT in an analysis by Loo [15] at Stanford. In this hypothetical situation specifying treatment to the axillae and the mediastinum, a much more conformal plan is achieved with IMRT, with sparing of tissues especially in the posterior mediastinum. The corresponding DVHs show very dramatically the ability to decrease the doses to the esophagus and heart. However, an interesting phenomenon can be seen in the lungs that has a potentially very significant impact: there is a large volume of lung that receives a low dose of radiation, even though the volume of lung that receives a high dose is smaller. DVHs for the breasts of women show the same phenomenon. This element of IMRT, that organs at risk can receive low doses of radiation to much larger volumes, is a serious concern for such patients, often with long life expectancies. For instance, doses as low as 4 Gy to the breast have been associated with a risk of secondary breast cancer in women (especially young women) treated for Hodgkin's disease.

Integration of PET Imaging

PET scanning has greatly helped to define the sites of disease more clearly, especially in noncontiguous areas such as in the cardiophrenic angle area during mediastinal radiotherapy. Hutchings et al. [16] evaluated the impact of PET/CT on radiation therapy planning. In two thirds of the patients, there were no notable changes to therapy plans, but such scans necessitated increases in the field sizes in 7 patients and decreases in 2.

What steps does one take to integrate this useful information into the planning process? They are very similar to the planning steps required for head and neck cancer patients:

(a) One should obtain a PET/CT prior to chemotherapy. Ideally, this should be a PET/CT simulation in the treatment position for future therapy planning. Certainly one cannot wait until after chemotherapy to obtain the first PET because most patients become PET negative early-on during chemotherapy.

(b) Prior to chemotherapy, a diagnostic CT is also needed to evaluate for enlarged nodes that are *not* PET-avid, since PET scanning is not perfect.

(c) One should repeat the simulation after chemotherapy, primarily to define anatomy for initiating radiotherapy. Optimally, this would be another PET/CT simulation, enabling assessment of response.

(d) The GTV definition is based upon the information from these studies combined. It is important to include the PET-negative areas that were identified on CT.

(e) If there are any residual PET-positive areas, one can define a separate GTV boost area for those patents.

(f) For patients who have been treated with chemotherapy, one may assume that the chemotherapy will treat the clinical target volume (CTV) expansion that would normally be incorporated into the treatment field. For these patients, the CTV can equal the GTV, and one can simply add margin for the planning target volume (PTV). Also the lateral, left-to-right contours may be reduced, based on the response to chemotherapy.

(g) One then adds appropriate expansion to obtain the PTV. At Stanford, this is approximately 1 cm for lymphoma patients.

There may be practical difficulties in implementing these steps. Often the initial staging PET/CT, obtained by the referring oncologist, is not performed in the treatment position, and a second study may not be authorized by the insurance carriers. Therefore, one may be required to merge the prechemotherapy diagnostic PET/CT with the postchemotherapy CT simulation for radiotherapy planning, even though the patient may be in somewhat different positions and there have been anatomic changes between the studies. Problems encountered with imperfect registration may necessitate the use of appropriately larger fields.

Clinical Trials Utilizing PET

Field Size

Experience to date has been encouraging regarding the accuracy and usefulness of PET imaging in the lymphomas. Using PET imaging, can we have confidence in reducing the size of treatment fields even further? For instance, if there is a single positive axillary node, perhaps one does not need to treat the entire axillary lymph node region. This concept is now being tested. Investigators in the EORTC are looking carefully at patients with prechemotherapy PET/CT scans performed in the treatment position, and comparing them with postchemotherapy PET/CT simulations also obtained in the treatment position. Using these 2 studies, they are designing fields that incorporate only the initially involved nodes with margin. In some instances, they have been able to demonstrate dramatic reductions in field size using this approach. The EORTC has adopted this as routine in their current clinical trials. On the HD17 trial, the German Hodgkin Study Group is randomizing patients who have stage I-II Hodgkin's disease with unfavorable factors to combined modality therapy that utilizes involved node irradiation rather than the more conventional involved field irradiation.

PET Response and Radiotherapy

There are 2 interesting European clinical trials testing the use of PET in lymphoma treatment planning. These are both evaluating the effect of PET imaging on the utility of involved nodal irradiation.

An EORTC trial is evaluating patients with favorable stage Hodgkin's disease. The control group receives 2 cycles of ABVD chemotherapy, then imaging with PET. This is followed by involved node irradiation to 30 Gy with a 6-Gy boost to any residual abnormality on PET. The investigational arm receives 2 cycles of ABVD, then PET imaging. The PET is then used to adapt the therapy. If the PET scan is negative after 2 cycles of ABVD, patients receive 2 additional cycles of ABVD (for a total of 4) only, without radiotherapy. If the PET scan is positive, patients receive an intensification of their chemotherapy, followed by radiotherapy. For the unfavorable patients, there is a similar algorithm: in the control arm, all patients receive irradiation after chemotherapy; in the investigational arm, whether patients receive radiotherapy is dependent on the PET activity.

The German Hodgkin's Study Group has a slightly different trial. They also initiate therapy with 2 cycles of ABVD followed by PET imaging. Their control arm includes involved field radiation to 30 Gy. In their investigational arm, if the PET scan is negative after 2 cycles of ABVD, the patients receive no further therapy, but if the PET scan is positive, then they receive radiotherapy.

Guidelines for Clinical Practice

- Prior to chemotherapy, one should obtain a PET/CT. Ideally, this should be a *PET/CT simulation* in the treatment position for future therapy planning.
- Prior to chemotherapy, a diagnostic CT is also needed to evaluate for enlarged nodes that are not PET-avid.
- After chemotherapy, repeat the PET scan, optimally the same PET/CT simulation. This assesses response and also obtains current anatomy for planning. Alternatively, a diagnostic PET and a CT simulation are obtained.
- The GTV definition is based upon the combined information from these studies. It is important to include the PET-negative areas that were identified on CT.
- If there are any residual PET-positive areas, one can define a separate GTV boost area for those patents.
- For patients who have been treated with chemotherapy, one may assume that the chemotherapy will treat the CTV expansion that would normally be incorporated into the treatment field. For these patients, the CTV can equal the GTV, and one can simply add margin for the PTV. Also the lateral, left-to-right contours may be reduced, based on the response to chemotherapy, in the mediastinum, para-aortic nodes and discrete nodal masses.
- One then adds appropriate expansion to obtain the PTV. At Stanford, this is approximately 1 cm for lymphoma patients.

References

1 Bangerter M, Moog F, Buchmann I, et al: Whole-body 2-[^{18}F]-fluoro-2-deoxy-D-glucose positron emission tomography (FDG-PET) for accurate staging of Hodgkin's disease. Ann Oncol 1998;9:1117–1122.

2 Partridge S, Timothy A, O'Doherty MJ, et al: 2-Fluorine-18-fluoro-2-deoxy-D glucose positron emission tomography in the pretreatment staging of Hodgkin's disease: influence on patient management in a single institution. Ann Oncol 2000;11:1273–1279.

3 Munker R, Glass J, Griffeth LK, et al: Contribution of PET imaging to the initial staging and prognosis of patients with Hodgkin's disease. Ann Oncol 2004;15:1699–1704.

4 Naumann R, Beuthien-Baumann B, Reiss A, et al: Substantial impact of FDG PET imaging on the therapy decision in patients with early-stage Hodgkin's lymphoma. Br J Cancer 2004:90:620–625.

5 Rigacci L, Vitolo U, Nassi L, et al: Positron emission tomography in the staging of patients with Hodgkin's lymphoma. A prospective multicentric study by the Intergruppo Italiano Linfomi. Ann Hematol 2007;86:897–906.

6 Wirth A, Foo M, Seymour JF, et al: Impact of [18F] Fluorodeoxyglucose positron emission tomography on staging and management of early-stage follicular non-Hodgkin lymphoma. Int J Radiat Oncol Biol Phys 2008;71:213–219.

7 Petersen PM, Gospodarowitz M, Tsang R, et al: Long-term outcome in stage I and II follicular lymphoma following treatment with involved field radiation therapy alone. Proc ASCO 2004; 23:561.

8 Vaughan Hudson B, Vaughan Hudson G, MacLennan KA, et al: Clinical stage 1 non-Hodgkin's lymphoma: long-term follow-up of patients treated by the British National Lymphoma Investigation with radiotherapy alone as initial therapy. Br J Cancer 1994;69:1088–1093.

9 MacManus MP, Hoppe RT: Is radiotherapy curative for stage I and II low-grade follicular lymphoma? Results of a long-term follow-up study of patients treated at Stanford University. J Clin Oncol 1996;14:1282–1290.

10 de Wit M, Bohuslavizki KH, Buchert R, et al: 18FDG-PET following treatment as valid predictor for disease-free survival in Hodgkin's lymphoma. Ann Oncol 2001;12:29–37.

11 Advani R, Maeda L, Lavori P, et al: Impact of positive positron emission tomography on prediction of freedom from progression after Stanford V chemotherapy in Hodgkin's disease. J Clin Oncol 2007;3902–3907.

12 Gallamini A, Hutchings M, Rigacci L, et al: Early interim 2-[18F]fluoro-2-deoxy-D-glucose positron emission tomography is prognostically superior to international prognostic score in advanced-stage Hodgkin's lymphoma: a report from a joint Italian-Danish study. J Clin Oncol 2007;25:3746–3752.

13 Barton MA, Rose A, Lonergan D, et al: Mantle planning: report of the Australasian Radiation Oncology Lymphoma Group film survey and consensus guidelines. Australas Radiol 2000;44:433–438.

14 Stromberg JS, Sharpe MB, Kim LH, et al: Active breathing control (ABC) for Hodgkin's disease: reduction in normal tissue irradiation with deep inspiration and implications for treatment. Int J Radiat Oncol Biol Phys 2000;48:797–806.

15 Loo BW Jr, Hoppe RT: Lymphoma: overview; in Mundt A, Roeske J (eds): Intensity Modulated Radiation Therapy – A Clinical Perspective. Hamilton, BC Decker, 2005, pp 535–548.

16 Hutchings M, Loft A, Hansen M, et al: Clinical impact of FDG-PET/CT in the planning of radiotherapy for early-stage Hodgkin lymphoma. Eur J Haematol 2007;78:206–212.

Prof. Richard Hoppe
Department of Radiation Oncology, Stanford University Medical Center
875 Blake Wilbur Drive, Rm CC-G224
Stanford, CA 94305-5847 (USA)
Tel. +1 650 723 5510, Fax +1 650 498 6922, E-Mail rhoppe@stanford.edu

Prostate Cancer

Image-Guided, Adaptive Radiotherapy of Prostate Cancer: Toward New Standards of Radiotherapy Practice

Patrick Kupelian[a] · John L. Meyer[b]

[a]Department of Radiation Oncology, University of California School of Medicine, Los Angeles, Calif., and [b]Department of Radiation Oncology, Saint Francis Memorial Hospital, San Francisco, Calif., USA

Abstract

The development and acceptance of new image-guided radiotherapy (IGRT) technologies have often been initiated with the treatment of prostate cancer. Imaging and tracking of the prostate during a treatment course has yielded a great deal of information about the motion and deformation of the gland during radiotherapy, and has led the way toward the development of more accurate treatment methods including dose-guided and adaptive strategies. Now, there is long-term experience with the use of fiducials and electromagnetic implantable beacons that give high-quality tracking of prostate motion. From analyzing these extensive tracking data sets, a clear understanding of prostate motion and its dosimetric significance has developed. This knowledge can now be used to define current expectations and guidelines for clinical care. The random nature of prostate motion requires daily localization if treatment is to be delivered with small margins. Interfraction motion can have a significant impact on prostate gland dosimetry, and even more of an impact on the seminal vesicles and possibly intraprostatic tumor areas. The dosimetric impact on normal structures (bladder/rectum) is less clear, and there are significant individual variations. Interfraction and intrafraction rotations and deformations of the prostate are routinely detected. The dosimetric impact of these motions of the prostate gland is minimal when daily localization is used, even when the treatment margins are small. However, deformations of the seminal vesicles, rectum and bladder are much more pronounced. The dosimetric impact of deformation of the rectum and bladder is highly variable among patients, and the clinical consequences remain unclear. Daily volumetric imaging and dosimetry may become quite important for these volumes. Due to the random nature of motion/deformation during prostate radiotherapy, adaptive radiotherapy ideally would be performed as an on-line process. On-line adaptive radiotherapy requires robust deformable registration and replanning programs. These are beginning to emerge in useful clinic applications.

Copyright © 2011 S. Karger AG, Basel

Prostate Radiotherapy: A Model for Adaptive Image-Guided Radiotherapy

Most image-guided radiotherapy (IGRT) technologies have proven their utility first for prostate cancer therapy, and have extended their applications to other disease sites based on these results. From evaluating current approaches to prostate radiotherapy, one can gain perspectives on where the field is moving with regard to the most efficient and reliable solutions for IGRT and how the more far-ranging issues of adaptive radiotherapy may be approached. New standards of radiotherapy practice are often established first for the management of prostate cancer.

Why is prostate therapy a good model? The gland itself has a relatively simple geometry, and well-defined treatment plans can be generated for the primary site. Deformation is not as large a problem as for many other organ sites, change in prostate size during treatment is minimal and respiratory motion is typically absent. This allows clinicians to focus on issues of coverage of the target and surrounding anatomy. Clinical studies have yielded a great deal of information about the motion and deformation of the prostate, as well as the tissues around it, including normal organs and regions of potential tumor extension. Now, there is long-term experience with fiducials and electromagnetic implantable transponders that give high-quality tracking of intrafraction motion. From analyzing these extensive tracking data sets, a clear understanding of prostate motion and its dosimetric significance has developed. This knowledge can now be used to define current expectations and guidelines for clinical care.

Two major challenges must be overcome before accurate radiation coverage for prostate and other cancers can be achieved. Perhaps the solutions for prostate treatment will become guides for radiotherapy practice in general. The first is the relative movement among the different targets in the field. If both the seminal vesicle and lymph node volumes are to be treated, new strategies must be developed since these targets move and deform independently of each other. Simple tracking solutions are not sufficient.

The other challenge is that prostate motion is random and not periodic. Unlike breathing, progressive weight loss or progressive tumor shrinkage, the sources of movement of the prostate and other pelvic structures are unpredictable. To correct for these variations most accurately, dosimetry fields need to be adapted. The best adaptive solutions would be *on-line* methods to immediately account for the anatomy that is imaged before each treatment. These immediate processes would eliminate the need for later *off-line* adaptations that attempt to adjust for dose already given. There are many issues in bringing robust adaptive therapy solutions into clinical practice; however, there are many advantages of initiating this work for prostate cancer patients.

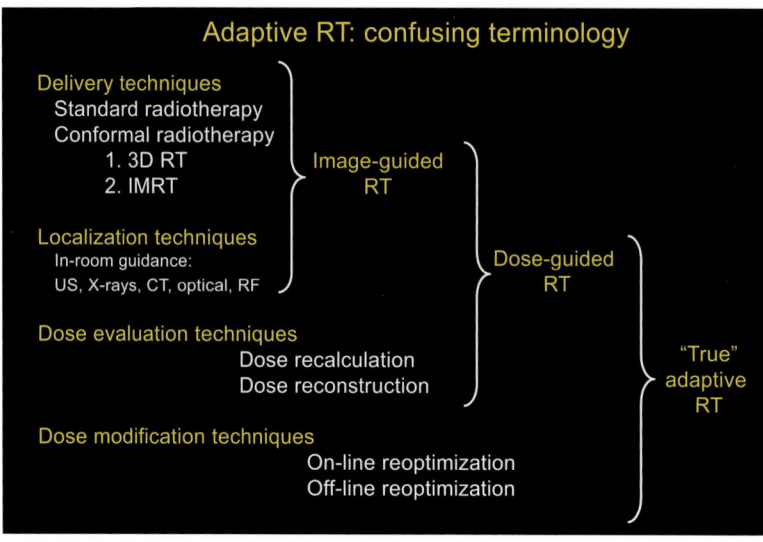

Fig. 1. Expanding approaches to radiation dose delivery.

It is important to keep the progress in improving radiotherapy accuracy in perspective. As figure 1 indicates, new technologies in radiation dose delivery (intensity-modulated radiotherapy; IMRT) have been coupled with new technologies for guiding them (IGRT). Image guidance methods have been developed that add an overlay of the dose distribution and this can be used to assist the setup registration. With volumetric image guidance, one can even recalculate the doses to be delivered based on the anatomy imaged at treatment. These dosimetric enhancements of IGRT might be called 'dose-guided radiotherapy'. Yet these processes fall short of the goal of true adaptive radiotherapy, which integrates methodologies to re-contour the changing internal anatomies and adapt treatment plans to adjust for these changes.

Dosimetric Impact of Motion in Prostate Radiotherapy

What are the results of motion tracking, and what is the resulting difference in the daily and cumulative dosimetry? How much benefit do current guidance interventions contribute? The answers depend on the frequency and amount of internal motion, and these have been studied extensively. Motion can be introduced between treatments (interfraction) or during treatment delivery (intrafraction); both are clinically important, and will be discussed here (see online supplementary material).

Interfraction Motion

Daily Setup Accuracy

Adaptive radiotherapy attempts to bring the delivered radiation dose distribution to the planned distribution. In addition to sophisticated methods of dose recomputation at treatment, this goal can be approached through many practical improvements in the delivery process. The first is simply to perform accurate daily setup. Currently, image guidance allows x-y-z (and even rotational) alignment of the targets, which is critical to accurate treatment. The degree of benefit depends on how accurately one understands where the target is at each treatment fraction. If one uses external skin marks to represent the target, often the target is partially missed. Use of visualized bone anatomy adjacent to the target, provides somewhat better results. More precise methods of aligning to the tumor, by directly visualizing the prostatic soft tissues and/or markers within them, improve the results further and represent the current standard of care in most centers. The ability to setup to deformed registration volumes will be the most accurate, and is a realistic goal [1].

It is also important to eliminate systematic errors made during therapy planning. For instance, there are two series illustrating the importance of eliminating rectal distension during treatment planning, which can introduce error throughout therapy. There is documentation that failure to eliminate rectal distension in planning is associated with lower tumor control rates, which is not surprising since such systematic errors will introduce the most consequential shifts [2, 3]. Recent work at the M. D. Anderson Cancer Center Orlando (MDACCO) has shown that the use of daily image guidance can significantly reduce this concern [4].

Need for Daily Image Guidance

Is interfraction motion sufficient to require imaging before every treatment for prostate irradiation, or can imaging be performed less often? Could a program be developed that would eliminate some of these daily images without altering the accuracy of therapy? At MDACCO, the daily imaging records of 74 prostate cases (2,252 total fractions) were retrospectively analyzed to address these questions. Many different scenarios for imaging were considered, such as imaging every other day, once per week, and so forth. Since imaging *was* obtained every day in these patients, the difference that the daily imaging contributed could be assessed. 'Residual errors' were calculated for each day from the difference between the known setup corrections and those that would have been made if a different protocol were followed. How significant were these errors? Regardless of the imaging protocol, the residual errors were considered excessively large compared with the accuracy of daily imaging. For instance, every-other-day imaging still resulted in residual errors of ≥3 mm in about 40% of fractions, and ≥5 mm in about 25% of fractions. This work indicates that daily imaging is indeed necessary to avoid these sizeable uncertainties [5].

Significance of Interfraction Motion

To evaluate the dosimetric impact of interfraction motion, investigators at the University of Utrecht reviewed treatment data from a large number of patients (n = 217). Each had daily imaging, with shifts based on implanted fiducials, for 35 fractions per patient. They compared the daily-imaged prostate anatomy relative to the plan, with or without shifts applied ('corrected' or 'uncorrected'). What would be the dosimetric implication of the changing anatomy, assessed by dose recalculation and dose accumulation? For the areas of interest, they evaluated (a) the prostate plus seminal vesicles as the target, (b) the prostate gland by itself and (c) the bladder and rectum. They also considered (d) a tumor 'proxy' for the likely location of the tumor by modeling a peripheral zone. They then evaluated the dosimetric impact of interfraction motion on all of these regions of interest (fig. 2) [6].

Figure 2 shows the average dose and the $D_{99\%}$ isodose coverage for 3 different scenarios: the initial plan (static), at treatment before image-guided correction (uncorrected) and after daily image guidance (corrected), for each for the anatomic areas shown. Consider first the middle panels for the *prostate only*. The upper middle panel shows the mean dose, and dots below the dotted line show cases beyond the 10th percentile. One can see that some patients become significant dosimetric 'outliers' when no correction is made, though the mean target dose was acceptable for most patients when using an 8-mm margin. When evaluating a more sensitive metric, the $D_{99\%}$, one sees even more outliers, and daily image guidance does not eliminate this in all cases, perhaps because of deformation. Inside the prostate, the *peripheral zone* target, there are many dosimetric outliers without correction for either the mean or 99% dose measure (red circles). For the *prostate and seminal vesicles*, correction eliminates the outliers less often because of more pronounced deformation.

The study also considered bladder and rectal high-dose volumes. On average, the bladder and rectum received acceptable doses, but there were outlier cases that potentially received high doses to large volumes of the bladder and the rectum. From these results, it is apparent that not all patients will benefit equally from daily image guidance, but it is critically important for some patients. In the treatment of prostate cancer, with cure rates already high and toxicity rates low, it is important to provide optimal therapy to every patient if further improvements in therapeutic gain are to be achieved.

Current Findings, Remaining Challenges

To summarize our knowledge about interfraction motion, we now understand that the motion of the prostate is random rather than cyclic. The random nature of the motion requires daily localization. This is especially true as treatment margins have become tighter; image guidance with dose escalation has brought margins below 8 mm in most institutions. Also, systematic errors will introduce the

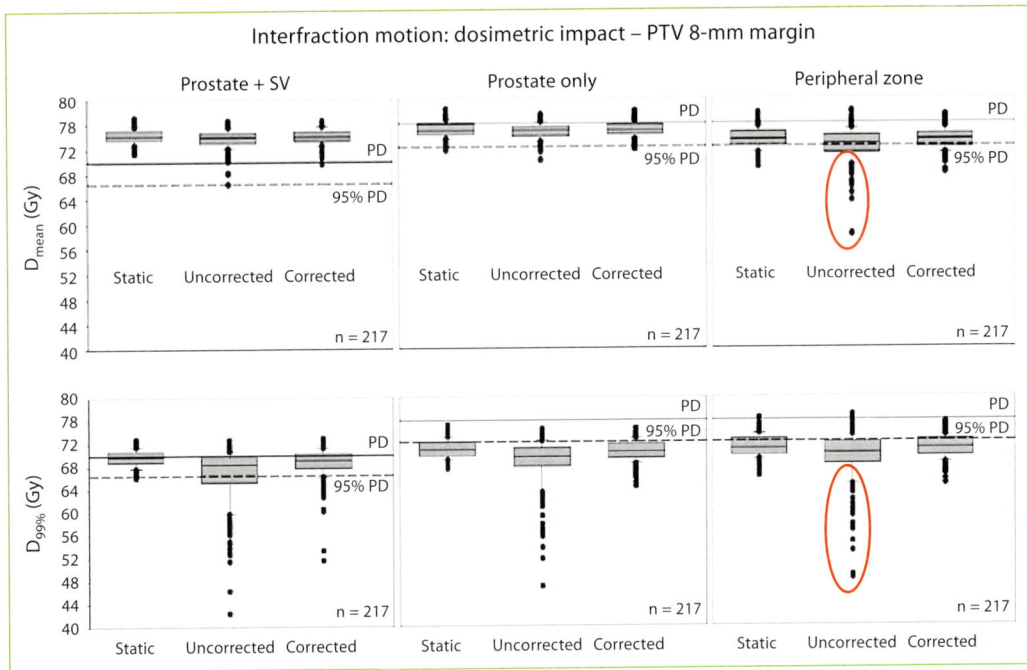

Fig. 2. Dosimetric impact of interfraction motion and image-guided correction. Dosimetry of original plan ('static') is compared with daily dosimetry without or with image guided translational shifts ('correction'). D_{mean} = Average dose; $D_{99\%}$ = dose to 99% of CTV; PD = prescription dose; SV = seminal vesicles. Shaded boxes = Median (line), 25th and 75th percentiles; whiskers above and below boxes are 10th and 90th percentiles. Dots represent outliers [6].

most consequential errors, and care must be taken in avoiding these in each step of the planning process.

The dosimetric impact of motion will differ from area to area within the anatomic treatment field. When one evaluates the locale of the prostate tumor (peripheral zone), prostate as a whole, seminal vesicles and lymph nodes, the motion of each is different. It is a challenge to manage this relative anatomical change. Currently, treatments are essentially designed to align to the prostate gland itself. Overall, the average doses to most organs will be acceptable, but one must be aware that dosimetric outliers will occur. The likelihood of this increases with greater distance from the prostate setup point. The dosimetric impact of interfraction motion on the prostate gland can be significant, but the impact on seminal vesicles, lymph nodes and even intraprostatic areas can be even more significant. This remains a major challenge and indicates the need for adaptive planning methods. The dosimetric impact of interfraction motion on normal structures, especially the bladder and rectum, is less clear though significant individual variations do occur and require vigilance.

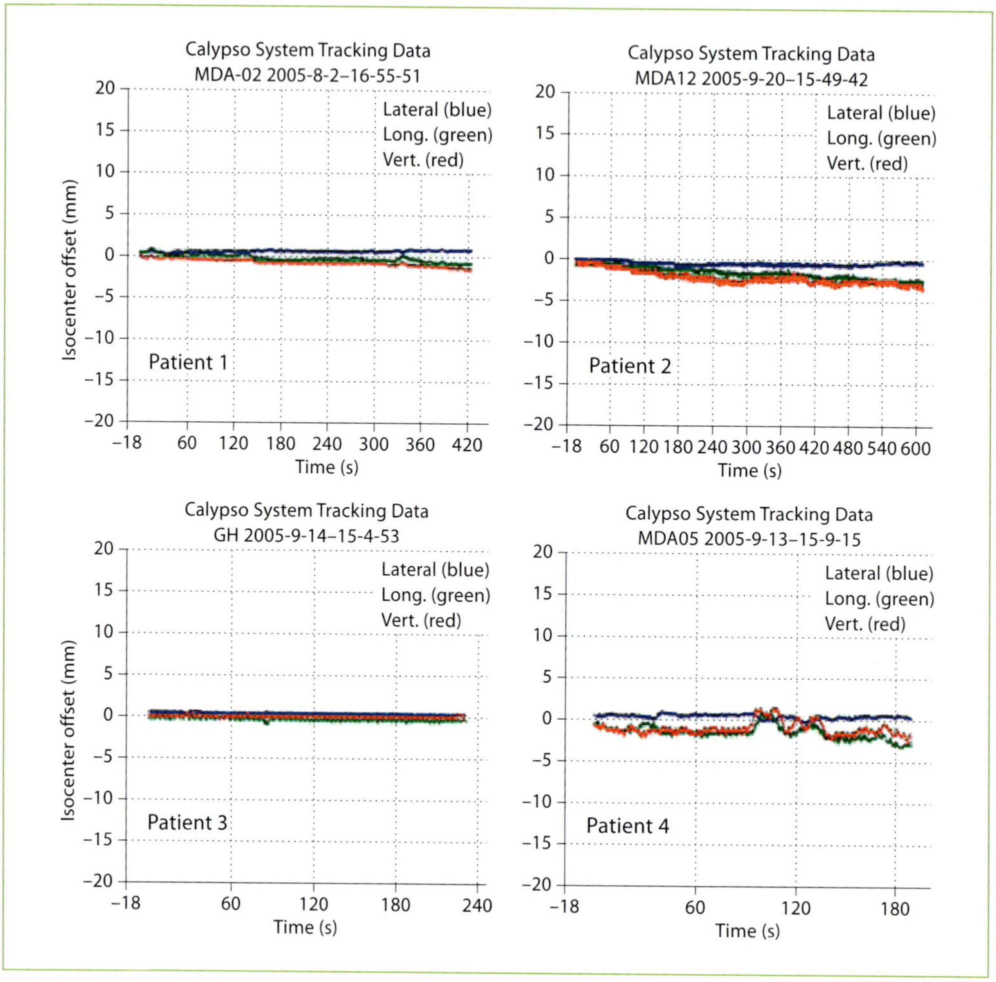

Fig 3. Electromagnetic beacon monitoring showing prostate motion during radiotherapy. Colors indicate the beacon positions in 3 dimensions (blue: lateral, green: longitudinal, red: vertical). Some patients had significant positional drift during treatment, while others had large momentary changes [7]. See online supplementary videos for active demonstrations.

Intrafraction Motion

Tracking during Treatment

Electromagnetic tracking of the prostate (Calypso System) has provided a rich source of data on its movement during radiotherapy; few other organ sites have been so carefully studied for intrafraction motion. Figure 3 shows results from four different patients, showing the positional shifts during treatment in the vertical, lateral and longitudinal dimensions. For most patients, little change occurs in

Table 1. Percentage of total tracking time that implanted prostate transponder was beyond a given distance during treatment delivery

	Percentage of total tracking time beyond distance (3D vector)			
	>10 mm	>7 mm	>5 mm	>3 mm
Average	0.2%	0.6%	2.4%	12.5%
Median	0.1%	0.4%	1.8%	9.8%
Individual patients				
Min.	0.0%	0.0%	0.0%	0.40%
Max.	0.9%	1.8%	7.7%	33.8%

Distance = 3D vector from original position [7].

the prostate's position. Very occasionally, shifts occur that could be significant, reflecting ongoing changes that likely occur silently in other patients during radiotherapy.

A large amount of information about prostate intrafraction motion has now been acquired. How clinically significant is this motion? It is difficult to define a method that best quantifies this. One method used at MDACCO has been to assess the percentage of the total tracking time that the prostate was located beyond a certain distance – 1 cm, 7 mm, 5 mm, 3 mm, and so forth – measured as the 3D vector from the original position. Table 1 shows the mean and median values for 17 patients, as well as the value range for individual patients. Prolonged intrafraction motion, such as for an entire treatment delivery, was uncommon. But when evaluating the results in individual patients, one can observe a drift of >3 mm for 33.8% of the monitoring time, and a drift of >5 mm for 7.7% of the time. Such events can also be documented on fluoroscopic X-rays. For example, gas bubbles that appear in the rectum may cause a shift of the fiducial markers (and the prostate) of >1 cm.

Clinical Use of the Tracking Data

What is the practical impact of intrafraction motion? The overall implications are being investigated using these data. For now, difficulties observed in individual patients should lead to reconsideration of the treatment margins, delivered doses and clinical management based on the degrees of motion. The data obtained from implanted transponders lead to better understanding of images of implanted fiducial markers obtained for delivery guidance. When fiducials are used for guidance at MDACCO, second and even third sets of X-rays midway through the course of therapy delivery may be taken to make certain that treatment is still on target. These repeat images have shown significant motion in some patients.

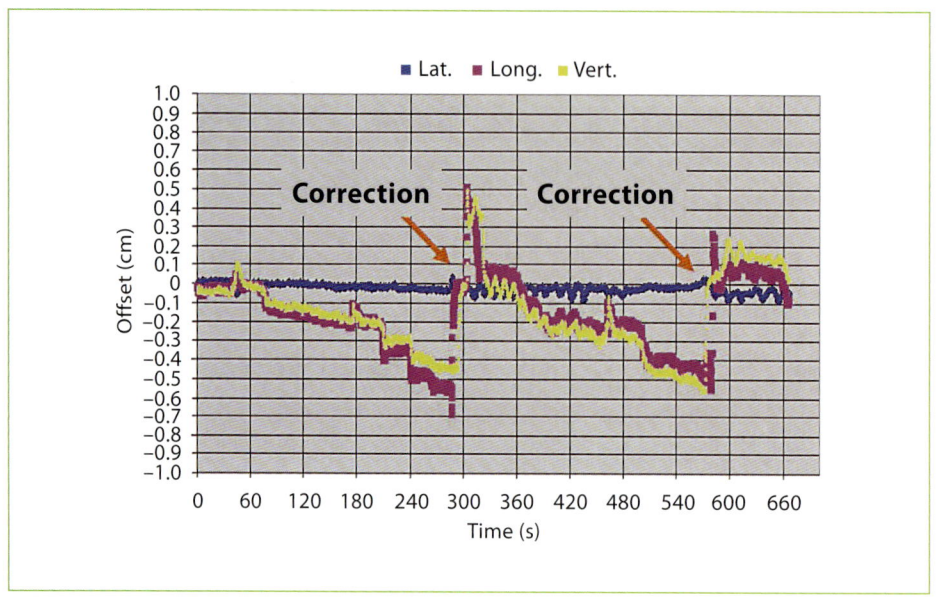

Fig 4. Adjustment during therapy using electromagnetic tracking. Treatment is interrupted and adjusted if position change exceeds predefined limits. Lat. = Lateral; Long. = longitudinal; Vert. = vertical.

When tracking the prostate using the Calypso System, a 3-mm threshold has been used at MDACCO (fig. 4). When this threshold is exceeded, treatment is stopped for realignment between treatment beams; this avoids interruption of therapy during delivery of individual beams. Table 2 shows how often this has been required. We evaluated nearly 1,000 treatments in 29 patients, and found that realignment during treatment was required in 25% of the fractions. Of importance, the range for individual patients was enormous, and some patients required intervention in 85% of their treatment fractions. In some cases, shifts were observed before therapy delivery was even initiated, indicating patient movement after positioning. This shows another role for real-time monitoring to assure treatment quality.

By monitoring prostate position actively during therapy, one can reduce the margins of treatment required (table 3). These are significant results, which have implications for improved normal tissue protection, dose escalation and potential tumor control. Technologies for active tracking during therapy will undoubtedly be developed for use at other disease sites, such as lung and liver, and may dramatically alter management programs and potentially the results of treatment with radiotherapy for these tumors.

Table 2. Frequency of intervention needed to realign beams for motion >3 mm

Events	Monitoring intrafraction motion	
	mean frequency	range (indiv. patient)
No motion >3 mm, no intervention	59	10–100
Motion >3 mm, transient, no intervention	14	0–42
Motion >3 mm, realignment between beams	25	0–85
MD disagrees with therapist intervention (interuser variability)	1	0–8

Values are percentages. Clinical protocol using prostate transponders, 3-mm threshold for realignment between beams in 29 patients, 963 total fractions. Using a 3-mm threshold, realignment was required in 25% of the fractions [8].

Table 3. Treatment margins needed to encompass intrafraction motion (University of Michigan)

	Treatment margins and intrafraction motion		
	treatment margins		
	lateral	ant./post.	sup./inf.
Skin marks			
Ignore intrafraction motion	8.0 mm	7.3 mm	10.0 mm
With intrafraction motion	8.2 mm	10.2 mm	12.5 mm
Daily marker alignments	1.8 mm	5.8 mm	7.1 mm
Interbeam adjustments	0.4 mm	2.3 mm	1.8 mm
Intrafraction tracking and correction (3-mm threshold)	0.3 mm	1.5 mm	1.5 mm

Intrafraction tracking and correction using a 3-mm threshold required the smallest margins [9].

Dosimetric Consequences of Tracked Motion

Data from active tracking has been analyzed through 4D dose calculation engines, to evaluate the dosimetric impact of the observed motion. We have evaluated the dosimetric consequences for helical tomotherapy delivery [10, 11] and similar work has been reported for static field IMRT delivery [12, 13].

Figure 5 shows a case in which the prostate drifted anteriorly during therapy. For this fraction, one can document the dosimetric differences that resulted. Using the 3D tracking information, one can *anatomically* place the dose differences, and show regions that received dose more or less than that intended, including by

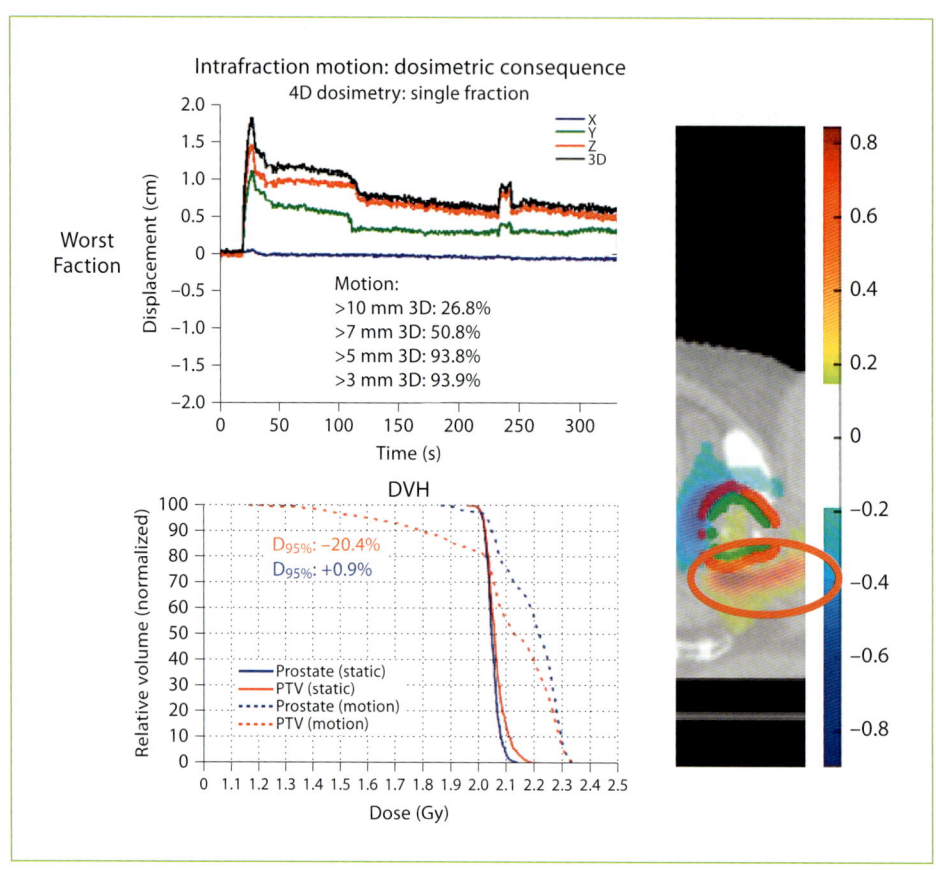

Fig. 5. Dosimetric consequences of prostate motion can be anatomically localized using dose recalculations based on tracking data [11].

how much. For an entire treatment course, one can have an accurate 3D picture of the exact dose delivered to the therapy volume for each fraction – including the regions of concern for over- or underdosing. This dosimetric information can provide the basis for adaptive replanning, which could compensate for dose inaccuracies in subsequent treatments.

Since treatment courses involve multiple fractions, it is useful to accumulate the dose differences that occur over the entire therapy course. Figure 6 shows this information for 16 patients over their treatment courses. For most patients, motion on individual days did not change the cumulative dosimetry significantly, since the dose patterns homogenized over time. For a few patients, however, the delivered doses never met the intended ones. There were also large variations in delivered doses for individual fractions. For hypofractionation, such daily dose changes caused by motion will have much greater impact.

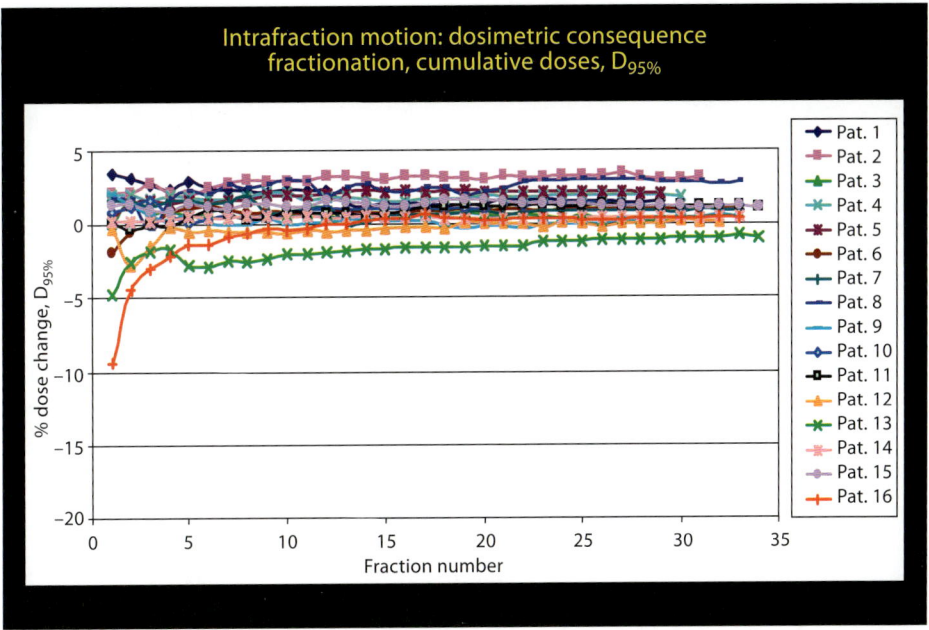

Fig. 6. Dosimetric consequence of monitored dose differences in the $D_{95\%}$ for 16 patients, by increasing numbers of treatments. Note that differences of individual treatments tend to reduce over a fractionated course, but not entirely [8].

Prostate Deformation/Rotation and Its Dosimetric Consequences

Interfraction Deformation/Rotation
It is common to see changes in the 3D geometries of treatment volumes from fraction to fraction. As a result, aligning beams 'correctly' may still fail to encompass the intended volumes. This problem is frequently confronted in prostate therapy since the seminal vesicles can move differently from the prostate due to daily variations in rectal filling (fig. 7). Perfect alignment of the prostate may still result in inadequate coverage of the seminal vesicles.

If one recontours the bladder and rectal volumes for every treatment of a therapy course, one discovers wide variations in the daily geometries of these normal tissues [14]. It is very unlikely that one could determine the rectal or bladder dosimetry throughout the course of treatment by any predictive method, since there are no specific trends in the daily changes of the rectal and bladder contours. These daily dosimetric changes can be large (fig. 8), and these issues pose difficult problems conceptually. However, the work on prostate tracking dosimetry encourages a thorough approach: daily volumetric imaging, and accumulating the rectal and bladder dosimetry based on the actually observed daily contours.

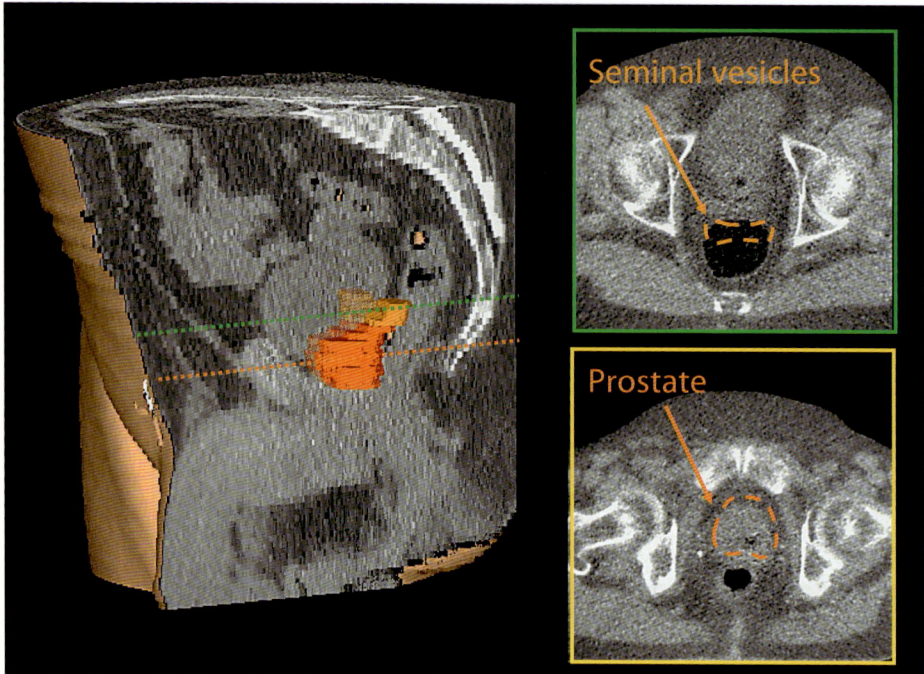

Fig. 7. Motion and deformation of the prostate and seminal vesicles may be different and cause misalignment of one or the other organs despite image guidance. See online supplementary video.

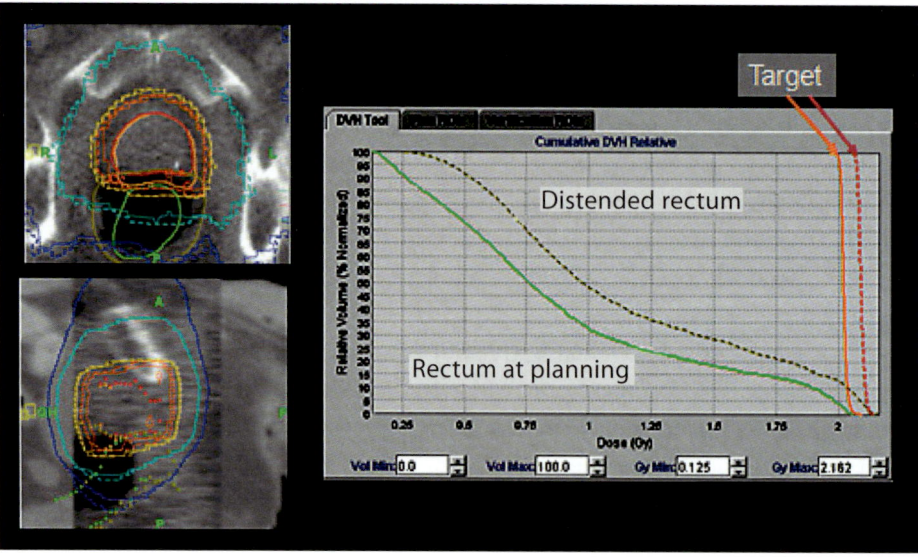

Fig. 8. Effect of daily deformation on rectal dose, based on dose recalculation on volumetric imaging at treatment. Note the large effect on the rectal DVH with rectal distention.

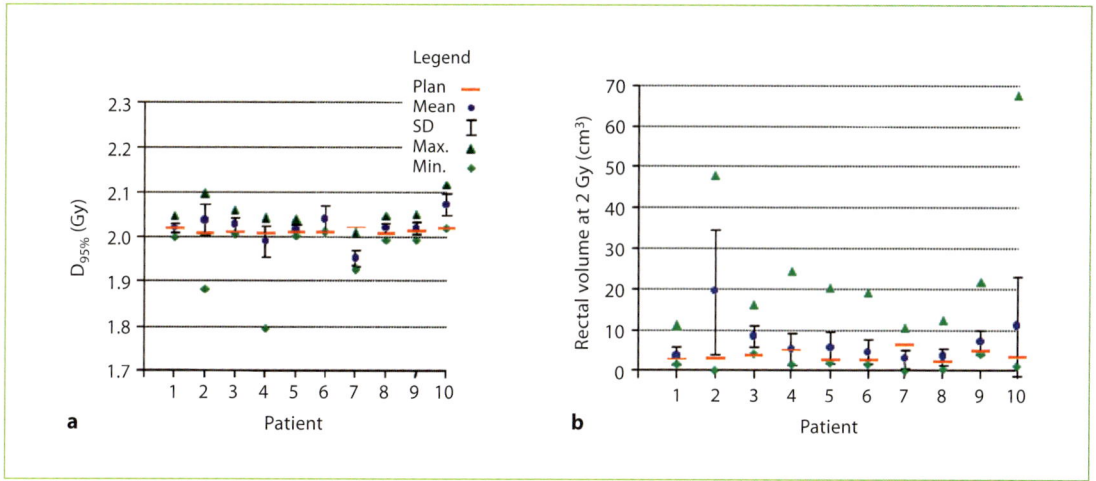

Fig. 9. Dosimetric impact of interfraction motion on the prostate and rectum. 10 patients were imaged daily; the prostate, rectum and bladder were recontoured, and doses recalculated for the anatomy imaged at treatment. **a** Prostate dose, $D_{95\%}$. Alignment was on markers, and showed excellent coverage of the prostate overall despite possible prostate deformation. **b** Absolute rectal volume receiving 2 Gy. In many patients, the rectal dose was substantially higher than planned. Red line = Intended dose per fraction; error bars = standard deviation in actually delivered doses for that patient. Mean, minimum and maximum values are also shown [15].

With daily volumetric imaging, especially with helical tomotherapy, it is relatively easy to acquire images and recalculate doses based on the data. If one recontours the volumes, one can obtain the dose-volume histogram (DVH) for that fraction day. This was performed for 10 patients, with recontouring of the bladder, rectum and prostate [15]. Patients were treated using 6-mm margins, except 4-mm posteriorly. All of the treatments were performed with alignment on markers. The $D_{95\%}$ for each fraction was evaluated, for all 35 fractions for these 10 patients (fig. 9). With little exception, the prostate target doses were met by aligning to the markers (fig. 9a). The results for the rectum were different, and large individual variations occurred in the daily rectal dose delivered (fig. 9b). Therefore, deformation/rotation may have much less impact on the prostate treatment than on the normal tissue doses especially for the rectum.

Intrafraction Deformation/Rotation

So far, we have considered interfraction deformation/rotation. These effects can also be observed on cine CT and MRI scans during the treatment session itself, representing intrafraction geometrical changes [1] (see online supplementary video). Analysis of cine MRIs from William Beaumont Hospital shows that these

can be used to quantify different shifts for individual points of interest throughout the treatment volume [16].

What significance does intrafraction deformation/rotation have? At MDACCO, CT scans were obtained before and after prostate therapy in 46 patients [17]. The first was used for alignment, and dose distributions were calculated on both. The dose differences were evaluated for different anatomical regions. Treatment plans used only 3-mm margins – was this adequate for the prostate gland? They found that intrafraction deformation did not have significant impact on the final dosimetry of the prostate gland itself since nearly all patients had better than 90% coverage of the prostate on the second scan. However, when they evaluated the seminal vesicles for motion and coverage, they found a very different answer. Continued coverage of the seminal vesicles varied greatly between patients. In a few, more than 50% coverage of the region was lost during therapy. The impact of intrafraction deformation was clear.

Intrafraction Deformation/Rotation and Proton Therapy
Protons interact in tissue differently than photons, and recent reports focus on the effects of motion and deformation during treatment for proton delivery. At MDACCO, investigators compared proton and IMRT treatments and found no differences in this regard [18]. A group in Korea delivered proton doses in hypofractionated schedules [19]. They found that small target movements could significantly reduce target proton doses. From the University of Florida, investigations have shown that the dosimetric impact is small as long as the prostate motion is <5 mm [20]. These are important issues for prostate treatment using protons, and are more fully discussed in the chapters devoted to proton radiotherapy.

Conclusion: Deformation/Rotation and Prostate Dosimetry
In summary, intrafraction and interfraction rotations and deformations of the prostate are routinely detected. While it is easy to routinely image these differences, recontouring of the target volumes on each image is arduous and impractical for all patients. However, studies on limited numbers of patients show consistent results. For treatments using accurate daily localization, the dosimetric impact of such motion on the prostate itself is minimal, even when margins as small as 3 mm are used. This may not be true of other tissue volumes. Deformations of the seminal vesicles, rectum and bladder are more pronounced, and their dosimetric impact is highly variable between patients. In some patients the dosimetric differences are large. The clinical consequences of these differences are still unclear, since practical methods of contouring and evaluating the volumes are typically not available. Ultimately, regular volumetric imaging and determination of the actual doses delivered to the bladder and rectum may be quite important for some patients, and clearly needs further investigation.

Solutions to Motion and Deformation in Prostate Image-Guided Radiotherapy

Reduction in Motion

How can these dosimetric effects of motion be managed effectively? The simplest solution might be to reduce or eliminate the motion. Intrarectal balloons are used at some centers with this intent. It is not clear whether immobilization is always achieved, since MRI and CT cine studies continue to show some degree of motion despite placement of the balloon. Comparative evaluations of fiducial marker positions, with and without balloons, do not always show advantage in reducing marker movement [21].

It does appear that an intrarectal balloon can improve rectal dosimetry [22] and reduce rectal bleeding rates. However, the gland may not be the same shape or in the same position after each placement. Since the particular shape of the balloon-flattened gland may change with each placement, irregular deformation problems are introduced. Also, one may be required to increase the length of rectum that is exposed superiorly, since the balloon can push and distend the gland in this direction. This is especially true if an attempt is made to cover the seminal vesicles. At MDACCO, intrarectal balloons are used for patients entering trials using hypofractionated stereotactic body radiotherapy in 5 sessions. It is apparent that posterior rectal sparing is easily achieved, but there can be a greater length of anterior wall exposed.

To increase the separation of the prostate from the rectum, spacer materials are being developed that can be injected transperineally. One is hyaluronic acid, which persists for about 10–12 months. Other compounds are being investigated as well, and conceptually these could be used for other anatomical sites. Perhaps one could increase the separation of esophagus cancers from the spine, for instance. This is a new development and we await more information and clinical experience.

Adaptive Radiotherapy

To improve the dosimetric accuracy during a treatment course, radiotherapy dosimetry can be adapted to the anatomy imaged at the time of treatment. Traditionally, this has been performed at an arbitrary point in the treatment course by obtaining a new planning CT, representing a snap-shot of the anatomy during the treatment course (fig. 10). This is still commonly used for replanning lung cancer therapies, for instance. However, volumetric image guidance allows acquisition of anatomy data on a daily basis, which can be used for replanning in 2 ways: (a) replanning may be performed between treatment fractions *(off-*

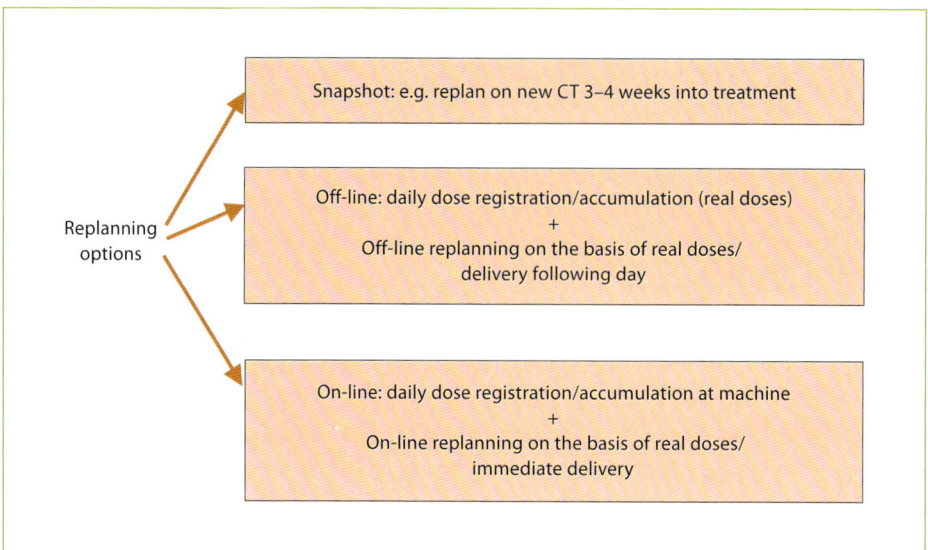

Fig. 10. Adaptive radiotherapy options. See online supplementary material.

line), or (b) integrated into an ongoing treatment fraction prior to delivery *(on-line)*. In both cases, new contours must be defined to produce target and normal tissue DVHs.

Both off-line and on-line methods can reduce systematic errors through dose recalculation during a treatment course. Only on-line adaptive radiotherapy will capture and adjust for daily, random anatomic variations. As yet, on-line approaches are difficult to implement since fast, efficient and reliable deformable registration methodologies are still needed for quick recontouring. While these are under development, a few centers have brought them into the clinic.

Daily Dose-Volume Histogram and Patient Positioning
Since the soft tissues can be imaged every day, contours can be obtained for specific tissues manually or with automated tools. Dose distributions can be recalculated, and one can obtain a 'DVH of the day' (fig. 11). In the simplest form of dose guidance, this could be used for daily repositioning of the patient. For instance, if the rectal doses were judged unacceptable, a shift could be modeled virtually to improve the rectal dose and a second DVH could be generated. If the new DVH is judged acceptable, the appropriate couch offsets could be applied. In this approach, a full reoptimization plan would not be performed, but rather the dose-volume information would be used only to adjust the patient position.

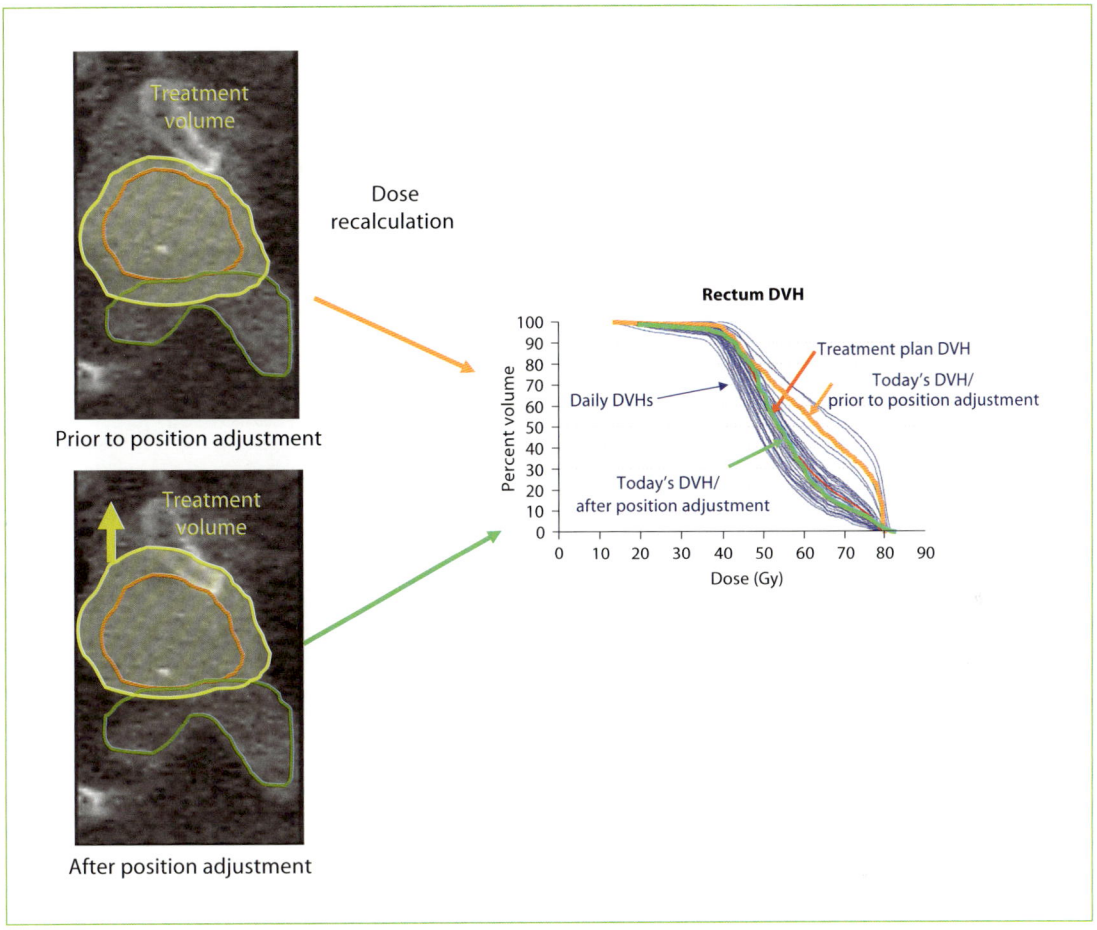

Fig. 11. Rectal dose DVHs. Daily DVHs can be obtained for areas of interest such as the rectum. If the DVH for that day appears atypical, patient repositioning could be performed and a repeat DVH obtained, as shown on the images to the left.

While such daily DVH information is undoubtedly useful, and might be used for daily couch adjustments, its importance for a series of treatments cannot become clinically meaningful until there are better 3D tools for accumulating the dosimetric information. The only way to process several dose distributions on different images is to develop reliable and useable deformable registration methods, the basis for rapid off-line and on-line adaptive guidance.

Deformable Registration
Anatomic changes recorded on daily CTs imply that structures are being deformed, and these changes can result in dosimetric changes. The process of translating

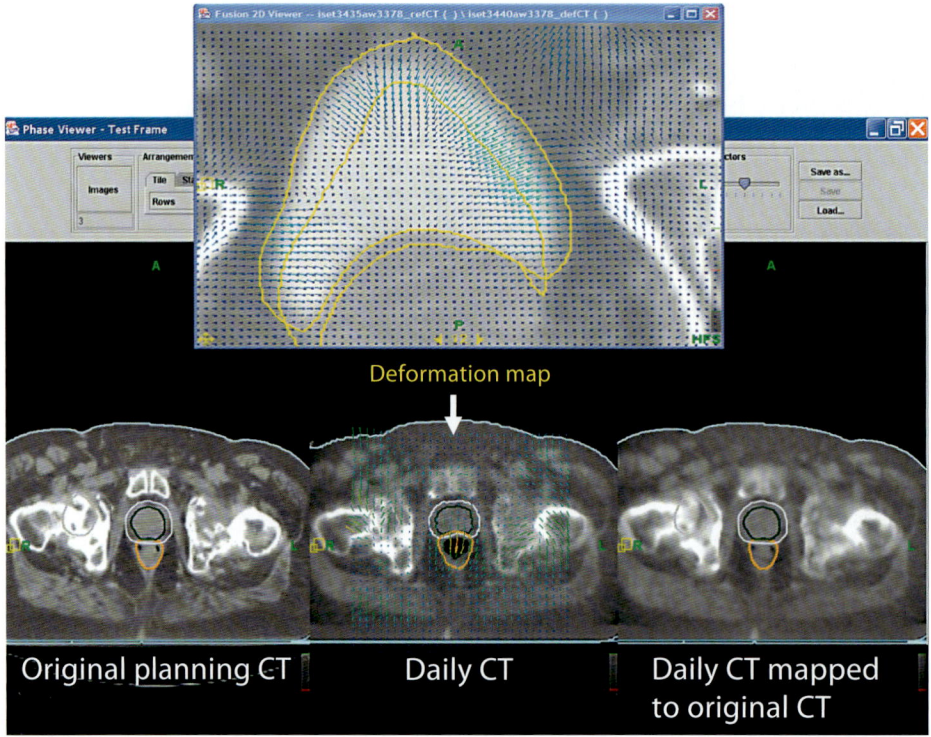

Fig. 12. Deformable registration. The corresponding anatomy of 2 scans is identified pixel by pixel and a correspondence map generated, enabling accumulation of dose on the scans.

these anatomic changes into a cumulative view coupled with a dosimetric pattern is called deformable dose registration. The general idea is as follows. Assume that the prostate appears different on 2 CT scans – the original CT and the daily treatment CT. With a deformation algorithm, the CT voxels of one scan can then be individually adjusted until they correspond to the other. This is called generating a deformation map. Once that map is generated, the doses delivered to those voxels can follow, creating an overall dose pattern (fig. 12). Rapid, automated deformable image registration methods are currently under development at several centers.

Early clinical applications of deformable registration have been tested in head and neck sites, as for evaluation of parotid volumes. In the pelvis, the large daily variations in the anatomical volumes, especially the rectum, make automated deformable registration procedures more difficult. A sophisticated approach, even accounting for daily changes in rectal filling, was reported from the University of North Carolina [1]. MDACCO has also reported a useable registration process [23]. These and other algorithms for automated deformation need to be validated clinically, confirmed to be anatomically correct, and confirmed as being efficient

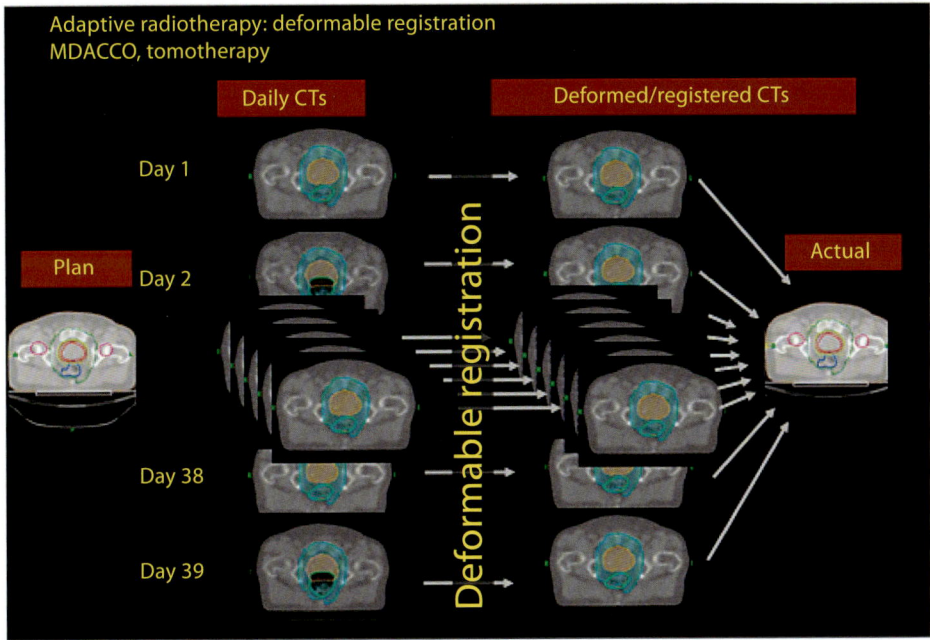

Fig. 13. Deformable registration can be used to cumulate actually delivered doses through a treatment course. This allows comparison of the actual dose delivered to the original plan dose.

and reliable [24]. It will be challenging to achieve a program fast enough to be used for the patient on the treatment table, and still achieve results equal to careful off-line planning.

Off-Line Adaptive Radiotherapy
The off-line approach requires accumulating all of the actually delivered doses, and adjusting the plan accordingly for a modified delivery for later therapy fractions. This might be organized as follows (fig. 13). Daily DVHs are acquired from daily imaging. Using deformable registration, the cumulative DVHs are determined and compared with the intended DVH from the planning CT. A new plan is then modeled that adjusts for the difference between the two. This could be performed on a daily basis, but might be effective at a given point during the therapy course. It might be performed approximately halfway through treatment, as in the example given in figure 14. After 26 fractions, the cumulative rectal DVH showed that this organ was receiving a higher dose than planned. On the basis of all the actual doses from those prior 26 fractions, a second plan could be generated, adjusting dose distributions to result in a final dose profile that closely matches the originally planned distribution.

One version of off-line adaptive therapy has been investigated in a prospective phase II trial from 1997 to 2003 at William Beaumont Hospital [25–28]. 512 patients with tumors T1–3 and with PSA ≤30 were enrolled. Larger, generic margins (1 cm) were used for the initial fractions. Daily CTs were obtained during the first 5–6 fractions, essentially obtaining an internal target volume for the targets over time. The median prescribed dose was 75.6 Gy (70.2–79.2). The reported clinical outcomes were excellent. GI and GU chronic toxicities, seen in about 15% of cases, were mostly grade 2, and high biochemical control rates were obtained for all risk groups. Similar offline adaptive processes have been undertaken at other institutions. In one, daily cone-beam CTs were obtained for 6 initial sessions using broad 1-cm margins [29]. From these, average CTVs and rectal volumes were obtained for the definitive planning of the remaining course with smaller target margins.

A more sophisticated off-line adaptive approach has recently been described in a preliminary report [30]. In a hypofractionated program (60 Gy given in 20 fractions), this group calculated the doses actually delivered to points during the initial 15 fractions. The final 5 fractions were replanned based on the actually delivered doses, with the dose-volume constraints determined by subtracting the already delivered doses from the original plan. As an example, the mean rectal volume at 60 Gy was higher than expected by 47% in one patient, and the dose to the highest dose 10 cm^3 of rectum was greater by 24%. They could adjust the later dosimetry, accounting for these unanticipated normal tissue doses. This shows a general approach, a way of accumulating the daily doses and then at one point intervening, specifically by adjusting for the dose accumulation.

On-Line Adaptive Radiotherapy
On-line adaptive radiotherapy offers the advantage of immediately adjusting the dosimetry to the imaged anatomy and is a work-in-progress awaiting practical deformable registration methods. Several approaches are now being investigated in the clinic, as discussed above.

Fig. 14. Off-line adaptive replanning based on cumulative DVHs for the rectum. **a** Daily CTs are obtained, registered using deformation programs, and the actual dose delivered up to that point in the therapy course determined. **b** The remainder of the course can be replanned to adjust for the actual delivered doses, bringing the final actual dose to the originally planned dose.

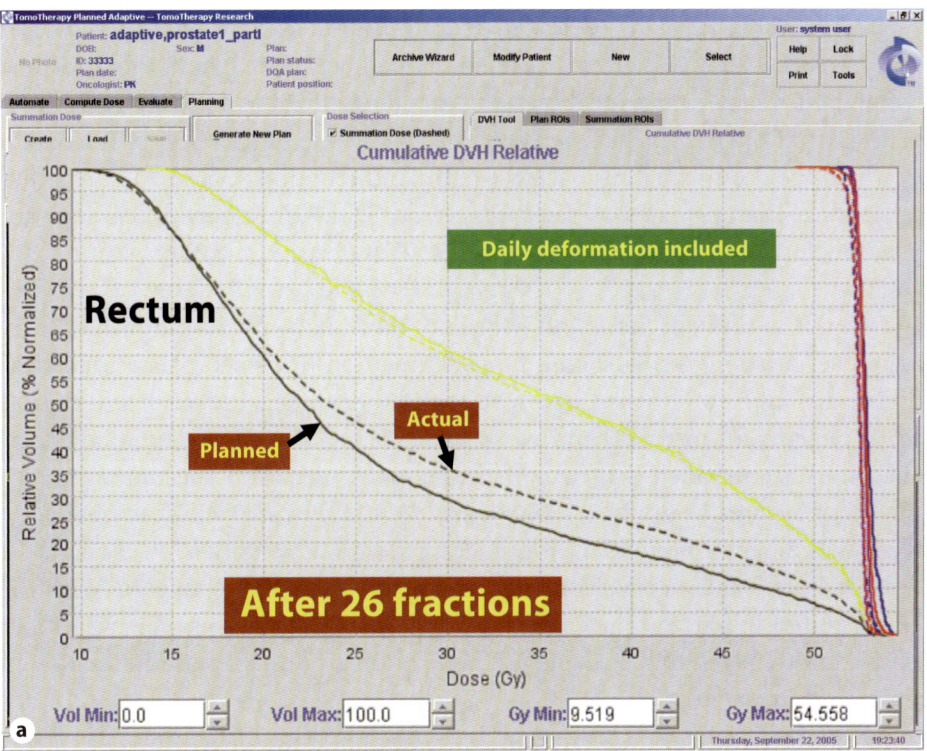

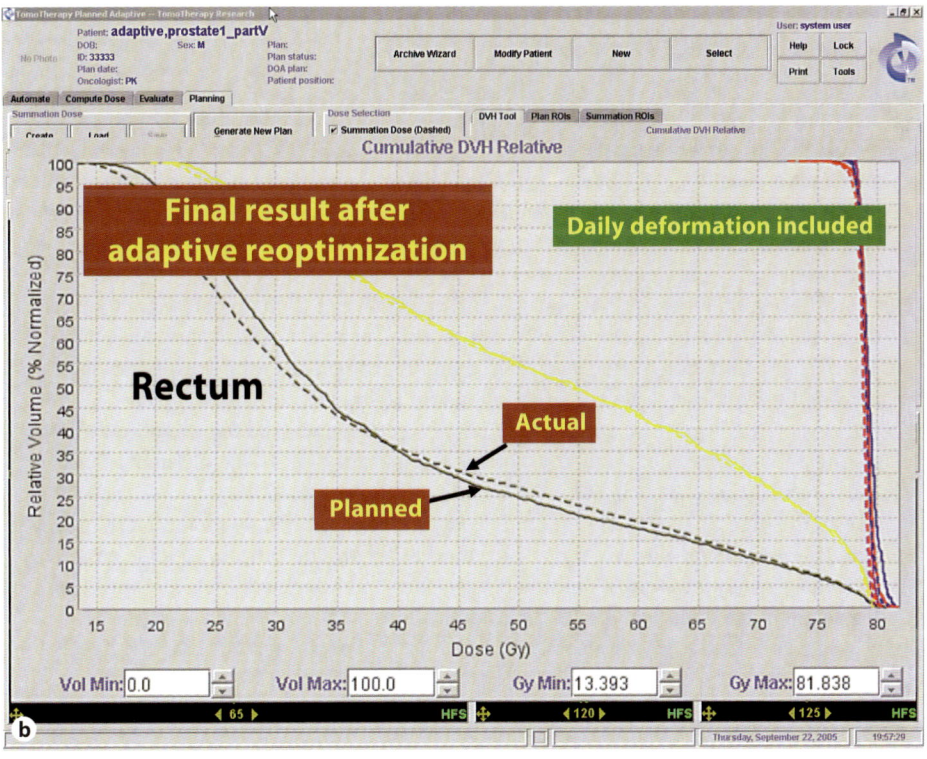

Other approaches using on-line plan optimization are now being designed or implemented. The UCSF group has reported methods to compensate for prostate/lymph node relative movement by selecting on a daily basis from a series of pre-plans [31]. Also, a process has been described by the Medical College of Wisconsin in which they can recontour within 10 min and reoptimize doses [32]. Another process has been reported by researchers from Duke University, where they applied a deformable registration process that allowed reoptimization within 2 min [33].

Due to the random nature of deformation of many areas within the pelvis, adaptive radiation ideally should be performed on-line for prostate cancer radiotherapy. If the on-line issues can be solved for this site, the result will be dramatic benefits for many other disease sites. We are still dependent on deformable registration, which is difficult in the pelvis. As these tools become available, they may become as useful as IGRT itself. While they may clinically benefit a limited number of individuals, it is uncertain who those patients will be until a method is available to efficiently and routinely assess all treatments. With the expansion of hypofractionated therapy approaches, it will become even more important to have rapid, reliable methods of on-line adaptive radiotherapy in the near future.

Guidelines for Clinical Practice

- The random nature of prostate motion requires daily localization if treatment is to be delivered with small margins.
- Interfraction motion can have a significant impact on prostate gland dosimetry, and even more of an impact for the seminal vesicles and possibly intraprostatic tumor areas. The dosimetric impact on normal structures (bladder/rectum) is less clear, and there are significant individual variations in these.
- Interfraction and intrafraction rotations and deformations of the prostate are routinely detected. The dosimetric impact of these motions of the prostate gland are minimal when daily localization is used, even with margins as small as 3 mm. However, deformations of the seminal vesicles, rectum and bladder are much more pronounced.
- The dosimetric impact of deformation of the rectum and bladder is highly variable among patients, and the clinical consequences remain unclear. Daily volumetric imaging and dosimetry may become quite important for these volumes.
- Due to the random nature of motion/deformation during prostate radiotherapy, adaptive radiotherapy ideally would be performed as an on-line process. On-line adaptive radiotherapy requires robust deformable registration and replanning programs. These are beginning to emerge in useful clinic applications.

References

1 Foskey M, Davis B, Goyal L, et al: Large deformation image registration in image-guided radiation therapy. Phys Med Biol 2005;50:5869.
2 de Crevoisier R, Tucker SL, Dong L, et al: Increased risk of biochemical and local failure in patients with distended rectum on the planning CT for prostate cancer radiotherapy. Int J Radiat Oncol Biol Phys 2005;62:965–973.
3 Heemsbergen WD, Hoogeman MS, Witte MG, et al: Increased risk of biochemical and clinical failure for prostate patients with a large rectum at radiotherapy planning: results from the Dutch trial of 68 versus 78 Gy. Int J Radiat Oncol Biol Phys 2007;67:1418–1424.
4 Kupelian PA, Willoughby TR, Reddy CA, et al: Impact of image guidance on outcomes after external beam radiotherapy for localized prostate cancer. Int J Rad Oncol Biol Phys 2008;70:1146–1150.
5 Kupelian PA, Lee C, Langen KM, Zeidan OA, Mañon RR, Willoughby TR, Meeks SL: Evaluation of image-guidance strategies in the treatment of localized prostate cancer. Int J Radiat Oncol Biol Phys 2008;70:1151–1157.
6 van Haaren PMA, Bel A, Hofman P, et al: Influence of daily setup measurements and corrections on the estimated delivered dose during IMRT treatment of prostate cancer patients. Radiother Oncol 2009;90:291–298.
7 Langen KM, Willoughby TR, Meeks SL, et al: Observations on real-time prostate gland motion using electromagnetic tracking. Int J Radiat Oncol Biol Phys 2008;1084–1090.
8 Langen KM, Shah AP, Willoughby TR, Meeks SL, Kupelian PA: Real-time position adjustment during treatment for localized prostate cancers: observations on clinical use and acute toxicity. Int J Radiat Oncol Biol Phys 2009;75:S294.
9 Litzenberg DA, Balter JM, Hadley SW, et al: Influence of intrafraction motion on margins for prostate radiotherapy. Int J Rad Oncol Biol Phys 2006:65:548–553.
10 Langen KM, Lu W, Willoughby TR, et al: Dosimetric effect of prostate motion during helical tomotherapy. Int J Radiat Oncol Biol Phys 2009;74:1134–1142.
11 Langen KM, Lu W, Willoughby TR, et al: Correlation between dosimetric effect and intrafraction motion during prostate treatments delivered with helical tomotherapy. Phys Med Biol 2008;53:7073.
12 Pierburg BA, Parikh PJ, Roy M, et al: Dosimetric consequences of intrafraction prostate motion on patients enrolled on multi-institutional hypofractionation study. Int J Radiat Oncol Biol Phys 2007;69:S23–S24.
13 Li HS, Chetty IJ, Enke CA, et al: Dosimetric consequences of intrafraction prostate motion. Int J Radiat Oncol Biol Phys 2008;71:801–812.
14 Frank SJ, Dong L, Kudchadker RJ, et al: Quantification of prostate and seminal vesicle interfraction variation during IMRT. Int J Radiat Oncol Biol Phys 2008;71:813–820.
15 Kupelian PA, Langen KM, Zeidan OA, et al: Daily variations in delivered doses in patients treated with radiotherapy for localized prostate cancer. Int J Radiat Oncol Biol Phys 2006;66:876–882.
16 Ghilezan MJ, Jaffray DA, Siewerdsen JH, et al: Prostate gland motion assessed with cine-magnetic resonance imaging (cine-MRI). Int J Radiat Oncol Biol Phys 2006;62:406–417.
17 Melancon AD, O'Daniel JC, Zhang L, et al: Is a 3-mm intrafractional margin sufficient for daily image-guided intensity modulated radiation therapy of prostate cancer? Radiother Oncol 2007;85:251–259.
18 Zhang X, Dong L, Lee AK, et al: Effect of anatomic motion on proton therapy dose distributions in prostate cancer treatment. Int J Radiat Oncol Biol Phys 2007;67:620–629.
19 Yoon M, Kim D, Shin DH, et al: Inter-and intrafractional movement-induced dose reduction of prostate target volume in proton beam treatment. Int J Radiat Oncol Biol Phys 2008:71:1091–1102.
20 Vargas C, Wagner M, Mahajan C, et al: Proton therapy coverage for prostate cancer treatment. Int J Radiat Oncol Biol Phys 2008:70:1492–1501.
21 van Lin EN, van der Vight LP, Witjes JA, et al.: The effect of an endorectal balloon and off-line correction on the interfraction systematic and random prostate variations. Int J Radial Oncol Biol Phys 2005;61:278–288.
22 Patel RR, Orton N, Tome WA, Chappell R, Ritter MA: Rectal dose sparing with a balloon catheter and ultrasound localization in conformal radiation therapy for prostate cancer. Radiother Oncol 2003;67:285–294.
23 Wang H, Dong L, Lii MF, et al: Implementation and validation of a three-dimensional deformable registration algorithm for targeted prostate cancer radiotherapy. Int J Radiat Oncol Biol Phys 2005;61:725–735.
24 Wang H, Garden AS, Zhang L, et al: Performance evaluation of automatic anatomy segmentation algorithm on repeat or four-dimensional computed tomography images using deformable image registration method. Int J Radiat Oncol Biol Phys 2008;72:210–219.

25 Brabbins D, Martinez A, Yan D, et al: A dose-escalation trial with the adaptive radiotherapy process as a delivery system in localized prostate cancer. Int J Radiat Oncol Biol Phys 2005:61:400–408.
26 Martinez AA, Yan D, Lockman D, et al: Improvement in dose escalation using the process of adaptive radiotherapy combined with three-dimensional conformal or intensity-modulated beams for prostate cancer. Int J Radiat Oncol Biol Phys 2001;50:1226–1234.
27 Harsolia A, Vargas C, Yan D, et al: Predictors for chronic urinary toxicity after the treatment of prostate cancer with adaptive three-dimensional conformal radiotherapy. Int J Radiat Oncol Biol Phys 2007;69:1100–1109.
28 Vargas C, Yan D, Kestin LL, et al: Phase II dose escalation study of image-guided adaptive radiotherapy for prostate cancer. Int J Radiat Oncol Biol Phys 2005;63:141–149.
29 Nikamp J, Pos FJ, Nuver TT, et al: Adaptive radiotherapy for prostate cancer using kilovoltage cone-beam computed tomography: first clinical results. Int J Radiat Oncol Biol Phys 2008;70:75–82.
30 Jacob R, Wu X, Dobelbower CM, et al: Daily variations in rectal and bladder doses with image guided prostate radiotherapy and the feasibility of adaptive radiotherapy planning. Int J Radiat Oncol Biol Phys 2008;72:S348.
31 Ludlum E, Mu G, Weinberg V, et al: An algorithm for shifting MLC shapes to adjust for daily prostate movement during concurrent treatment with pelvic lymph nodes. Med Phys 2007;34:4750–4756.
32 Ahunbay EE, Peng C, Chen G-P, et al: An on-line replanning scheme for interfractional variations. Med Phys 2008;35:3607–3615.
33 Wu JQ, Thongphiew D, Wang X, et al: On-line re-optimization of prostate IMRT plans for adaptive radiation therapy. Phys Med Biol 2008;53:673–691.

Patrick Kupelian, MD
Department of Radiation Oncology, University of California School of Medicine
200 UCLA Medical Plaza, Suite B265
Los Angeles, CA 90095-6951 (USA)
Tel. +1 310 825 9775, Fax +1 310 794 9795, E-Mail pkupelian@mednet.ucla.edu

III. SBRT: Stereotactic Body Radiotherapy

The Expanding Roles of Stereotactic Body Radiation Therapy and Oligofractionation: Toward a New Practice of Radiotherapy

Brian D. Kavanagh[a] · Robert Timmerman[b] · John L. Meyer[c]

[a]Department of Radiation Oncology, University of Colorado School of Medicine, Anschutz Medical Campus, Aurora, Colo., [b]Department of Radiation Oncology, University of Texas Southwestern Medical Center, Dallas, Tex., and [c]Department of Radiation Oncology, Saint Francis Memorial Hospital, San Francisco, Calif., USA

Abstract

The range of clinical applications for stereotactic body radiation therapy (SBRT) continues to expand based on clinical outcomes data from prospective trials and carefully analyzed institutional experiences. As a result of this strong scientific foundation, there has been burgeoning implementation of SBRT and other forms of hypofractionated radiation therapy in the practice of radiation oncology worldwide. In spite of the clinical successes achieved thus far – or, perhaps, because of them – fundamental questions about SBRT remain and have come into greater focus. Where and when is SBRT optimally integrated into the range of evolving modern multidisciplinary cancer treatment programs? What scientific insights (biological, technical and medical) might lead to further improvements in the efficacy of SBRT? What efficiencies are needed to achieve greater availability of SBRT? These and many other questions, fueled by the clinical accomplishments of SBRT to date, provide compelling directions for further exploration in scientific and clinical studies and further contributes to discoveries already transforming the practice of radiation oncology.

Copyright © 2011 S. Karger AG, Basel

Stereotactic body radiation therapy (SBRT) epitomizes the trend toward hypofractionation, a fundamental shift in the clinical approach to radiotherapy for the treatment of cancer. The 5-or-fewer-treatment regimen of SBRT is the most densely compressed example of extracranial hypofractionated radiotherapy, a spectrum of treatment schedules with increasing validation in clinical trials, including the shortened schedules of treatment now proven appropriate for many patients with early-stage breast cancer [1]. Whereas SBRT is used exclusively in the realm of definitive rather than postoperative radiotherapy, the efficiency and potency of SBRT open new opportunities to broaden the role that radiotherapy plays in cancer management. The ultra-high-dose-per-treatment regimens of SBRT also offer new insights into the radiobiology of tumor and normal tissue responses. The accumulating observations may challenge traditional concepts of radiation effects, but radiation oncologists and radiobiologists should be alert to the new and possibly unexpected opportunities that these findings may provide.

The Radiobiology of Stereotactic Body Radiation Therapy

Traditional radiobiological understanding of cancer treatment largely surrounds the differences in the damaging effects occurring from chromosomal changes within tumor and normal cells, in both cases resulting from the relatively homogeneous dose exposures of conventional radiotherapy. By comparison, the large dose-per-fraction irradiation associated with SBRT would be expected to cause dramatic DNA damage within any tissue exposed to such a prescription dose. DNA damage to tumor cells is likely to be a major mechanism by which SBRT exerts tumor control and causes toxicity risks in normal tissue. As such, much greater care is exercised in the conduct of SBRT to geometrically partition the dose levels encountered by tumor and normal tissues. Furthermore, SBRT dose distributions are characteristically heterogeneous, showing large variations of delivered dose between tumor, immediately adjacent normal tissues and more separated normal tissues. This variability of dose exposure further complicates comparisons between SBRT and conventionally fractionated radiotherapy.

Is the radiobiology of SBRT simply a high-dose extrapolation of well-understood principals derived from decades of conventional radiation dose delivery? Alternatively, are there dose-threshold effects, i.e. effects that do not occur below a specified dose but may occur beyond the dose levels traditionally used in radiotherapy clinics? Evidence is growing that SBRT-related tissue effects follow unique mechanisms not previously appreciated with conventional small daily dose delivery. While DNA damage is still likely to be the predominant mode of radiation effect, alternate pathways of injury may be associated with treatment programs using large doses per fraction given for few sessions. These highly compressed dose schedules, called oligofractionation, may provide important opportunities for both improving tumor control and avoiding toxicity.

High-Dose Radiation and Endothelial Apoptosis

Radiation therapy induces tumor cell death directly through a variety of mechanisms [2]. Among the pathways to tumor cell clonogenic death, radiation-induced apoptosis is generally assumed to be of less clinical importance for solid tumors than for hematopoietic malignancies. However, a growing body of preclinical and clinical evidence suggests that radiation-induced apoptosis may play an important role in the success of SBRT in controlling solid tumors. The mechanism, though, is very possibly of an indirect nature; it may result from ablation of the tumor microcirculation rather than an effect on tumor cells themselves.

Nearly a century ago, the eminent physician-scientist and cancer expert extraordinaire, James Ewing, proposed that for certain radioresistant tumors, the predominant mode of action of radiotherapy appeared to be devascularization, whereby capillary cells were killed, thus choking off the supply of oxygen and nutrients to the tumor itself [3]. Microvascular effects are now recognized, and in this era of bevacizumab and other vessel-targeted interventions, the argument could be made that ionizing radiation was the original antiangiogenic agent. In reality, though, relatively little attention was paid to this notion throughout the rest of the 20th century.

Leading the current resurgence in interest in the devascularizing effects of radiotherapy is the group from the Memorial Sloan-Kettering Cancer Center. In 2003, they reported studies using a xenograft model. In mice deficient in enzymes that enable apoptosis, tumors were significantly more radioresistant than the same tumors grown in apoptosis-competent mice [4]. The underlying reason for the observed differences in growth delay in the melanoma and fibrosarcoma tumors appeared to be a large difference in radiation-induced apoptosis in endothelial cells. Interestingly, there was a suggestion of a dose threshold in the observations reported: below a single exposure dose of 11 Gy, the percent of apoptotic endothelial cells was relatively small, on the order of only a few percent or less. With doses of 11 Gy or higher, there was a marked increase in the percent of apoptotic endothelial cells (from roughly 20% to more than 60%). In later clinical work involving the use of single fraction high-dose spinal SBRT, investigators from the same institution illustrated their analysis with an image of the strongly pronecrotic response seen after doses in the range of 18–24 Gy, a radiographic change consistent with a devascularizing effect [5].

New Pathways of Repopulation

Every student of radiobiology learns the traditional 4 *R*s describing the response of tumor cells and normal tissues to radiation therapy: *R*epair of injury, *R*edistribution into other phases of the cell cycle, *R*eoxygenation of hypoxic regions within the tumor and *R*epopulation of tumor cells following initial cytotoxic effect. It is at first intuitive to consider that with SBRT, given the extremely brief schedules of treatment, there is no need for concern about the 4th *R*, repopulation, at least within the tumor target. However, considering the aforementioned proapoptotic pathways of special interest with SBRT, certain new observations about the secondary effects of cellular apoptosis must now be considered.

Investigators at the University of Colorado recently described the so-called 'phoenix rising' pathway of tissue regeneration that is triggered specifically by apoptosis [6]. This group demonstrated that mice lacking caspase 3 or caspase 7, key enzymes that regulate apoptosis, are unable to mount a robust normal tissue regenerative response to cytotoxic injury. Arachidonic acid is an essential molecule in the signaling pathway. Radiation was shown to induce arachidonic acid and consequently upregulate prostaglandin E2, which directly stimulates stem cell proliferation. The direct significance of this pathway in response to SBRT is not yet known, but future studies to consider its relevance are warranted given the expectation that SBRT likely induces more apoptosis in tumor and normal tissues than conventionally fractionated radiotherapy.

Deconstructing and Reconstructing the Radiation Cell Survival Curve

Lea and Catcheside's [7] original application of a linear-quadratic (LQ) formalism to describe the relationship between radiation dose and incidence of chromosomal translocations is generally well remembered, and has served as the basis for modeling radiation dose effects ever since. Less well remembered from the same publication, though, is their description of a condition that would apply to high single-dose exposure. Based on their observations and understanding, there would need to be a correction applied to the LQ model to account for a decaying effect related to the prolonged time required to deliver high doses, presumably related to a component of single strand break recombination without translocation. In essence, the result would be a straightening of the dose-log response curve at high doses that would depart significantly from the constantly curving βD^2 predictions.

LQ modeling has worked quite well for decades without any sort of high-dose correction factors, presumably because there is little departure from simple LQ estimates in the conventional dose range of roughly 2 Gy per fraction so commonly used in radiotherapy. The advent of SBRT-style doses per fraction, however, has prompted numerous groups to take a closer look at the high-dose range of the curve to consider one or another form of mathematical correction that would account for the apparent log-linear appearance of the radiation dose-survival curve at doses upwards of 8–10 Gy.

One of the first attempts was that of Guerrero and Li [8], who proposed a modified LQ model based largely on the lethal and potentially lethal model of Curtis [9]. They proposed a new term, δ, to characterize the repair kinetics at high doses, though the expectation of a measurable dose-rate effect remained. More recently, Park et al. [10] at the University of Texas-Southwestern have offered a 'universal survival curve', which hybridizes the LQ model using a multitarget model that applies in the range of doses above a transition dose, D_T. A conceptually similar contribution to the literature was published around the same time by Astrahan [11].

In summary, emerging evidence suggests that distinct mechanisms of cell death, likely including a major contribution from vascular epithelial cell apoptosis, might play a major role in the success of SBRT in controlling tumors. However, studies geared toward maximizing this effect should also be designed with the cautionary awareness that tissue regenerative processes might also be triggered as a result. It is not yet clear whether there would be selective regeneration only in normal tissues as opposed to tumor cells. Finally, the net effects of these high-dose-induced pathways and counter-pathways might not be reliably predicted by the mathematical models that have proven useful for conventional low-dose per fraction effects.

Stereotactic Body Radiation Therapy Applications

Growth Trends in Stereotactic Body Radiation Therapy Publications and Hypofractionated Treatment Regimens

The number of technical and clinical reports published annually on the use of SBRT has more than quadrupled over the past 5 years. While national data on utilization are not readily available, the percent of patients treated with SBRT in some radiation oncology departments that provide comprehensive radiation treatment services is now on the order of 12–15%, equaling or exceeding the percent of patients treated with cranial radiosurgery.

SBRT represents just one form of hypofractionated radiotherapy. If one considers the number of radiotherapy cases treated with 10 or fewer fractions with the goal of curative treatment, then considerable changes in our patterns of practice are rapidly developing. It is estimated that more than 25,000 patients with prostate cancer might qualify for hypofractionated treatment each year in the USA. A similar number of early stage lung cancer cases could potentially qualify for SBRT treatment programs, and the range of applications to other sites continues to expand as well. For example, postlumpectomy accelerated partial breast irradiation is now acknowledged to be a reasonable option for many patients with early stage

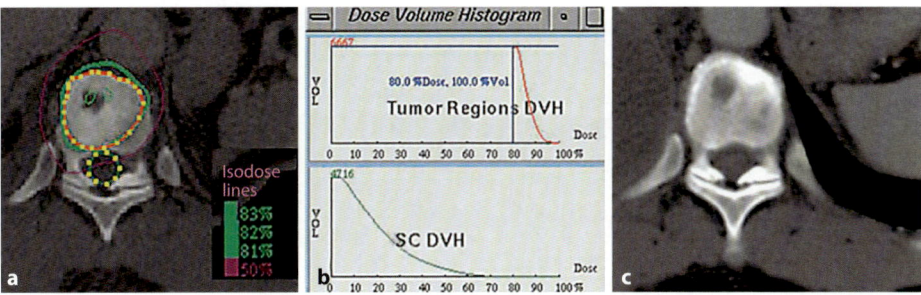

Fig. 1. a Axial slice from the radiosurgery treatment plan with superimposed prescription isodose line (green) and 50% isodose line (purple) showing the rapid falloff of dose. **b** The dose volume histograms (DVH) for the target and spinal cord (SC). **c** A posttreatment axial CT slice at 15 months shows the stability of the bone. Case example courtesy of Iris Gibbs MD, Stanford University.

breast cancer, and achieves a high rate of breast conservation [12]. This form of hypofractionated radiotherapy might ultimately prove to be the single most common clinical application, with potential relevance for upwards of 50,000 patients in the USA alone.

Examples of Stereotactic Body Radiation Therapy Applications: 2 Case Studies

The accuracy of SBRT in approaching critical structures is its recognized advantage. The case example shown in figure 1 illustrates the very high-dose gradients that can be achieved, in this case protecting the spinal cord while treating the vertebral body. The palliative application of treatment to visceral sites is described in case 2 (fig. 2), illustrating the focal high-dose targeting of hepatic lesions with SBRT.

Case Study 1
Spinal Stereotactic Body Radiation Therapy
Patient. A 55-year-old man with metastatic lung cancer, previously treated with neoadjuvant chemotherapy followed by lobectomy, presented with a persistent, isolated T11 vertebral body spinal metastasis causing pain.

Treatment. 20 Gy was prescribed to the 82% isodose surface in 2 fractions (10 Gy per fraction) to the T11 vertebral body lesion of a volume of 6.67 cm^3 (modified conformity index = 1.53; maximum dose = 24.4 Gy; maximum point dose to the spinal cord = 17.9 Gy in 2 fractions). The volume of the spinal cord receiving a biologically equivalent dose of 10 Gy in a single fraction was <0.004 cm^3.

Result. The patient was alive at 15 months after treatment with complete back pain relief, though he had other areas of systemic metastases (fig. 1).

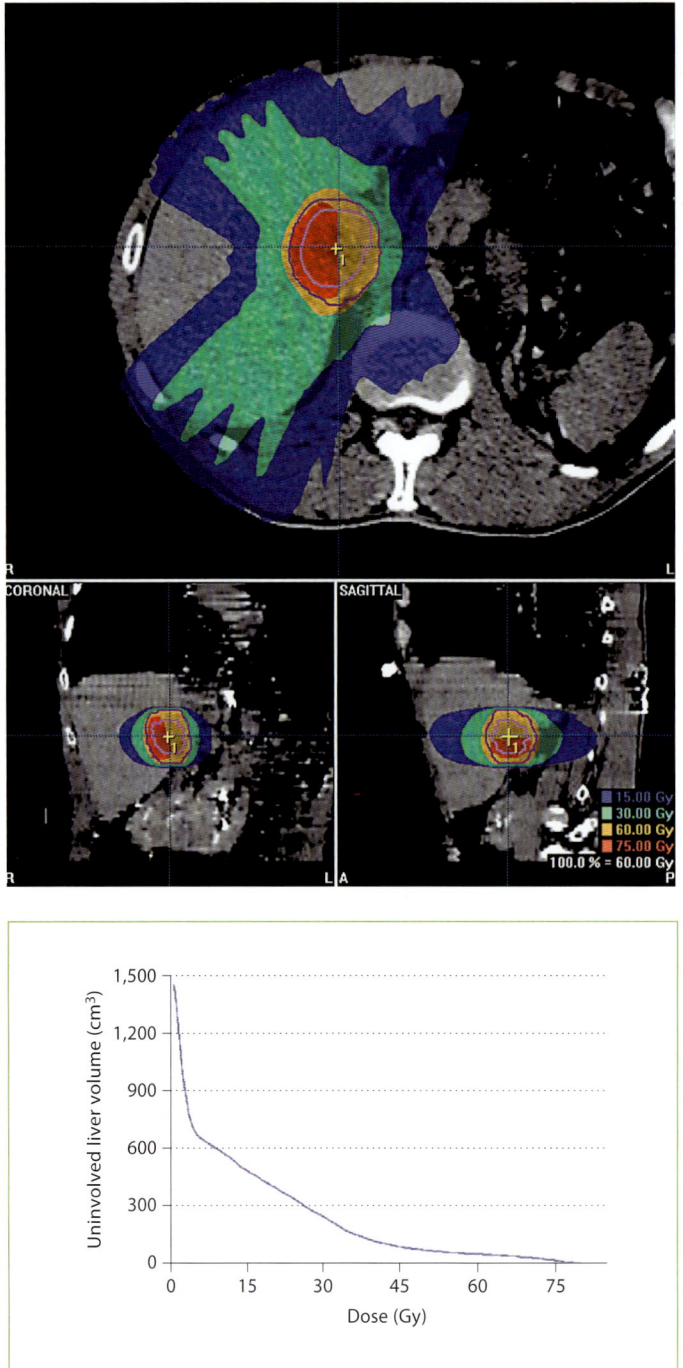

Fig. 2. The figure includes axial, coronal and sagittal views of the isodose distribution superimposed on CT images. Also, the dose-volume histogram of uninvolved liver is shown.

Case Study 2
Hepatic Stereotactic Body Radiation Therapy
Patient. A 51-year-old man who had undergone neoadjuvant chemoradiation 2 years previously and subsequent esophagectomy for a clinical T3N1M0 adenocarcinoma of the esophagus later developed liver metastases and achieved a good response from chemotherapy. However, the patient then presented with clinical and radiographic evidence of a solitary 2.5-cm brain metastasis and a 3-cm liver metastasis.
Treatment. The patient elected stereotactic radiosurgery for the brain metastasis, deferral of any whole-brain radiotherapy and enrollment on a phase II study of SBRT for the solitary liver metastasis. A dose of 60 Gy was given in 3 fractions using dynamic conformal arcs.
Result. From a pre-SBRT level of 450 ng/ml, the carcinoembryonic antigen decreased to 273 ng/ml 1 month following SBRT, indicating a favorable early response. No acute SBRT-related complications were observed. The patient will continue periodic surveillance (fig. 2).

Stereotactic Body Radiation Therapy Applications – On Protocol or Off Protocol? The Example of Prostate Stereotactic Body Radiation Therapy

Many applications of SBRT have become accepted into clinical practice, while other applications remain appropriate for continued study in clinical trials. For the treatment of the medically inoperable patient with early stage lung cancer, long-term data confirm that the control rates are convincingly higher (from 80% to more than 90%) than with conventionally fractionated therapy (control rates about 30%), and complication rates are acceptably low. These results are presented and discussed in the chapter by Timmerman et al. (pp. 395–411). Likewise, the application of SBRT as an ablative therapy for oligometastatic disease in the liver, lung and spine has become a well-established indication.

In other areas, such as SBRT for prostate cancer, a more cautious approach seems warranted. Many of these concerns exemplify types of issues that are important to introducing SBRT as definitive therapy for other primary tumors. For prostate cancer, we await long-term data on rates of both tumor control and toxicity. While early results with PSA response are encouraging, PSA results may not always reflect later metastasis-free survival. Late complication rates remain unknown, since some complications may occur only after several years, especially fibrotic reactions such as urethral stricture and related incontinence. Radiation cystitis may occur several years after therapy. Comparisons of complication rates from SBRT with conventional treatment will only be reliable after several additional years of accumulated follow-up.

While the prescription doses delivered to the prostate with SBRT are similar to those delivered with high-dose rate (HDR) brachytherapy, it remains uncertain whether the outcomes will be the same. The methods of delivery, dose gradients and instantaneous dose rates differ in potentially meaningful ways. HDR necessarily involves very heterogeneous exposure of intratumoral regions to doses much higher than those prescribed to the margins. HDR may also permit focal dose modifications, as for urethral sparing. As yet, it is unknown whether this sparing can be accomplished adequately using SBRT technologies.

There are other concerns with curative approaches using SBRT for prostate cancer. It is unknown whether the current SBRT programs use the optimal fraction sizes or total doses. Estimations of the doses required have been partly based (or justified) on linear quadratic formula analyses, though these results are not universally accepted for SBRT applications. This model was created to predict responses at lower doses per fraction, and it appears to be less accurate with higher SBRT fraction sizes. Also, the α/β estimations used in this modeling for prostate irradiation are difficult to confirm clinically. Before general acceptance, SBRT for prostate cancer will benefit from further analysis of empiric normal and cancer tissue responses through sequential prospective clinical evaluations.

Future Directions for Stereotactic Body Radiation Therapy

A New Gold Standard: Setting the Bar for Future Protocols in Stereotactic Body Radiation Therapy

The unprecedented safety and efficacy of SBRT for medically inoperable early stage non-small cell lung cancer (NSCLC) is epitomized by the results of the Radiation Therapy Oncology Group (RTOG) 0236 protocol [13]. This protocol has not only established SBRT as the preferred mode of treatment for patients in that category, but has also set a lofty goal for investigations of SBRT in other settings. While it might not be possible to replicate the nearly 100% durable primary tumor control achieved in RTOG 0236 in other settings, the study will nevertheless always serve as a paradigm for future investigations, insofar as it was a rigorously conducted multi-institutional trial with careful attention to detail and tight patient follow-up.

Cancer clinicians from around the world are investigating the potential expansion of the role of SBRT in lung cancer, particularly in operable patients. The results of the recently completed RTOG 0618 protocol in medically operable patients with early stage NSCLC are eagerly awaited. Augmenting the RTOG 0618 data will be the findings of 2 studies. The Radiosurgery or Surgery for Operable Early Stage Non-Small Cell Lung Cancer (ROSEL) Study, underway in the Netherlands, randomizes patients with operable stage IA NSCLC between SBRT and surgery. The American College of Surgeons Oncology Group (ACOSOG) and the RTOG are collaborating on their most ambitious prospective trial using SBRT to date. The 2 groups will offer patients ACOSOG 4099/RTOG 1021 in a phase III trial in high-risk operable patients comparing sublobar resection to SBRT. This population is technically operable, but have limited tolerance to the standard lobectomy surgical resection. This pivotal work is discussed further in the chapter by Timmerman et al. (pp. 395–411).

In a separate study, the RTOG is comparing the pain-relieving efficacy of 16-Gy single fraction spinal SBRT, also known as spine stereotactic radiosurgery, with 8-Gy conventional single fraction external beam radiotherapy (RTOG 0631).

Stereotactic Body Radiation Therapy and Oligometastases

Alongside the development of SBRT, there have been numerous noteworthy advances in other disciplines in clinical oncology. Surgical techniques have improved and become generally less invasive in many instances, and traditional multiagent chemotherapy regimens have been refined and combined with molecular agents targeting tumor vasculature (e.g. bevacizumab) or key growth factor-mediated pathways (e.g. erlotinib and cetuximab). When the best combinations of local and systemic therapies are applied for patients with localized disease, a high percentage enjoy lengthy disease-free survivorship. Unfortunately, many patients will still eventually manifest distant disease recurrence. However, with modern imaging techniques these recurrences are often detected at a point where the overall disease burden is still very limited. The ultimate question for SBRT is whether it can play a meaningful role in these patients with so-called oligometastatic disease, especially when they maintain a good performance status overall.

Hellman and Weichselbaum [14] first coined the term *oligometastases* in the mid-1990s. In the next few years, the University of Rochester group initiated a series of clinical trials specifically designed to consider whether it might be possible to achieve long-term survivorship in patients with oligometastic disease. One of the reports from this work was an analysis of 40 patients with breast cancer treated with hypofractionated radiotherapy with curative intent [15]. Most had multiple [2–5] metastatic lesions, but good performance status. Chemotherapy and/or hormonal therapies were given to most patients before and/or after treatment. Remarkably, the 4-year overall survival rate was 59%, with rates of progression-free survival of 38% and lesion local control of 89%. Although the systemic agents used likely contributed to the success in the Rochester experience, one may compare a recent large multi-institutional study of first-line docetaxel with or without bevacizumab (without consolidative SBRT) that showed a median survival in all groups of only 30–32 months [16].

A strategy that appears attractive for oligometastatic NSCLC is to act upon the expected pattern of failure after first-line systemic therapy in a 'consolidative' approach. A series reported from the University of Colorado included 64 consecutive patients with metastatic NSCLC whose full extent of disease would have been treatable by SBRT [17]. The observed patterns of failure were categorized as local (L) for lesions known prior to treatment or distant (D) for new lesions. Among SBRT-eligible patients, first failure was L-only in 68%, D-only in 14% and L + D in 18%. In these patients, the time to first progression was 3.0 months in those with L-only failure versus 5.7 months in those with any D failure. It was concluded that the predominant pattern of failure in advanced NSCLC after first-line systemic therapy is local-only, suggesting that the strategic use of SBRT in this setting could improve time to progression in a substantial proportion of patients. The North Central Cancer Treatment Group has opened a study for the consolidative treatment of NSCLC in stage IV disease, and it will be interesting to see whether the results are consistent with those predicted from the University of Colorado retrospective review.

Conclusions

SBRT has a firmly established role in the treatment of patients with medically inoperable early stage lung cancer. From this starting point, there has been a proliferation of SBRT applications for a wide range of other cancer problems. Encouraging results for primary prostate, pancreas and hepatocellular cancers have all been reported [18–22]. For patients with limited spine, lung or liver metastases, durable local control on the order of 90% or higher can be achieved [5, 23–24]. To continue the process of refining patient selection, treatment technique and fractionation schedules, it will be important to conduct and analyze even larger new studies and to follow the existing studies longer. Furthermore, as for all sophisticated medical technologies, physicians must be sensitive to the relative cost-effectiveness of different modalities in comparison with each other. The major advantage for SBRT and other hypofractionated radiotherapy regimens, obviously, is the chance for reduced cost by virtue of the use of fewer total treatments, an important factor to add to the outcome benefits of local control and therapeutic gain.

References

1 Whelan TJ, Pignol JP, Levine MN, et al: Long-term results of hypofractionated radiation therapy for breast cancer. N Engl J Med 2010;362:513–520.
2 Eriksson D, Stigbrand T: Radiation-induced cell death mechanisms. Tumour Biol 2010;31:363–372.
3 Ewing J: Factors determining radioresistance in tumors. Radiology 1930;14:186–190.
4 Garcia-Barros M, Paris F, Cordon-Cardo C, et al. Tumor response to radiotherapy regulated by endothelial cell apoptosis. Science 2003;300:1155–1159.
5 Yamada Y, Bilsky MH, Lovelock DM, et al: High-dose, single-fraction image-guided intensity-modulated radiotherapy for metastatic spinal lesions. Int J Radiat Oncol Biol Phys 2008;71:484–490.

6 Li F, Huang Q, Chen J, et al: Apoptotic cells activate the 'phoenix rising' pathway to promote wound healing and tissue regeneration. Sci Signal 2010;3(110):ra13.
7 Lea DE, Catcheside DG: The mechanism of the induction by radiation of chromosome aberrations in *Tradescantia*. J Genetics 1942;44:216–245.
8 Guerrero M, Li XA: Extending the linear-quadratic model for large fraction doses pertinent to stereotactic radiotherapy. Phys Med Biol 2004;49:4825–4835.
9 Curtis SB: Lethal and potentially lethal lesions induced by radiation – a unified repair model. Radiat Res 1986;106:252–270.
10 Park C, Papiez L, Zhang S, et al: Universal survival curve and single fraction equivalent dose: useful tools in understanding potency of ablative radiotherapy. Int J Radiat Oncol Biol Phys 2008;70:847–852.
11 Astrahan M: Some implications of linear-quadratic-linear radiation dose-response with regard to hypofractionation. Med Phys 2008;35:4161–4172.
12 Smith BD, Arthur DW, Buchholz TA, et al: Accelerated partial breast irradiation consensus statement from the American Society for Radiation Oncology (ASTRO). Int J Radiat Oncol Biol Phys 2009;74:987–1001.
13 Timmerman R, Paulus R, Galvin J, et al: Stereotactic body radiation therapy for inoperable early stage lung cancer. JAMA 2010;303:1070–1076.
14 Hellman S, Weichselbaum RR. Oligometastases. J Clin Oncol 1995;13:8–10.
15 Milano MT, Zhang H, Metcalfe SK, et al: Oligometastatic breast cancer treated with curative-intent stereotactic body radiation therapy. Breast Cancer Res Treat 2009;115:601–608.
16 Miles DW, Chan A, Dirix LY, et al: Phase III study of bevacizumab plus docetaxel compared with placebo plus docetaxel for the first-line treatment of human epidermal growth factor receptor 2-negative metastatic breast cancer. J Clin Oncol 2010;28:3239–3247.
17 Rusthoven KE, Hammerman SF, Kavanagh BD, et al: Is there a role for consolidative stereotactic body radiation therapy following first-line systemic therapy for metastatic lung cancer? A patterns-of-failure analysis. Acta Oncol 2009;48:578–583.
18 Madsen BL, Hsi RA, Pham HT, et al: Stereotactic hypofractionated accurate radiotherapy of the prostate (SHARP), 33.5 Gy in five fractions for localized disease: first clinical trial results. Int J Radiat Oncol Biol Phys 2007;67:1099–1105.
19 Katz AJ, Santoro M, Ashley R, et al: Stereotactic body radiotherapy for organ-confined prostate cancer. BMC Urol 2010;10:1.
20 King CR, Brooks JD, Gill H, et al: Stereotactic body radiotherapy for localized prostate cancer: interim results of a prospective phase II clinical trial. Int J Radiat Oncol Biol Phys 2009;73:1043–1048.
21 Mahadevan A, Jain S, Goldstein M, et al: Stereotactic body radiotherapy and gemcitabine for locally advanced pancreatic cancer. Int J Radiat Oncol Biol Phys 2010;78:735–742.
22 Tse RV, Hawkins M, Lockwood G, et al: Phase I study of individualized stereotactic body radiotherapy for hepatocellular carcinoma and intrahepatic cholangiocarcinoma. J Clin Oncol 2008;26:657–664.
23 Rusthoven KE, Kavanagh BD, Cardenes H, et al: Multi-institutional phase I/II trial of stereotactic body radiation therapy for liver metastases. J Clin Oncol 2009;27:1572–1578.
24 Rusthoven KE, Kavanagh BD, Burri SH, et al: Multi-institutional phase I/II trial of stereotactic body radiation therapy for lung metastases. J Clin Oncol 2009;27:1579–1584.

Brian D. Kavanagh, MD, MPH
Department of Radiation Oncology, University of Colorado School of Medicine
Anschutz Medical Campus
1665 N. Ursula Street, Suite 1032, Box 6510, Mail Stop F706
Aurora, CO 80045 (USA)
Tel. +1 720 848 0154, Fax +1 720 848 0222, E-Mail Brian.Kavanagh@ucdenver.edu

Stereotactic Body Radiation Therapy: Normal Tissue and Tumor Control Effects with Large Dose per Fraction

Robert Timmerman · Michael Bastasch · Debabrata Saha · Ramzi Abdulrahman · William Hittson · Michael Story

Department of Radiation Oncology, University of Texas Southwestern Medical Center, Dallas, Tex., USA

Abstract

Stereotactic body radiation therapy (SBRT) is a potent noninvasive means of administering high radiation doses to demarcated tumor deposits in extracranial locations. The treatments use image guidance and related advanced treatment delivery technologies for the purpose of escalating the radiation dose to the tumor, while sharply minimizing the radiation doses to surrounding normal tissues. The local tumor control outcomes for SBRT have been higher than any previously published for the radiotherapy of frequently occurring carcinomas. In addition, the pattern, timing and severity of the toxicities have been very different than from those seen with conventional radiotherapy. These issues pose challenges to our understanding of the radiobiological mechanisms and the optimal uses of SBRT. In this review, the clinical characteristics and outcomes of SBRT are presented in the context of their possible underlying mechanisms. While some of these considerations remain theoretical, they may outline at least qualitative understandings of the observed clinical effects, and motivate continuing research into the effects of SBRT that guide its most effective use in the clinic.

Copyright © 2011 S. Karger AG, Basel

Stereotactic body radiation therapy (SBRT) is defined by its conduct. It employs a limited number of very large dose per fraction treatments (with a biologically equivalent dose of at least 75–100 as a minimum or even higher), fields only slightly larger than gross tumor targets, high confidence in precision even for moving targets (with 95% probability of including the entire target with margins of 0.5–1.0 cm) and dosimetry constructed to be very conformal with sharp gradients from high- to low-dose areas [1]. At the completion of the delivery of this very potent and biologically damaging dose, dramatic tissue effects may occur, both in normal

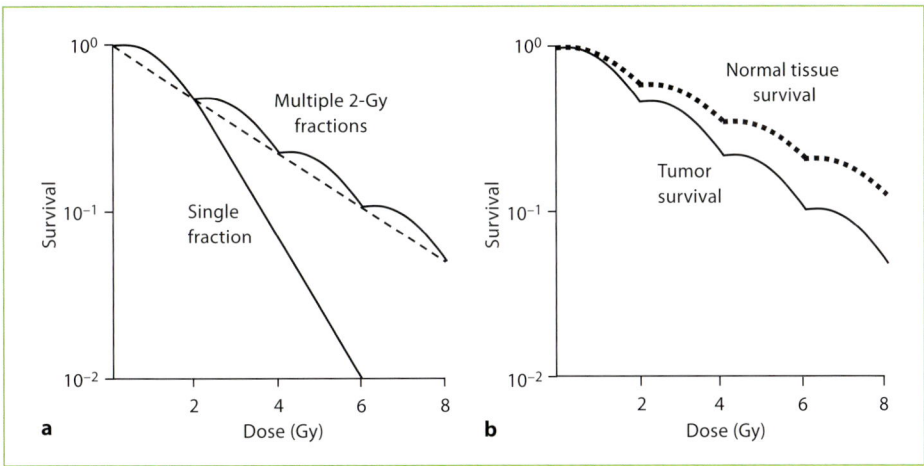

Fig. 1. a Typical cell survival curve shape comparing single-dose and multiple-dose radiation exposure. **b** Cell survival differs between normal tissues and tumor tissues for multiple-fraction exposure.

tissues and within the intended tumor target. Irrespective of the dose and fractionation, tissue responses are dependant on 3 basic characteristics: dose delivered (including dose per fraction), volume exposed and inherent tissue radiosensitivity. To some degree, these characteristics are not independent, but rather, interdependent. For example, at a constant level of dose, volume effects in normal tissue may demonstrate a spectrum from no noticeable effect to dramatic dysfunction. Furthermore, the type of toxicity and its manifestations in the person may change throughout this spectrum. We know that radiosensitivity varies among various tumor types, even within the same patient. It also varies with age, getting less sensitive as the host matures and organ volumes enlarge. Throughout the history of radiation oncology, biologists have attempted to understand these processes and discover their underlying mechanisms. They have done this mostly in the context of delivering the only type of radiation generally available – conventionally fractionated radiotherapy (CFRT). With the advent of technology facilitating SBRT, the lessons learned with CFRT investigation may or may not apply.

Traditional Radiobiology

Traditional radiobiology generally describes survival curves and tries to define models that would account for the observed data. As demonstrated in figure 1a, these curves initially have a curving portion then become more linear, in plots of log of surviving fraction versus dose. The initial curving portion of the curve is called the 'shoulder' and is said to represent a damage/repair process called 'sublethal' damage. With CFRT, a typical daily dose of around 2 Gy per fraction is given, which for most tissues would not reach the linear portion, and we repeat that fraction many times as shown in figure 1. Each time we give this rather small daily dose, the cumulative dose is never as potent as an equivalent single-fraction dose. This is still potentially beneficial as there often is a difference in the survival characteristics of the tumor on that curvy portion versus the normal tissues resulting in a therapeutic ratio that may be positive for CFRT as shown in figure 1b.

The shoulder, or the curvy portion of the survival curve, exists because of the 4 'Rs' of radiobiology that have been engrained into radiation oncology residency training for decades. There is *repair* of sublethal damage, which is beneficial for dose spilling to normal tissues. The normal tissues will *repopulate* during the course of a CFRT. Since CFRT is protracted in time, tumor cells will proceed through their cell cycle. Through this process of *reassortment,* tumor cells initially in a resistant phase will proceed eventually into a sensitive phase when exposed to damaging radiation. Finally, as tumors shrink, the central necrotic core becomes closer to the peripheral blood supply allowing *reoxygenation* which makes all cells more sensitive. If the story stopped here, we would conclude that CFRT is always better and not explore SBRT-type dose fractionation schemes. However, the 4 Rs realistically also work against ideal therapeutic outcome. In other words, tumors can *repair* and may *repopulate* as well during the course of protracted CFRT resulting in local tumor progression. Also, normal tissues may become more sensitive during a protracted course of radiation, leading to more toxicity.

So why do we fractionate? Many would casually convey that it is simply the right thing to do based on long experience (tradition) within the field of radiation oncology. Indeed, when hypofractionated schedules similar to what are now being used for SBRT were explored 100 years ago during the advent of radiotherapy, toxicity was prohibitive. Others have gone even further and said that hypofractionated radiation is simply not as effective as fractionated radiation at controlling tumors, and have cited historical clinical comparisons [2–4]. Overall, the dogma championing CFRT has mostly been justified by implying better tumor control. Survival curves should promptly dismiss this notion as tumor kill; total dose for total dose shows hypofractionation is always more effective for any tumor cells that have a shoulder. The real reason we use CFRT is in fact because of the normal tissues, not the tumor. Because of our lack of dose delivery technology capable of limiting normal tissue exposure, normal tissue volumes in classic CFRT delivery have been typically much larger than the tumor volume itself. In such a scenario, the only option for delivering high tumor dose is to use CFRT. Therefore, the use of CFRT during the past 80 years in the field of radiation oncology draws from a compromise, not an inherent advantage.

The survival curve also shows us that for CFRT, small increments in dose per fraction really do not change the surviving fraction very much. The arithmetic slope of the curve within the shoulder is small, such that for any increase in dose, or the 'run', you get a modest decrease in tissue survival, or the 'rise'. CFRT operates in this shoulder region. But at large doses per fraction, for a small increment in dose you get a rather dramatic increase in effect. That effect may be improved tumor kill, which would be favorable. However, this is a double-edged sword. The dramatic change in effect with modest increase in dose may also be enhanced toxicity, particularly if radiation is spilled to the uninvolved normal tissues.

Modeling the Survival Curve

Mathematicians can easily show that any real function (i.e. one that only has 1 y-value for each x-value) can be characterized by a geometric power series. A power series has a progression of terms starting with a constant, followed by a linear term, quadratic term, cubic term, and so forth as depicted in figure 2a. Now, if the series goes out to infinity, it will perfectly match the shape of any real function (in our case the survival curve). If we truncate that power series, say after 3 terms, it may approximate the curve, particularly in a certain range, but it will not match perfectly. In the case of the radiation survival curve, the constant is theoretically zero, because at zero doses there is 100% survival and the log of 1 is zero. So for a truncated power series with 3 terms, we end up with the linear term and the quadratic term. If we call the coefficients *alpha* and *beta,* we get the 'linear-quadratic' model [5]. The physical meanings of the linear and quadratic terms have been linked to single-strand or double-strand DNA breaks. While this may be partly true, the most precise definition is that they are mathematical coefficients describing physical data.

Since the alpha-beta model describing the survival curve is not the entire power series, it does not exactly fit the curve. When fitting data to a truncated power series to determine the coefficients, the goodness of fit will be best in the range of the measured input data. In the case of CFRT and the application of the alpha-beta model, most of the data have been collected, appropriately, in the range of the shoulder where actual clinical practice is carried out. Since very large dose per fraction treatments historically could not be carried out, due to prohibitive toxicity, there was no compelling reason to fit the data well after 10 Gy per fraction. At very low dose per fraction, even below what is used in CFRT, the alpha-beta model's survival values are dominated by the alpha (linear) term because its coefficient is larger than that of the beta (quadratic) term. At higher doses per fraction within the shoulder, the beta term dominates since it is applied to the dose squared. The quadratic term changes the slope of the survival curve, giving the shoulder portion its characteristic shape. Eventually, however, the quadratic term would overwhelmingly dominate the representation of the curve by constantly curving

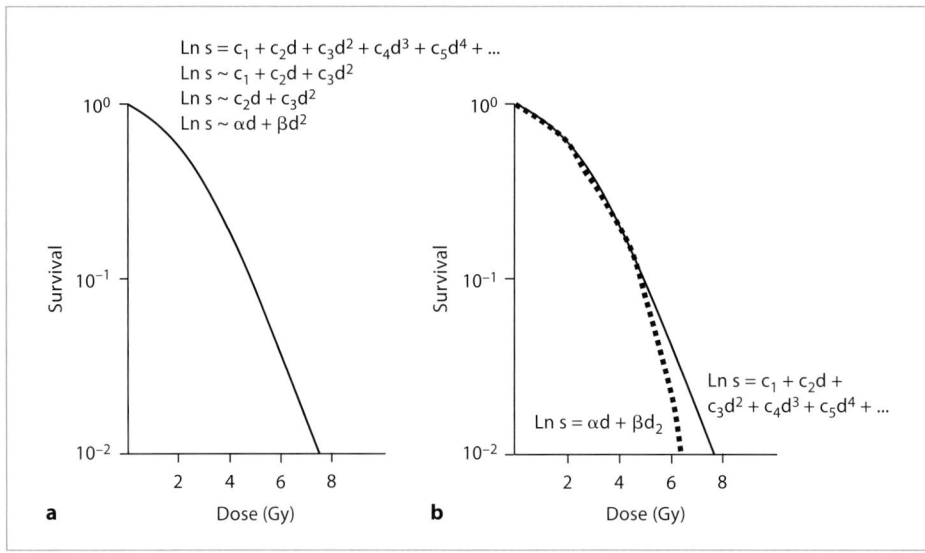

Fig. 2. a Real functions such as the shape of the cell survival curve can be expressed as a power series. A theoretical perfect match can be obtained when the number of terms is taken to infinity; however, approximate correlation can be obtained with truncated series, such as the linear-quadratic model where the coefficients correspond to alpha and beta. **b** While the data may fit the truncated series well over a low-dose range where most measurements are taken, the truncated linear-quadratic series overpredicts cell kill compared with the full series and actual results due to dominance by the quadratic term constantly curving the slope downward.

downward, implying greater fractions of cell kill as shown in figure 2b. In reality, though, the actually measured curve becomes linear after the shoulder (not quadratic), which is a fact not handled well by the linear-quadratic formalism. In effect, the alpha-beta model overpredicts cell kill at large doses per fraction.

Biologically Equivalent Dose

Professor Fowler helped us use the alpha-beta or linear-quadratic model prudently for CFRT [5]. He derived the formalism of biologically equivalent dose, allowing us to compare the potency of various fractionation regimens so long as we had reliable information about survival characteristics for the applied range (e.g. alpha-beta ratios). In the preceding paragraphs, we argued that the alpha-beta formalism breaks down when applied to treatment schedules with very large doses per fraction. While we maintain this position, it is still convenient and cautiously useful to apply this formalism in these extrapolated circumstances. It might be useful, for example, in assigning the starting dose for a phase I study or making estimates of normal tissue tolerance for protocols. However, do not forget the limitations of the linear-quadratic model. At large doses, it becomes a poor fit, with overestimation of the effect of the treatment. Furthermore, it does not account for the duration of the treatment, variation in the tissue architecture or function (e.g. lung vs. spinal cord), nor look at the impact of nonchromosomal mechanisms of radiation damage like apoptosis or vascular injury. The net effect and most important observation is that the tumor control conversion will be overstated and the normal tissue conversion will be understated. By using the alpha-beta conversion without reservation, clinicians will think that SBRT at a given dose is both more effective at controlling tumor as well as safer than it is in reality. This can be a very dangerous fallacy for the patient.

Organ/Tissue Function and Microscopic Architecture

Wolbarst et al. [6] described the concept of functional subunits in a model of cell structure. They said that there were 2 types of functional subunits, structurally defined and structurally undefined. The structurally defined subunits had a very demarcated architecture, the undefined subunits were monotonous. They stated that organ damage would result from constituent damage to the functional subunits composing the organ.

In order to understand this model, it is convenient to consider some simplified examples. One example is the lung which is structurally defined at the level of the alveolus and capillary since there is a clear distinction between each adjacent capillary/alveolus complex capable of exchanging gases as shown in figure 3a. A contrasting example is the structurally undefined esophagus where we have a monotonous tube which functions as a conduit as shown in figure 3b. However, both cases are similar in that a cellular layer including a population of stem cells (capable of repopulating this layer) lies over a basement membrane. Consider a dose of irradiation capable of eradicating all the mucosal cells and their stem cells in a given volume. For the tissue to 'heal', stem cells from somewhere beyond the volume irradiated must migrate into the functional subunit to 'rescue' by way of repopulation of the mucosa. In the case of the structurally defined lung, if you irradiate a specific alveolus/capillary functional subunit to the point where all stem cells within the subunit are eradicated, neighboring adjacent subunits cannot rescue the affected subunit because stem cell migration is blocked by the intervening basement membranes. Even though there may be numerous stem cell clonogens in a neighboring alveolus, they cannot physically help their neighbor as shown in figure 4a. For the structurally undefined tissue of the esophagus, however, stem cells are free to migrate to help repair damage. Structurally undefined stem cell migration is only limited by a reasonable dis-

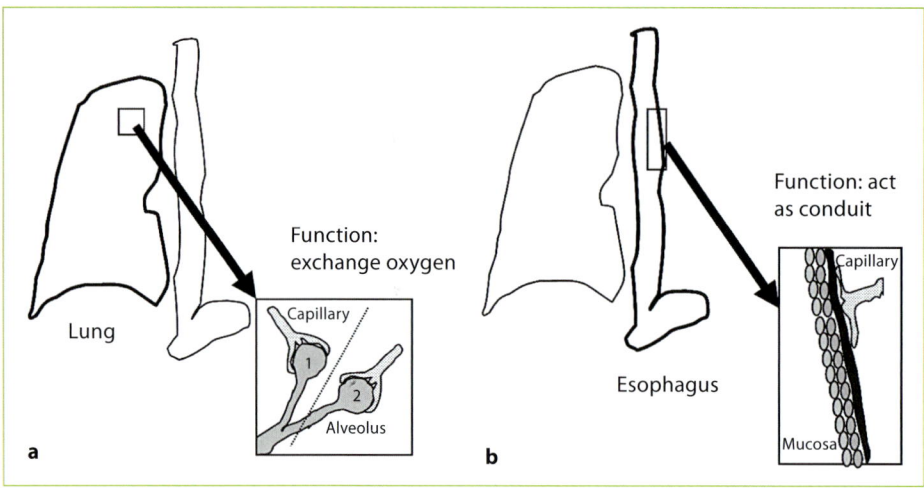

Fig. 3. a Structurally defined normal tissues like the peripheral lung exhibit a distinct demarcation between the functional subunits 1 and 2. The functional subunits independently perform the tissue function, exchanging oxygen. **b** Structurally undefined normal tissues like the mucosa of the esophagus show no distinct separation between functional subunits, which in the esophagus act in concert to perform the overall function of acting as a conduit for passage of food to the stomach.

tance, nothing anatomical as indicated in figure 4b. This would explain the observation that the radiation tolerance per cubic centimeter of the lung is significantly lower than that of the esophagus.

Tissues that are made up mostly of structurally defined subunits like the peripheral lung are called *parallel functioning* tissues. Tissues that are made up of the monotonous undefined subunits are called *serially functioning* tissues. The parallel tissues are in the peripheral lung, peripheral liver, peripheral kidney and acini regions of glands. The serial tissues would include the body's more linear or hollow structures like the spinal cord, esophagus, bowels and ducts. Parallel functioning tissues are typically very large and have inherent functional redundancy or reserve. After radiation dose injury to the parallel tissues, which may occur at quite low levels, the toxicity will be mostly related to volume. That is why for things like pneumonitis, we talk about the 'V20' (volume receiving more than 20 Gy) rather than a toxic dose. For serially functioning tissues, it is the dose that is the more important predictor of toxicity. We use metrics like 'TD 5/5' (tolerance dose that causes 5% of patients to have radiation injury within 5 years) for spinal tolerance or esophageal tolerance. The trouble with these models is that most organs are actually a mixture of parallel and serially functioning tissues.

In the 1980s, Yeas and Kalend [7] came up with the notion of the *critical volume model*. The critical volume defined the organ volume exceeding the threshold dose

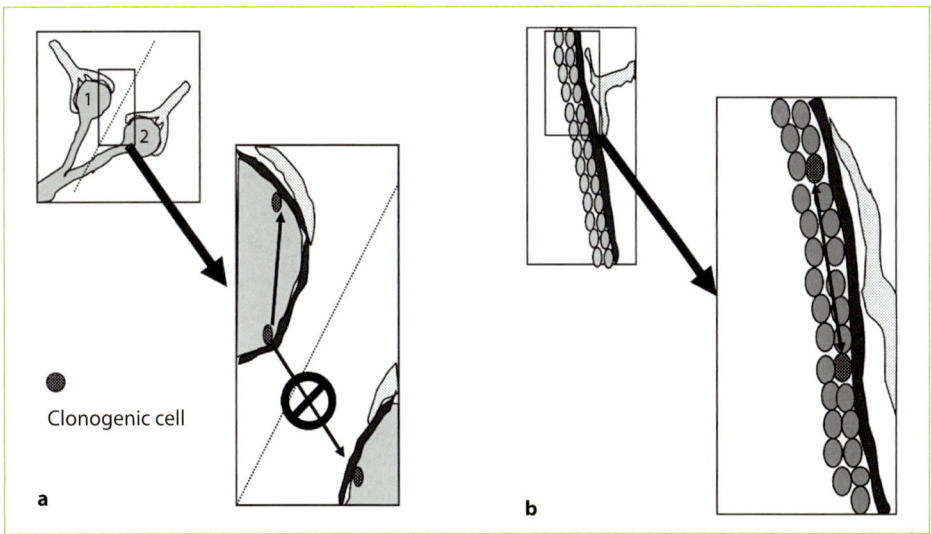

Fig. 4. a Migration of clonogenic stem cells within structurally defined tissues is unhindered within a particular functional subunit (e.g. within a specific alveolus). However, clonogenic cells even adjacent to another functional subunit cannot rescue a damaged neighbor since they cannot migrate across basement membranes. **b** Migration of clonogenic stem cells within structurally undefined tissues is unhindered by the basement membrane and is only limited by practical distance for stem cell movement. As such, adjacent functional subunits can be rescues from injury given time and the ability of neighboring stem cells to migrate.

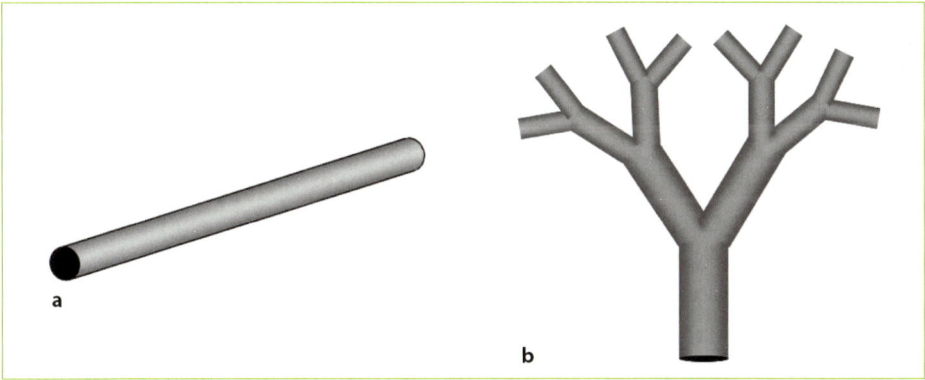

Fig. 5. a Serially functioning tissue arranged as a straight or curved line without branch points is referred to as a linear structure. Ablative injury to the circumference of a linear tubular structure at any point along its length will eventually totally disrupt the function of the organ. In the case of the linear tubular small intestine, such an injury would result in a bowel obstruction. **b** Serially functioning tissue arranged akin to a tree or bush with a larger trunk dividing into major limbs and finally smaller limbs and branches is referred to as a branching structure. Ablative injury to a branch will primarily cause dysfunction downstream from the injury, not the organ as a whole unless the major trunk itself is damaged. The airways in the lung and duct in the liver follow a branching tubular structure.

to wipe out all stem cells within all irradiated functional subunits. For tumors near serially functioning tissues, radiation tolerance would be relatively high and even higher with protracted radiation. In order to give a tumoricidal dose while still maintaining the possibility to rescue adjacent functional subunits, it made sense to fractionate the dose, more like conventional treatments, or to use radiosensitizers. For tumors near parallel functioning tissues, even small doses would eliminate the functional subunits with little chance of rescue. As such, it would be best in terms of tumor kill to give the dose all at once, and then do a lot of measures to exclude volume because volume was the problem with toxicity. So, for these structures, it was most important to find radiation methods to limit moderate- and high-dose irradiation of normal tissues. Finally, we introduced our need for the techniques of SBRT. Now we have many vendors that offer the technologies to help us do these SBRT treatments. Indeed this technology is very useful if you are in parallel functioning tissues to improve the dose distribution and minimize the high-dose regions. But for treatments next to a serially functioning tissue, these technologies will only be of limited benefit.

Serially Functioning Tissue Macroscopic Architecture

Our bodies have two basic types of linearly arranged serially functioning hollow lumen tissues as indicated in figure 5. One is a *linear tubular* structure like the esophagus or the gastrointestinal tract. The other is the *branching tubular* structure like the hepatic ducts and the bronchus of the lungs, extending out toward the parallel functioning tissues in the periphery [8].

The function of a linear tubular organ is simply to transmit something downstream. The 'something' may include nutrients, secretions or waste products. Within the walls or channels of linear tubular structures, specialized structures may aid this process. In the case of the hollow viscus or bowels, for instance, the clonogens are monotonously layered on a basement membrane throughout the lumen. They have a vascular supply that starts from two main arterial feeders following the tube longitudinally. These vessels are connected to arcades of vessels that go around providing vascular supply to the circumference of the walls.

In contrast, while branching tubular organs function for the same reasons, they have branching conduits that split akin to the branches of a tree or bush. Their blood supply follows along those same branches out to the periphery from the central trunk. In this arrangement, conduits beyond a split point are no longer transmitting the same material even though the structure of two adjacent branches may be identical.

Severe damage to any segment of a linear tubular structure like the small intestine (as might be caused by SBRT) results in total and catastrophic dysfunction of the organ. This might include a bowel obstruction that totally disrupts effective digestion. In contrast, with severe damage to a branching tubular structure like the pulmonary airways, the severity of the dysfunction depends to a great degree on the position of the damage within the branching structure. If the damage occurs at a distal branch, only the relatively small volume of tissue supplied by that particular branch will become dysfunctional. However, if the damage occurs in the larger more proximal trunks, a severer dysfunction will occur as many downstream branches are cut off. In either case, damage to tubular structures tends to manifest in the patient as 'late' toxicity, often not appearing for 4–6 months after therapy.

Stereotactic Body Radiation Therapy and Ablation

The main objective of therapeutic radiation is to disrupt clonogenicity, and there are many genes involved with clonogenicity. Stem cells are typically fairly radiosensitive, so often it does not take much dose to affect them. Cellular division or clonogenicity is a very complicated process regulated by a multitude of genes. The entire process can be stymied by radiation damage to any one of these genes, which may occur at relatively modest doses. Cellular function, like the secretion of a hormone, is usually only coded for by one or a few genes, and therefore radiation must damage every one of those genes in every functioning cell to disrupt the function. As such, it takes a very high dose of radiation to disrupt cellular function. As an illustration, growth control of a pituitary adenoma is achieved at a relatively modest dose, but to stop the secretion of a hormone from a hormone-secreting pituitary adenoma requires a much higher dose. Another example can be borrowed from the use of radioactive iodine to treat thyroid cancer. The therapy is so well targeted that massive doses reach both the thyroid cancer and the residual functioning thyroid gland. These doses both disrupt clonogenicity and cellular function. Such a treatment has been termed *ablative*. In turn, typical dose regimens used with SBRT are also often ablative to the tissues seeing dose levels near the target dose.

The assumptions with SBRT, at least in the region of the planning target volume, are that we will destroy clonogenicity and that tumors will be totally disrupted. More than likely, we will also destroy tissue function (e.g. transmitting air through an involved bronchus). In the shell of normal tissue just outside of the planning target volume, the dose is so high that it will effectively disable all of the clonogenicity and function of that portion of the organ. So in that regard, SBRT is almost like a surgeon's knife, effectively cutting the tissues and leaving them totally disabled both at the point of contact and downstream. Nonetheless, from a tumor control point of view, ablative dose levels are still desirable since in this range the likelihood of disrupting clonogenicity in every tumor cell is very high.

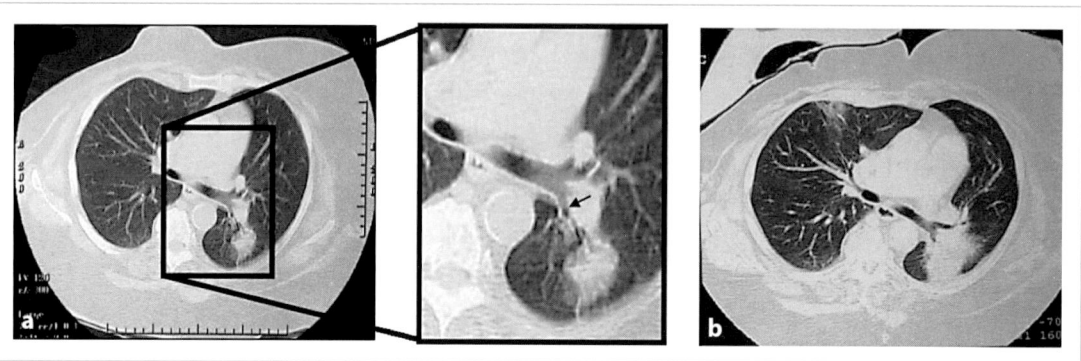

Fig. 6. a Early-stage lung cancer patient prior to ablative dose SBRT. There is a small branching bronchus designated by the arrow oriented toward the tumor. This airway was situated inside the planning target volume and received the ablative dose. **b** Two-year post-treatment CT scan showing dense consolidation distal to the small airway indicated in **a**. The consolidation was not avid on PET and is most consistent with postradiation atelectasis.

Normal Tissue Consequences of Ablative Dose Delivery

With SBRT, radiotherapy can act almost like a knife. As such, lessons can be learned from surgeons. The surgeon must achieve homeostasis, remove all devascularized tissues and restore continuity of a hollow viscus. If a surgeon removes a tumor of the esophagus, he/she has to reconnect the ends or the patient would suffer. If a surgeon removes a lung tumor, he/she does not just enucleate the tumor. Instead, the surgeon must take out all downstream lung which otherwise would necrose and lead to postoperative complications. So if we use SBRT in these same examples and we treat a tumor of a hollow viscus like an esophageal cancer, what we can expect is that the function of that structure, i.e. the tubular function, will be destroyed. If we do not restore it, the patient will have a severe problem. It would be foolish, as things stand, to treat esophageal tumors with ablative doses using SBRT, as you will end up with a debilitating stenosis. In the case of the branching tubular structures like the lung, SBRT will affect all tissues downstream from airways exposed to ablative dose. We will end up with a collapsed, atelectatic tissue downstream. The overall post-treatment function of the organ depends heavily on the volume of downstream lung. Obviously, lesions treated in the lung periphery will be less prone to problems as compared with lesions near the main stem bronchus.

For the branching tubular organs, the lung has been the most prominently treated site. Downstream collapse from bronchial injury is commonly observed as shown in figure 6 [9–14]. The liver has a similar architecture as the lung, but it has the capacity to regenerate [15]. This capacity will make little difference, however, if the draining bile ducts are obliterated. Whether stenting can prevent the effects of radiation-induced biliary sclerosis is yet unclear. As long as we do not set the patient back too far during the course of an SBRT treatment, the liver will likely be a preferred site to treat.

The normal tissue tolerances to SBRT remain unclear. Large doses per fraction are associated with increased late effects. Late effects take more time to observe. Therefore, we must collect toxicity data years after therapy to quantify risk and define tolerance. The Radiation Therapy Oncology Group has included tolerances in its 0236 protocol for SBRT in medically inoperable lung cancer. These were agreed upon by extrapolation, limited data and committee discussions. In all likelihood, they will have to be modified over time with more adequate follow-up.

Tumor Control Dose

SBRT uses anything from 1 to 5 fractions. Different centers have used different dose levels. Some have argued that they have found adequate dose levels, but these are typically based on retrospective testing. Prospective testing is now ongoing, but results are still immature. At any rate, it is clear that better tumor control is achieved when using ablative dose ranges. This would require at least 10 Gy per fraction for 3–5 fractions. The only prospective dose escalation study testing multiple levels up to a maximum-tolerated dose was carried out at Indiana University. It showed that 20 Gy per fraction times 3 (60 Gy total) was tolerable in most patients with early-stage medically inoperable lung cancer. It also showed that doses of 16 Gy times 3 or less had higher rates of local tumor progression [9]. In a previous section, we made an argument that alpha-beta conversions overpredict dose potency in this range. At any rate, in cancer therapy development, the first objective of clinical studies is to determine a potent dose in phase I studies. That way, the therapy will be in its most potent form for phase II and III trials for testing. It would be feasible, if needed, to reduce the dose because of excessive toxicity with late follow-up. Again, ideally the therapy should afford tumor control probabilities near 90%. Such probabilities are attainable with SBRT, provided an adequate dose is delivered.

Conclusions

What are the true dose-response characteristics of SBRT? What are the mechanisms of both tumor and normal tissue injury? To date, the answers to these questions are unclear. It can already be appreciated, however, that lessons learned from the long history of CFRT do not necessarily apply to SBRT. The biggest obstacle to delivering SBRT surrounds the treatment of tumors near serially functioning tissues. Technology cannot solve this problem. The future of SBRT will require a considerable amount of work in the realm of classic radiobiology. The encouraging results in parallel functioning organs may very well change the patterns of care for many diseases, including early-stage lung cancer, as trials mature. However, the true potential of SBRT will not be realized until chemical modifiers or dose schedules can be developed to use this same ablative therapy in and about serially functioning tissue. This challenge should be viewed as a true opportunity.

References

1 Potters L, Steinberg M, Rose C, Timmerman R, Ryu S, Hevezi JM, Welsh J, Mehta M, Larson DA, Janjan NA, American Society for Therapeutic Radiology and Oncology, American College of Radiology: American Society for Therapeutic Radiology and Oncology and American College of Radiology practice guideline for the performance of stereotactic body radiation therapy. Int J Radiat Oncol Biol Phys 2004;60:1026–1032.
2 Rosenthal DI, Glatstein E: We've got a treatment, but what's the disease? Or a brief history of hypofractionation and its relationship to stereotactic radiosurgery. Oncologist 1996;1:1–7.
3 Glatstein E: Personal thoughts on normal tissue tolerance, or, what the textbooks don't tell you. Int J Radiat Oncol Biol Phys 2001;51:1185–1189.
4 Goffman TE, Glatstein E: Hypofractionation redux? J Clin Oncol 2004;22:589–591.
5 Douglas BG, Fowler JF: The effects of multiple small doses of X-rays on skin reactions in the mouse and a basic interpretation. Radiat Res 1976;66:401–426.
6 Wolbarst, AB, Chin, LM, Svensson, GK: Optimization of radiation therapy: integral-response of a model biological system. Int J Radiat Oncol Biol Phys 1982;8:1761–1769.
7 Yeas RJ, Kalend A: Local stem cell depletion model for radiation myelitis. Int J Radiat Oncol Biol Phys 1988;14:1247–1259.
8 Timmerman RD, Kavanagh BD: Stereotactic body radiation therapy. Curr Probl Cancer 2005; 29:120–157.
9 Timmerman R, Papiez L, McGarry R, Likes L, DesRosiers C, Frost S, Williams M: Extracranial stereotactic radioablation: results of a phase I study in medically inoperable stage I non-small cell lung cancer. Chest 2003;124:1946–1955.
10 Aoki T, Nagata Y, Negoro Y, Takayama K, Mizowaki T, Kokubo M, Oya N, Mitsumori M, Hiraoka M: Evaluation of lung injury after three-dimensional conformal stereotactic radiation therapy for solitary lung tumors: CT appearance. Radiology 2004;230:101–108.
11 Takeda T, Takeda A, Kunieda E, Ishizaka A, Takemasa K, Shimada K, Yamamoto S, Shigematsu N, Kawaguchi O, Fukada J, Ohashi T, Kuribayashi S, Kubo A: Radiation injury after hypofractionated stereotactic radiotherapy for peripheral small lung tumors: serial changes on CT. AJR Am J Roentgenol 2004;182:1123–1128.
12 Yamada Y, Shiomi H, Sumida I, Suzuki O, Isohashi F, Oh RJ, Tanaka E, Inoue T, Nakamura H: Quantitative evaluation of changes in irradiated lung fields after stereotactic irradiation by the Polygon Method. Radiat Med 2004;22:98–105.
13 Slotman BJ, Solberg T (eds): Extracranial Stereotactic Radiosurgery and Radiotherapy, ed 1. Hamilton, BC Decker Inc, in press.
14 Kavanagh B, Timmerman RD (eds): Stereotactic Body Radiation Therapy. Baltimore, Lippincott Williams and Wilkins, 2005.
15 Herfarth KK, Hof H, Bahner ML, Lohr F, Hoss A, van Kaick G, Wannenmacher M, Debus J: Assessment of focal liver reaction by multiphasic CT after stereotactic single-dose radiotherapy of liver tumors. Int J Radiat Oncol Biol Phys 2003; 57:444–451.

Dr. Robert Timmerman
Department of Radiation Oncology
University of Texas Southwestern Medical Center
5801 Forest Park Road
Dallas, TX 75390-9183 (USA)
Tel. +1 214 645 7651, Fax +1 214 645 7624
E-Mail robert.timmerman@utsouthwestern.edu

Stereotactic Body Radiation Therapy for Thoracic Cancers: Recommendations for Patient Selection, Setup and Therapy

Robert Timmerman[a] · John Heinzerling[a] · Ramzi Abdulrahman[a] · Hak Choy[a] · John L. Meyer[b]

Departments of Radiation Oncology, [a]University of Texas Southwestern Medical Center, Dallas, Tex., and [b]Saint Francis Memorial Hospital, San Francisco, Calif., USA

Abstract

Advanced technologies have facilitated the development of stereotactic body radiation therapy (SBRT) programs capable of delivering ablative radiation doses for the control of lung cancers. To date, experience with these programs has been highly favorable, as reflected in the results of careful clinical trials. The medically inoperable lung cancer patient, lacking more effective options, has served as the initial clinical base to test SBRT; the therapeutic outcomes have confirmed a significant role for this approach. For many patient groups, SBRT may become a non-invasive alternative to some thoracic surgeries, especially ones with more limited therapeutic goals such as wedge resection. Despite these results, long-term evaluation of the cases treated is required to allow greater understanding of the limitations and contributions of this new modality. The successful delivery of SBRT requires the development of a comprehensive, specialized clinical program providing advanced technology and the technical expertise of physicians, physicists and therapists specially trained in SBRT applications. To achieve successful clinical outcomes, careful patient selection and attention to therapy design and delivery are required since exacting clinical procedures are involved. This chapter will outline many details essential for establishing an effective SBRT program in clinical practice.

Copyright © 2011 S. Karger AG, Basel

Early-stage lung cancer has been the primary clinical model for the testing of stereotactic body radiation therapy (SBRT) [1–3], also known as stereotactic ablative radiotherapy. Since first introduced in the mid-1990s [4], SBRT has already distinguished itself as an attractive treatment that combines high rates of tumor control, tissue tolerance and patient survival with a convenience that is unique among curative therapies for most cancers. SBRT is a potent, oligofractionated (oligo- mean-

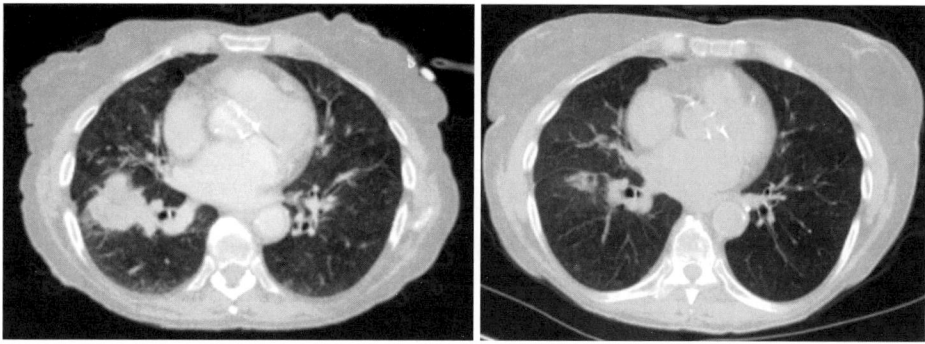

Fig. 1. Patient with a T2, N0, M0 non-small cell lung cancer prior to and 3 months after SBRT.

ing 'few'), ablative therapy that can be delivered noninvasively in 1–5 outpatient treatments given daily or every other day. Each treatment takes 20–60 min to set up and deliver. The total treatment duration is only 1–2 weeks, and most patients can return to their normal activities immediately.

Distinct from conventionally fractionated radiotherapy (daily doses of 1.8–2.5 Gy), or hypofractionated radiotherapy (daily doses of 3–6 Gy, often used for palliation at end of life), SBRT uses very high ablative-range daily doses of 8–30 Gy facilitated by sophisticated image guidance technologies. These dose levels overwhelm the tumor repair mechanisms that cause other therapies to fail. The danger exists, however, for these potent doses to cause late effects that may be intolerable to the patient. The keys to success are found in the careful conduct of the treatment process and in a proactive approach toward supporting patient/organ health. In this chapter, we will briefly outline the evidence justifying the use of SBRT for lung cancer and then describe selective clinical approaches for its planning and delivery.

Prospective Testing of Stereotactic Body Radiation Therapy

SBRT was first used to treat thoracic tumors in the early 1990s, by pioneering groups in Sweden [4] and Japan [5] working nearly simultaneously. Initial reports were mostly retrospective and uncontrolled, yet the dramatic early tumor responses (typical of SBRT) showed that it was biologically unique from conventional radiotherapy (fig. 1). Complete response rates for SBRT were reported to be 50% or better [6], well beyond the 20% complete response rates for conventional radiotherapy alone [7, 8]. Such high rates of response indicated not only the capacity of SBRT to disrupt tumor proliferation, but also apparently to preserve the competence of the patient's immune function to promote tumor reduction through its surveillance/attack processes.

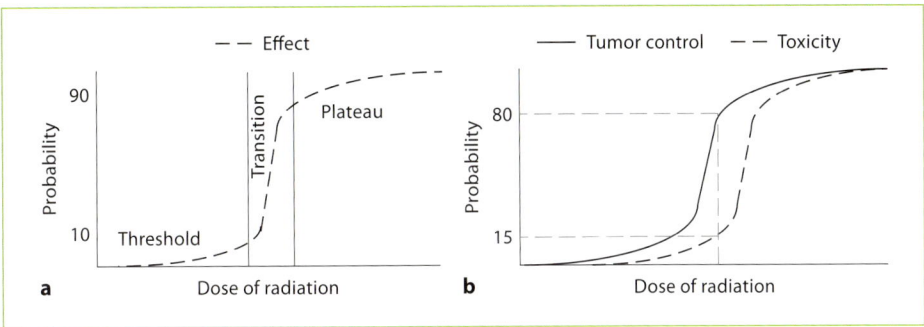

Fig. 2. a Three ranges of sigmoid dose response: threshold, transition and plateau. **b** The therapeutic window between tumor control and toxicity. There is a limited range of radiation doses that permit differential responses.

The rapid scientific acceptance of SBRT, and its clinical acceptance into practice even at the community level, reflect the integration of several technical advances into an effective treatment strategy and its confirmation in outcome evaluations. Based on careful hypothesis testing through well-designed prospective trials, the new technologies yielded significant clinical results. The goals of the early prospective testing of SBRT were to test and validate the new technologies for managing image guidance, repositioning accuracy and motion control. Subsequently, efforts were made to formulate dose-response relationships between toxicity and local control – critical steps in assessing value for any new treatment approach.

Finding a clinically relevant dose-response range is critical for optimizing a proposed therapy. Most dose-response relationships have a sigmoid shape with 3 regions: (1) at lower doses, there is little effect with increasing dose initially *(threshold)*; (2) at some point, a transition zone is entered where additional dose causes more dramatic change in effect *(transition)*; and (3) at higher doses, further dose shows little additional response *(plateau;* fig. 2a).

Most chemotherapy phase I trials specifically search for the maximum-tolerated dose (MTD) of a new agent by escalating its dose in small cohorts of only 3–6 patients. Toxicities, mostly early or short-term effects, are observed at each dose level. By this, the MTD can be reasonably characterized while exposing only a small number of patients to blatantly subtherapeutic doses, and even fewer to doses beyond the MTD. However, these types of trials are grossly underpowered to assess the range of beneficial effects in tumor control. They are often rationalized by the dogma that a new agent will not be very effective unless given at its most potent dose level (the MTD), which must be determined first. Clinical phase I testing of SBRT has been more instructive than this, since its clinical benefit is substantial. The optimized dose for its clinical implementation is considerably below its MTD [9, 10]. It would be a mistake to consider only toxicities and MTD in phase I studies using highly effective therapies like SBRT since the transition zone for therapeutic effect is likely to be substantially below that for toxicity. Instead, phase I studies of SBRT permitted early assessment of the ideal therapeutic window (fig. 2b), balancing benefit and harm, and began to focus on dose selection for given patient/tumor characteristics.

Stereotactic Body Radiation Therapy for Medically Inoperable Lung Cancer: Early Trials

The first North American multicenter trial for medically inoperable early-stage lung cancer patients was RTOG 0236 [6]. The prescription dose was 20 Gy × 3 = 60 Gy, calculated as if in water density. This was equivalent to about 18 Gy × 3 = 54 Gy if properly accounting for tissue heterogeneity [11]. RTOG 0236 also prescribed careful technical procedures for carrying out SBRT, rigorous dose constraints for normal tissues and quality management programs that included a central accreditation process [12]. The trial served the unusual role of providing instruction for many centers on how to develop reasonable SBRT programs using the clinical guidance of RTOG. The trial opened in 2004, enrolled 57 patients and completed accrual in 2006. The results were respectable: at 3 years, the primary tumor control rate was 98% and the survival rate was 56% [6]. With the outcomes of this trial now mature, the treatment dose of 18 Gy × 3 has become a standard for future RTOG trials involving SBRT for peripheral tumors in medically inoperable patients. RTOG 0236 excluded central lung tumors because of the complications seen in the Indiana University experience [13]. It is now recognized that tubular structures, like the airways, hilar vessels and esophagus, are much more prone to toxicity with ablative treatments [3].

The RTOG is currently conducting a phase I study in patients with centrally located tumors (RTOG 0813, principal investigator: Andrea Bezjak). This is a phase I/II dose escalation study starting at 10 Gy × 5 = 50 Gy total, and tests the hypothesis that slightly increasing the number of fractions to deliver the dose will yield a reduction of toxicity. The goal is to define a potent dose schedule that can be carried forth into future trials. It requires proper dosimetric accounting of tissue heterogeneities and use of IMRT to avoid parts of tubular structures like the trachea and esophagus.

The trial conducted in Indiana was one of the earliest to show a dose-response effect in lung cancer local control, and probably has longer follow-up data than any other prospective trial [9, 14]. While local control rates improved with increasing dose (fig. 3), there were no differences in survival according to dose. Why? With surgically treated patients, improvement in local control does appear to improve survival [15]. However, retrospective data indicate that nearly one half of untreated medically inoperable patients will eventually die of comorbidities rather than

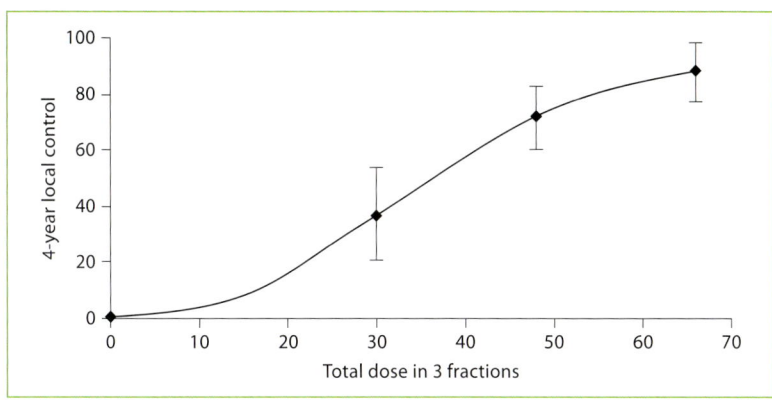

Fig. 3. Dose response for local control after 3-fraction SBRT treatment. Indiana University prospective phase I trial for medically inoperable patients with stage I non-small cell lung cancer [9, 14].

their lung cancers. The Indiana results likely reflected the selection of patients for SBRT who were medically unfit for surgery and more infirm generally.

What has greater impact on survival in patients with medically inoperable lung cancers: avoiding death from non-small cell lung cancer or from aggravated comorbidities? RTOG 0915, a randomized phase II trial for peripheral tumors, is a first step in answering this reasonable question. This trial compares 2 published regimens, both using less aggressive treatment doses than used in RTOG 0236, and will primarily measure toxicity outcomes. Stopping rules based on efficacy will be used. Following the completion of RTOG 0915, the intent is to compare its less toxic arm with the now standard RTOG 0236 regimen in a phase III trial. Given the 98% 3-year control rate already achieved in this standard regimen, the RTOG 0915 winner is not expected to improve local control further, but may have merit for this frail population. It represents a trial design that may not be appropriate for operable cases, where it is already established that improved tumor eradication improves survival.

Stereotactic Body Radiation Therapy for Medically Operable Patients

The results of SBRT in medically inoperable patients show that it is highly effective in eradicating selected primary tumors, and reasonably well tolerated even in this medically high-risk population. This has led to speculation on its applicability in operable patients. In Japan, healthier patients refusing surgery have already been treated with SBRT. The results have been fairly similar to those of surgical resection, even with lobectomy, so long as a reasonable radiation dose potency has been delivered [16].

The RTOG has moved forward with a pilot study, RTOG 0618, testing SBRT in operable patients with the dose standardized in RTOG 0236. Adequate radiation dose is a key to success in this population because the control rates would need to be competitive with established surgical approaches and lobectomy can control 90% or more of similar lesions. Indeed, the primary objective of this trial will be 2-year local control with a target rate of 90% and with secondary objectives being the rates of survival and toxicity. Clearly, the local control rate is a most important prognostic determinant for these patients, but if the control rate in RTOG 0618 proves to be similar to that in RTOG 0236 after the identical treatment, this requirement should be met. Finally, 2 trials have been organized that randomize operable patients to treatment with either surgery or SBRT. These are active trials, but still early in their enrollment [17, 18].

A Guide to Stereotactic Body Radiation Therapy: Procedures and Practice

The prototypical stereotactic radiosurgical tool, the Gamma Knife, integrates many radiation beams, each relatively weak and causing little entry damage. The beams converge to produce a potent and damaging cumulative dose at the target. We mimic this paradigm with SBRT by designing a high radiation dose convergence at the target site using external beam equipment. In order to confidently hit this target and miss the innocent normal tissues, sophisticated image guidance and motion control techniques must be used. It is crucial for the treatment team to learn how to properly select patients, design treatment plans and operate this equipment prudently in order to deliver the biological doses that define SBRT. In this section, we will describe the current treatment process at the University of Texas Southwestern Medical Center (UTSW).

Immobilization

Step one in the process of SBRT for lung cancer is to effectively immobilize the patient. Our technique uses a body frame. Whatever method is selected, the patient must be comfortable for these longer treatment sessions and the setup must be highly reproducible. This is important to avoid any large geometric corrections at treatment and the uncertainties that they may introduce. We have patients lie on a very large vacuum bag integrated within the frame. To introduce them into the frame, we use regimented steps bringing them from standing, to sitting, and then to lying positions. Patients are asked to make final adjustments to their position such that they establish a comfortable alignment within the vacuum pillow. When fully deflated, it makes a cast of the patient on 3 sides to allow a high surface

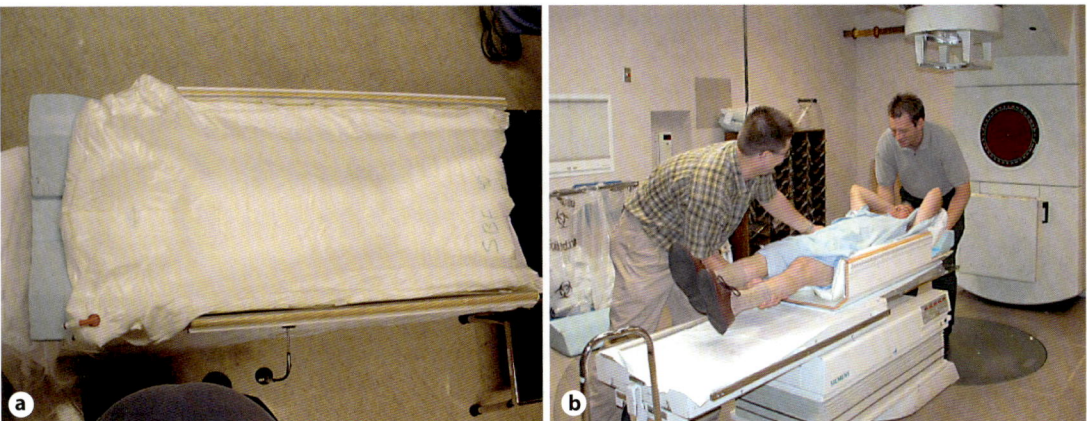

Fig. 4. a The vacuum pillow shown prior to introducing the patient. It will be molded to the patient's contours on 3 sides. **b** The patient, referenced to the frame and comfortably immobilized, is moved into position for scanning or treatment.

area of contact for better immobilization, reproducibility and comfort (see fig. 4a). Each patient is carefully referenced to the frame; these values are recorded into the medical records so that the position can be accurately reproduced in subsequent treatments. Figure 4b shows how a patient is slid on the treatment couch to the starting position for stereotactic localization by the operators.

Motion Assessment and Management

It is important to accurately assess the degree of respiratory motion of the treatment target and to make a deliberate decision whether motion control intervention is necessary. This step is critical to adequately minimize treatment margins. At UTSW, we use real-time fluoroscopy to initially assess motion. This is followed by 4D CT, and we use a 10-interval phase binning method. It is also important to understand that fluoroscopic and 4D CT assessments of motion are prone to errors related to artifacts, predominantly due to irregularity of breathing. Some centers perform separate inspiration and expiration CT scans. We would caution against this, as it is an artificial type of breathing that usually does not represent endpoints of respiration at the time of treatment. Also, this method may lead to excessive exposure to normal tissues, and incorrect representation of normal tissue doses by overrepresentation of the respiratory extremes in the planning process.

    A great deal of information about tumor motion can be assessed in the various phase representations from the 4D scan. For example, place a line on the superior and the inferior points of the tumor as it moves in cranial and caudal directions

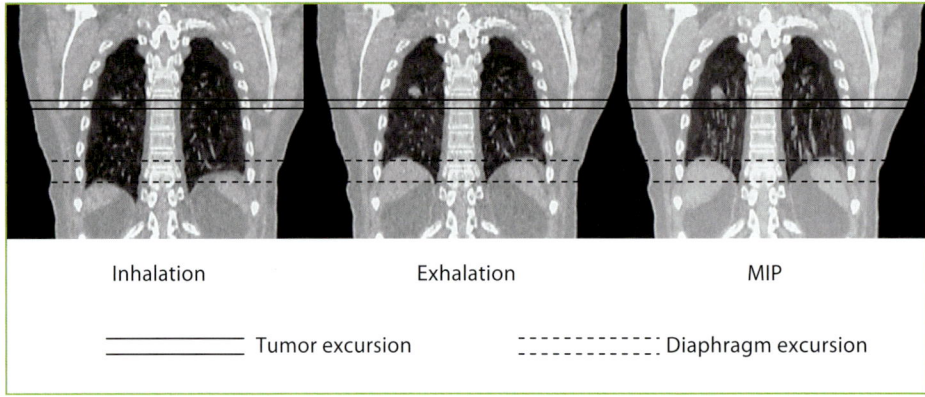

Fig. 5. The diaphragm may be a reasonable surrogate for the phase of the tumor motion, but not the amplitude (lines indicate motion boundaries, inhale and exhale). MIP = Maximum intensity projection. See online supplementary material for demonstration.

with respiration (fig. 5). This constitutes the magnitude of the respiratory movement throughout a cycle. The 4D scan will generate an artificial reconstruction called the maximum intensity projection image that provides this same information in one sequence. Relative motion can also be assessed on a 4D CT. Notice in figure 5 that total tumor movement (depicted by the lines at inspiration and expiration) show considerably less displacement than the diaphragm. Obviously, the diaphragm is not necessarily a reliable surrogate for the *amplitude* of the motion of lung tumors. Nonetheless, the diaphragm could be a reasonable surrogate for the tumor's *phase* characteristics.

Motion management is introduced according to the initial fluoroscopic real-time assessment. If the motion is more than a half centimeter superior and a half centimeter inferior, abdominal compression is used to control that motion. After motion is dampened under compression to within these range limits, the 4D CT is obtained; it is only carried out under adequately controlled motion, and confirms that the motion has been controlled, as shown in figure 6.

Abdominal compression is perhaps a 'low-tech' solution, but has a practical utility used and confirmed as effective in centers around the world. In the resting breathing cycle, the inspiratory force is typically generated primarily by the diaphragm. With heavy exertion, the chest wall musculature can add a powerful inspiratory force that dramatically increases the tidal volume. It is typically recruited when needed, but can be consciously used at any time.

Diaphragm contraction generates force vectors upon the pulmonary parenchyma mostly in the cranial-caudal direction, leading to significant tumor displacements. In contrast, chest wall expansion lifts the chest wall in all directions, resulting in force vectors that mostly cancel one another throughout the pulmonary

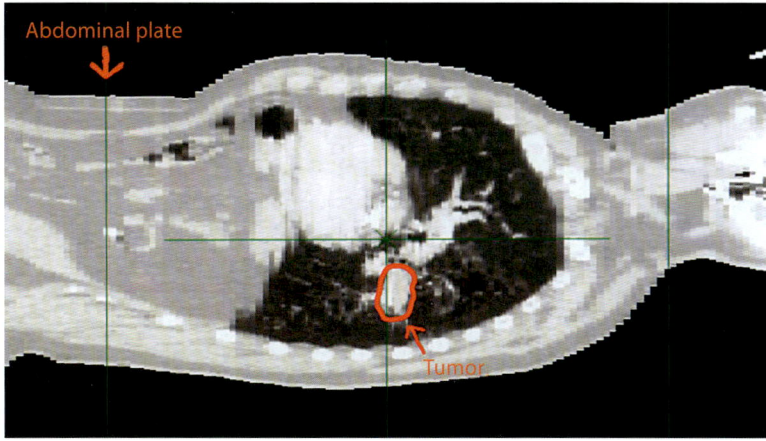

Fig. 6. Abdominal compression. See online supplementary material for demonstration.

parenchyma. With the abdominal compression technique, the goal is to persuade patients to inflate their lungs with chest wall musculature instead of diaphragm motion, resulting in significantly less tumor motion. Active coaching of patients to consciously use chest wall rather than abdominal musculature is essential to the success of this approach. Abdominal compression is best used as an accessory device, a deliberate guide to encourage inspiration using chest wall expansion.

Treatment Planning (Dosimetry)

Step 3 of the process is to construct 'stereotactic' dose distributions, characterized by both high-dose and high conformality of dose delivered to the target, and steep-dose gradients surrounding the target. Proper target delineation is critically important to this process (see online supplementary material for step-by-step demonstration). The radiation oncologist should define the gross tumor volume (GTV) on pulmonary CT windows. Ideally, highest-quality CT imaging and image fusion should be employed to achieve maximum confidence in defining the GTV. In SBRT, the GTV should be very close to the clinical target volume (CTV). This volume will move with every breath, and the full extent of this movement becomes the internal target volume. Finally, additional margin is added for treatment uncertainties such as setup errors, creating the planning target volume (PTV). Each expansion should be no more than is justifiably required. The actual treatment volume, the PTV, is considerably larger than the GTV. As the PTV increases, the final treatment volume increases by the cube root of its radius, and greatly increases the potential for toxicity as it does. Therefore, the determination of the GTV,

CTV and PTV through careful steps is critical for minimizing toxicity and maximizing tumor control.

In planning, one must create a very compact dosimetry that results in a high level of dose/target conformality (see online supplementary material for step-by-step demonstration). With current planning systems, it has become relatively easy to achieve this goal at the target, but much more challenging to reduce the regions of intermediate dose around it to keep the gradients as compact as possible. This is the most difficult aspect of the SBRT planning process, and distinguishes a good from a poor plan since intermediate levels of dose may be spilled in any direction. The key is to spill them in places that will likely cause less toxicity. Finally, the stereotactic paradigm trades large volumes treated to low doses in exchange for concise volumes treated to intermediate and high doses. In stereotactic treatments, this is understood and accepted, and considered desirable at least in nonpediatric patients.

Dose Constraints

Objective dose constraints for all normal tissues are needed so that safe SBRT dose distributions can be planned reliably. While constraints have been reasonably well validated for conventionally fractionated radiotherapy, equivalent ones have not been established for SBRT, though the basic data are now being accrued. One of the most important roles of the RTOG Advanced Technology Integration Committee's activities has been to systematically collect SBRT treatment plans together with their resulting outcomes. The risks for specific normal tissue volumes at given radiation doses can then be quantified. While we provide estimates regarding normal tissue constraints in this chapter, the reader should understand that these constraints are not fully validated with long-term follow-up data.

The concept of the critical volume, first introduced in the mid-1980s, is especially important in carrying out SBRT [19–21]. A critical volume is the absolute volume of an organ that must be spared in order to avoid its dysfunction after treatment. The evidence for this critical model derives from experience with surgical resections of metastases. A surgeon must preserve enough functioning normal organ tissue for the healthy survival of the host. This is the critical volume for that organ, its required residual capacity. In the liver, for example, about two thirds can be resected; if more is removed, basic hepatic function is not sustained.

This critical volume concept applies primarily to parallel tissues. These tissues, like the liver and lung parenchyma, are characterized by a redundancy of monotonous functional subunits. These converge and connect to the organism as a whole at the organ's hilum, composed of its ducts, blood vessels and nerves. Importantly, the critical volume concept – as it relates to parallel tissues – describes the amount or volume of tissue that can be damaged, not the degree of this damage. In reality, the threshold doses to disable function of parallel tissues are typically quite low, since the body insures the protection of these organs by means of redundancy, not their capacity to repair. As always, the dose limits for each organ will also depend on the context of each treatment, such as the number of fractions and the use of systemic agents, as well as the total radiation dose and baseline organ condition.

The strategies for planning SBRT are different depending on whether the therapy involves parallel or serial tissues. For parallel tissues, toxicities are primarily dependent on the organ volume treated, not on the dose potency used, whether for ablative or conventional radiotherapy. In parallel tissues, the planning goal for SBRT is to keep a *critical volume* of tissue below its threshold dose and functionally intact.

In contrast, serial tissue damage is very dependent on the radiation *maximum dose* potency delivered. These organs, like the esophagus, nerves and airways, are often highly active in repairing damage. This is critical because even small, subsegmental damage can be potentially catastrophic to the function of the entire organ. As a result, the safest planning approach with respect to serial organs is to simply avoid them whenever possible. When it is not, SBRT must avoid any dose that is beyond the capacity for tissue repair, even to the smallest volume. Unlike parallel tissues, serial tissues are preserved by limiting the maximum doses in the intermediate- and high-dose volumes.

Table 1 lists the dose constraints that are currently being used at UTSW, and is an update of values that we previously published [22]. Our planning algorithms use these constraints. However, there are times when they cannot be met because of proximity of the target. In these circumstances, we first insure that the dose distribution has the high dose conformality and steep gradients of a quality stereotactic plan, and then decide how much we can compromise on specific constraints in the given clinical circumstance. Aside from the spinal cord constraints, which should be generally obeyed in patients with reasonably good prognosis, the constraints in table 1 should be considered as guidelines, not firm requirements, until they can be modified and validated with long-term data.

Tissue Heterogeneity

It is very important that the correct dose heterogeneity algorithm is used in the planning process for SBRT. We have previously appreciated [23], and subsequently confirmed by dose measurements from the Radiologic Physics Center, that pencil beam and Clarkson methods can result in errors of 10–20% (and even up to 40%) for low-density materials because they do not adequately account for secondary scattering events [11]. Even performing the calculations in water density, as we

Table 1. SBRT dose constraints currently used at UTSW

1 Fraction

Serial tissue	Volume	Volume max., Gy	Max. point dose, Gy[b]	Endpoint (≥grade 3)
Optic pathway	<0.2 cm^3	8	10	neuritis
Cochlea			9	hearing loss
Brainstem (not medulla)	<0.5 cm^3	10	15	cranial neuropathy
Spinal cord and medulla	<0.35 cm^3	10	14	myelitis
	<1.2 cm^3	7		
Spinal cord subvolume (5–6 mm above and below level treated per Ryu)	<10% of subvolume	10	14	myelitis
Cauda equina	<5 cm^3	14	16	neuritis
Sacral plexus	<5 cm^3	14.4	16	neuropathy
Esophagus[a]	<5 cm^3	11.9	15.4	stenosis/fistula
Brachial plexus	<3 cm^3	14	17.5	neuropathy
Heart/pericardium	<15 cm^3	16	22	pericarditis
Great vessels	<10 cm^3	31	37	aneurysm
Trachea and large bronchus[a]	<4 cm^3	10.5	20.2	stenosis/fistula
Bronchus-smaller airways	<0.5 cm^3	12.4	13.3	stenosis with atelectasis
Rib	<1 cm^3	22	30	pain or fracture
Skin	<10 cm^3	23	26	ulceration
Stomach	<10 cm^3	11.2	12.4	ulceration/fistula
Duodenum[a]	<5 cm^3	11.2	12.4	ulceration
	<10 cm^3	9		
Jejunum/ileum[a]	<5 cm^3	11.9	15.4	enteritis/obstruction
Colon[a]	<20 cm^3	14.3	18.4	colitis/fistula
Rectum[a]	<20 cm^3	14.3	18.4	proctitis/fistula
Bladder wall	<15 cm^3	11.4	18.4	cystitis/fistula
Penile bulb	<3 cm^3	14	34	impotence
Femoral heads (right and left)	<10 cm^3	14		necrosis
Renal hilum/vascular trunk	<2/3 volume	10.6		malignant hypertension

Parallel tissue	Critical volume, cm^3	Critical volume dose max., Gy	Endpoint (≥grade 3)
Lung (right and left)	1,500	7	basic lung function
Lung (right and left)	1,000	7.4	pneumonitis
Liver	700	9.1	basic liver function
Renal cortex (right and left)	200	8.4	basic renal function

did in RTOG 0236, provides more accurate dose representation to the margin of the PTV than these faulty algorithms. Monte Carlo, collapsed cone (superposition) and AAA algorithms adequately account for secondary scattering and electronic equilibrium. At UTSW, we perform our planning using the collapsed cone algorithm on the average intensity projection dataset, an approach that we have confirmed to be accurate with phantom measurements.

Table 1 (continued)

3 Fractions

Serial tissue	Volume	Volume max., Gy (Gy/fx)	Max. point dose, Gy[b] (Gy/fx)	Endpoint (≥grade 3)
Optic pathway	<0.2 cm³	15.3 (5.1)	17.4 (5.8)	neuritis
Cochlea			17.1 (5.7)	hearing loss
Brainstem (not medulla)	<0.5 cm³	18 (6)	23.1 (7.7)	cranial neuropathy
Spinal cord and medulla	<0.35 cm³	18 (6)	21.9 (7.3)	myelitis
	<1.2 cm³	12.3 (4.1)		
Spinal cord subvolume (5–6 mm above and below level treated per Ryu)	<10% of subvolume	18 (6)	21.9 (7.3)	myelitis
Cauda equine	<5 cm³	21.9 (7.3)	24 (8)	neuritis
Sacral plexus	<5 cm³	22.5 (7.5)	24 (8)	neuropathy
Esophagus[a]	<5 cm³	17.7 (5.9)	25.2 (8.4)	stenosis/fistula
Brachial plexus	<3 cm³	20.4 (6.8)	24 (8)	neuropathy
Heart/pericardium	<15 cm³	24 (8)	30 (10)	pericarditis
Great vessels	<10 cm³	39 (13)	45 (15)	aneurysm
Trachea and large bronchus[a]	<4 cm³	15 (5)	30 (10)	stenosis/fistula
Bronchus-smaller airways	<0.5 cm³	18.9 (6.3)	23.1 (7.7)	stenosis with atelectasis
Rib	<1 cm³	28.8 (9.6)	36.9 (12.3)	pain or fracture
Skin	<10 cm³	30 (10)	33 (11)	ulceration
Stomach	<10 cm³	16.5 (5.5)	22.2 (7.4)	ulceration/fistula
Duodenum[a]	<5 cm³	16.5 (5.5)	22.2 (7.4)	ulceration
	<10 cm³	11.4 (3.8)		
Jejunum/ileum[a]	<5 cm³	17.7 (5.9)	25.2 (8.4)	enteritis/obstruction
Colon[a]	<20 cm³	24 (8)	28.2 (9.4)	colitis/fistula
Rectum[a]	<20 cm³	24 (8)	28.2 (9.4)	proctitis/fistula
Bladder wall	<15 cm³	16.8 (5.6)	28.2 (9.4)	cystitis/fistula
Penile bulb	<3 cm³	21.9 (7.3)	42 (14)	impotence
Femoral heads (right and left)	<10 cm³	21.9 (7.3)		necrosis
Renal hilum/vascular trunk	<2/3 volume	18.6 (6.2)		malignant hypertension

Parallel tissue	Critical volume, cm³	Critical volume dose max., Gy (Gy/fx)	Endpoint (≥grade 3)
Lung (right and left)	1,500	10.5 (3.5)	basic lung function
Lung (right and left)	1,000	11.4 (3.8)	pneumonitis
Liver	700	17.1 (5.7)	basic liver function
Renal cortex (right and left)	200	14.4 (4.8)	basic renal function

Treatment Verification and Delivery

The final step for lung SBRT is to verify the accuracy of the beam targeting and therapy delivery. In our experience, cone-beam CT reconstructions provide excellent information for image guidance at the time of treatment. They are particularly useful for lung targets, where the large density changes between solid tumor

Table 1 (continued)

4 Fractions

Serial tissue	Volume	Volume max., Gy (Gy/fx)	Max. point dose, Gy[b] (Gy/fx)	Endpoint (≥grade 3)
Optic pathway	<0.2 cm³	19.2 (4.8)	21.2 (5.3)	neuritis
Cochlea			21.2 (5.3)	hearing loss
Brainstem (not medulla)	<0.5 cm³	20.8 (5.2)	27.2 (6.8)	cranial neuropathy
Spinal cord and medulla	<0.35 cm³	20.8 (5.2)	26 (6.5)	myelitis
	<1.2 cm³	13.6 (3.4)		
Spinal cord subvolume (5–6 mm above and below level treated per Ryu)	<10% of subvolume	20.8 (5.2)	26 (6.5)	myelitis
Cauda equine	<5 cm³	26 (6.5)	28 (7)	neuritis
Sacral plexus	<5 cm³	26 (6.5)	28 (7)	neuropathy
Esophagus[a]	<5 cm³	18.8 (4.7)	30 (7.5)	stenosis/fistula
Brachial plexus	<3 cm³	23.6 (5.9)	27.2 (6.8)	neuropathy
Heart/pericardium	<15 cm³	28 (7)	34 (8.5)	pericarditis
Great vessels	<10 cm³	43 (10.75)	49 (12.25)	aneurysm
Trachea and large bronchus[a]	<4 cm³	15.6 (3.9)	34.8 (8.7)	stenosis/fistula
Bronchus-smaller airways	<0.5 cm³	20 (5)	28 (7)	stenosis with atelectasis
Rib	<1 cm³	32 (8)	40 (10)	pain or fracture
Skin	<10 cm³	33.2 (8.3)	36 (9)	ulceration
Stomach	<10 cm³	17.6 (4.4)	27.2 (6.8)	ulceration/fistula
Duodenum[a]	<5 cm³	17.6 (4.4)	27.2 (6.8)	ulceration
	<10 cm³	12 (3)		
Jejunum/ileum[a]	<5 cm³	18.8 (4.7)	30 (7.5)	enteritis/obstruction
Colon[a]	<20 cm³	24 (6)	33.2 (8.3)	colitis/fistula
Rectum[a]	<20 cm³	24 (6)	33.2 (8.3)	proctitis/fistula
Bladder wall	<15 cm³	17.6 (4.4)	33.2 (8.3)	cystitis/fistula
Penile bulb	<3 cm³	26 (6.5)	46 (11.5)	impotence
Femoral heads (right and left)	<10 cm³	26 (6.5)		necrosis
Renal hilum/vascular trunk	<2/3 volume	21 (5.25)		malignant hypertension

Parallel tissue	Critical volume, cm³	Critical volume dose max., Gy (Gy/fx)	Endpoint (≥grade 3)
Lung (right and left)	1,500	11.6 (2.9)	basic lung function
Lung (right and left)	1,000	12.4 (3.1)	pneumonitis
Liver	700	19.2 (4.8)	basic liver function
Renal cortex (right and left)	200	16 (4)	basic renal function

and aerated lung can be exploited. Our objective is to completely eliminate interfraction errors by obtaining pretreatment cone-beam CTs and making the appropriate couch adjustments. Furthermore, we try to quantify the intrafraction error by recording shifts at the *end* of a daily treatment so that we can have ongoing assessment of our institutional PTV margins.

Table 1 (continued)

5 Fractions

Serial tissue	Volume	Volume max., Gy (Gy/fx)	Max. point dose, Gy[b] (Gy/fx)	Endpoint (≥grade 3)
Optic pathway	<0.2 cm³	23 (4.6)	25 (5)	neuritis
Cochlea			25 (5)	hearing loss
Brainstem (not medulla)	<0.5 cm³	23 (4.6)	31 (6.2)	cranial neuropathy
Spinal cord and medulla	<0.35 cm³	23 (4.6)	30 (6)	myelitis
	<1.2 cm³	14.5 (2.9)		
Spinal cord subvolume (5–6 mm above and below level treated per Ryu)	<10% of subvolume	23 (4.6)	30 (6)	myelitis
Cauda equina	<5 cm³	30 (6)	32 (6.4)	neuritis
Sacral plexus	<5 cm³	30 (6)	32 (6.4)	neuropathy
Esophagus[a]	<5 cm³	19.5 (3.9)	35 (7)	stenosis/fistula
Brachial plexus	<3 cm³	27 (5.4)	30.5 (6.1)	neuropathy
Heart/pericardium	<15 cm³	32 (6.4)	38 (7.6)	pericarditis
Great vessels	<10 cm³	47 (9.4)	53 (10.6)	aneurysm
Trachea and large bronchus[a]	<4 cm³	16.5 (3.3)	40 (8)	stenosis/fistula
Bronchus-smaller airways	<0.5 cm³	21 (4.2)	33 (6.6)	stenosis with atelectasis
Rib	<1 cm³	35 (7)	43 (8.6)	pain or fracture
Skin	<10 cm³	36.5 (7.3)	39.5 (7.9)	ulceration
Stomach	<10 cm³	18 (3.6)	32 (6.4)	ulceration/fistula
Duodenum[a]	<5 cm³	18 (3.6)	32 (6.4)	ulceration
	<10 cm³	12.5 (2.5)		
Jejunum/ileum[a]	<5 cm³	19.5 (3.9)	35 (7)	enteritis/obstruction
Colon[a]	<20 cm³	25 (5)	38 (7.6)	colitis/fistula
Rectum[a]	<20 cm³	25 (5)	38 (7.6)	proctitis/fistula
Bladder wall	<15 cm³	18.3 (3.65)	38 (7.6)	cystitis/fistula
Penile bulb	<3 cm³	30 (6)	50 (10)	impotence
Femoral heads (right and left)	<10 cm³	30 (6)		necrosis
Renal hilum/vascular trunk	<2/3 volume	23 (4.6)		malignant hypertension

Parallel tissue	Critical volume, cm³	Critical volume dose max., Gy (Gy/fx)		Endpoint (≥grade 3)
Lung (right and left)	1,500	12.5 (2.5)		basic lung function
Lung (right and left)	1,000	13.5 (2.7)		pneumonitis
Liver	700	21 (4.2)		basic liver function
Renal cortex (right and left)	200	17.5 (3.5)		basic renal function

Validation of these constraints requires long-term follow up of clinical outcomes, and clinical caution should be used.
[a] Avoid circumferential irradiation.
[b] 'Point' defined as 0.035 cm³ or less.

Guidelines for Clinical Practice

- The first North American multicenter trial for medically inoperable early-stage lung cancer patients was RTOG 023 [6]. At 3 years, the primary tumor control rate was 98% and the survival rate was 56%.
- In this trial, the prescription dose was equivalent to approximately 18 Gy x 3 = 54 Gy if properly accounting for tissue heterogeneity. With the outcomes of this trial now mature, the treatment dose of 18 Gy!3 has become a standard for future RTOG trials involving SBRT for peripheral tumors in medically inoperable patients.
- In Japan, healthier patients refusing surgery have already been treated with SBRT. The results have been fairly similar to those of surgical resection, even with lobectomy, so long as a reasonable radiation dose potency has been delivered [16].
- Step 1 in the process of SBRT for lung cancer is to effectively immobilize the patient. Our technique uses a body frame. Whatever method is selected, the patient must be comfortable for these longer treatment sessions, and the setup must be reproducible.
- It is important to accurately assess the degree of respiratory motion of the treatment target, and to make a deliberate decision whether motion control intervention is necessary. At UTSW, we use real-time fluoroscopy to initially assess motion. This is followed by 4D CT. If the motion is more than a half centimeter superior and a half centimeter inferior, abdominal compression is used to control that motion.
- The determination of the GTV, CTV and PTV through careful steps is critical for minimizing toxicity as well as tumor control. The strategies for planning SBRT are different when therapy involves parallel or serial tissues. Table 1 lists the normal tissue dose constraints that are currently being used at UTSW.
- The successful delivery of SBRT requires the development of a comprehensive, specialized clinical program providing advanced technology and the technical expertise of physicians, physicists and therapists specially trained in SBRT applications.

References

1. Lo SS, Fakiris AJ, Chang EL, et al: Stereotactic body radiation therapy: a novel treatment modality. Nat Rev Clin Oncol 2009;7:44–54.
2. Timmerman R, Abdulrahman R, Kavanagh BD, Meyer JL: Lung cancer: a model for implementing stereotactic body radiation therapy into practice. Front Radiat Ther Oncol 2007;40:368–385.
3. Timmerman RD, Park C, Kavanagh BD: The North American experience with stereotactic body radiation therapy in non-small cell lung cancer. J Thorac Oncol 2007;2:S101–S112.
4. Blomgren H, Lax I, Naslund I, Svanstrom R: Stereotactic high dose fraction radiation therapy of extracranial tumors using an accelerator. Clinical experience of the first thirty-one patients. Acta Oncol 1995;34:861–870.
5. Uematsu M, Shioda A, Tahara K, et al: Focal, high dose, and fractionated modified stereotactic radiation therapy for lung carcinoma patients: a preliminary experience. Cancer 1998;82:1062–1070.
6. Timmerman R, Paulus R, Galvin J, et al: RTOG 0236: stereotactic body radiation therapy (SBRT) to treat medically inoperable early stage lung cancer patients. American Society for Therapeutic Radiology and Oncology (ASTRO) Annual Meeting; 2009 November 2, 2009; Chicago, IL. Int J Radiat Oncol Biol Phys 2009;75(suppl):S3.
7. Perez CA, Bauer M, Edelstein S, Gillespie BW, Birch R: Impact of tumor control on survival in carcinoma of the lung treated with irradiation. Int J Radiat Oncol Biol Phys 1986;12:539–547.

8 Perez CA, Bauer M, Emami BN, et al: Thoracic irradiation with or without levamisole (NSC #177023) in unresectable non-small cell carcinoma of the lung: a phase III randomized trial of the RTOG. Int J Radiat Oncol Biol Phys 1988;15: 1337–1346.

9 Timmerman R, Papiez L, McGarry R, et al: Extracranial stereotactic radioablation: results of a phase I study in medically inoperable stage I non-small cell lung cancer. Chest 2003;124:1946–1955.

10 Schefter TE, Kavanagh BD, Timmerman RD, Cardenes HR, Baron A, Gaspar LE: A phase I trial of stereotactic body radiation therapy (SBRT) for liver metastases. Int J Radiat Oncol Biol Phys 2005;62:1371–1378.

11 Xiao Y, Papiez L, Paulus R, et al: Dosimetric evaluation of heterogeneity corrections for RTOG 0236: stereotactic body radiotherapy of inoperable stage I–II non-small-cell lung cancer. Int J Radiat Oncol Biol Phys 2009;73:1235–1242.

12 Timmerman R, Galvin J, Michalski J, et al: Accreditation and quality assurance for Radiation Therapy Oncology Group: multicenter clinical trials using stereotactic body radiation therapy in lung cancer. Acta Oncol 2006;45:779–786.

13 Timmerman R, McGarry R, Yiannoutsos C, et al: Excessive toxicity when treating central tumors in a phase II study of stereotactic body radiation therapy for medically inoperable early-stage lung cancer. J Clin Oncol 2006;24:4833–4839.

14 McGarry RC, Papiez L, Williams M, Whitford T, Timmerman RD: Stereotactic body radiation therapy of early-stage non-small-cell lung carcinoma: phase I study. Int J Radiat Oncol Biol Phys 2005;63:1010–1015.

15 Ginsberg RJ, Rubinstein LV: Randomized trial of lobectomy versus limited resection for T1 N0 non-small cell lung cancer. Lung Cancer Study Group. Ann Thorac Surg 1995;60:615–622, discussion 22–23.

16 Onishi H, Araki T, Shirato H, et al: Stereotactic hypofractionated high-dose irradiation for stage I non-small cell lung carcinoma: clinical outcomes in 245 subjects in a Japanese multi-institutional study. Cancer 2004;101:1623–1631.

17 M. D. Anderson Cancer Center to lead study of surgery vs. CyberKnife Radiosurgery for operable lung cancer patients. http://www.accuray.com/PressReleases.aspx?release=676 (October 26, 2007).

18 Hurkmans CW, Cuijpers JP, Lagerwaard FJ, et al: Recommendations for implementing stereotactic radiotherapy in peripheral stage IA non-small cell lung cancer: report from the Quality Assurance Working Party of the randomised phase III ROSEL study. Radiat Oncol 2009;4:1.

19 Yaes RJ, Kalend A: Local stem cell depletion model for radiation myelitis. Int J Radiat Oncol Biol Phys 1988;14:1247–1259.

20 Wolbarst AB, Chin LM, Svensson GK: Optimization of radiation therapy: integral-response of a model biological system. Int J Radiat Oncol Biol Phys 1982;8:1761–1769.

21 Wolbarst AB: Optimization of radiation therapy II: the critical-voxel model. Int J Radiat Oncol Biol Phys 1984;10:741–745.

22 Timmerman RD: An overview of hypofractionation and introduction to this issue of seminars in radiation oncology. Semin Radiat Oncol 2008; 18:215–222.

23 Mayer R, Williams A, Frankel T, et al: Two-dimensional film dosimetry application in heterogeneous materials exposed to megavoltage photon beams. Med Phys 1997;24:455–460.

Dr. Robert Timmerman
Department of Radiation Oncology
University of Texas Southwestern Medical Center
5801 Forest Park Road
Dallas, TX 75390-9183 (USA)
Tel. +1 214 645 7651, Fax +1 214 645 7624
E-Mail robert.timmerman@utsouthwestern.edu

Stereotactic Body Radiation Therapy for Gastrointestinal Malignancies

A. Yuriko Minn · Albert C. Koong · Daniel T. Chang

Department of Radiation Oncology, Stanford University Medical Center, Stanford, Calif., USA

Abstract

Stereotactic body radiotherapy (SBRT) is an emerging treatment for pancreas cancer and liver tumors. Early data suggest excellent control rates for locally advanced pancreas cancer. However, due to the close proximity of the duodenum and stomach, steps to effectively minimize toxicities must be taken through image guidance of treatments. SBRT for liver tumors has also shown high rates of local control with low risks for hepatic toxicity. Careful selection of cases for SBRT is essential to achieve disease control and to minimize toxicity for patients. In treatment, attention must be paid to minimizing exposure of nearby normal tissues, including ribs, skin and bowel as well as the functioning organs surrounding the tumors. There is no accepted standard for the SBRT dose/fractionation schedule for these cases and the optimal strategy will likely depend on the size, number and location of lesions for each patient. However, the published data seem to suggest an overall dose-response effect. To realize the clinical potential of SBRT for these tumors, investigations are needed to determine optimum fractionation schedules and to integrate its use with systemic chemotherapy programs.

Copyright © 2011 S. Karger AG, Basel

Stereotactic radiosurgery is a technique that has been used for intracranial tumors for over 4 decades but only recently applied to tumors elsewhere in the body. Improved technologies in tumor imaging and image guidance have allowed safer delivery of large doses per fraction of radiotherapy, and have led to stereotactic body radiotherapy (SBRT) applications in many sites including the abdomen. The major advantages to SBRT compared with conventionally fractionated radiotherapy are intensified treatment to the primary tumor and shortened overall treatment course time. The concise duration of therapy increases its practical applications since potential interference with other ongoing therapies such as chemotherapy programs is reduced. The hypofractionated schedules of SBRT deliver a higher biologically effective dose. This dosing may overcome mechanisms that contribute to radio-resistance to standard radiation fractionation such as tumor hypoxia. Tumors that are small, localized and reasonably well demarcated from their normal tissue surroundings are ideal for SBRT [1]. SBRT has been applied to certain gastrointestinal cancers, specifically pancreatic and hepatic tumors. This chapter will review this work with emphasis on the treatment programs at Stanford University.

Stereotactic Body Radiotherapy for Locally Advanced Pancreatic Cancers

Pancreatic cancer is the fourth leading cause of cancer death in the United States [2]. Surgical resection offers the only chance for long-term survival; however, only a minority of patients have potentially resectable disease at presentation. The overall 5-year survival rate for all stages remains less than 5% [3].

Combined chemoradiotherapy has played a significant role in managing locally advanced, unresectable pancreatic cancer. Two decades ago, the Gastrointestinal Tumor Study Group reported a survival benefit in patients treated with 5-fluorouracil and radiotherapy compared with radiotherapy alone [4]. Another Gastrointestinal Tumor Study Group study revealed improved survival in pancreatic cancer patients treated with chemoradiotherapy compared with chemotherapy alone [5].

Despite treatment of initially nonmetastatic pancreatic disease, local cancer progression is common, with rates of 24–58% reported from clinical trials [6]. While systemic progression is the primary cause of disease-related mortality, local progression can be a significant source of morbidity that impacts the quality of life of patients. For this reason, investigators at Stanford University have pursued a strategy of SBRT as a method of radiation dose intensification [7–9] (table 1).

Clinical Trials Using Stereotactic Body Radiotherapy for Pancreatic Cancer

In a phase I dose escalation study, Koong et al. [9] treated 15 patients with locally advanced pancreatic cancer with single fractions of 15, 20 or 25 Gy. The investigators observed minimal acute toxicity, and the maximum tolerated dose was not reached. The study was stopped at 25 Gy because the clinical objective of local control was met. A local control rate of 100% was observed, and the overall survival was 11 months. Nevertheless, all patients developed distant metastases. Most patients experienced clinical benefits including pain reduction and weight increase. Three patients (20%) developed grade 2 toxicities including treatment-related abdominal pain (n = 2) and diarrhea (n = 1).

Table 1. SBRT trials at Stanford University for locally advanced pancreatic cancer

Study	Number of patients	Treatment	Local control	Overall survival
Koong et al., phase I [9]	15	15 Gy → 20 Gy → 25 Gy × 1	100%	11 months
Koong et al., phase II [8]	16	45 Gy IMRT + 5-FU then 25 Gy × 1	94%	33 weeks
Schellenberg et al., phase II [7]	16	Gemcitabine then 25 Gy × 1 then gemcitabine	81%	11.4 months

Gy = Gray, → dose-escalation, IMRT = intensity-modulated radiation therapy, 5-FU = 5-fluorouracil.

Given these encouraging results, Koong et al. [8] began in 2005 a phase II study combining the use of conventionally fractionated radiotherapy using intensity-modulated radiotherapy followed by an SBRT boost to the primary tumor. The primary tumor and regional lymph nodes were treated using a standard intensity-modulated radiotherapy technique to 45 Gy at 1.8 Gy per fraction with concurrent 5-fluorouracil chemotherapy. Within a month of completing standard chemoradiation therapy, the primary tumor received a 25-Gy single-fraction boost using CyberKnife (Accuray, Sunnyvale, Calif., USA). Nineteen patients were enrolled, but 3 patients (16%) had distant progression of their disease before the boost and did not receive the planned SBRT. This study showed a local control rate of 94%, but toxicity rates appeared to be increased: 4 patients experienced grade 2 (21%) and 2 patients (10%) had grade 3 gastrointestinal toxicities.

In another phase II study, Schellenberg et al. [7] reported the outcomes for patients who received a single fraction of 25-Gy SBRT between full-dose gemcitabine chemotherapy treatments. Sixteen patients received 3 weekly doses of gemcitabine (1,000 mg/m^2) followed by a 2-week break. They then received SBRT and resumed the same chemotherapy following an additional 2-week recovery period. They remained on gemcitabine until their disease progressed or they experienced toxicity. In that study, 19% had local progression, and the median survival was 11.4 months. This study was the first to integrate SBRT with standard systemic chemotherapy. While results continued to show efficacy, toxicity rates seemed to increase: 5 patients (31%) developed grade 2 bowel ulcers, 1 patient had a grade 3 stenosis, and 1 patient had a grade 4 bowel perforation.

In a pooled analysis of these 3 trials plus additional patients treated off protocol, 77 patients who had received a single fraction of 25 Gy were analyzed [10]. Most patients (96%) received chemotherapy, typically gemcitabine-based. Freedom from local progression was 84% at 1 year, and the 1-year freedom from isolated local progression was 95%. There were 9 regional recurrences, but all occurred in the setting of a local or distant failure. Sixty-five patients (84%) had distant recurrences, and the 12-month progression-free survival rate was 9%. A median overall survival of 6.4 months was observed when calculated from the date of SBRT. It is

Table 2. Univariate analysis of dosimetric endpoints

Variable[†]	Cutoff cm³[††]	Incidence of grade 2–4 duodenal toxicity, %[†††]	log-rank p value
V5	<25	28	0.39
	≥25	31	
V10	<16	15	0.015
	≥16	46	
V15	<9.1	11	0.002
	≥9.1	52	
V20	<3.3	11	0.002
	≥3.3	52	
V25	<0.21	12	0.010
	≥0.21	45	

Courtesy of *Int J Radiat Oncol Biol Phys*, Murphy et al. [12].
[†] V5 refers to the volume of duodenum receiving 5 Gy. [††] Cutoff refers to the median value. [†††] Actuarial incidence at 12-months.

important to note, however, that some patients had received prior chemotherapy and radiotherapy, making the overall survival rate difficult to compare with those from other studies that included only de novo patients. The median survival from the time of diagnosis was 11.9 months.

Toxicities were significant: 14 patients (18%) experienced a grade 2 or higher acute or late toxicity. Ten of 18 (55%) developed stomach or small bowel ulcers, 3 patients had biliary strictures, 1 had a duodenal stricture and 1 patient had a duodenal perforation. The 6- and 12-month actuarial rates of toxicity (grade 2 or higher) were 11 and 25%, respectively [10].

In 2005, Hoyer et al. [11] published their experience with SBRT for locally advanced pancreas cancer. Of 22 patients who received 45 Gy in 3 fractions, 6 (27%) developed local progression as a component of failure and only 1 patient had isolated local tumor progression. The investigators reported significant gastrointestinal toxicity, including 5 patients (22%) with severe mucositis, ulceration or perforation, as well as worsened pain, nausea and performance status after treatment [11]. However, because larger margins of expansion were used, these investigators treated significantly larger volumes than patients would receive under the protocols described by Chang et al [10]. This difference most likely explains the increased toxicities observed in the Hoyer study.

Toxicity
The adjacent duodenum and stomach remain the critical structures at risk when treating the pancreas with large doses of radiation and have been dose limiting in

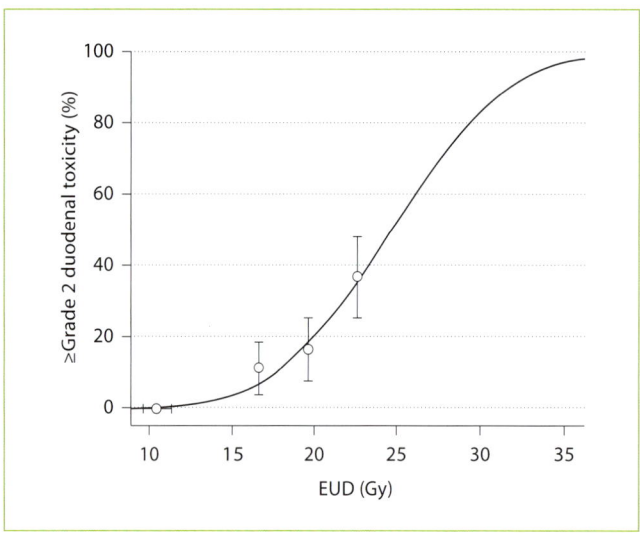

Fig. 1. Lyman NTCP model. The curve represents the estimated NTCP model as a function of equivalent uniform dose (EUD) fit from methods described in the Methods section. Circles represent actual patient data divided into quartiles; error bars represent the standard error. Courtesy of *Int J Radiat Oncol Biol Phys*; Murphy et al. [12].

treatment programs. With the location of the tumor often in very close proximity to, and sometimes intimately associated with these structures, irradiating portions of these organs is unavoidable. Therefore, strategies to reduce margins of expansion for setup uncertainties and respiratory motion are critical. Yet, despite these efforts, the rates of toxicity observed in the previously mentioned studies are not trivial.

As experience is gained with SBRT in the abdomen, understanding bowel tolerance with hypofractionation is vital to further refining dosing strategies. Murphy et al. [12] analyzed the outcomes of 73 previously unirradiatiated patients who had been treated with 25 Gy in a single fraction at Stanford University. Overall, 12 patients (16.4%) experienced grade 2 or higher toxicity, and dose-volume histogram parameters were correlated with toxicity. The authors found that V10 (the volume that received 10 Gy or more), V15, V20, V25, and the maximum dose (D_{max}) all correlated with the development of ≥grade 2 toxicity (table 2). In addition, application of the Lyman-Kutcher-Burman model for normal tissue complication probability (NTCP) [13, 14] revealed a strong correlation between equivalent uniform dose to the duodenum and the risk of toxicity (fig. 1). A multivariate analysis revealed that the NTCP model alone correlated with this risk (table 3).

Table 3. Multivariate analysis for risk of toxicity

Variable	Hazard ratio (95% CI)	p value
NTCP	1.06 (1.02, 1.10)	0.001
Age	0.96 (0.91, 1.01)	0.15
Gender	0.93 (0.20, 4.4)	0.93
Tumor location	0.5 (0.06, 4.1)	0.51
Time from neoadjuvant chemotherapy to radiation	0.76 (0.44, 1.3)	0.31
Treatment machine	0.41 (0.05, 3.7)	0.43
Bypass surgery	0.75 (0.13, 4.3)	0.75
Biliary stent	0.77 (0.22, 2.7)	0.68

Courtesy of *Int J Radiat Oncol Biol Phys*, Murphy et al. [12].

These data, the first to model duodenal tolerance in the setting of hypofractionation, confirm that the risk of toxicity is dose- and volume-related. Small portions of small bowel and stomach can likely tolerate larger doses of irradiation, but further studies are needed to characterize these relationships. In addition, while these data apply to single fraction SBRT, they do not address cases in the setting of multiple fraction courses, which presumably have very different tolerance profiles depending on the fraction sizes.

Treatment Planning and Delivery of Stereotactic Body Radiotherapy for Pancreas Cancer

SBRT is not constrained to any particular radiation machine, and the only requirement is the ability to use image guidance and respiratory motion management. A variety of motion-management techniques are available, including respiratory-gating, breath-hold and respiratory-tracking, and one must decide upfront which strategy will be utilized as treatment planning will differ for each. At Stanford University, respiratory-gating is typically used when treating with SBRT.

Target Tracking with Gold Fiducial Seeds

The complex respiratory motion of upper abdominal organs, with their close proximity to the diaphragm, would necessitate excessive planning target volume (PTV) margins of expansion unless some form of motion management was introduced. One method to address this uncertainty and reduce treatment margins is to place 3–5 gold fiducial seeds directly into the tumor, for tracking using kV image guidance. While fiducial placement can be accomplished by CT-guided intervention, laparoscopy or laparotomy, our preferred approach has been through the endoscopic route, using intraluminal ultrasound to minimize tissue invasiveness and possible complications. This procedure is performed 5–7 days before obtaining the treatment planning scans in order to allow the seeds to stabilize after any migration following placement.

Treatment Planning Using Respiratory-Gating

Multiple scans are taken during CT simulation for optimal target delineation and treatment delivery. For treatment positioning, patients are placed supine with their arms above their heads and immobilized using an Alpha Cradle (Smithers Medical Products, North Canton, Ohio, USA). When possible, FDG-PET (fluorodeoxyglucose positron emission tomography) is recommended to assist in target delineation and to survey for occult metastases. Intravenous contrast and a small amount of oral contrast are administered before the CT image acquisitions. Both arterial and portal venous-phase CT scans are obtained at a 1.25-mm slice thickness to accurately and thoroughly evaluate the extent of the tumor. Finally, a 4D CT scan is performed for each patient using the Real-Time Position Management System (RPM, Varian Medical Systems, Palo Alto, Calif., USA) to account for respiratory motion.

The gross tumor volume is contoured on axial slices of the biphasic CTs with assistance from the 4D CT and PET-CT images. The use of the 4D CT is dependent on the chosen strategy to account for respiratory motion. At Stanford University, the preferred strategy has been respiratory gating. Pancreatic tumor motion is visually inspected throughout the respiratory cycle, and a gating window is centered on the end-expiration phase since patients tend to spend most of their breathing cycle time in this phase. This phase has more stability in the tumor position and allows the largest window of time for treatment delivery. The biphasic CT images are co-registered with the end-expiration phase of the 4D CT, as well as the phases from each end of the selected gating window.

The gross tumor volume consists of the grossly visible tumor seen on contrast-enhanced CT and PET. The internal tumor volume is determined by expanding the gross tumor volume to include changes in tumor position seen on the selected 4D CT phases representing the gating window. The PTV is defined as the internal tumor volume plus 2–3 mm for setup error. The regional lymph nodes are not included in the target volume.

The dose is prescribed to the isodose line that covers 95% of the target. The prescription dose for single fraction SBRT is 25 Gy. Normal tissue constraints for treatment planning are shown in table 4. However, given the concern of normal tissue toxicity, consideration should be given to fractionate the dose over 2–5 fractions. The current protocol being tested in a multi-institution study is 33 Gy in 5 fractions.

Table 4. Normal tissue constraints

Organ	Constraint
Prescription dose	25 Gy to the isodose line covering >95% of PTV
Tissue	
Liver	50% of volume <5 Gy
	70% <2.5 Gy
Kidney	75% of volume of each kidney <5 Gy
Spinal cord	maximum dose <5 Gy
Stomach	<4% of volume >22.5 Gy
	the 50% isodose line should not reach the distal (nonadjacent) wall of the lumen
Duodenum	<5% of volume >22.5 Gy
	<50% of volume >12.5 Gy
	the 50% isodose line should not reach the distal (nonadjacent) wall of the lumen
Other bowel	maximum dose <21 Gy
	<5% of volume >20 Gy

Courtesy of *Cancer*, Chang et al. [10].

Tumor Motion and Treatment Delivery Using Respiratory Gating
With normal respiration, pancreatic tumors can move as much as 2–3 cm from inspiration to expiration [15, 16]. Fluoroscopic recordings of pancreas and liver tumors have shown average fiducial movements of 7.4 mm in the superior-inferior direction and 3.8 mm in the anterior-posterior direction [16]. The movement is complex and can occur in any dimension, but appears greatest in the superior-inferior direction. It is essential to account for this respiratory-associated movement. In a study of patients with pancreatic cancer treated with SBRT, significant movement of the pancreas was observed in the superior-inferior direction (mean up to 12.7 mm), left-right direction (mean up to 9.4 mm) and anterior-posterior direction (mean up to 5.5 mm) [17]. Arbitrary expansion of the treatment margin to account for this movement will result in treatment of larger normal tissue volumes and probable increases in toxicity. On the other hand, inadequate margins may lead to marginal misses of tumor cells and to local failures. Recognizing the exact extent of tumor movement in each direction allows clinicians to place an 'intelligent', nonuniform margin around the tumor volume.

On the day of treatment, before radiation delivery, orthogonal kV portal imaging is used to align the observed bone anatomy and the fiducial markers to their target positions based on the digitally reconstructed radiographs (DRR) from CT planning. After this initial alignment, fluoroscopy through orthogonal isocenter angles and each treatment angle is used to confirm the seed positioning during the

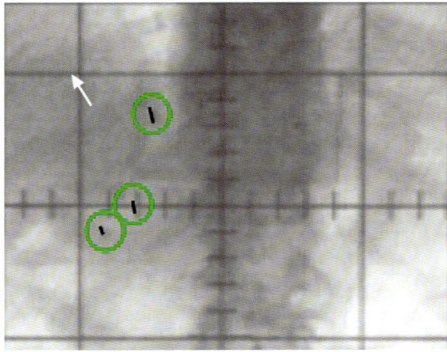 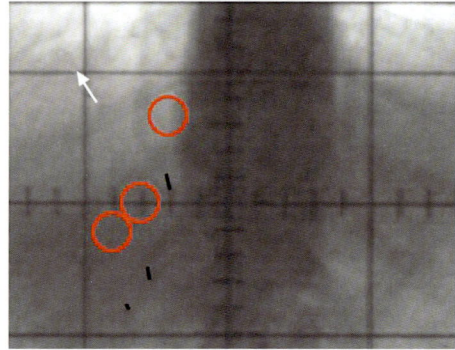

Fig. 2. Adjustments to the gating window. Adjustments are made to ensure that the radiation beam is 'on' when fiducials are within 2–3 mm of their position on the digitally reconstructed radiographs and 'off' when the seeds move outside of that region. Courtesy of Lippincot, Williams & Wilkins; Chang et al. [38]. See online supplementary material.

gating window. This window is adjusted to ensure that the radiation beam is 'on' when the fiducials are within 2–3 mm of their position on the digitally reconstructed radiographs, and 'off' when the seeds move outside of that region (fig. 2).

Current Indications and New Investigations in Stereotactic Body Radiotherapy for Pancreas Cancer

While there are no established criteria for eligibility to receive SBRT, the ideal patient should have unresectable or marginally resectable disease, no lymph node involvement, and no distant metastases. Ideally, the tumor size should be less than 5 cm, but more importantly, the anatomy with respect to the adjacent critical structures should be favorable. Generally this means that there should be some separation between the bowel/stomach and the tumor. Tumors that invade the duodenum or cause gastric outlet obstruction anatomically prevent sparing of these critical normal structures, and these patients are less favorable candidates for this treatment approach.

Due to rapid systemic progression, not all patients will benefit from the local control offered by SBRT. However, for those whose disease is more indolent or well controlled with systemic therapy, local control will take a higher priority since local progression could impact their quality of life and survival. For this reason, it currently appears reasonable to offer gemcitabine chemotherapy for approximately 2–4 months to patients with locally advanced, inoperable disease as initial therapy. Imaging studies may then be repeated to determine treatment response. SBRT would not be suitable for patients with progressive disease, but could be considered for those whose disease is stable or improved. This approach is supported by a recent German study, which found that radiotherapy benefitted patients who did not have systemic disease progression after 3–6 months of initial gemcitabine chemotherapy [18]. This design could be incorporated into future studies utilizing SBRT.

In addition, given the substantial toxicity seen in the clinical data using single fraction SBRT, more fractionated schedules (perhaps 2–5 fractions) should be investigated. If the adjacent bowel and stomach cannot be spared, it is likely that 25 Gy in a single fraction carries too high a risk of toxicity, and smaller doses per fraction may reduce the risk of morbidity. A better understanding of bowel tolerance to a range of hypofractionated schedules may allow a risk-adapted approach for each patient and a strategy for improving the therapeutic ratio generally.

Finally, it is extremely important that SBRT for pancreatic tumors be performed using an advanced treatment planning process. Optimized imaging studies are required for tumor visualization, respiratory motion must be accounted for and controlled, and treatment delivery must be verified with image guidance. SBRT is not recommended unless all of these criteria can be satisfied.

SBRT for pancreatic cancer has given us a greater understanding of the benefits and limitations of treating this disease. Overall, the clinical experience to date shows superior rates of local control, but no significant changes in patient survival compared with historical data. This highlights the overwhelming role that systemic disease spread plays in determining the clinical course of these patients, and the need to integrate SBRT with optimized chemotherapy as the most appropriate treatment strategy.

Stereotactic Body Radiotherapy for Hepatic Malignancies

Malignant tumors of the liver, both primary and metastatic, are important causes of cancer-related morbidity and mortality. Advances in systemic therapy for metastatic colorectal cancer have increased the median survival of patients to approximately 2 years [19]. As their life expectancy increases, control of coexisting liver disease becomes increasingly important to prevent organ failure and related morbidities. When possible, surgical resection remains the standard of care; however, these tumors often present as unresectable lesions.

Radiation therapy has not been a standard treatment for liver tumors; other focal therapies such as radiofrequency ablation, cryoablation, chemoembolization and ethanol injection are more established approaches. However, each of these treatments has limitations, often dictated by tumor location, size and underlying liver dysfunctions, as well as their risks for local failure [20–22]. Recently, clinical data have shown that SBRT is a potential therapy for treating liver tumors effectively and quickly.

Table 5. Selected studies using SBRT to treat liver lesions

Study	Lesions n	Patients/lesions, n	Follow-up	Dose (Gy/fraction)	Motion	CTV→PTV expansion	Prescription	Local control
Herfarth et al. 2001 [37]	1–3	37/60 (4 primary, 56 mets.)	5.7 month	14 Gy/1 → 26 Gy/1	abdominal compression	0.6 cm axial 1 cm SI	Isocenter 80% IDL around PTV	6-month: 75% 12-month: 71% 18-month: 67%
Schefter et al. 2005 [33]	1–3	18/24 (mets.)	7.1 months	36 Gy/3 60 Gy/3	abdominal compression or ABC	0.5 cm axial 1 cm SI	80–90% IDL around PTV	
Hoyer et al. 2006 [27]	1–6	64/141 (mets.)	4.3 years	45 Gy/3		0.5 cm axial 1 cm SI	Isocenter	24-month: 79%
Wulf et al. 2006 [26]		44/56 (5 primary, 51 mets.)	15 months	30 Gy/3 or 28 Gy/4 37.5 Gy/3 or 26 Gy/1	abdominal compression	0.5 cm axial 0.5–1 cm SI	65–80% IDL around PTV	1-year: 92% 2-year: 66% (primary – 100%, 15-month median f/u)
Mendez Romero et al. 2006 [29]	1–3	25/45 (11 primary, 34 mets.)	12.9 months	37.5 Gy/3 or 25 Gy/5 or 30 Gy/3	abdominal compression	0.5–1 cm, but now individualizing	65% IDL around PTV	1-year: 94% 2-year: 82%
Kavanagh et al. 2006 [28]	1–3	36 (mets.)	19 months	60 Gy/3	abdominal compression or breath-hold	0.5 cm axial 1 cm SI	80–90% IDL around PTV	18-month: 93%
Katz et al. 2007 [23]	1–6	69/174 (mets.)	14.5 months	30–55 Gy/ 7–20	breath-hold	0.7 cm transversal 1 cm SI	80% IDL around PTV	10-month: 76% 20 month: 57%
Rusthoven et al. 2009 [24]	1–3	47/63 (mets.)	16 months	36 Gy/3 → 60 Gy/3	abdominal compression breath-hold	0.7 cm axial 1.5 cm SI 0.5 cm axial 1 cm SI	max. IDL covering PTV	1-year: 95% 2-year: 92% (<3 cm 100%)
Lee et al. 2009 [25]	1–8	68 (mets.)	6 months	27.7 Gy 60 Gy/6	abdominal compression or breath-hold	min. 0.5 cm	max. IDL covering PTV	1-year: 71%
Goodman et al. 2009 [31]	1–5	26/40 (7 primary, 19 mets.)	17.3 months	18 Gy/1 → 30 Gy/1	synchrony	0.3–1 cm	80% IDL around PTV	1-year: 77%
Tse et al. 2008 [32]		41 (primary)	17.6 months	24–54 Gy/6	abdominal compression or breath-hold	min. 0.5 cm		1-year: 65%

CTV = Clinical target volume; mets. = metastases; SI = superior-inferior; IDL = isodose line; ABC = active breathing coordinator; f/u = follow-up.

Studies Using Stereotactic Body Radiotherapy for Hepatic Tumors

Table 5 summarizes recent studies of SBRT for liver lesions. The treatment schedules and techniques have been heterogeneous, but the resulting local control rates, ranging from 71 to 100%, have been promising [23–30].

In a dose escalation study at Stanford University, 26 patients with 33 lesions were treated with SBRT [31]: 19 patients (73%) had metastases and 7 (27%) had primary hepatic tumors, either hepatocellular carcinoma (HCC) or cholangiocarcinoma. The dose was escalated from 18 to 30 Gy in a single fraction at increments of 6 Gy. No patient developed dose-limiting acute toxicity even at the highest dose. However, 1 patient, treated to a porta-hepatis mass developed a late duodenal perforation. A total of 5 local failures were observed, and the 1-year cumulative incidence of failure was 23%.

A prospective study from Princess Margaret Hospital (Toronto, Ontario) enrolled 68 patients with metastatic liver tumors [25]. A 6-fraction SBRT regimen was used, and the fraction size was individualized based on the risk of liver toxicity (determined by NTCP modeling). Patients had up to 8 liver metastases, and the median dose delivered was 41.8 Gy (range: 27.7–60 Gy) over the 6 fractions. No radiation-induced liver disease, serious liver toxicity or other dose-limiting toxicities were observed. Six patients (9%) had acute grade 3 toxicities, including gastritis (n = 2), nausea (n = 2), lethargy (n = 1) and thrombocytopenia (n = 1), and 1 patient had a grade 4 thrombocytopenia. Five patients (7%) had significant late toxicities including grade 2 dyspepsia (n = 1), grade 2 rib fracture (n = 2), grade 4 duodenal bleeding (n = 1) and a grade 5 small bowel obstruction (n = 1). The 1-year local control rate was 71% [25], and the median survival was 17.6 months. Similarly promising results have been reported by Rusthoven et al. [24].

Much less is known about SBRT for primary liver cancer. For HCC, most cases present with unresectable disease. The largest SBRT series has been reported by the Princess Margaret group in a phase I trial. The median dose used was 36 Gy over 6 fractions (range: 24–54 Gy), though this was individualized as in their treatments for metastatic lesions [32]. Forty-one patients were enrolled with unresectable disease, 31 (76%) with HCC and 10 (24%) with cholangiocarcinoma. The investigators did not find any radiation-induced liver disease or grade 4–5 toxicities within 3 months following therapy. The 1-year local control rate was 65%, and most failures occurred outside of the treated volumes. The median survival was 11.7 months for patients with HCC and 15 months for patients with cholangiocarcinoma.

Treatment Planning and Delivery of Stereotactic Body Radiotherapy for Hepatic Tumors

Treatment Fractionation and Toxicity

Based on the available data, it is difficult to draw conclusions on the optimal dose schedule, given the wide range of dose schemes used in existing reports. Also, there was heterogeneity in the number of tumors treated, their size, histology and other factors. A critical issue in determining any SBRT dosing schedule is the tolerance of the normal tissues with hypofractionation, in this case, the liver. While the liver appears able to receive high doses of radiation as long as a sufficient volume of healthy liver tissue is spared, the exact dose-volume relationship is not well understood. Investigators from the University of Colorado have used a critical volume model where at least 700 cm^3 of healthy liver tissue should receive less than 15 Gy over 3 fractions [28, 33]. Other investigators have limited the V15 and V21 (normal liver tissue receiving 15 Gy and 21 Gy, respectively) to less than 50 and 30%, respectively, in 3 fractions [34]. As mentioned above, the investigators from Princess Margaret Hospital have used NTCP modeling to determine the prescription dose [32, 35].

In addition to liver tolerance, however, more attention should also be paid to structures like the chest wall and bowel, which can often be in close proximity depending on the location of the lesion. Pettersson et al. [36] performed a dosimetric analysis to determine the dose-volume relationship for rib fractures in 68 patients treated with SBRT for lung tumors. They found that keeping the dose received by 2 cm^3 of rib (D_{2cm3}) to less than 21 Gy in 3 fractions results in a near 0% risk of fracture. If the D_{2cm3} was greater than 27.3 or 49.8 Gy in 3 fractions, the resultant risk of fracture was 5 and 50%, respectively.

Treatment Planning

Treatment techniques have varied among studies. Investigators have used different methods to control for respiratory motion, including abdominal compression to limit the amount of diaphragmatic excursion and breath-holding. The preferred treatment approach at Stanford University has been to use inserted gold fiducials and respiratory gating, similar to the techniques used for pancreatic SBRT. Treatment planning has generally been based on IV contrast-enhanced thin-cut CT scans, in both the arterial and portal venous phases to optimally visualize the lesion(s). FDG-PET has also been used for planning when tumors have demonstrated FDG avidity, and 4D CT scans have been obtained to capture the respiratory motion of the tumor.

The optimal clinical target volume margin to account for microscopic tumor extension is not well established and is likely to depend on the tumor histology being treated; however, most studies have used only small margins of expansion beyond the visible gross lesion. The PTV margins for setup and respiratory uncertainty are dependent on the technique used for treatment. When patients are free-breathing with abdominal compression, a larger margin in the superior-inferior axis (1 cm) and a smaller margin in the axial direction (0.5–0.7 cm) have typically been used [24, 26–28, 33, 37]. At Stanford University, with the use of fiducials and respiratory gating, the PTV margin expansion typically ranges between 0.3 cm and 0.5 cm.

Current Indications for Stereotactic Body Radiotherapy for Hepatic Malignancy

As experience with SBRT for liver tumors has evolved, the eligibility criteria for treatment of metastatic disease have continued to change as well. There has been no consensus on the maximum size or number of lesions suitable for SBRT since other factors such as tumor location and biologic aggressiveness may be equally or more important. However, the most important consideration must be the clinical benefit that the patient is likely to derive from this therapy. Therefore, careful patient selection is required. In general, SBRT should only be considered after maximally effective systemic therapy has been completed. The anticipated PTV volume should also be taken into careful consideration to ensure that normal tissue constraints are likely to be satisfied.

For primary tumors, patients suitable for SBRT should have single lesions and adequate liver reserve. It is not clear whether SBRT should be used in patients with multifocal disease. It is important to remember that the evolution of therapy for HCC has occurred without radiotherapy as a standard treatment modality. Most of the treatment programs now involve surgery, chemotherapy, radiofrequency ablation or chemoembolization. Therefore, the role of SBRT is less clear and further studies should focus not only on maximizing efficacy, but also on determining how SBRT should be used in the context of these previously established therapies.

Guidelines for Clinical Practice

Pancreatic Cancer
- SBRT for locally advanced pancreatic cancer appears to provide high rates of local disease control.
- The vast majority of patients develop metastatic disease, with or without local control of their disease.
- An ideal candidate should have unresectable or marginally resectable disease, no lymph node involvement, and no distant metastases.
- Gastrointestinal toxicity is a significant complication of SBRT for pancreatic cancer.

- Advanced localization and treatment planning technologies and procedures are essential to performing SBRT for pancreatic cancer.
- Further investigations into optimizing SBRT fractionation schedules and their timing with chemotherapy are needed.

Liver Tumors
- SBRT can achieve promising local control rates for metastatic lesions in the liver.
- The role of SBRT for HCC is less clearly defined.
- The optimal dose and fractionation scheme is unknown.
- A better understanding of liver tolerance with hypofractionation is needed.
- The risk of radiation-induced liver disease is greater in cirrhotic livers.

References

1 Timmerman RD, Kavanagh BD, Cho LC, et al: Stereotactic body radiation therapy in multiple organ sites. J Clin Oncol 2007;25:947–952.
2 Jemal A, Siegel R, Ward E, et al: Cancer statistics, 2009. CA Cancer J Clin 2009;59:225–249.
3 Sener SF, Fremgen A, Menck HR, et al: Pancreatic cancer: a report of treatment and survival trends for 100,313 patients diagnosed from 1985–1995, using the National Cancer Database. J Am Coll Surg 1999;189:1–7.
4 Moertel CG, Frytak S, Hahn RG, et al: Therapy of locally unresectable pancreatic carcinoma: a randomized comparison of high dose (6,000 rads) radiation alone, moderate dose radiation (4,000 rads + 5-fluorouracil), and high dose radiation + 5-fluorouracil: The Gastrointestinal Tumor Study Group. Cancer 1981;48:1705–1710.
5 Treatment of locally unresectable carcinoma of the pancreas: comparison of combined-modality therapy (chemotherapy plus radiotherapy) to chemotherapy alone. Gastrointestinal Tumor Study Group. J Natl Cancer Inst 1988;80:751–755.
6 Willett CG, Czito BG, Bendell JC, et al: Locally advanced pancreatic cancer. J Clin Oncol 2005; 23:4538–4544.
7 Schellenberg D, Goodman KA, Lee F, et al: Gemcitabine chemotherapy and single-fraction stereotactic body radiotherapy for locally advanced pancreatic cancer. Int J Radiat Oncol Biol Phys 2008;72:678–686.
8 Koong AC, Christofferson E, Le QT, et al: Phase II study to assess the efficacy of conventionally fractionated radiotherapy followed by a stereotactic radiosurgery boost in patients with locally advanced pancreatic cancer. Int J Radiat Oncol Biol Phys 2005;63:320–323.
9 Koong AC, Le QT, Ho A, et al: Phase I study of stereotactic radiosurgery in patients with locally advanced pancreatic cancer. Int J Radiat Oncol Biol Phys 2004;58:1017–1021.
10 Chang DT, Schellenberg D, Shen J, et al: Stereotactic radiotherapy for unresectable adenocarcinoma of the pancreas. Cancer 2009;115:665–672.
11 Hoyer M, Roed H, Sengelov L, et al: Phase-II study on stereotactic radiotherapy of locally advanced pancreatic carcinoma. Radiother Oncol 2005;76:48–53.
12 Murphy JD, Christman-Skieller C, Kim J, et al: A dosimetric model of duodenal toxicity after stereotactic body radiotherapy for pancreatic cancer. Int J Radiat Oncol Biol Phys 2010;78:1420–1426.
13 Kutcher GJ, Burman C: Calculation of complication probability factors for non-uniform normal tissue irradiation: the effective volume method. Int J Radiat Oncol Biol Phys 1989;16:1623–1630.
14 Lyman JT: Complication probability as assessed from dose-volume histograms. Radiat Res Suppl 1985;8:S13–S19.
15 Bussels B, Goethals L, Feron M, et al: Respiration-induced movement of the upper abdominal organs: a pitfall for the three-dimensional conformal radiation treatment of pancreatic cancer. Radiother Oncol 2003;68:69–74.
16 Gierga DP, Chen GT, Kung JH, et al: Quantification of respiration-induced abdominal tumor motion and its impact on IMRT dose distributions. Int J Radiat Oncol Biol Phys 2004;58:1584–1595.
17 Minn AY, Schellenberg D, Maxim P, et al: Pancreatic tumor motion on a single planning 4D-CT does not correlate with intrafraction tumor motion during treatment. Am J Clin Oncol 2009; 32:364–368.

18 Huguet F, Andre T, Hammel P, et al: Impact of chemoradiotherapy after disease control with chemotherapy in locally advanced pancreatic adenocarcinoma in GERCOR phase II and III studies. J Clin Oncol 2007;25:326–331.
19 Hurwitz H, Fehrenbacher L, Novotny W, et al: Bevacizumab plus irinotecan, fluorouracil, and leucovorin for metastatic colorectal cancer. N Engl J Med 2004;350:2335–2342.
20 Berber E, Pelley R, Siperstein AE: Predictors of survival after radiofrequency thermal ablation of colorectal cancer metastases to the liver: a prospective study. J Clin Oncol 2005;23:1358–1364.
21 Machi J, Oishi AJ, Sumida K, et al: Long-term outcome of radiofrequency ablation for unresectable liver metastases from colorectal cancer: evaluation of prognostic factors and effectiveness in first- and second-line management. Cancer J 2006;12:318–326.
22 Siperstein AE, Berber E, Ballem N, et al: Survival after radiofrequency ablation of colorectal liver metastases: 10-year experience. Ann Surg 2007; 246:559–565, discussion 565–557.
23 Katz AW, Carey-Sampson M, Muhs AG, et al: Hypofractionated stereotactic body radiation therapy (SBRT) for limited hepatic metastases. Int J Radiat Oncol Biol Phys 2007;67:793–798.
24 Rusthoven KE, Kavanagh BD, Cardenes H, et al: Multi-institutional phase I/II trial of stereotactic body radiation therapy for liver metastases. J Clin Oncol 2009;27:1572–1578.
25 Lee MT, Kim JJ, Dinniwell R, et al: Phase I study of individualized stereotactic body radiotherapy of liver metastases. J Clin Oncol 2009;27:1585–1591.
26 Wulf J, Guckenberger M, Haedinger U, et al: Stereotactic radiotherapy of primary liver cancer and hepatic metastases. Acta Oncol 2006;45: 838–847.
27 Hoyer M, Roed H, Traberg Hansen A, et al: Phase II study on stereotactic body radiotherapy of colorectal metastases. Acta Oncol 2006;45:823–830.
28 Kavanagh BD, Schefter TE, Cardenes HR, et al: Interim analysis of a prospective phase I/II trial of SBRT for liver metastases. Acta Oncol 2006;45: 848–855.

29 Mendez Romero A, Wunderink W, Hussain SM, et al: Stereotactic body radiation therapy for primary and metastatic liver tumors: a single institution phase I-II study. Acta Oncol 2006;45:831–837.
30 Nishioka T, Nishioka S, Kawahara M, et al: Synchronous monitoring of external/internal respiratory motion: validity of respiration-gated radiotherapy for liver tumors. Jpn J Radiol 2009;27: 285–289.
31 Goodman KA, Wiegner EA, Maturen KE, et al: Dose escalation study of single fraction stereotactic body radiotherapy for liver malignancies. Int J Radiat Oncol Biol Phys 2010;78:486–493.
32 Tse RV, Hawkins M, Lockwood G, et al: Phase I study of individualized stereotactic body radiotherapy for hepatocellular carcinoma and intrahepatic cholangiocarcinoma. J Clin Oncol 2008; 26:657–664.
33 Schefter TE, Kavanagh BD, Timmerman RD, et al: A phase I trial of stereotactic body radiation therapy (SBRT) for liver metastases. Int J Radiat Oncol Biol Phys 2005;62:1371–1378.
34 Dawson LA, Ten Haken RK: Partial volume tolerance of the liver to radiation. Semin Radiat Oncol 2005;15:279–283.
35 Dawson LA, Eccles C, Craig T: Individualized image guided iso-NTCP based liver cancer SBRT. Acta Oncol 2006;45:856–864.
36 Pettersson N, Nyman J, Johansson KA: Radiation-induced rib fractures after hypofractionated stereotactic body radiation therapy of non-small cell lung cancer: a dose- and volume-response analysis. Radiother Oncol 2009;91:360–368.
37 Herfarth KK, Debus J, Lohr F, et al: Stereotactic single-dose radiation therapy of liver tumors: results of a phase I/II trial. J Clin Oncol 2001;19: 164–170.
38 Chang DT, Schellenberg DS, Koong AC: Nonhepatic gastrointestinal malignancies; in Timmerman RD, Xing L (eds): Image Guided and Adaptive Radiation Therapy. Philadelphia, Lippincott Williams & Wilkins, 2010, pp 225–232.

Daniel T. Chang, MD
Department of Radiation Oncology, Stanford University Medical Center
875 Blake Wilbur Drive, Rm CC-G229
Stanford, CA 94305-5847 (USA)
Tel. +1 650 724 3547, Fax +1 650 725 8231, E-Mail dtchang@stanford.edu

Stereotactic Body Radiotherapy for Prostate Cancer: Current Results of a Phase II Trial

Christopher King

Department of Radiation Oncology, University of California School of Medicine, Los Angeles, Calif., USA

Abstract

The hypofractionation of stereotactic body radiotherapy (SBRT) for prostate cancer has become a broad topic, and there are many aspects to consider before accepting this treatment into our clinics. Among the considerations are the data from the Stanford phase II trial, a seminal investigation into this area, which will be presented and reviewed here. A single-arm, prospective phase II trial was initiated at Stanford in December of 2003. This trial uses SBRT as monotherapy for 'low-risk' prostate cancer patients, and 69 patients have been entered to date. We have analyzed the patient data for the first 5 years of this study. For study entry, patients were required to have clinical stage T1c or T2a disease, prostate-specific antigen (PSA) ≤10 and a Gleason score of 3 + 3 (or 3 + 4 if the higher grade portion was of small volume, usually <25% of the cores involved). No prior treatment was permitted, including the use of transurethral resections or androgen deprivation therapies. A low urinary IPSS score of <20 was required for study entry as well. The prescription dose was 7.25 Gy for 5 fractions for a total dose of 36.25 Gy. This was normalized to cover ≥95% of the planning target volume with 100% of the prescription dose. Patients were treated using CyberKnife technology. To date, excellent PSA responses have been observed in patients with lower-risk disease selected for treatment and receiving 36.25 Gy in 5 fractions. To date, sexual quality of life outcomes have also been approximately comparable to other radiotherapy approaches. Rates of late GI and GU toxicity have been relatively low and generally comparable to dose-escalated approaches using conventional fractionation.

Copyright © 2011 S. Karger AG, Basel

Phase II Trial for Hypofractionated Radiotherapy of Prostate Cancer

Rationale

There are some unique aspects of the biology of prostate cancer that may favor its treatment approach with hypofractionation as compared with conventional fractionation. By taking advantage of this radiation biology, there is potential for higher control rates and for lower adverse effects, resulting in an enhancement of the therapeutic ratio. An additional advantage is simply the shorter duration of the treatment course.

It is well known that the curve for the single-dose/response cell survival relationship can be described by a linear quadratic model [1] of the form $\exp(-\alpha d - \beta d^2)$. The surviving fraction for a multifraction regimen is given by $[\exp(-\alpha d - \beta d^2)]^n$. The α/β ratio is a measure of tissue sensitivity to dose-fraction size. The uniqueness of prostate cancer can be found in the α/β ratio. It has been suggested that the α/β ratio for prostate cancer is approximately 1.5, and that it is significantly lower than that for all other cancers [2–5] (fig. 1). This encourages us to consider treating prostate cancer on a different fractionation scheme than has been used to treat other cancers.

However, many factors come into play beyond radiobiological formulas when deciding on a hypofractionation protocol. There is additional clinical evidence that prostate hypofractionation is an acceptable treatment regimen for prostate cancer from experience with high-dose rate brachytherapy and even from an external beam regimen of 6 Gy × 6 fractions [6].

Patient Selection

A single-arm prospective phase II trial was initiated at Stanford in December of 2003. This trial has used stereotactic body radiotherapy (SBRT) as monotherapy for 'low-risk' prostate cancer patients, and 69 patients have been entered to date. We have analyzed the patient data for the first 5 years of this study [7]. For study entry, patients were required to have clinical stage T1c or T2a disease, prostate-specific antigen (PSA) ≤10 and a Gleason score of 3 + 3 (or 3 + 4 if the higher grade portion was of small volume, usually <25% of the cores involved). No prior treatment was permitted, including the use of transurethral resections or androgen deprivation therapies. A low urinary IPSS score of <20 was required for study entry as well.

Treatment Planning

The treatment program involved planning for intensity-modulated therapy using image guidance similar to other approaches: 3 solid gold fiducials were placed using transrectal ultrasound, and a CT scan at a slice width of 1.25 mm was obtained through the treatment region. Anatomic contouring of the prostate, seminal vesicles, rectum, bladder, penile bulb and femoral heads were performed. The PTV expansion margin was 3 mm in the posterior and 5 mm in the other directions. The prescription dose was 7.25 Gy for 5 fractions for a total dose of 36.25 Gy. This was normalized to cover ≥95% of the PTV with 100% of the prescription dose.

Patients were treated using CyberKnife technology (Accuray Inc., Sunnyvale, Calif., USA). Using this system, planning typically required prescribing to the 89% isodose line on average. The frequency of treatment evolved during the trial period, and therapy was given every other day in the later study period. To derive and approve an adequate plan, the most critical aspect was considered to be the rectal dose-volume histogram. Plans were considered acceptable if they achieved a rectal dose-volume histogram similar to those commonly obtained in prostate intensity-modulated radiotherapy programs employing high-dose escalation. Our rectal dose-volume histogram goals were: <50% rectal volume receiving 50% of the prescribed dose, <20% receiving 80% of the dose, <10% receiving 90% of the dose and <5% receiving 100% of the dose.

Patients were immobilized with an alpha cradle. Additional external immobilization devices to fix the external surface of the patient were not used, since the goal in using the CyberKnife was to track the prostate movement, which is due to internal organ motion.

Typical CyberKnife treatments consisted of about 200 individual beams for each patient. Planning time on the CyberKnife system takes about 1 h to develop an initial plan. Typically, 2–4 iterations are required, but this depends on the experience of the treatment planner. The dose distributions produced by the CyberKnife are similar in terms of any differences that one would see with a conventional intensity-modulated radiotherapy planning and delivery system.

From the linear quadratic equation, one can derive the equivalent biologic dose when given at 2 Gy per fraction (EQD2) from that of a hypofractionated course for any tissue or tumor type by the simple relationship: $EQD2 = D[\alpha/\beta + d]/[\alpha/\beta + 2]$, where D is the total dose given at dose d, the dose per fraction. Our hypofractionated dose regimen corresponds to a tumor EQD2 of 90.6 Gy (assuming an α/β of 1.5 Gy), a normal tissue late effect EQD2 of 74.3 Gy (assuming an α/β of 3 Gy) and an acute toxicity EQD2 of 52.2 Gy (assuming an α/β of 10 Gy).

Image-Guided Treatment Delivery

Throughout this volume, there are many excellent discussions on management of inter- and intrafraction organ motion. For the prostate, the dominant uncertainty in prostate position is due to interfraction motion, that is, the day-to-day setup uncertainty of the patient. However, there are various sources of movement that can occur during treatment delivery, as discussed by Drs. Kupelian and Meyer in this volume.

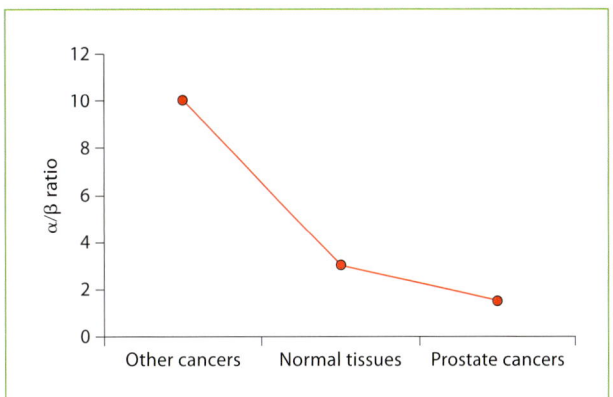

Fig. 1. The α/β ratio of prostate cancer compared with normal tissues and other cancers.

The primary reason to use the CyberKnife for this prostate protocol was its image guidance system. One can place fiducials inside a target and the CyberKnife has the ability to automatically track the motion of the fiducials. At this time, no other commercially available external radiation delivery system has the ability to automatically track fiducial markers by orthogonal kV imaging. This tracking is done in 'real-time'. Localization of the prostate (approximately every 30–90 s) based on 3 gold fiducial seeds placed in the prostate and the position of the robot is automatically adjusted based on any detected changes of prostate position.

Treatments were given over 5 consecutive days for the first 21 patients and 3 times per week subsequently. Total treatment time is about 40 min from the patient on the table to the last beam delivered. Imaging is done every 5–7 beams, and that corresponds to around 30–90 s.

Patient Evaluation

Endpoints included early and late urinary and rectal toxicities, quality of life (QOL) measures, and patterns of PSA response. The patients continue to be followed with serial PSA tests and QOL-validated questionnaires using the IPSS (International Prostate Symptom Score) and EPIC (Expanded Prostate Cancer Index) every 3 months for the first 2 years, and every 6 months thereafter.

Special Considerations

One issue that has not been well addressed is the potential for incidental dose to the testes. This is particularly important for those patients treated with the CyberKnife, as it is a very non-coplanar beam delivery, though it could apply to oth-

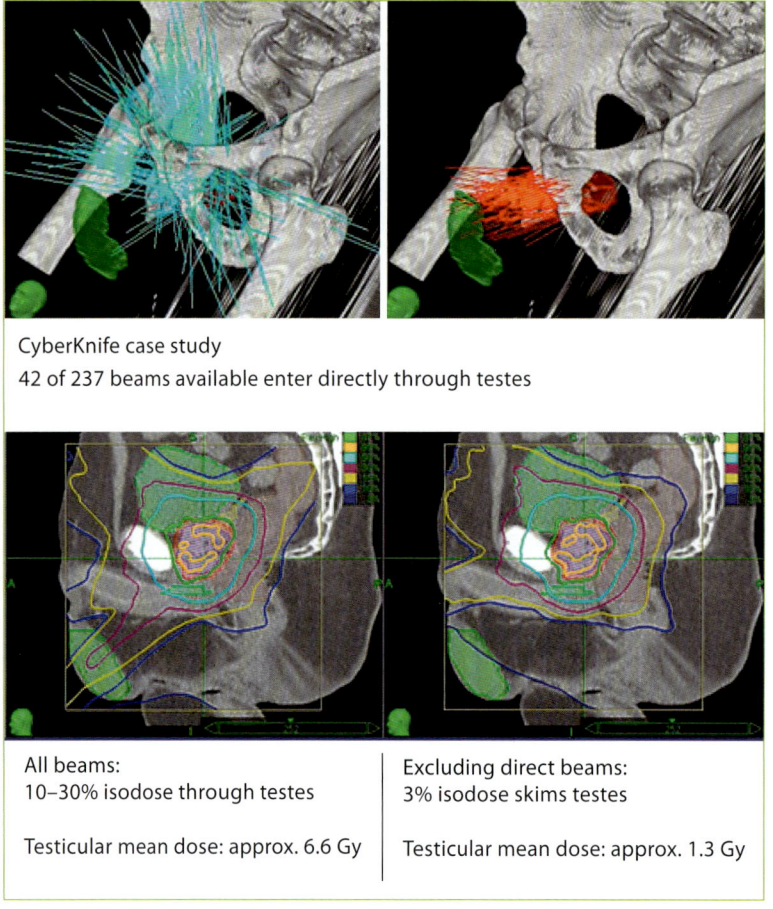

Fig. 2. Testicular dose resulting from non-coplanar therapy. The left panel shows incident beams and isodose profile resulting from planning without delineating testis as an avoidance region. The right panel shows possible dose reduction to testis with testicular avoidance.  See online supplementary material for additional displays.

er approaches as well. Figure 2 shows the testicular volume in green, and the beam entries relative to this volume [8]. In a plan using all of the available CyberKnife beams, 42 of the available 237 beams would be incident through the testes and are shown highlighted in red. This would result in the 10–30% isodose line being delivered directly through the testis, for a mean testicular dose of 6.6 Gy given in the hypofractionated dose.

A review of the literature estimates the LD50 of the testicular Leydig cells to be about 10 Gy. Therefore, a dose of 6 Gy would be a significant normal tissue exposure. Not only would it cause hypogonadism, with all of its attendant consequences, but it would also confound any PSA measurement of treatment effect for years. Through planning, one can exclude all of the CyberKnife incident beams through

Table 1. PSA response in evaluable trial patients

Initial results	PSA response			
	PSA nadir	1 year	2 years	3 years
41 evaluable patients	≤1 ng/ml	53	70	93
Median follow-up of 33 months (6–45)	≤0.6	31	70	87
Median age of 66 years	≤0.4	19	53	67
Median PSA of 5.6 ng/ml (0.7–10)	≤0.2	9	6	40
T1c (30 patients), T2a (10 patients), T2b (1 patient)				
3 + 3 (29 patients), 3 + 4 (12 patients)				

Patient characteristics before treatment are shown in the left column ('initial results'). The columns to the right show the percent of patients achieving the given nadir value.

the testes and reduce this dose to about 1.3 Gy or less, without significant degradation of the plan quality. Our review shows less than a 1% difference in any dose statistic between the 2 plans.

Trial Results

PSA Response

Forty-one evaluable patients at the time of analysis in 2008 had been followed for a median of 33 months (table 1). Figure 3 shows the initial PSA response as a function of time. There is a gradual and progressive decline in the PSA, with a lower PSA nadir as a function of time. There is an initial quick decline in PSA over the first 6 months, followed by a slower decline. A high proportion of patients (78%) with 12 or more months of follow-up achieved a low PSA nadir of 0.4 ng/ml. As years elapse, a greater percentage of patients achieve a lower and lower PSA value. This indicates that prostate cancer, unlike many other cancers, responds quite slowly to irradiation. The median PSA overall was 0.32 at the time of analysis. There is now 1 PSA failure in this group, with biopsy showing persistent (and high-grade) disease. This patient has undergone high-dose rate brachytherapy for salvage treatment.

As in other forms of irradiation for prostate cancer, one can see a PSA 'bounce' following treatment. This phenomenon can be defined in various ways. If one considers a PSA rise of 0.2 over the nadir, then it has occurred in 29% of trial patients. Three such cases are shown in figure 4, showing PSA fluctuations over time. The median time to this bounce is 18 months, and the median PSA increment is 0.4 (with a range of 0.2–2.4).

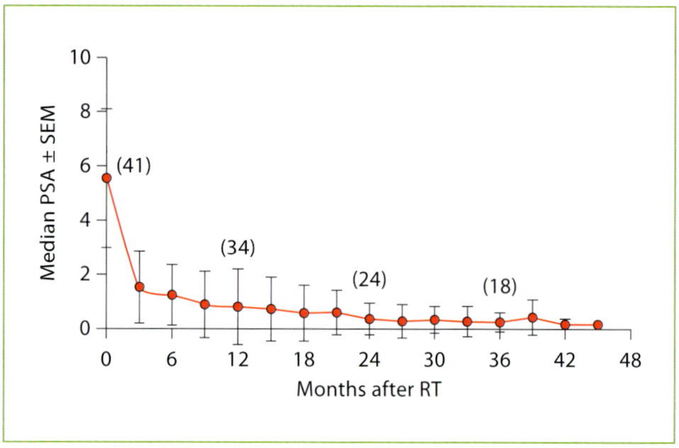

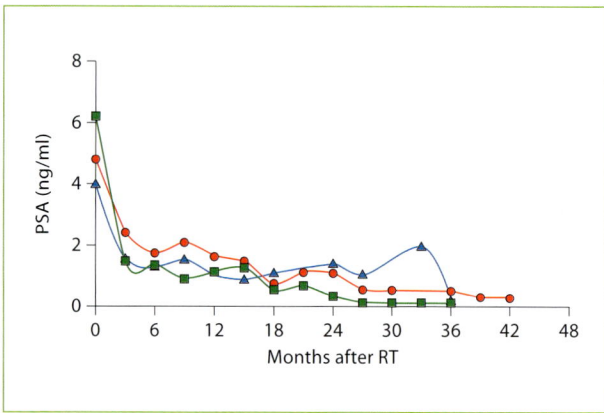

Fig. 3. PSA response after treatment. Error bar: 95% confidence around the mean; figures in parentheses are the evaluable patients at that time point. Median PSA nadir: 0.32 ng/ml; no PSA failures. RT = Radiotherapy.

Fig. 4. PSA fluctuations over time in 3 patients. 12 of 41 patients (29%); median time of 18 months (12–33); median PSA height of 0.4 ng/ml (0.2–2.4). RT = Radiotherapy.

Toxicity

The late urinary and rectal toxicities in these patients are shown in table 2. These data are compared with results from the M.D. Anderson dose escalation trial using 78 Gy and 3D conformal techniques [9]. In the current SBRT trial, late grade III urinary effects were seen in 2 patients: one with a stricture that resolved following cystoscopy, and the other with persistent dysuria. However, our follow-up period is not as long as in the M.D. Anderson series, and it is possible that additional effects will be observed with SBRT over longer times. We have not observed grade III or IV rectal toxicities, and grade I-II rectal toxicities have been seen with a frequency similar to that in the M.D. Anderson dose escalation series.

The first 20 patients in the trial received therapy on 5 consecutive days (QD). The dose regimen was subsequently changed to treatment delivery every other day (QOD). Comparing these 2 programs, 4 of 21 (19%) of the QD patients reported

Table 2. Late urinary and rectal toxicity rates following SBRT

RTOG grade	0	I	II	III	IV
Late urinary toxicity					
Number of patients	11	15	9	2	–
Patients, %	30	41	24	5	–
Number of patients	*114*	*21*	*11*	*5*	
Patients, %	*76*	*14*	*7*	*3*	*–*
Late rectal toxicity					
Number of patients	20	13	6	–	–
Patients, %	51	33	15	–	–
Number of patients	*71*	*42*	*28*	*10*	
Patients, %	*47*	*28*	*19*	*7*	*–*

These results are compared with results (in italics) from the M.D. Anderson dose escalation trial (3DCRT 78 Gy) [9].

Table 3. Late rectal toxicity rates following SBRT, comparing 2 treatment schedules

'Moderate' or 'big' problem	QD vs. QOD (late rectal)		
	QD	QOD	p
Any item[1]	38% 8/21	0% 0/20	0.0035
Overall[2]	24% 5/21	0% 0/20	0.048

[1] Urgency, frequency, control, bloody stool, pain.
[2] Bowel habits (at 6 months or beyond).

Table 4. Sexual QOL rates following radiotherapy

Series	Age, years	% Δ in potency rate
EB (13 reports)	68–73	15–54 (mean 40)
HDR boost (2 reports)	65–68	40–47 (mean 44)
LDR mono (2 reports)	64	22–36 (mean 29)
SBRT (this study)	68	53

Published literature was surveyed for studies using objective QOL assessment instruments. Hormonal therapy was not used in these series. EB = External beam radiotherapy; HDR boost = high-dose-rate brachytherapy following 50 Gy EB; LDR = low-dose-rate brachytherapy.

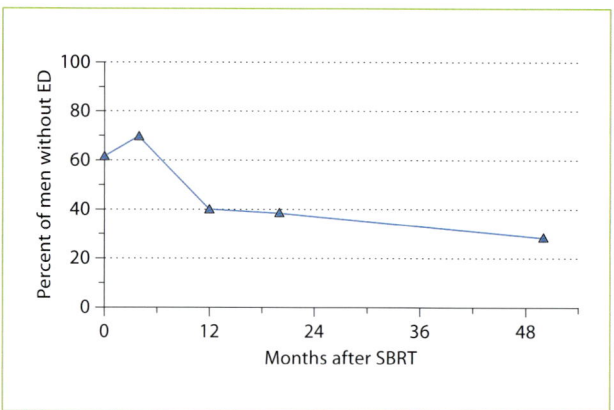

Fig. 5. Percent of men without ED after SBRT.

dissatisfaction on their QOL questionnaire compared with 1 of 20 (5%) of the QOD patients (table 3). This difference was more pronounced when comparing late rectal symptoms in the EPIC scores for the 2 groups. In the QD regimen, there were a significant number of patients reporting distressing rectal symptoms following treatment. There were no significant differences in genitourinary toxicity between the QD and QOD dose regimens. Currently, we are proceeding with the QOD approach as our standard.

We have also evaluated the EPIC-based assessment of sexual function in these patients, and show these results in figure 5 [10]. At baseline, 62% of patients were free of erectile dysfunction (ED), which is typical for men in this age group without prostate cancer. This percentage declined over the subsequent 48 months to a value of 33%, translating to a 53% decline in the rate of erectile function at 4 years after SBRT.

The use of ED medications in these patients has increased gradually over this time, and currently about 25% of patients are using these. There is now extensive literature showing that ED is a multifactorial and complex issue in these patients, but strongly a function of age. Evaluating ED as a function of patient age in this trial, a strong correlation is apparent: 9 of 15 patients (60%) are potent at age <70, compared with only 2 of 17 (12%) at age ≥70, suggesting that age is one of the primary factors affecting erectile function in these patients.

There are several reports of ED frequency following different radiotherapy-alone approaches (without androgen deprivation) for prostate cancer, and these are summarized in table 4. One must also take into account the use of ED medications and the differences in age (and other potential confounding factors) between these studies. It does appear that SBRT may be comparable to other treatment approaches in ED outcome measures.

Guidelines for Clinical Care

- Hypofractionation has the potential to biologically dose-escalate radiotherapy for prostate cancer.
- To date, excellent PSA responses have been observed in patients with lower risk disease selected for treatment, and receiving 36.25 Gy in 5 fractions.
- Rates of late GI and GU toxicity have been relatively low and generally comparable to dose-escalated approaches using conventional fractionation.
- To date, sexual QOL outcomes have also been approximately comparable to other radiotherapy approaches.
- The every-other-day fractionation is possibly better than daily fractionation for rectal outcome measures, though further study of this is needed.
- Longer follow-up and more patients are needed for more definitive conclusions, and these results warrant further prospective trials before being generally accepted.

References

1 Hall EJ: Radiobiology for the Radiologist, ed 4. Philadelphia, J.B. Lippincott Company, 1994.
2 Brenner DJ, Hall EJ: Fractionation and protraction for radiotherapy of prostate carcinoma (comment). Int J Radiat Oncol Biol Phys 1999;43: 1095–1101.
3 King CR, Fowler JF: A simple analytic derivation suggests that prostate cancer alpha/beta ratio is low (comment). Int J Radiat Oncol Biol Phys 2001; 51:213–214.
4 Fowler JF, Chappell R, Ritter M: Is alpha/beta for prostate tumors really low? (comment). Int J Radiat Oncol Biol Phys 2001;50:1021–1031.
5 Brenner DJ, et al: Direct evidence that prostate tumors show high sensitivity to fractionation (low alpha/beta ratio), similar to late-responding normal tissue (comment). Int J Radiat Oncol Biol Phys 2002;52:6–13.
6 Lloyd-Davis RW, Collins CD, Swan AV: Carcinoma of prostate treated by a radical external beam radiotherapy using hypofractionation: 22 years experience (1962–1984). Urology 1990;36:107–111.
7 King CR, Brooks JD, Gill H, Pawlicki T, Cotrutz C, Presti JC: Stereotactic body radiotherapy for localized prostate cancer: interim results of a prospective phase II clinical trial. Int J Radiat Oncol Biol Phys 2009;73:1043–1048.
8 King CR, Lo A, Kapp DS: Testicular dose from prostate cyberknife: A cautionary note. Int J Radiat Oncol Biol Phys 2009;73:636–637.
9 Kuban DA, Tucker SL, Dong L, et al: Long-term results of the M.D. Anderson randomized dose-escalation trial for prostate cancer. Int J Radiat Oncol Biol Phys 2008;70:67–74.
10 Wiegner EA, King CR: Sexual function after stereotactic body radiotherapy for prostate cancer: results of a prospective clinical trial. Int J Radiat Oncol Biol Phys 2010;78:442–448.

Christopher King, MD, PhD
Department of Radiation Oncology, University of California School of Medicine
200 UCLA Medical Plaza, Suite B265
Los Angeles, CA 90095-6951 (USA)
Tel. +1 310 825 9775, Fax +1 310 794 9795, E-Mail crking@mednet.ucla.edu

IV. Proton Beam Radiotherapy

Techniques and Technologies

Proton Therapy: Clinical Gains through Current and Future Treatment Programs

Radhe Mohan[a] · Thomas Bortfeld[b]

[a] M.D. Anderson Cancer Center, Houston, Tex., and [b] Francis H. Burr Proton Therapy Center, Massachusetts General Hospital, Boston, Mass., USA

Abstract

Proton beams can provide a substantial dosimetric advantage because of their unique depth-dose characteristics, which can be exploited to achieve significant reductions in normal tissue doses proximal and distal to the target volume. These may allow escalation of tumor doses, potentially improving local control and survival while at the same time reducing toxicity and improving quality of life. While many of the steps in proton and photon treatment planning processes are similar, there are also significant differences. Some of these arise from the unique physical characteristics of protons, while others are the result of their greater vulnerability to uncertainties, especially from inter- and intrafractional variations in anatomy. These factors must be considered in designing margins and field-shaping devices, as well as in designing treatment plans as a whole and in evaluating them. Ongoing research is aimed at better estimation of these uncertainties and their impact on proton therapy, and reducing these uncertainties through image guidance, adaptive radiotherapy and the development of novel imaging devices and dose computation algorithms. For proton therapy delivery, intensity modulation techniques are already in use, and will continue to be developed and utilized increasingly. The advantages include greater flexibility in dose shaping for improved target coverage and reduced normal tissue dose, potential improvement in plan robustness, and improvement in clinical efficiency. A spectrum of imaging techniques can now be used to assist our understanding of proton dosimetry in the patient, and PET imaging is the one that is furthest developed toward the goal of in vivo dose imaging. To decrease the cost of proton therapy and increase its availability, many technical improvements and practical delivery technologies are being developed, including compact proton machines that will soon become clinically available.

Copyright © 2011 S. Karger AG, Basel

The use of protons for radiation treatment of cancers can provide a substantial dosimetric advantage compared with the use of conventional forms of radiation. The clinical potential of protons stems from their unique depth-dose characteristics illustrated in figure 1. Protons interact continuously as they traverse a medi-

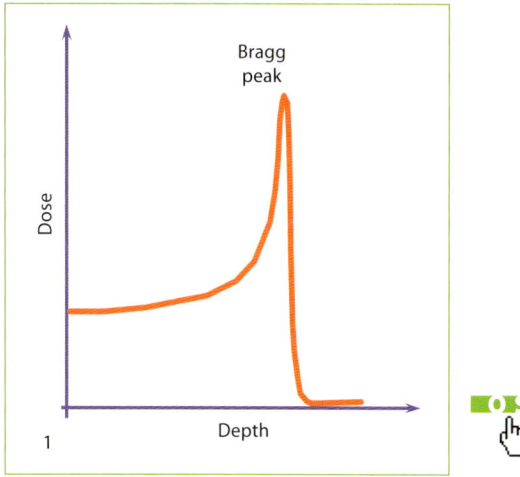

Fig. 1. Characteristic depth-dose distribution of a monoenergetic proton beam. The pristine Bragg peak is due to a pencil beam of protons in a homogeneous phantom. See online supplementary material for additional information.

um, losing energy along the way until they come to a stop. They deposit the greatest amount of energy per unit path length near the end of their range. The peak of highest dose is called the Bragg peak. Such characteristics can be exploited to achieve significant reduction in normal tissue doses proximal and distal to the target volume and to allow escalation of tumor doses, potentially improving local control and survival while at the same time reducing toxicity and improving quality of life.

Protons have been used for therapeutic applications since the early 1960s in facilities intended primarily for physics experiments. However, since about 1990, a number of hospital-based proton therapy facilities have come into operation. Currently, there are approximately 30 particle therapy centers in operation worldwide (8 in the USA) and 22 others are either under construction or in the planning stage (at least 2 in the USA; see Particle Therapy Cooperative Group, http://ptcog.web.psi.ch). This does not include several single-room facilities under construction or consideration.

Over the decades, protons have been shown to be effective for the treatment of several types of cancers including uveal melanomas, chordomas and chondrosarcomas of the base of skull and axial spine, small lung and liver tumors, paranasal sinus cancers, and central nervous system and pediatric tumors. In pediatric cancers in particular, proton therapy leads to much lower dose outside the target volume and to lower integral dose. These factors are important to decrease the risks of late effects, especially skeletal abnormalities and second malignancies. Currently, prostate patients constitute the largest fraction of patients undergoing proton therapy, at least in the USA. Previous experience with proton therapy of cancers of the thorax has been limited and has not yet shown a clear advantage over photon therapy in clinical outcomes. In general, the current mode of proton ther-

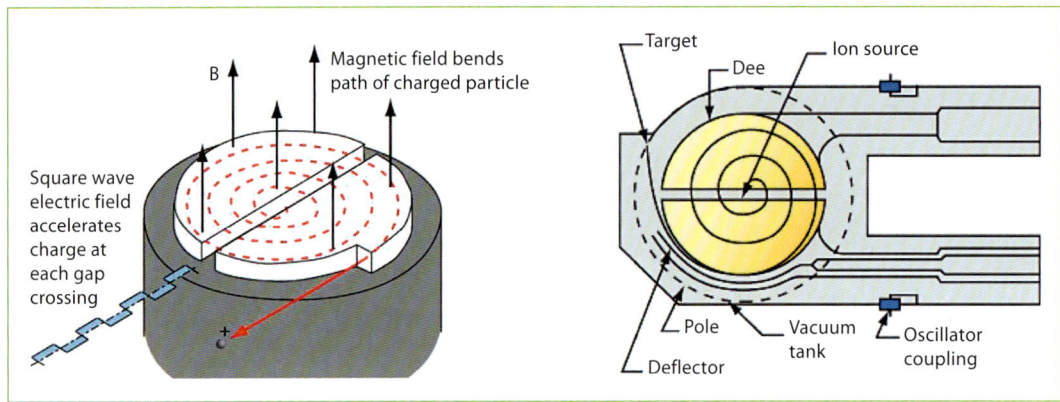

Fig. 2. Acceleration of protons in a cyclotron. A fixed magnetic field bends the path of protons, and they are accelerated by a square wave electric field applied between gaps of 2 D-shaped regions (known as 'Dees'). As energy increases, the radius of the proton path increases until the designated maximum is reached and protons are extracted.

apy using passively scattered technologies may offer advantages over conventional photon 3D conformal therapy, though the advantage over photon intensity-modulated radiation therapy is being debated and may be specific to the clinical situation.

While the state of photon therapy has advanced significantly over the last 2 decades, the high cost of proton facilities and the small number of such facilities engaged in research have been impediments in the corresponding advances in proton therapy. Considerable additional research is necessary to improve proton therapy and to conduct trials to evaluate its true potential compared with the best forms of photon therapy. Some of such research and trials are in progress jointly by the M.D. Anderson Cancer Center (MDACC) and Massachusetts General Hospital (MGH) with support from the National Cancer Institute.

The State of the Art in Proton Therapy

Proton Beam Production

Protons of energies appropriate for therapeutic applications (from 70 to 250 MeV) are typically accelerated using a cyclotron or a synchrotron. Of the two, cyclotrons (fig. 2) are more compact. They have higher beam intensities and produce a continuous stream ('DC') of protons which are accelerated to a maximum energy, e.g. 250 MeV. The clinically desired initial energies are then achieved by inserting degraders in the proton stream ('beam line') before it enters the treatment room.

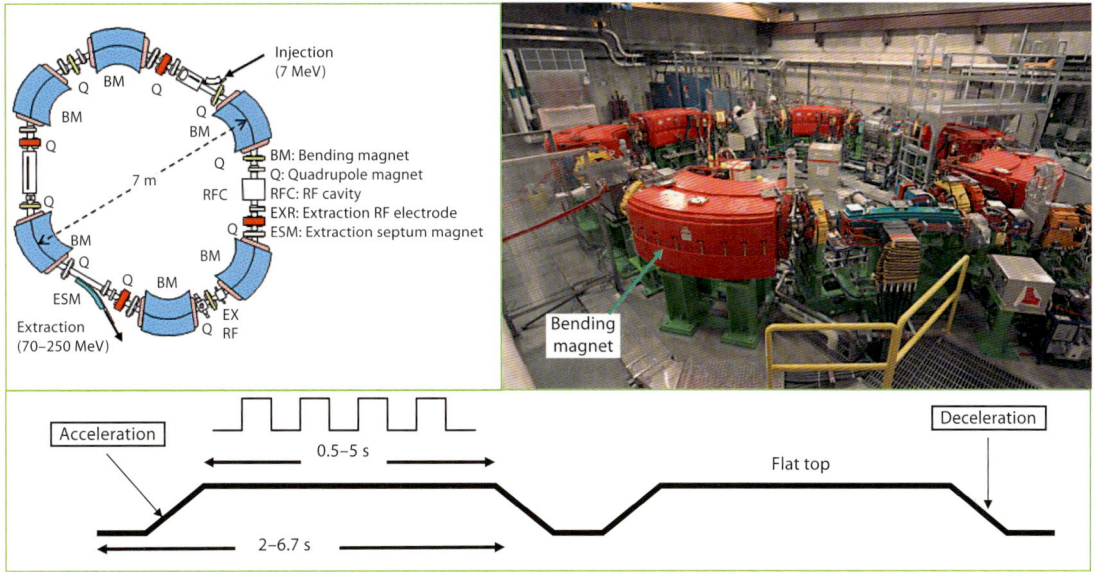

Fig. 3. The synchrotron at MDACC. A batch of protons is initially accelerated by a linear accelerator to a low energy (7 MeV) and injected into the synchrotron. Protons, as they are accelerated by the successive application alternating electric field, are constrained to move in a fixed circular path by increasing the magnetic field. When the batch of protons has reached the specified energy, it is extracted and transmitted to one of the treatment rooms.

Synchrotrons (fig. 3), on the other hand, accelerate pulses of protons to the desired energy levels. Protons gain in energy with each revolution through the accelerator. As soon as a batch has reached the desired energy, it is extracted and steered via the beam line to the treatment room for delivery. The extraction may occur over a variable period of time (anywhere from 0.5 to 5 s), depending on the application. The accelerator then goes through deceleration and reset steps before it starts the next cycle. Each cycle may produce protons of different energies. Generally, synchrotrons have greater energy flexibility, smaller energy spread and lower power consumption.

A typical proton accelerator serves multiple rooms. In the newer designs, the beam is switched automatically from one room to the next based on the order of request and priority. The narrow beam of protons produced by the accelerator has a relatively small initial energy and intensity spread. After extraction from the accelerator, the beam is directed magnetically through the beam line to the gantry and into the 'nozzle'. The term nozzle is used in particle therapy to describe the treatment head that the beam passes through after leaving the evacuated beam line. Figure 4 shows the treatment floor configuration at MDACC.

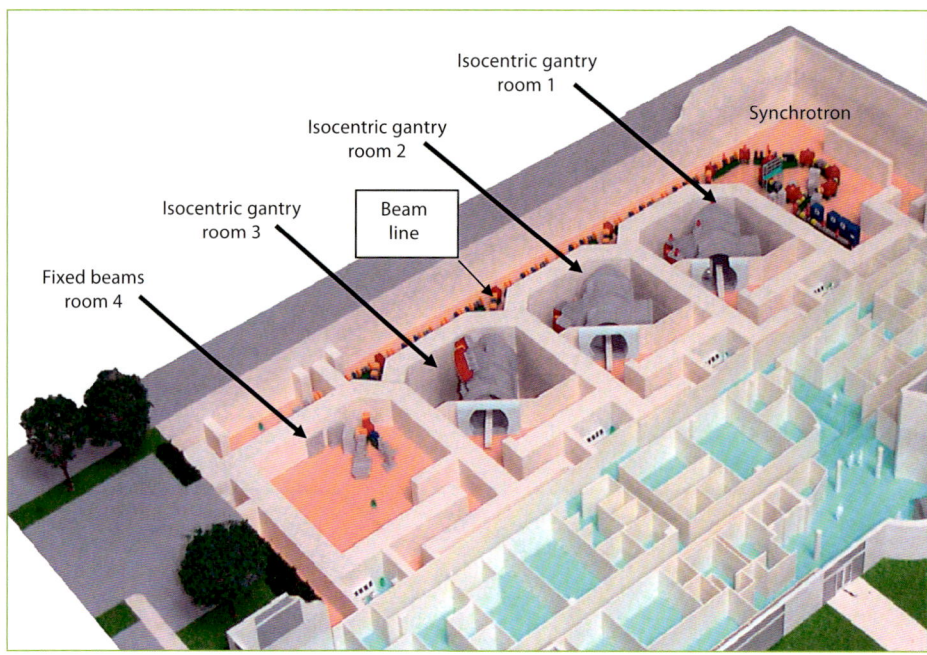

Fig. 4. Treatment room configuration at MCACC, which is similar to other multiroom facilities. Accelerated protons are carried to one of the treatment rooms via the beam line.

Therapeutic Proton Beams

A beam of protons entering the nozzle is focused, circularly symmetric and pencil-thin. The protons have maximum ranges (in tissue equivalent media) typically from 4 to 30 cm. As the proton beam traverses the air and monitoring devices, it spreads out slightly by the time it reaches the surface of the medium (patient, phantom, compensator, etc.). Once it enters the medium, the rate of interactions increases substantially. A typical depth-dose distribution in water from such a beam is shown figure 1. To be clinically useful, such distributions must be spread out laterally and longitudinally, i.e., along the direction of the beam. This can be achieved either by (a) *passively scattering* protons with a combination of a rotating modulation wheel (RMW) and a scattering foil, or (b) varying the energy of the protons before they enter the nozzle and *actively scanning* the protons with magnets. These two approaches will be discussed in greater depth throughout this chapter, as they have many implications.

For passively scattered proton therapy, protons are generated to one of a small number of discrete initial energies, spanning the therapeutic range, and are directed into the gantry. For instance, the MDACC passively scattered proton ther-

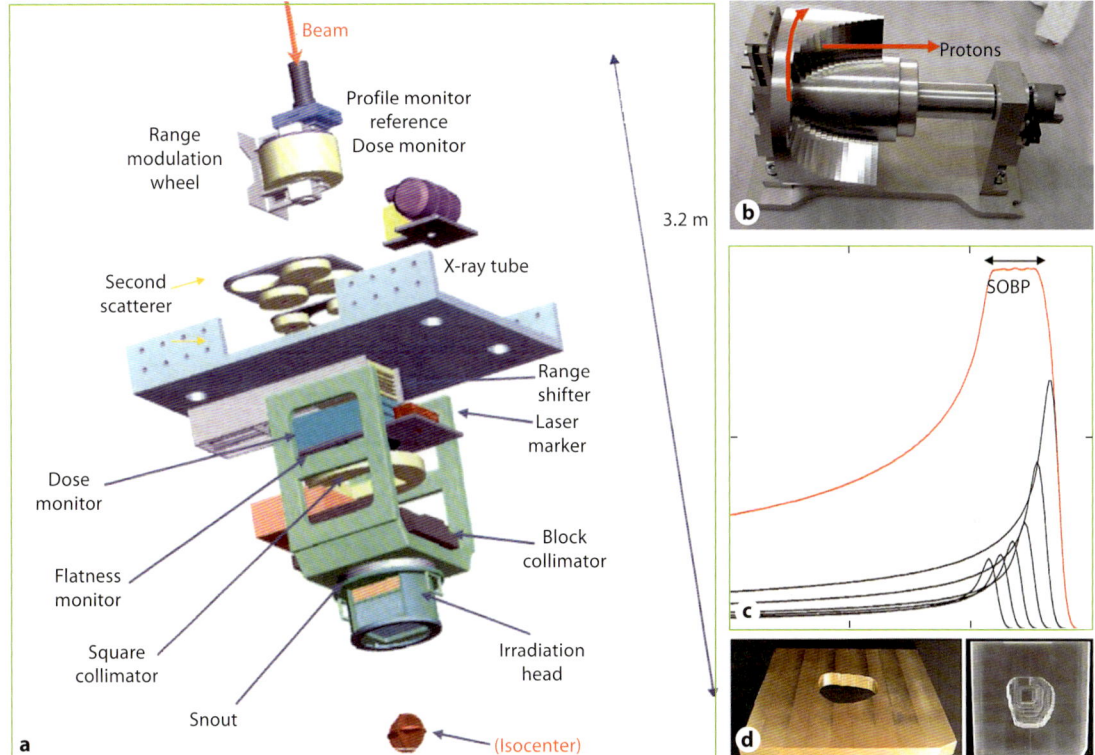

Fig. 5. a A passive scattering nozzle. The beam entering the nozzle is spread longitudinally by the RMW and spread laterally by the scatterers. The inset (**b**) illustrates how the RMW interposes steps of different thickness to produce Bragg peaks of different ranges and intensities, creating the SOBP (shown in **c**). The proton beam is turned on at the thinnest step. The width of the Bragg peak can be changed by gating the beam off at an appropriate step. **d** A brass aperture on the left and a compensator on the right. See online supplementary material for additional information.

apy system consists of 8 initial energies from 100 to 250 MeV. In this mode, an initially monoenergetic beam of desired penetration is modulated using an RMW to produce protons of a *sequence* of lower energies. These form the required depth-dose distribution, called the spread-out Bragg peak (SOBP).

Figure 5 shows the design of a Hitachi RMW along with the schematic of a Hitachi passive scattering nozzle. The beam is also scattered mechanically to achieve lateral spreading. In the current designs, 2 scatterers made of high Z materials are used to spread the beam laterally and make it flat within the region of interest. Often, the first scatterer is built into the RMW. For most patients treated with proton therapy currently, the dose distributions are produced by passive scattering.

For scanning beam applications, the proton energies entering the nozzle are much larger in number and at finer steps (e.g. 94 energies from 72 to 222 MeV in the case of MDACC Hitachi system). A scanned beam is a composite of pencil beams (or 'beamlets') of a range of needed energies. These beamlets are positioned laterally magnetically to produce the desired dose distribution. Magnetic scanning of beamlets provides greater flexibility and control for creating an optimal conformal proton dose distribution. In addition, the elimination of mechanical shaping devices (such as apertures and compensators) saves the time required for the insertion of these devices, obviates the need to enter the treatment room between fields and makes the treatments more efficient. Scanned beams are used to plan and deliver intensity-modulated proton therapy (IMPT), as discussed later in this chapter.

Passively Scattered Proton Therapy

Field Shaping
When a passively scattered beam is used to irradiate a water phantom, it produces a dose distribution with the depth-dose characteristics depicted in figure 5c. The lateral distribution of dose is axially symmetric and tapers off gradually beyond the maximum radius of the intended flat region. To conform the dose distribution laterally to the shape of the target volume in the patient (plus appropriate margins for inter- and intrafractional anatomic variations and penumbra) a custom *aperture* (fig. 5d) is used. This is typically made of brass of sufficient thickness (2–8 cm) to absorb incident protons of highest energy. The field aperture fits in a 'snout,' a component at the end of the nozzle (fig. 5a) designed to hold field-shaping devices.

To create a dose distribution that conforms to the distal shape of the target, the SOBP of the passively scattered beam is shaped using a *range compensator* (fig. 5d). A compensator is usually made of a nearly water-equivalent material such as Lucite. It reduces the beam energy ray-by-ray by appropriate amounts so that the distal edge of the beam conforms to the shape of the target plus a suitable margin for range uncertainty. The compensator is placed just below the aperture and is the final element in the nozzle. The air gap between the patient and the compensator is minimized to reduce the penumbra by moving the snout as close to the patient as practical. The aperture and compensator for each beam are designed by the planning system, and the design information is used to fabricate these devices.

Since the SOBP width for a passively scattered beam is designed to be constant across the entire field, passive scattering provides no control over dose distribution proximal to the target. For a target with a highly irregular distal edge, this may lead to a substantial excess volume of high dose in the region proximal to the target. In contrast, for scanning beams, distal, proximal and lateral field-shaping is achieved by limiting the positions of the spots to within the target regions only. Therefore, scanned beams can deliver lower doses in the proximal regions as well.

Treatment Planning for Passively Scattered Proton Therapy

Treatment planning determines the number and directions of proton beams and the parameters of the beams to achieve optimum dose distribution. Beam parameters include distal and proximal margins, lateral margin, beam energy (or the maximum proton range), modulation width, and the characteristics of the aperture and compensator. Energy and modulation width are subsequently used to select the RMW and its SOBP width parameter and the second scatterer.

A typical proton planning system computes dose distributions expected to be received by the patient over the course of radiotherapy. The plans are evaluated using dose distributions and dose-volume histograms (DVHs). While many of the steps in the proton and photon treatment planning processes are similar, there are significant differences. Some of these differences arise due to the physical characteristics of protons, while others are the result of the greater vulnerability of protons to uncertainties due to such factors as inter- and intrafractional variations in anatomy, the approximations in computation of dose distributions, etc. These uncertainties create an uncertainty in knowing the exact point where a beam of protons stops. In addition, anatomic inhomogeneities in the patient degrade the distal field edge, which may not be nearly as sharp as expected.

These factors must be considered in designing margins and field-shaping devices, as well as in designing treatment plans as a whole and in evaluating them. They challenge the conventional concepts described in ICRU Reports 50 and 62 [1, 2], for instance, the validity of planning target volume (PTV) for proton therapy. One consequence is that, in contrast with photon therapy, margins for proton therapy are not isotropic. Lateral margins depend on the anatomic variations orthogonal to the beam direction and are generally determined in the same manner as the PTV margins for photons. To these margins is added the width of the penumbra from 50% to, say, 90 or 95% to define the aperture margin. The penumbra margin is a function of maximum proton energy. To account for uncertainty in range, field-specific distal and proximal margins must be assigned to the clinical target volume (CTV) to ensure coverage of the CTV along the path of each beam. These margins have been estimated empirically to be of the order of 3.5% of the depth of the target.

Compensators designed to conform the distal edge of the proton dose distribution with the distal edge of the target assume that the patient's anatomy is invariant over the course of radiotherapy and remains aligned with the compensator inter- and intrafractionally. However, positioning uncertainties and changes in inhomogeneities in the path of a proton beam to the target volume affect the depth of penetration and the conformity. To ensure coverage of the target in the presence of misalignment of the compensator with the anatomy and the variations of anatomic structures of significantly different densities, compensators are 'smeared' [3]. The smearing process, illustrated in figure 6, essentially reduces the width of

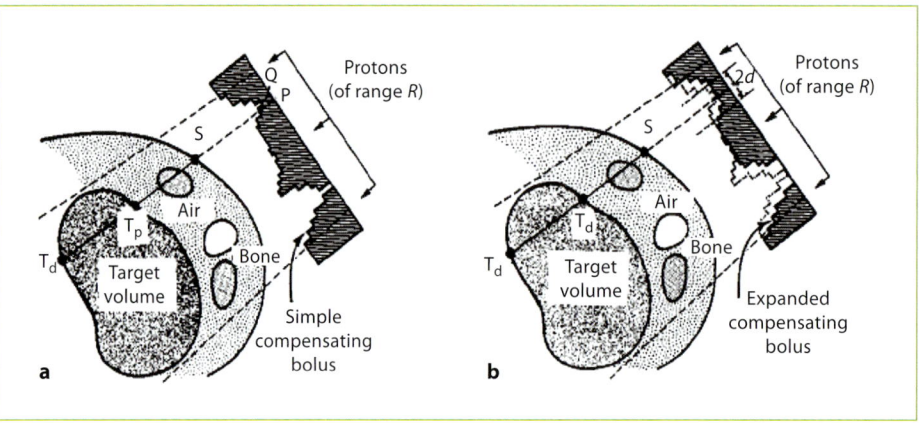

Fig. 6. Compensator smearing. **a** Compensator designed assuming perfect alignment. **b** Smeared (expanded) compensator to account for misalignment of compensator with anatomy and anatomic structures relative to each other. Reprinted with permission of Urie et al. [3].  See online supplementary material for additional information.

the higher thickness regions of the compensator, to allow protons to penetrate more deeply even when adjacent higher density tissues move into their path. Smearing may necessitate an increase in the modulation width to ensure that dose to the proximal edge is not compromised.

 Targets that move due to respiration pose additional challenges for passively scattered proton therapy planning. To incorporate respiration-induced motion into proton planning, 4D CT data sets, comprising a sequence of 3D CT data sets (one set for each phase of the respiratory cycle) may be used to design the compensator. One strategy currently being used requires the definition of a motion-integrated internal target volume (ITV) (see online supplementary material for demonstration). The compensator design is then based on the so-called maximum intensity projection image of the ITV in which each voxel within the ITV is set to its maximum intensity in any one of the phases [4]. In another approach [5], one compensator is designed for each phase of the 4D CT. The union of all the compensators represents the final composite compensator. Either strategy ensures that protons penetrate to sufficient depths to cover the target adequately regardless of its position. However, actual dose distribution behind the target may be significantly higher compared with what is seen on the resulting treatment plan.

While designing compensators to incorporate motion ensures adequate coverage of targets, motion management with breath-hold techniques and respiratory gating of proton beams should lead to significant sparing of normal tissues. These are being introduced gradually into proton clinical practice.

Treatment Plan Evaluation

Can proton treatment plans be evaluated in the same manner as photon plans? In practice, there are important distinctions to be made because of the effect of various uncertainties on the depth of proton penetration. For protons, anatomic variations perturb the depths of penetration along beams from different directions, creating range uncertainties unique to each beam. Therefore, *beam-specific* distal and proximal margins need to be assigned. As alluded to above, this brings into question the validity of the concept of plan evaluation based on a *composite* dose distribution display of all beams, and their PTV coverage or DVHs. In the current practice, the passive scattering plan evaluation process involves making certain that, for each beam individually, the distance between the lateral, distal and proximal edges of the CTV and the prescription isodose surface are equal to or greater than the assigned lateral, distal and proximal margins. A similar process is used to ensure that normal critical structures are at sufficient distance from high-dose isodose surfaces and are adequately spared. Even though the concept of PTV and the corresponding normal tissue volumes are unsuitable for protons, in the absence of suitable efficient tools, their DVHs continue to be used for plan evaluation and for comparison of proton and photon plans.

An improvement is a plan evaluation method suggested originally by Goitein [6], extended by Tony Lomax in the 1990s and, more recently, by others by incorporating respiratory motion [presented by R. Mohan at PTCOG 44 (2006) in Zurich]. Multiple dose distributions for a given plan are computed for the nominal image and for each of the images obtained with +/– AP, lateral and SI displacements of the body; plus and minus changes in CT numbers proportional to range uncertainty and, if applicable, end-inhale and end-exhale. A composite 'worst case' dose distribution is then obtained by assigning to each voxel inside the CTV the minimum dose on any of the plans, and to voxels outside the CTV the maximum dose on any of the plans. The resulting dose distributions and DVHs, representing the worst-case scenario, are then used to evaluate the plan. The computation of the large number of plans required makes this approach resource-intensive!

Radiobiological Issues

In the practice of proton therapy, it is assumed that protons (a low LET radiation) do not differ significantly from photons from a biological standpoint. Based on numerous in vitro and animal experiments, it has been found that the relative biological effectiveness (RBE) of protons is *on average* 1.1 relative to cobalt-60 radiation. Physical proton dose distributions are multiplied by 1.1 to obtain values in a unit known as 'Cobalt Gray Equivalent' or CGE. More recently, this terminology has been changed to *Gy(RBE)*. The dose distributions thus obtained are used to compare proton and photon plans, or even to add proton and photon plans.

However, the biological interactions of protons may be more complex than what has been assumed so far in clinical practice. For instance, almost all of the experiments to date have been carried out in the middle of the SOBP of a flat homogeneous dose distribution. Most of the in vivo experiments have used single fractions of high doses. However, it is known that proton RBE at low doses is *higher* than at high doses. Furthermore, RBE is also a function of tissue type. In addition, the biological interactions of protons with cells and tissues in the presence of chemotherapy or molecular agents may be different compared with the corresponding interactions with photons. Such variations in RBE are ignored in current practice. For a passively scattered beam, this might mean that the biologically effective dose near the end of the range is uncertain – and possibly higher than what is seen on a treatment plan. Clinical evidence to date does not provide sufficient clarity on radiobiological uncertainties. For this reason, and because of uncertainties in range, aiming the beams toward critical normal structures is avoided.

Uncertainties in Proton Therapy: Current Challenges, Future Directions

The finite range and sharp distal fall-off of protons are physical characteristics that give them significant potential to produce superior dose distributions and to improve the therapeutic ratios over photon therapy. However, the characteristics that give protons their therapeutic advantage are also their Achilles heel, rendering their dose distributions vulnerable to uncertainties to a greater extent than for photons. Examples of these uncertainties include the following:

(a) Day-to-day positioning uncertainties.
(b) Intrafractional motion due to respiration, and interfractional changes due to tumor shrinkage, weight loss and other anatomic changes.
(c) Approximate values of CT numbers, artifacts on CT images due to high atomic number materials and approximations in the conversion of CT numbers to proton stopping powers required for accurate dose calculations.
(d) Approximations in the current dose computation algorithms.
(e) Uncertainties in RBE.

Similar uncertainties also exist in conventional radiotherapy modalities; however, consequences for protons are more serious. For instance, protons do not stop exactly where they are expected to and the dose distributions actually delivered to a patient may be substantially different from what is seen on the treatment plan.

In addition, passage of protons through complex heterogeneities, such as those in the head and neck and lungs, degrades the distal edge of the proton beam so that the fall-off is not nearly as sharp as expected. The range uncertainties and distal edge degradation issues are particularly significant for tumors of the lung due to the longer range of protons in low-density media.

In the current state of the art, these uncertainties necessitate the use of undesirably large margins around the CTV that expose larger volumes of normal tissues to higher doses of radiation. Despite these margin expansions, there may be concerns about target miss. Thus, in many situations, the outcomes may be unsatisfactory compared with what is possible in principle, and may prevent or limit the use of protons for some important sites. Ongoing research is aimed at better estimation of these uncertainties and their impact on proton therapy, and at better management of these uncertainties through image guidance, adaptive radiotherapy and the development of novel imaging devices and dose computation algorithms. Methods are also being investigated to incorporate residual uncertainties into the planning process and to assess the robustness of the treatment plans. An important area of current research is the so-called robust optimization, which renders treatment plans less vulnerable to uncertainties. This is discussed later in this chapter.

Current and Future Developments in Clinical Proton Technology

In this section we will review 3 important areas of recent research and development in proton beam therapy, including their potential to resolve some of the above-mentioned challenges. We will first discuss the development of IMPT, where we will make a distinction between the use of pencil beam scanning and fully developed IMPT. The second research focus is on biological dose imaging for proton beam verification. Here the motivation comes from the fact that, even though proton stopping is the main physical advantage of proton therapy, in the clinical situation we do not know precisely where the beam stops always. The good news is that there are a number of unique opportunities for imaging using new technologies to visualize the proton distribution in the patient. These may help to reduce the proton range uncertainties. Finally, the primary concerns with proton therapy, and the reason why it is not available more broadly in clinics today, are the higher costs and footprints of the machines. The question is, if technological developments can yield new methods to accelerate protons, which of them are cheaper and more compact? We will discuss some techniques that have just appeared on the horizon, as well as others that are nearer to clinical implementation.

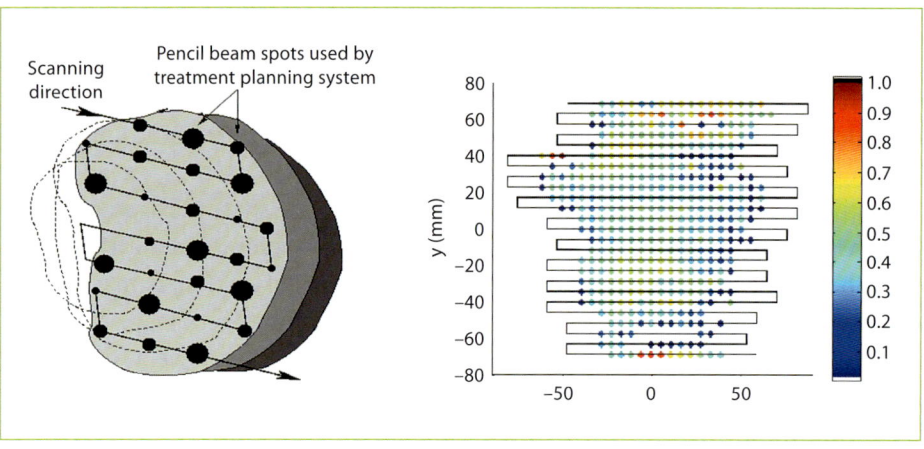

Fig. 7. Illustration of the pencil beam scanning technique. Each layer of the target volume is covered with dose in a zigzag pattern. Layers at different depths are covered by changing the energy of the beam [7]. See also online supplementary material.

The Promise of Beam Scanning and Intensity Modulation in Proton Therapy

Pencil beam scanning is a proton delivery technology alternative to the standard passive scattering technique reviewed above. Pencil beam scanning can be used to deliver uniform SOBP proton fields, like the passive scattering technique. It can also be used to deliver IMPT, where the dose delivered from each beam is intentionally modulated and nonuniform. We will first discuss pencil beam scanning and its clinical promise.

Pencil Beam Scanning
The basic idea behind pencil beam scanning is simple: instead of producing a broad beam and shaping it with apertures and compensators to match the target volume, one uses a narrow 'pencil' beam, then deflects it with magnets and scans it across the treatment field to cover just the area where the dose is wanted (the basic principle is quite similar to how an image is produced on a standard TV screen). Figure 7 shows an illustration of how pencil beam scanning works. The target is typically treated layer by layer, starting with the most distal layer, shown in dark grey. Each layer is irradiated in a zigzag manner either in a step-and-shoot mode, or in a dynamic continuous mode. After finishing a layer, the energy is reduced and the next layer is 'painted' until the whole target volume is covered with dose.

The pencil beam technique has some advantages even when it is used to mimic the passive scattering technique, i.e. by delivering uniform (total) dose to each layer. The first advantage is proximal dose shaping. One limitation of the older

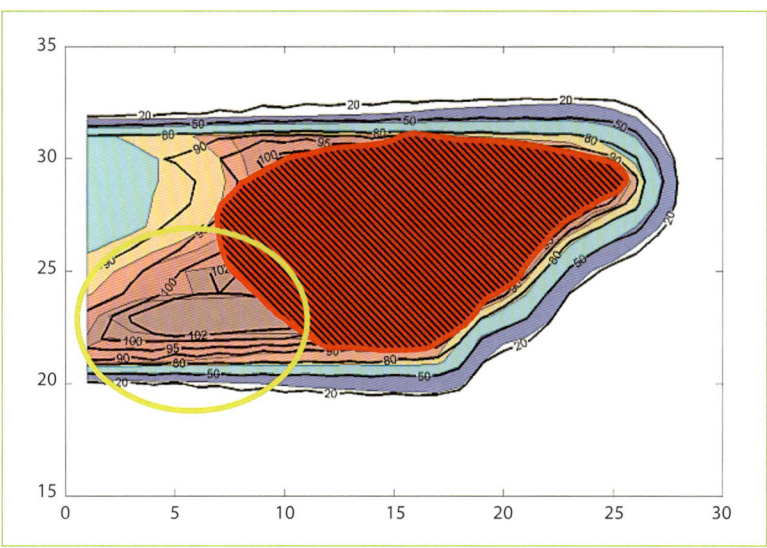

Fig. 8. Dose distribution achievable with the passive scattering technique (see isodose lines and color-wash; target volume: shaded red). Because the SOBP is the same everywhere across the field, the area circled in yellow is normal tissue that unnecessarily receives the full dose. With beam scanning this problem can be avoided by 'painting' the target volume with dose.

passive scattering technique is that the width of the high-dose region (the SOBP) is the same everywhere across the treatment field. Consequently, for most irregularly shaped target volumes there is a normal tissue volume proximal to the target that receives full therapeutic dose unnecessarily. Figure 8 shows an example. With pencil beam scanning, this problem can be avoided. However, the practical benefit of this is less than might be initially expected. With passive scattering, the problem is already reduced by using multiple fields, as are typical of these plans, since the proximal high-dose regions for each field occur in different locations and are spread out. With beam scanning, the problem is not fully corrected since the proximal dose can still be substantial, and it is certainly never zero. However, beam scanning significantly improves proximal normal tissue sparing.

The second advantage of beam scanning is the reduction of secondary neutrons. In an editorial, Hall [8] raised the issue of neutron dose contamination in proton therapy and suggested that this could be a significant problem for proton therapy centers using passive scattering technologies. Subsequent letters and follow-up papers have questioned this full assessment, but it is certainly true that protons do interact with the beam scatterers and produce neutrons in the passive scattering methods. These externally produced neutrons then 'flood' the patient, and add to the small number of neutrons produced inside the patient. With pencil beam scanning, on the other hand, magnets are used to deflect the beam and there

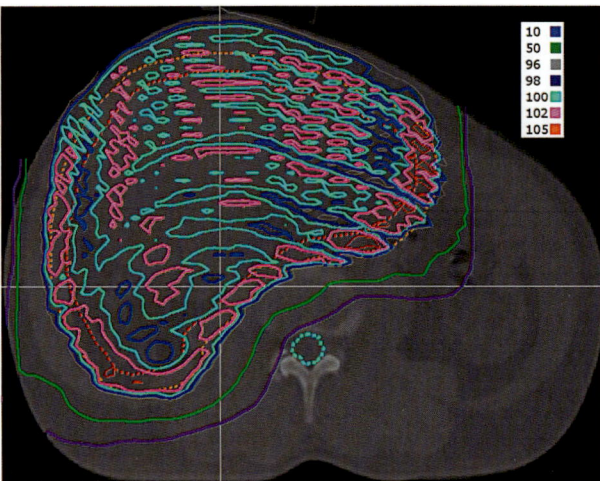

Fig. 9. Pencil beam scanning enables coverage of large target volumes efficiently and uniformly, as shown in this example of a very large abdominal sarcoma [9].

is no interaction with scatterers, so neutrons are not produced in the beam line. The neutron dose to the patient with beam scanning is lower than with passive scattering techniques, though the doses are not large with either.

The third advantage of beam scanning is its increased efficiency. In the passive scattering technique, the field size is typically limited because for larger fields, the field uniformity tends to suffer. For example, at MGH the maximum field size is 25 cm in diameter. If larger fields are required, they need to be 'patched' together. With beam scanning, on the other hand, much larger fields can be delivered without patching. To give a specific clinical example, the first patient who was treated with beam scanning at MGH had a large abdominal sarcoma (fig. 9). The patient was treated with beam scanning in 24 energy layers. The treatment time was only 20 min (11 min of beam-on time), and the treatment could be delivered with a single isocenter. To compare, passively scattered proton treatment was given for the initial fractions for the same patient. Because of the large target volume and because of the field size limitation, 8 fields had to be patched together, requiring the manufacture of 8 brass apertures. These needed to be switched during the treatment for the multiple fields, along with multiple settings of the isocenter, to cover the whole field uniformly. It resulted in a very complex and time consuming (40 min) treatment. Overall, beam scanning has the potential to make the whole planning and delivery process more efficient.

There are also a few potential disadvantages of pencil beam scanning. The first is its potentially broader penumbra. It is a technical challenge to create a very narrow pencil beam. In today's practice, proton pencil beams have 'standard deviations' (σ) of around 7 mm 'in air', even before the additional scattering within the patient. If one wishes to translate this into the full width (at half maximum) of the beam, one has to multiply this number by 2.4, which yields a width of 1.7 cm. Therefore, these beams are not really like pencils, but more like relatively broad brushes. One can easily imagine that this is not ideal for painting the fine details of target geometries. As a consequence, the penumbra in pencil beam scanning can be broader than that of a passively scattered beam which uses solid apertures to shape the field edges. That said, there are a few developments underway to remedy these potential issues with beam scanning. Physicists and engineers are working on the reduction of the beam width (σ) in air, apertures can also be used for beam scanning (even though this may reduce the efficiency advantage discussed above), and there are ways to sharpen the penumbra in beam scanning by increasing the intensities of the pencil beams at the field edges [10].

Another potential disadvantage of pencil beam scanning is in the treatment of moving targets. It has been shown that there can be an 'interplay' effect between the motion of the scanned beam and the motion of inner organs or the target volume that can result in unintentional dose heterogeneities, especially in the case of respiratory motion [11, 12]. This effect actually gets worse as the width of the pencil beam is reduced. However, it has also been shown that simply 're-painting' the target volume several times during treatment reduces this effect substantially just on basic probability reasons. It should also be mentioned that magnetic beam scanning theoretically allows one to track organ motion, i.e. to follow the organ motion with the beam [13]. While intriguing, this approach has not been implemented in the clinic yet.

Intensity-Modulated Proton Therapy

Though pencil beam scanning is becoming available as a clinical treatment modality at several centers, true IMPT has been in clinical use only at 1 center so far, the Paul Scherrer Institute in Switzerland [14]. By 'true' IMPT, we mean the delivery of deliberately nonuniform fields from different directions of incidence, such that the desired (uniform) target dose is achieved only after the superimposition of all fields (fig. 10). There are many ways to combine nonuniform individual fields to produce the desired overall dose distribution.

The gain in flexibility through IMPT can be exploited to accomplish several goals. One is to safely bend beams around critical structures, ensuring good sparing of these structures and at the same time guaranteeing target coverage. In the more conventional passive scattering approach, we use patch fields – multiple fields patched together along a match line. Obviously, this may create overdose or underdose along that match line, even when a feathering approach is used. With IMPT, this problem can be greatly reduced and the delivery simplified by delivering wedge-shaped fields. Figure 11 shows an example of 2 fields with wedge-shaped

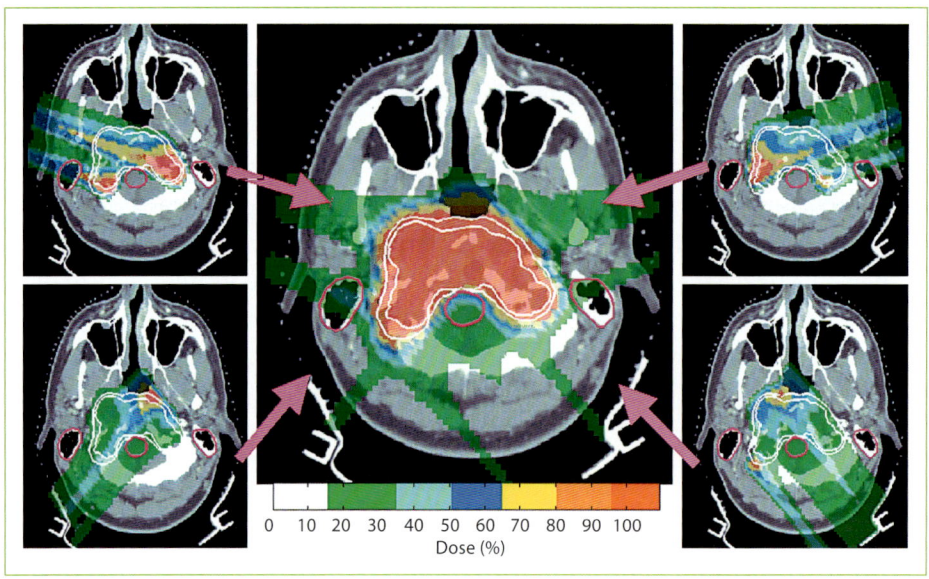

Fig. 10. An example of IMPT. Copyright held by A. Trofimov.

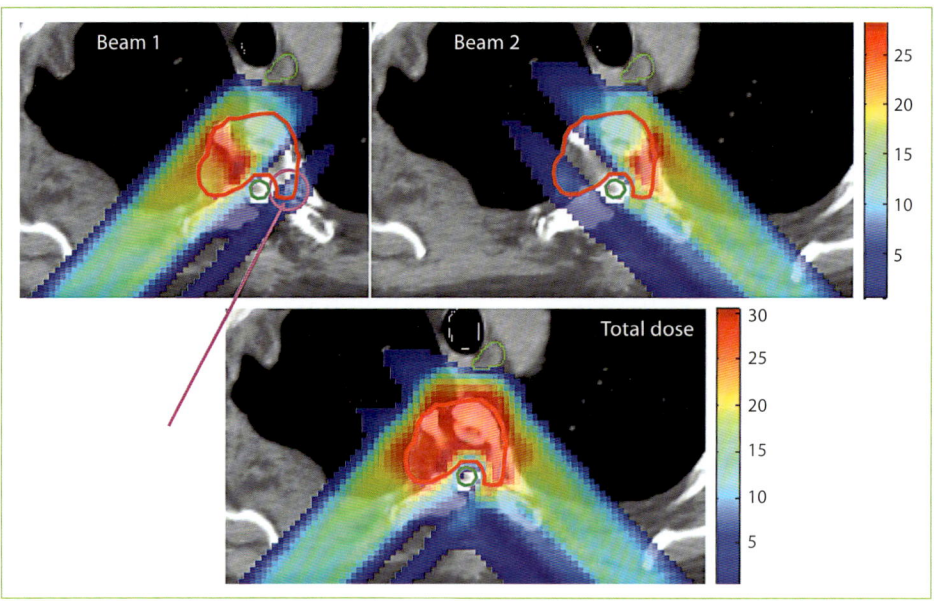

Fig. 11. Two advantages of IMPT can be seen in this paraspinal case. First, adding 2 beams with smooth wedge-shaped distal fall-off (shown in the upper half of the figure) improves the robustness of the total dose distribution (lower half). Second, dose can be boosted to small areas (arrow), which helps to avoid dose cold-spots in the target volume. Courtesy of S. Safai, Massachusetts General Hospital.

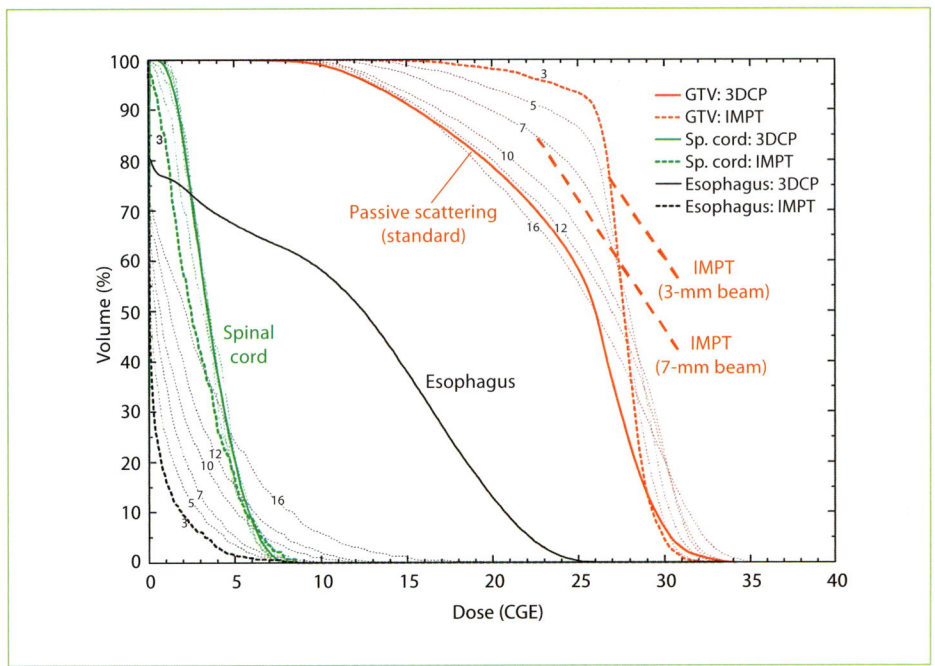

Fig. 12. Comparison of DVHs for passive scattering vs. IMPT with different pencil beams sizes [standard deviation (σ) in mm]. This is from the case also shown in figure 11, a retreatment of a recurrent tumor. Courtesy of S. Safai, Massachusetts General Hospital.

dose fall-off from the left and right posterior directions. Adding these fields together yields a uniform target dose, and the total dose distribution is quite insensitive to small variations of the range from the individual fields. So, one advantage of IMPT is robustness of the dose distribution, i.e. insensitivity to delivery errors. However, it is important to note that this robustness does not come automatically with IMPT. Rather, it has to be incorporated into the IMPT optimization algorithm [15]. Without that, IMPT plans may actually be more sensitive to delivery errors than passive scattering plans, which can be manually 'robustified' by experienced planners.

Another advantage of IMPT is better dose conformality and better dose coverage to the target volume in the most complex cases. This advantage depends greatly on the size of the pencil beams that can be produced by the proton system. If it could achieve a 3-mm (σ) beam this would be excellent, but realistically it may be about 7 mm (σ). Figure 12 shows dose distributions from IMPT using various pencil beam sizes compared with distributions from standard passive scattering techniques. Even with a 7-mm (σ) beam, IMPT is better (though not dramatically) than passive scattering technologies.

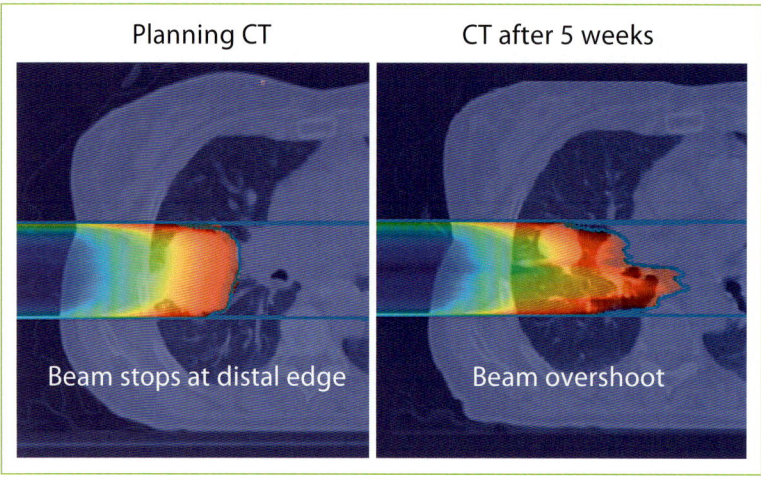

Fig. 13. Treatment planning example for a lung tumor. Left: initial treatment planning CT with correct proton dose coverage. Right: 5 weeks into treatment, repeat CT shows significant shrinkage of the tumor and dose overshoot into healthy tissue. Courtesy of S. Mori, NIRS. See online supplementary material for additional information.

Overall, the advantages of IMPT are: (a) greater flexibility in dose-shaping for improved target coverage or reduced normal tissue dose, (b) potential improvement in plan robustness (such as avoiding patching fields, which makes the therapy vulnerable to misalignments and range errors) and (c) possible use for dose painting, as might be used to boost dose in hypoxic areas.

Imaging where the Beam Stops

The second area of active development is imaging of the proton dose distribution, in order to 'see' where the beam actually stops in the patient. As discussed above, there are a number of reasons why the exact stopping point of the beam may be uncertain despite careful initial planning. Protons are more vulnerable to variations of the patient's anatomy than photons. One example is tumor shrinkage during radiation therapy of lung tumors (fig. 13). With a photon treatment technique, this shrinkage causes relatively minor dosimetric changes by itself. With proton therapy, it may result in significant overshooting into lung and increased lung doses. Range uncertainties are amplified in the lung because of its relatively lower density: a 1-cm range uncertainty in water equivalent tissue, for example, would have nearly a 4-cm range uncertainty in lung tissue simply because of its reduced density, an effect that could be clinically significant.

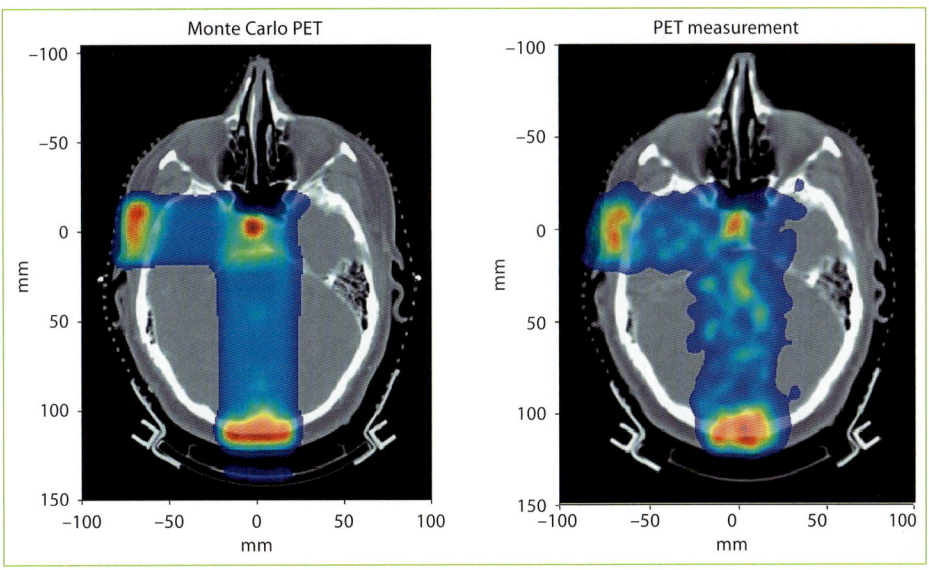

Fig. 14. PET/CT imaging provides a means to visualize the treated area in a proton treatment. Right: a measured PET/CT scan starting 20 min after the end of the treatment, with a scan time of 30 min [17]. Left: the expected PET image, based on the treatment plan.

PET/CT imaging is being investigated as a means to visualize the proton beam distribution inside the patient, and thereby reduce uncertainty about its range. Patients are scanned immediately after the treatment. We do not inject specific PET markers for these scans; rather, the proton beam directly activates O-15 and C-11 isotopes as it penetrates the patient [16, 17]. The resulting scans are compared with simulated PET scans based on the treatment plan. Figure 14 shows an example of such a patient scan, taken 20 min after treatment for a pituitary adenoma. Even though there is noise in this picture, one can see very clearly the 2 treatment beam paths. When compared with the Monte Carlo simulation of what is expected from this PET measurement, it agrees within a few millimeters at the distal edge, which is very reassuring even though it is not a direct measurement of the treatment dose.

Another way to obtain information about in vivo-delivered dose in proton therapy is to use measured 'prompt' gamma radiation produced from the interaction of protons with the patient [18, 19]. By measuring the gammas emitted laterally to the Bragg peak, one receives a signal that should be correlated closely with the dose distribution. This could be exploited to achieve information in real-time on the dose distribution, as opposed to using the offline PET method. This development is at an early stage; it has not been demonstrated in patients yet. Nevertheless,

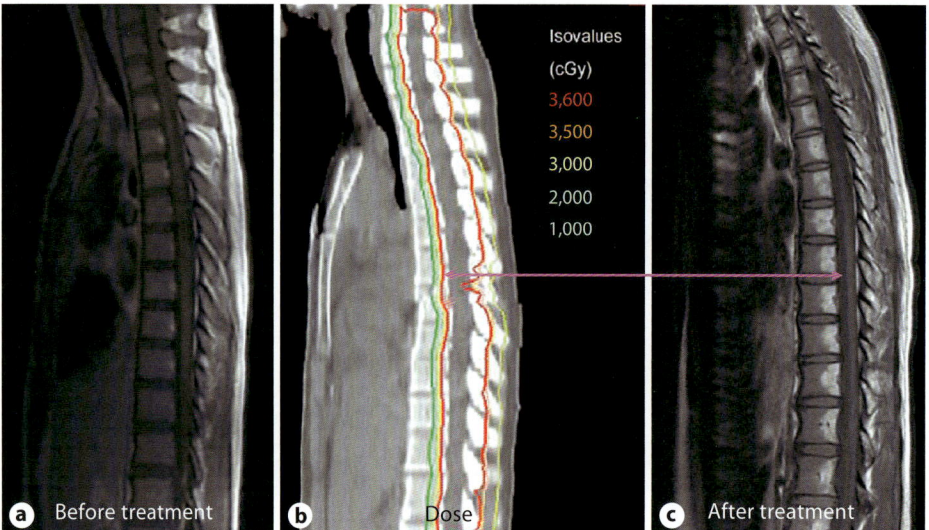

Fig. 15. Fatty replacement of the bone marrow as a consequence of irradiation can be visualized through T1-weighted MR imaging. **a** MRI scan taken before the treatment. **b** The planned dose distribution, and **c** an MRI scan taken after treatment [20].

theoretical and experimental work in Korea [18] and at MGH show a trend that there is a correlation (albeit not a perfect one) between the expected and measured gamma production and the proton dose distribution.

Except for physical phenomena, one can also exploit biochemical changes that occur as a consequence of irradiation. An example is fatty replacement of the bone marrow due to irradiation, which leads to a signal enhancement in T1-weighted MR images. These changes are less directly associated with the radiation dose and are not specific to proton irradiation. However, they are generally more useful in highly conformal types of irradiation such as proton therapy, because of the higher 'contrast' in the dose distribution. At MGH we have analyzed MRI scans taken after radiation therapy. Figure 15 shows MRI images of an adolescent treated for a medulloblastoma before and after treatment [20]. It shows a very clear signal enhancement in the treated area, an effect that appears to be a surrogate for the underlying dose distribution. Indeed we have been able to effectively translate the signal changes seen on MR images into delivered isodose curves, then to compare them with the calculated dose distributions [21]. We have found good agreement. This is another way of gaining confidence in our ability to deliver proton dose with millimeter accuracy (see double-arrow in fig. 15) and to effectively exploit the advantage of proton therapy, its finite range.

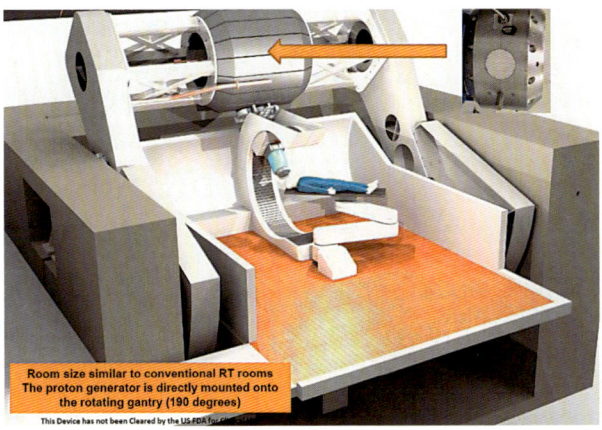

Fig. 16. Schematic drawing of a single-room proton machine with the cyclotron (arrow) mounted on the gantry (Still River Systems).

When and How Will Protons Become Cheaper?

Technologies that can reduce the size and cost of proton accelerators have received considerable attention recently. These include more compact versions of existing technologies yielding single-room proton machines, and also new acceleration mechanisms such as laser-accelerated protons and linear acceleration through dielectric wall acceleration.

Compact Proton Accelerator. There are a few commercial attempts to produce a proton accelerator that would support a single treatment room, rather than the 3 or 4 treatment room facilities that we have today. This could permit an overall reduction in the initial cost of the facility (but of course also the patient throughput). The Still River System features a compact cyclotron that is mounted on the gantry (fig. 16). There is no separate vault for the cyclotron; everything is integrated into a single treatment room (though somewhat larger than a typical photon therapy room). There are other developments by Accel (now Varian) that are following similar directions.

Wake Field Accelerator. Another approach that has generated interest, though less so recently, is the idea of accelerating protons with laser wake fields (fig. 17). It is very interesting from a physics point of view and has generated many journal articles. How does this work? The basic constituents are a laser beam and a thin foil. When the laser hits the foil, it produces plasma, a kind of hot gas of electrons and ions. The electrons are ejected very rapidly in the forward direction and produce an electrical field. Protons are sucked into the field, which is why it is called 'wake field acceleration'. The wake field produced by the electrons directly accelerates the protons that follow.

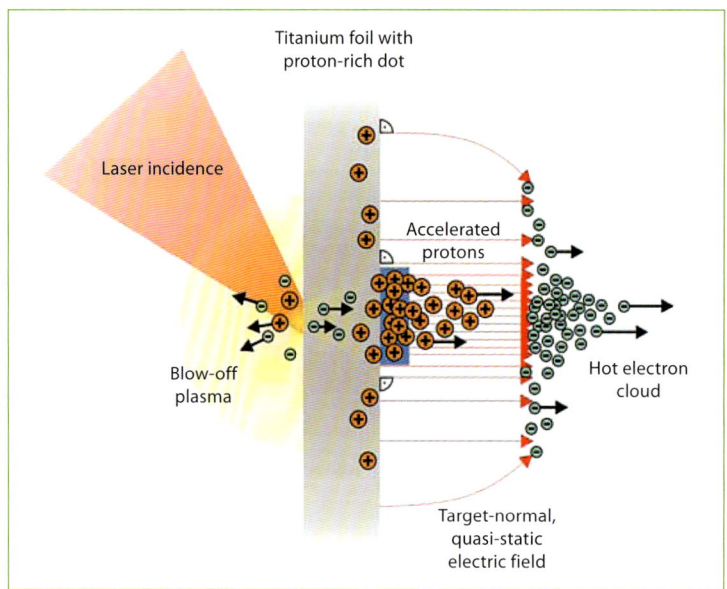

Fig. 17. The principle of laser wake field acceleration [22]. A laser beam incident on a thin foil produces a cloud of electrons on the other side of the foil, which pull protons in their direction. Reprinted with permission.

One of the problems with this approach is that, while it is possible to achieve high energies (150–180 MeV), the energy spectrum is not one that we would like to see. For proton therapy one needs a large number of protons with high energies, and a small number with lower energies. Unfortunately, the laser accelerated techniques typically produce the opposite spectrum. One would have to filter the spectrum greatly to obtain a clinically useful energy spectrum, which would result in low intensities and in treatment times that would be far too long (at least using current approaches). While physics investigations are ongoing, it is unclear whether or when clinically useful solutions will appear.

Proton Linear Accelerator. The idea of a proton linear accelerator is being promoted by TomoTherapy. Instead of accelerating protons in a cyclotron or synchrotron and then guiding and bending the beam with heavy magnets into the treatment room, one accelerates protons in a linear fashion with electric fields, but without heavy magnets, right on the gantry. The key elements are very good insulators to be able to produce high electric fields on a small length scale, and fast switches that allow the fields to be switched in phase with the accelerated protons, i.e. with the accelerating field at the right time and at the right location. The device would still be fairly large, of the order of 2 meters to produce 200-MeV protons, but it would not require either a separate cyclotron or synchrotron, or a separate beam line. Figure 18 shows a drawing of this device.

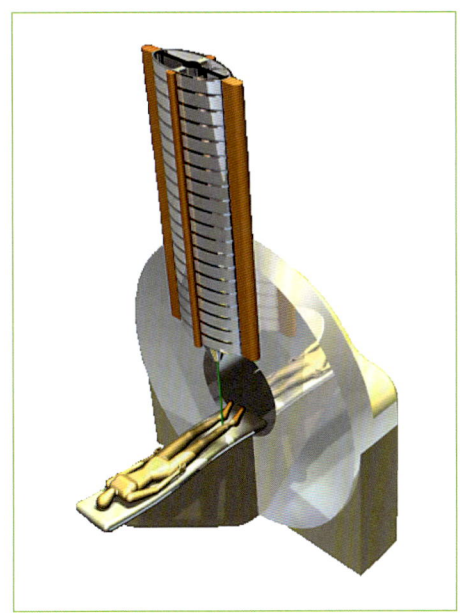

Fig. 18. Conceptual drawing of a treatment device using the dielectric wall acceleration technique, which is basically a proton linear accelerator. Courtesy of TomoTherapy.

Conclusions: Future Proton Technologies

For proton delivery, pencil beam scanning and IMPT techniques are already in use, and their further development and expanded utilization will undoubtedly continue. It should be understood that these approaches are not always better, and that further developments are necessary to exploit their benefits fully. Primarily, better methods are needed to focus the beams so that sharper penumbras are obtained.

A spectrum of fascinating imaging techniques can now be used to assist our understanding of where the proton beam stops exactly in the patient. PET imaging is the one that is furthest developed, allowing us to approach in vivo dose imaging. The question is whether we can use one or more of these techniques to perform dose-guided adaptive strategies, i.e. use the imaging information to drive the beam to the intended target position with millimeter accuracy and precision. This is not quite clear yet.

If one is considering proton therapy for long-term use, is it worthwhile to wait until new delivery technologies are developed? In our opinion, compact proton machines will become clinically available in the foreseeable future. They use tested technologies, and they need only to be 'put together'. This should occur within the next 2–3 years. Laser-generated and linearly accelerated technologies are not fully tested, and additional research and development are needed to achieve robust systems that are clinically useful; it seems unlikely that we will see clinical units based on such technologies within the next 5 years.

References

1 ICRU Report 50: Prescribing, Recording, and Reporting Photon Beam Therapy. Bethesda, International Commission on Radiation Units and Measurements, 1993.
2 ICRU Report 62: Prescribing, Recording and Reporting Photon Beam Therapy (supplement to ICRU report 50). Bethesda, International Commission on Radiation Units and Measurements, 1999.
3 Urie M, Goitein M, Wagner M: Compensating for heterogeneities in proton radiation therapy. Phys Med Biol 1984;29:553–566.
4 Kang Y, Zhang X, Chang JY, Wang H, Wei X, Liao Z, Komaki R, Cox JD, Balter PA, Liu H, Zhu SR, Mohan R, Dong L: 4D proton treatment planning strategy for mobile lung tumors. Int J Radiat Oncol Biol Phys 2007;67:906–914.
5 Engelsman M, Rietzel E, Kooy HM: Four-dimensional proton treatment planning for lung tumors. Int J Radiat Oncol Biol Phys 2006;4:1589–1595.
6 Goitein M: Calculation of the uncertainty in the dose delivered during radiation therapy. Med Phys 1985;12:608–612.
7 Trofimov A, Bortfeld T: Optimization of Beam Parameters and Treatment Planning for Intensity Modulated Proton Therapy. Technol Cancer Res Treat 2003;2:437–444.
8 Hall E: Intensity-modulated radiation therapy, protons, and the risk of second cancers. Int J Radiat Oncol Biol Phys 2006;65:1–7.
9 Kooy H, Clasie BM, Lu H-M, et al: A case study in proton pencil-beam scanning delivery. Int J Radiat Oncol Biol Phys 2010;76:624–630.
10 Pedroni E, Bacher R, Blattmann H, et al: The 200-MeV proton therapy project at the Paul Scherrer Institute: conceptual design and practical realization. Med Phys 1995;22:37–53.
11 Phillips MH, Pedroni E, Blattmann H, Boehringer T, Coray A, Scheib S: Effects of respiratory motion on dose uniformity with a charged particle scanning method. Phys Med Biol 1992;7:223–234.
12 Bortfeld T, Jokivarsi K, Goitein M, Kung J, Jiang SB: Effects of intra-fraction motion on IMRT dose delivery: statistical analysis and simulation. Phys Med Biol 2002;47:2203–20.
13 Bert C, Saito N, Schmidt A, Chaudhri N, Schardt D, Rietzel E: Target motion tracking with a scanned particle beam, target motion tracking with a scanned particle beam. Med Phys 2007;34:4768–4771.
14 Lomax J, Boehringer T, Coray A, et al: Intensity modulated proton therapy: a clinical example. Med Phys 2001;28:317–324.
15 Unkelbach J, Bortfeld T, Martin BC, Soukup M: Reducing the sensitivity of IMPT treatment plans to setup errors and range uncertainties via probabilistic treatment planning. Med Phys 2009;36:149–163.
16 Oelfke U, Lam GKY, Atkins MS: Proton dose monitoring with PET: quantitative studies in Lucite. Phys Med Biol 1996;41:177–196.
17 Parodi K, Paganetti H, Shih HA, Michaud S, Loeffler JS, DeLaney TF, Liebsch NJ, Munzenrider JE, Fischman AJ, Knopf A, Bortfeld T: Patient study of in vivo verification of beam delivery and range, using positron emission tomography and computed tomography imaging after proton therapy. Int J Radiat Oncol Biol Phys 2007;68:920–934.
18 Min C-H, Kim CH, Youn M-Y, Kim J-W: Prompt gamma measurements for locating the dose fall-off region in the proton therapy. Appl Phys Lett 2006;89:183517.
19 Polf JC, Peterson S, Ciangaru G, Gillin M, Beddar S: Prompt gamma-ray emission from biological tissues during proton irradiation: a preliminary study. Phys Med Biol 2009;54:731–743.
20 Krejcarek SC, Grant PE, Henson JW, et al: Physiologic and radiographic evidence of the distal edge of the proton beam in craniospinal irradiation. Int J Radiat Oncol Biol Phys 2007;68:646–649.
21 Gensheimer MF, Yock TI, Liebsch NJ, Sharp GC, Paganetti H, Madan N, Grant PE, Bortfeld T: In vivo proton beam range verification using spine MRI changes. Int J Radiat Oncol Biol Phys 2010;78:268–275.
22 Schwoerer H, Pfotenhauer S, Jäckel O, Amthor K-U, Liesfeld B, Ziegler W, Sauerbrey R, Ledingham KWD, Esirkepov T: Laser-plasma acceleration of quasi-monoenergetic protons from microstructured targets. Nature 2006;439:445–448.

Radhe Mohan, PhD
M.D. Anderson Cancer Center
1515 Holcombe Blvd., Unit 94
Houston, TX 77030 (USA)
Tel. +1 713 563 2505, Fax +1 713 563 2545, E-Mail rmohan@mdanderson.org

Medical Applications and Advances

Proton Therapy in the Clinic

Thomas F. DeLaney

Francis H. Burr Proton Therapy Center, Massachusetts General Hospital, Boston, Mass., USA

Abstract

The clinical advantage for proton radiotherapy over photon approaches is the marked reduction in integral dose to the patient, due to the absence of exit dose beyond the proton Bragg peak. The integral dose with protons is approximately 60% lower than that with any external beam photon technique. Pediatric patients, because of their developing normal tissues and anticipated length of remaining life, are likely to have the maximum clinical gain with the use of protons. Proton therapy may also allow treatment of some adult tumors to much more effective doses, because of normal tissue sparing distal to the tumor. Currently, the most commonly available proton treatment technology uses 3D conformal approaches based on (a) distal range modulation, (b) passive scattering of the proton beam in its x- and y-axes, and (c) lateral beam-shaping. It is anticipated that magnetic pencil beam scanning will become the dominant mode of proton delivery in the future, which will lower neutron scatter associated with passively scattered beam lines, reduce the need for expensive beam-shaping devices, and allow intensity-modulated proton radiotherapy. Proton treatment plans are more sensitive to variations in tumor size and normal tissue changes over the course of treatment than photon plans, and it is expected that adaptive radiation therapy will be increasingly important for proton therapy as well. While impressive treatment results have been reported with protons, their cost is higher than for photon IMRT. Hence, protons should ideally be employed for anatomic sites and tumors not well treated with photons. While protons appear cost-effective for pediatric tumors, their cost-effectiveness for treatment of some adult tumors, such as prostate cancer, is uncertain. Comparative studies have been proposed or are in progress to more rigorously assess their value for a variety of sites. The utility of proton therapy will be enhanced by technological developments that reduce its cost. Combinations of 3D protons with IMRT photons may offer improved treatment plans at lower cost than pure proton plans. Hypofractionation with proton therapy appears to be safe and cost-effective for many tumor sites, such as for selected liver, lung and pancreas cancers, and may yield significant reduction in the cost of a therapy course. Together, these offer practical strategies for expanding the clinical availability of proton therapy.

Copyright © 2011 S. Karger AG, Basel

Rationale for Protons

From a technical point of view, advances in radiotherapy begin with the principle that higher doses of radiation to the tumor increase rates of local tumor control. At the same time, higher doses to normal tissues increase the rates of normal tissue complications. Dr. Herman Suit has pointed out that complications do not occur in unirradiated tissues, and indeed normal tissue irradiation does not benefit the patient. We can optimize the therapeutic ratio by increasing the tumor dose and/or minimizing the normal tissue dose. The contribution of proton beam therapy to this is clear. The proton beam does not deliver dose beyond the distal edge of its Bragg peak, whereas photons continue to deposit dose in tissues beyond the target. The dose to normal tissues is less with protons.

To treat tumor volumes with passively scattered proton radiation therapy (the predominant form of beam delivery to date), the Bragg peaks of proton beams are spread out with range modulator wheels or ridge filters [1]. Lateral field-shaping is performed with custom-milled brass apertures. Distal field edge-shaping is accomplished using customized compensating boluses (see online supplementary material for illustrations).

The proton beam (because of its charge) can also be magnetically scanned within the patient and the energy modulated to deliver either uniform or intensity-modulated proton therapy; in time, it is expected that beam scanning will become the preferred treatment delivery system [2]. Figure 1 shows several important points. Photons deposit dose beyond the distal edge of the tumor, whereas protons do not. Photons, for any given beam, also deposit more dose proximal to the tumor. One subtle point is that with passively scattered protons, the skin dose with protons is higher than with photons. That can have some implications if treatment involves only a single field. Particularly when using concurrent chemotherapy, one will see more acute skin reactions if only a single proton field is employed.

Protons: Clinical Advantages and Special Concerns

Much of the original work with proton beam therapy was done at the Harvard Cyclotron Lab in conjunction with clinicians from the Massachusetts General Hospital (MGH). The facility had a fixed proton beam with 160 MeV maximum energy that could penetrate 15 cm in tissue. This limited the tumor sites that could be studied, but the clinical outcomes did provide the proof of principle for the use of protons.

The clinical advantage for protons is physical, not biological. We correct for the difference in relative biological effectiveness (RBE) of protons over photons by multiplying the physical dose in Gy by the recommended RBE correction factor of 1.1 to yield the biologic dose, expressed in units of *GyRBE* as recommended by the

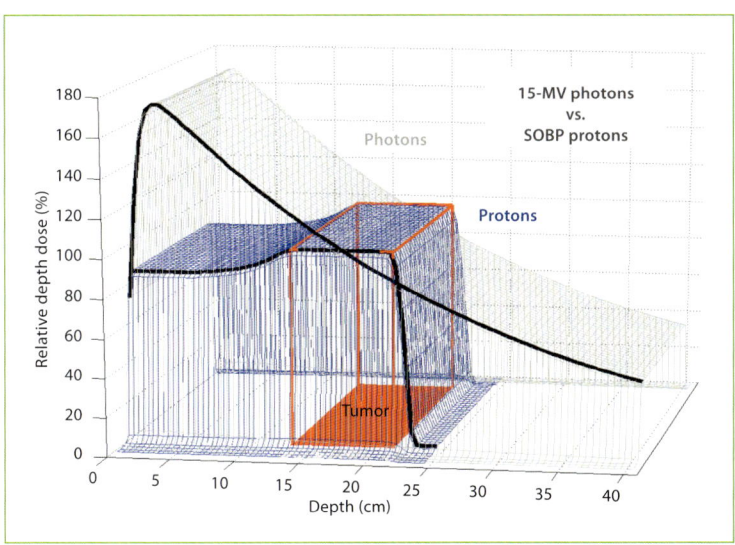

Fig. 1. Comparison of relative depth dose distributions of 15-MV photons versus spread-out Bragg peak (SOBP) protons. Courtesy of Alfred Smith, PhD.

ICRU [3]. Some of the heavier charged particles such as carbon may have additional biological benefit, such as a differential effect on hypoxic cells, although this remains to be proven in the clinic.

Michael Goitein [4] recently published an excellent review 'Radiation oncology, a physicist's eye view'. He made the point that there is a marked reduction in integral dose with proton therapy of roughly 60%, as a general rule. This is a useful concept. In the initial use of protons, there were improved dose distributions with protons compared with 2D photons or 3D conformal photons. This permitted dose escalation with acceptable normal tissue side effects. Ocular melanoma was one of the early sites treated (fig. 2), and in fact was one of the early anatomic sites that required the use of tumor motion management strategies: the patient was simply asked to fix his gaze on a light, and there was video confirmation of the globe position during treatment!

Table 1 shows some of the results achieved in the early proton studies. Those for ocular melanoma are impressive, with a 95% local control rate at 15 years. Chondrosarcomas, tumors that even some orthopedic texts still indicate are not sensitive to radiation, have been treated with high-dose photon/proton radiation when they arise in the skull base or spine. In the skull base, they have been locally controlled (defined as the absence of tumor growth) in the overwhelming majority of cases; they are not responsive in that they do not shrink, but they clearly are sensitive in their lack of progression long-term. Indeed, table 1 shows the result of base of skull treatment, with a 95% local control rate at 10 years.

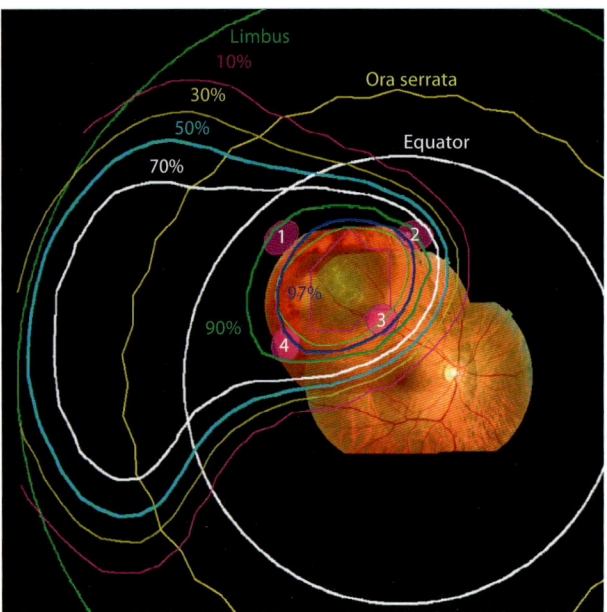

Fig. 2. Wide-angle fundus view of the eye model from the proton treatment planning program (displayed with a mosaic photo made of 3 narrow fundus photographs superimposed and aligned by matching the optic disc and the macula of the model to those in the mosaic). Four tantalum marker rings outlining the tumor are shown in magenta numbered 1–4. The base of the tumor on the fundus, shown in green, is seen to be enclosed by the 97% isodose line. The magenta line delineates the area receiving 100% of the dose [26].

Table 1. Some of the early data from selected anatomic sites treated with protons that demonstrated encouraging treatment results with the modality

Proton clinical results: local control

Ocular melanoma: 70 GyRBE in 5 fractions
    95% at 15 years (Harvard Cyclotron Lab) [26]
Chondrosarcoma skull base: 69.6 GyRBE in 37 fx
    95% at 10 years (Harvard Cyclotron Lab) [28]
Prostate cancer stages T1–T2B: 75 GyRBE in 46 fx
    88% PSA disease-free 5-year survival (Loma Linda University) [29]

The prostate data from Loma Linda University Medical Center indicate that this technology is transferable to other centers, and that one can achieve high rates of disease control with proton therapy – in this case, PSA disease-free survival. Many other proton facilities have now opened around the world and have demonstrated encouraging results, similar to those in the earlier studies, for a

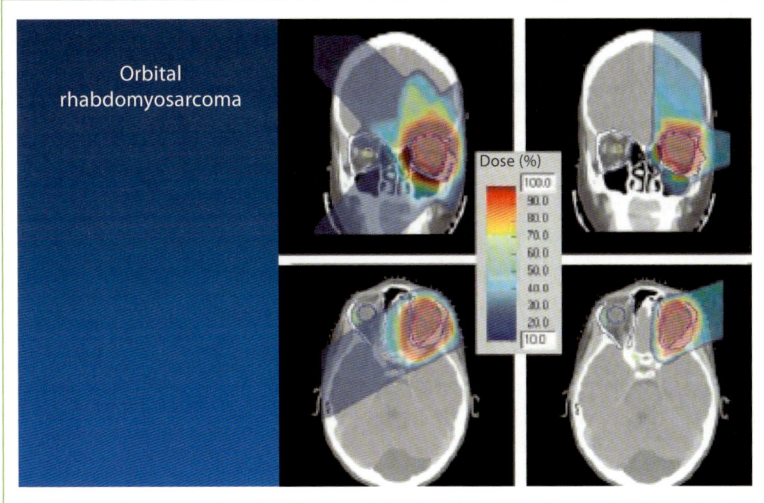

Fig. 3. Dose distributions for photons on the left compared with protons on the right for treatment of an orbital rhabdomyosarcoma [27]. See online supplementary material for this and an additional medulloblastoma case.

variety of tumor sites. Hospital-based facilities are now being constructed by commercial vendors, a development that is driving much of the current interest in the field. Our own facility was the cornerstone for the new cancer center, and similar facilities have recently been completed or are under construction at other sites.

The clinical advantages of proton therapy include the lower integral dose and absence of an exit dose. A smaller treatment volume is being irradiated. Certainly in some situations, this can provide substantial advantages for the patient. The reduction in normal tissue exposure may be important if one is combining radiotherapy with chemotherapy. Especially in the pediatric population, the reduction of radiation exposure to bone growth plates can spare the growth centers and eliminate arrest of bone growth. Figure 3 compares proton and photon fields for one of the common rhabdomyosarcomas, orbital. When using photons, with any given beam arrangement, one exits out through developing tissues, which creates an obvious advantage for protons.

One point that deserves consideration is the fact that the bigger the field, the bigger the advantage is for protons since the volume of normal tissue sparing increases. In the past, protons were used to treat small targets because the available proton beams limited the treatment possibilities. Current technologies permit the treatment of larger targets. For example, the advantage of proton therapy is clear for treatment of long craniospinal fields, such as for medulloblastomas [5].

Second Malignancies

At the 2008 ASTRO meeting, we presented our data on second malignancies following proton irradiation, and compared our results with matched SEER database photon results [6]. 503 Harvard Cyclotron patients were matched with 1,591 SEER patients, and both groups were evaluated for second malignancy occurrence. 6.4% of the proton patients (32 patients) developed a second malignancy versus 12.8% of photon patients (203 patients), resulting in a hazard ratio of 2.73 in favor of proton therapy (95% CI: 1.67–3.98, $p < 0.0001$). The median follow-up was 7.7 years for the proton group and 6.1 years for the photon group. These results are not entirely surprising; one is simply treating a smaller volume of tissue with protons.

Proton Radiotherapy Treatment Planning

Normal Tissue Sparing

Even for adult patients, one can think of advantages with proton therapy for several treatment sites. For instance, hepatocellular carcinomas typically arise in diseased livers, and organ sparing has greater importance for these patients. Comparing a 60-Gy field using helical tomotherapy with one using proton therapy, the latter provides absolute sparing of a much larger volume of liver. Currently, our group is working on a dose escalation study for these patients based on the dose delivered to the liver. In fact, the liver itself has not been the dose limiting organ in the study, but rather the normal tissues adjacent to the liver (personal communication: Theodore Hong MD, MGH).

Combining Proton and Photon Therapy

In practice, we often combine photons and protons in treatment courses in adult patients. After all, these are tools. In radiotherapy, we have always used modalities most suitable to the particular clinical situation, as when we combine electron and photon therapies routinely. One can also combine photon-based intensity-modulated radiotherapy (I will use the term IMRT exclusively for photon methods), and we frequently do this. Nearly all of the large-field Harvard Cyclotron results (excluding ocular melanomas) were achieved with at least 20% of the treatments from photons, as the facility was only available 4 days a week to patients for fractionated, large-field proton therapy. Currently, in order to maximize the number of patients that we can treat with at least a component of protons, many of our adult patients receive about 20–30 Gy of photon therapy. When treating with 3D protons, where one relies on the aperture edge and penumbra to tangent the target, the target dose distribution can also be improved by employing some component of IMRT. For example, one can use IMRT to conform the target dose around the spinal cord to reduce the potential spinal cord dose overall. Indeed, a recent study comparing 3D passively scattered protons, IMRT and a combination of the two concluded that the optimal plan (in terms of target coverage and integral dose) was a combination of 3D passively scattered protons and IMRT [7].

Normal Tissue Organ at Risk Avoidance

When delivering high doses to volumes immediately adjacent to the spinal cord – doses that well exceed cord tolerance – one needs to know where the cord is precisely. As we continue to explore hypofractionated dose escalation programs, it will be increasingly important to be able to anatomically locate avoidance structures with high degrees of certainty. While image fusion with MRI has been helpful, we have also employed intrathecal contrast at the time of CT simulation and daily pretreatment imaging for the high-dose portions of the treatment course. An example of the dose distributions for a patient with a C2 chordoma is shown in figure 4, demonstrating the use of intrathecal contrast to optimally define the spinal cord position precisely.

Target Definition

In the current practice of radiation oncology, we are often in the ironic situation of being able to deliver radiation dose with more precision than we know where the tumor is. As such, delineation of clinical target volumes can become a matter of professional interpretation to a degree. A variety of studies have demonstrated rather marked differences in the target volumes delineated even among radiation oncologists with subspecialty expertise [8]. There will be increasing need for precision in defining target volumes, especially for subclinical or microscopic tumor, and there are many ongoing efforts to improve the ability of diagnostic imaging to detect smaller volumes of tumor [9].

Dose Comparisons with Photon-Based Intensity-Modulated Radiotherapy

Because IMRT is more widely available and less costly than proton therapy, it is essential to know when protons can be advantageous. Treatment planning comparisons between the two modalities are important for clinicians and patients, and

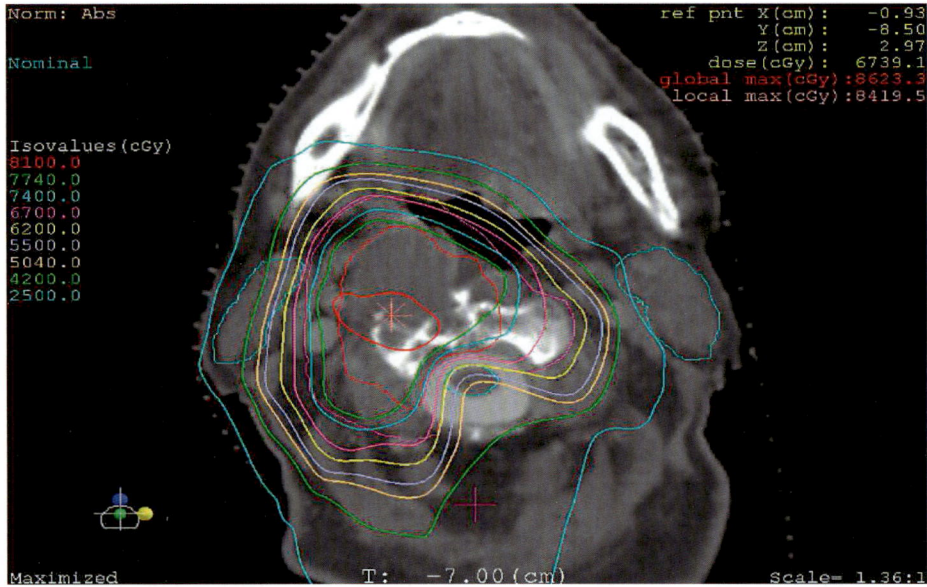

Fig. 4. Dose distribution for a patient treated definitively (following biopsy only) to 77.4 GyRBE with 19.8 Gy of IMRT photons and 57.6 GyRBE passively scattered 3D conformal photons for a chordoma involving the right side of the C2 vertebral body with extraosseous extension. Note the use of intrathecal contrast for spinal cord delineation and avoidance.

ultimately for health policy analysts. Figure 5 shows one such study performed by our genitourinary group [10]. It compares IMRT with 3D conformal protons, as actually implemented in the clinic using designed smearing procedures (the intentional smearing increases the effective range to offset any misalignment of the bone anatomy with the range compensator). In addition, comparisons were made with intensity-modulated proton therapy, which is just entering into clinical use. When these were evaluated, the IMRT gave obviously higher integral doses to the pelvis. However, in regions at risk for high dose such as the anterior wall of the rectum and the bladder neck, the doses were actually lower with IMRT than 3D conformal protons for the majority of the patients. This is illustrated by the dose-volume histogram curves in figure 5. This is important, since we know that clinical problems (at least in the short-term) will be in these two anatomic areas.

Overall, this study indicated that IMRT actually outperformed 3D conformal protons (both as clinically used at the time) in bladder and rectal dosimetry in the high-dose regions. Intensity-modulated protons were better than IMRT, but require better control of range than is currently available in the clinic (fig. 6). Hence, one can achieve better dosimetry with intensity-modulated protons than IMRT, but only if one understands and controls the range uncertainty associated with the

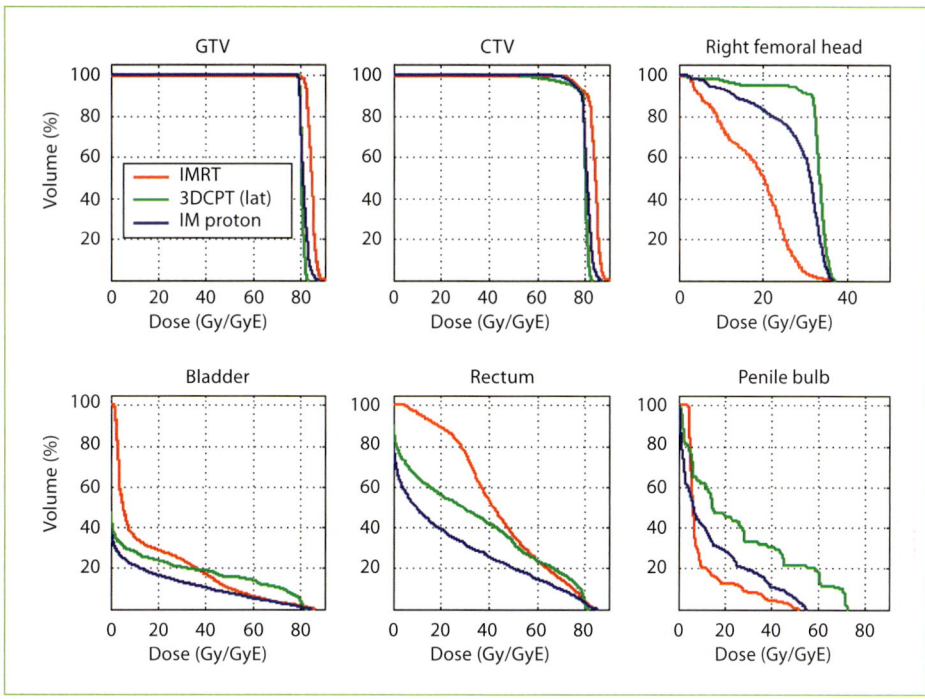

Fig. 5. Dose-volume histogram curves for 7-field IMRT, 3D conformal proton therapy with 2 parallel opposed proton fields, and intensity-modulated proton therapy with 2 parallel opposed proton fields (IM proton). The intensity-modulated proton therapy treatment plan here used 2 parallel opposed fields, which would not likely be an optimal arrangement; greater hip and penile bulb sparing would be achieved with several additional fields and other beam directions [10]. GTV = Gross tumor volume; CTV = clinical target volume; 3DCPT = 3D conformal proton therapy.

proton treatment. We recently advanced a research proposal for using a rectal probe dosimeter to assist us in defining this uncertainty in a practical way in the clinic and to facilitate implementing intensity-modulated proton therapy for prostate cancer.

Issues in the Utilization of Proton Radiotherapy

Cost of Proton Compared with Photon Therapy

The relative cost of proton to other therapies is particularly important, since the potential costs to society will increase with the number of proton centers. In many of the centers, a large percentage of patients are being treated for prostate cancer

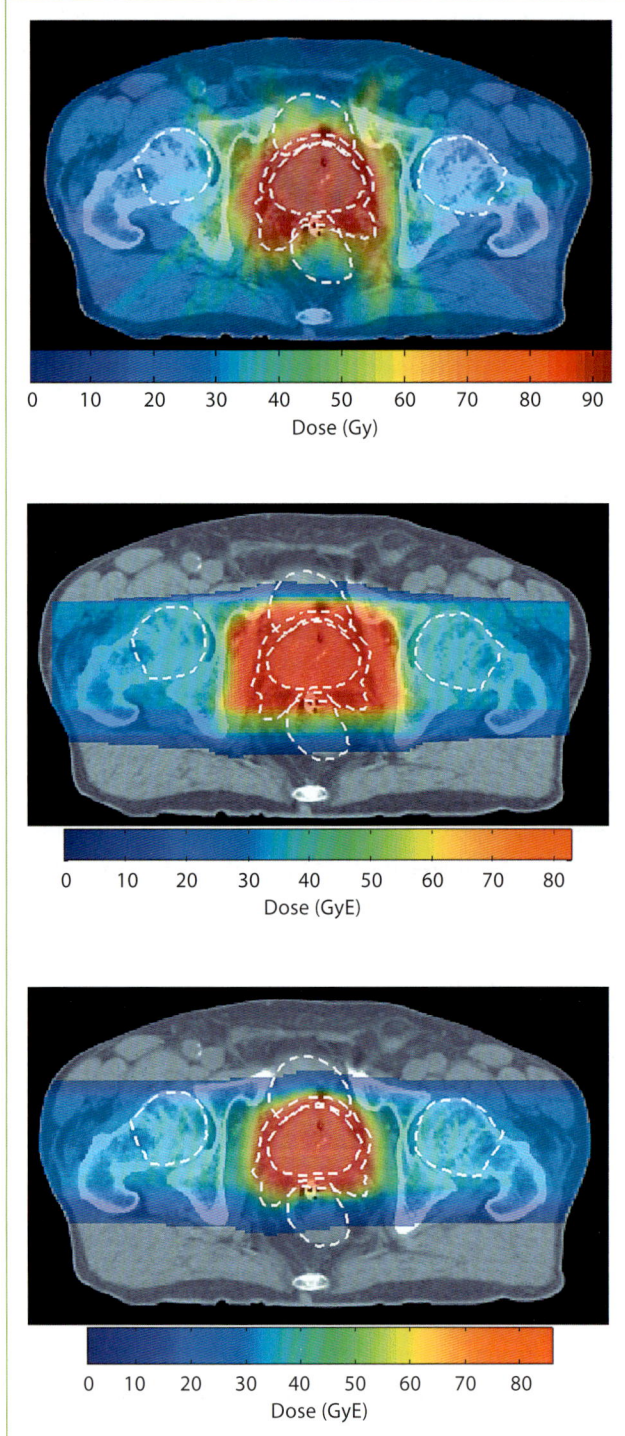

Fig. 6. Radiotherapy treatment plans for one patient included in the study by Trofimov et al. [10], 7-field IMRT (top), 3D conformal proton with 2 opposed lateral fields (middle), and intensity-modulated proton therapy with 2 opposed lateral fields (bottom).

although it has several alternative treatment options. Underlying this issue is the cost differential between IMRT and protons, and clearly the cost is higher for proton therapy. In one evaluation [11], the cost of intensity-modulated proton therapy was 2.4 times higher than IMRT. This was primarily related to the difference in the capital construction costs. The authors projected that this might decrease in time, with projected gains in the operating efficiencies of proton centers, to about 1.7 times. Also, if one were able to garner philanthropic support to offset the initial construction costs, this differential would be further diminished.

Hence, reduction in the cost differential between proton radiation therapy and IMRT should be a high priority in order to advance this technology. If protons and photons were equal in cost, discussion about the need for randomized proton versus photon studies would end. Since this is not the case, we need less expensive facilities and improvement of treatment efficiencies. Also, exploration of hypofractionation programs, especially in disease sites where fields are safely away from critical nerves and other structures, may yield significant reduction in the cost of a proton therapy course. Finally, one can combine proton and photon therapies to reduce cost and provide greater access to the currently limited number of proton treatment slots.

In our own group, we have several hypofractionation studies underway. One study is using neoadjuvant accelerated short-course proton beam radiation therapy with concurrent capecitabine for resectable pancreatic cancer. The phase I study for this has already been completed [12] and the phase II study is now in progress (http://clinicaltrials.gov/ct2/show/NCT00438256?term=proton+pancreas+hong&rank=1). The proton radiotherapy is 5 GyRBE × 5 fractions = 25 GyRBE in a single week. When we looked at the reimbursable charges for treatment recognized by the Center for Medicare Services (CMS) for 3D conformal or IMRT treatment to a standard dose of 50.4 Gy in 28 fractions over 5.5 weeks versus hypofractionated proton therapy of 25 GyRBE in 5 fractions over 1 week, the proton therapy was cheaper. This illustrates one opportunity to reduce proton treatment costs through hypofractionation made clinically possible by the lower integral doses.

Protons: Clinical Experience

There are some tumor sites where protons may already be cost-effective. For example, in the treatment of pediatric medulloblastomas, one may see a cost savings related to the reduction or elimination of certain late toxicities [13]. By avoiding these, protons were associated with an estimated cost savings of EUR 23,000 over the lifetime of the child as well as a 0.68 additional quality-adjusted life year (QALY) of survival per patient. Potential benefits were shown even for some common adult tumors, for instance left-sided breast cancer, although the cost for the QALY gained was EUR 67,000. Because this was due to avoidance of the heart, the cost per QALY gained would be lower in a population at risk for developing cardiac disease [14].

If proton therapy permits dose escalation, what incremental value (benefit to the patient divided by the cost) might be given? For instance, if proton therapy permitted a 10-Gy dose escalation over 81 Gy for adenocarcinoma of the prostate (91 GyRBE with protons, compared with 81 Gy with IMRT), what would be the cost-benefit to the patient? Using common criteria for health care interventions, Konski et al. [15] concluded that protons would not be cost-effective for most patients, based on available data. The author raised the question whether the number of proton facilities should be limited until more comprehensive evaluations of this modality, including its indications and cost effectiveness, are completed. Certainly, in an ideal world, medical interventions should meet an accepted minimum benefit for their cost, as can be reported by QALY gained for the cost. One potential alternative, however, would be to try to reduce the costs and increase the efficiencies of proton treatment.

My sense is that there will be clinical gains resulting from proton therapy for a number of tumor sites. The societal question remains, however, how to provide access to this technology to permit the greatest gain for the maximum number of patients? From my point of view, this would be approached if protons were used for patients not well treated by photons. For adult patients, this may involve combing proton with photon treatment, and there are even dosimetric advantages of combining IMRT and 3D protons in some sites [7]. Another efficiency would be the use of hypofractionation, already being performed for hepatocellular carcinomas and for medically inoperable, early-stage non-small cell lung cancers. Finally, it is anticipated that technological improvements will reduce the cost and complexity of proton treatment, making its availability more practical.

The Role of Clinical Trials

There is an emerging consensus that this technology may be an effective treatment with less acute and late morbidity than comparable external beam photon techniques. How do we objectively document the perceived or presumed gains with proton therapy? We obviously prefer to have the level 1 evidence of phase III trials. For pediatric patients, however, phase II studies are the only ones that can be undertaken from an ethical perspective because of the marked reduction in the integral doses with protons. Eligible children should be considered for referral to an appropriate center if felt to have a potential gain with protons. To accommodate all such patients, the number of proton centers would need to be expanded, as well

Table 2. V5, V10 and V20 percentages with 3D photon or IMRT photons to 60–63 Gy or protons to 74 GyRBE

Mean lung	V5	V10	V20
63 Gy 3D Photon	54.1%	46.9%	34.8%
74 GyRBE Proton	39.7%	36.6%	31.6% 3D (p = 0.002)
60–63 Gy IMRT	58.5%	45.3%	34.5%
74 GyRBE Proton	44.0%	39.3%	33.3%

Stage III. M.D. Anderson Cancer Center (from Chang et al. [18]).

as the professional expertise in proton therapy, pediatric oncology and pediatric anesthesia needed to manage these patients. As more proton centers come on-line, training for proton personnel at the new centers will become an additional issue. The expectation of proton use for pediatric patients also implies the development of new proton technologies with scanned beams so that the neutron doses associated with passively scattered proton beam lines can be reduced as well [16, 17].

For adult malignancies, randomized studies of protons versus photons raise the issue of *equipoise*. Are knowledgeable professionals truly uncertain of the benefits of proton therapy, making it worthy of scientific study? Clinicians and patients, aware of the dose advantages with proton therapy, may refuse randomization in phase III studies. However, further understanding can raise pertinent concerns and legitimate questions. For example, consider treatment of tumors in the head/neck regions. Initial therapy plans may be superior with protons, but the accumulation of mucus in the sinuses, target motion or changes in tumor or normal tissue density over the course of treatment may affect and degrade proton treatment plans to a greater degree than photon plans. Hence, one can make the case for randomization because of the greater uncertainties of dose distributions in some sites, and the possibility that target dose or normal tissue sparing might be compromised sufficiently with protons would make a randomized comparison with photons ethical and appropriate.

Lung Cancer

The M.D. Anderson group has reported comparisons of IMRT and 3D protons for therapy of stage III non-small cell lung cancers. They demonstrated that normal tissue doses were reduced with protons, even when tumor doses were escalated (74 GyRBE with protons vs. 63 Gy with photons). The lung doses were uniformly lower, as seen in table 2, which documents lower V5, V10 and V20 percentages with protons [18]. Even when comparing protons with IMRT, one could deliver higher

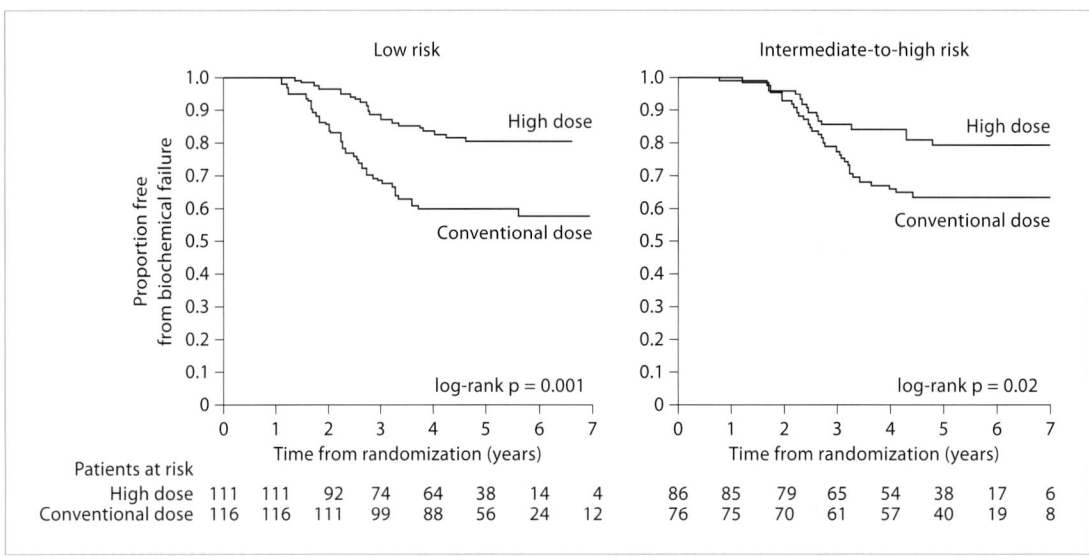

Fig. 7. Freedom from biochemical failure (ASTRO definition) following either conventional-dose (70.2 GyRBE) or high-dose (79.2 GyRBE) conformal radiation therapy (from Zietman et al. [19]). Analysis of these early cases is by risk subgroup. Low-risk patients have a PSA level <10 mg/ml, stage ≤T2a tumors and Gleason scores ≤6.

proton doses with comparable lung doses. Doses to the lung, spinal cord, heart and esophagus, and the integral dose were lower compared with IMRT; therefore, proton treatment appeared to reduce dose to normal tissues significantly, even with dose escalation.

The MGH and M.D. Anderson are currently conducting a randomized phase II study for patients undergoing chemoradiation for locally advanced non-small cell lung cancer (http://clinicaltrials.gov/ct2/show/study/NCT00915005?term=randomized+lung+cancer+proton&rank=1&show_desc=Y#desc). Patients whose 3D conformal or IMRT plans meet normal lung constraints will be randomized to 74 Gy photons versus protons. For those patients whose normal lung doses exceed the constraints, the protocol is being modified to randomize them to photons or protons at lower doses, either 66 Gy or 60 Gy depending on their lung doses.

Prostate Cancer

Because protons are a limited resource, they should be ideally used for patients not well treated by photons. Prostate cancer has a number of very good treatment options including surgery, external beam with IMRT or protons, brachytherapy or even observation in selected patients. Because prostate cancer patients constitute a significant number of the proton patients treated at some facilities, there is ongoing discussion about a registry or randomized study of IMRT versus 3D protons for prostate cancer (personal communication: Jason Efstathiou, MD, PhD, MGH).

For 3D conformal proton treatment, the most commonly used treatment plans specify lateral fields. Since there are range uncertainties for the distal edge of the beam, anterior and anterior oblique fields raise a level of uncertainty about where exactly the beam would stop in the rectum, and whether the posterior edge of the prostate would be adequately treated. This uncertainty makes these plans less robust than those with lateral fields.

The Proton Study Group, a collaboration of MGH and Loma Linda University Medical Center, incorporated proton radiotherapy into a randomized dose escalation study comparing 70.2 GyRBE with 79.2 GyRBE [19]. The dose above 50.4 Gy was delivered with protons. For both low-risk and intermediate-/high-risk patients, dose escalation to 79.2 Gy showed an improvement in biochemical disease-free survival without any increase in grade 3 toxicity in the high-dose arm, indicating that the protons were a good tool for safe dose escalation from 70.2 to 79.2 GyRBE (fig. 7). However, this study was designed to evaluate dose escalation rather than proton versus photon treatment, although the dose escalation was performed with protons.

Normal Tissue Effects

Since the initial period of development of proton therapy, photon IMRT has been introduced and widely implemented. IMRT often allows delivery of doses similar to protons and in some cases with better conformality than with 3D conformal protons [10]. Nevertheless, the integral dose with photon IMRT is always higher than with proton therapy [10]. While one selectively spares some normal tissues with photon IMRT, it is done at the cost of increased dose to other tissues. We do not know as yet what the late toxicity to the normal tissues will be from the low-to-moderate 'dose bath', which is part of IMRT.

Proton intensity modulation can now be implemented [20]. We have performed comparison dosimetric studies of photon versus proton IMRT radiotherapy for thoracic paraspinal sarcomas [21]. With the proton approach, one can almost completely eliminate the dose to the lungs anterior to the volume, offering a considerable normal tissue-sparing advantage.

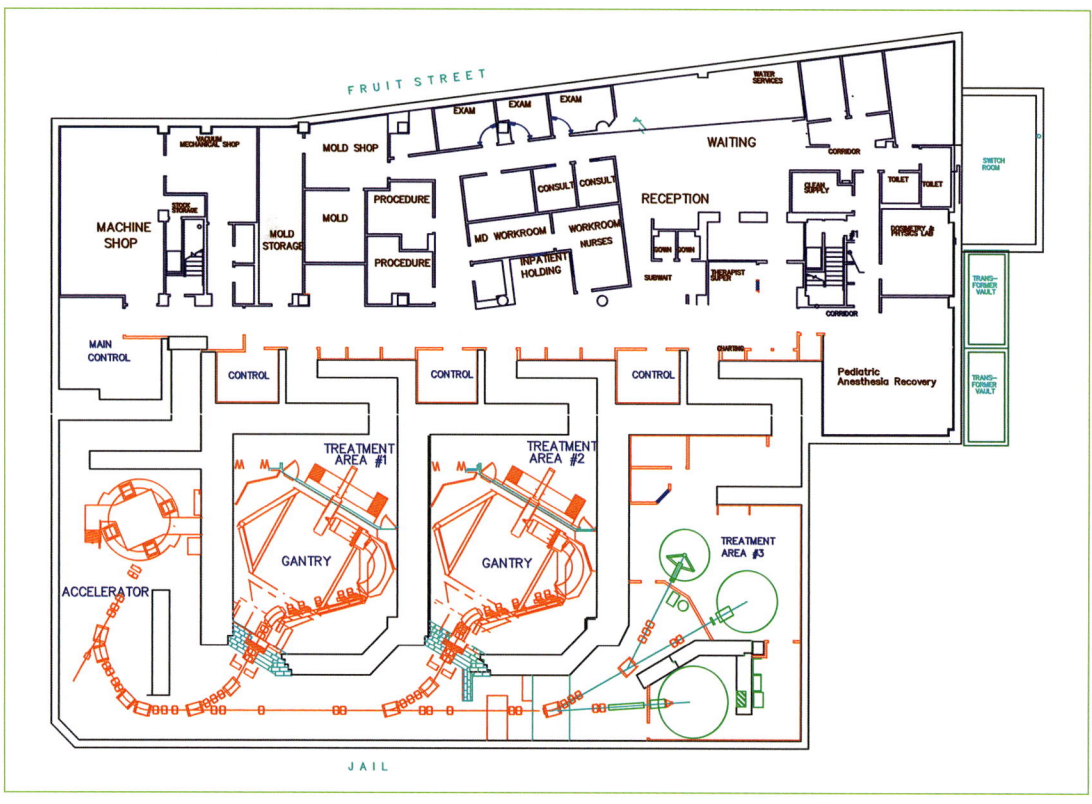

Fig. 8. The layout of the Francis H. Burr Proton Therapy Center at MGH. Beam lines transport protons from the cyclotron (round figure at lower left) into 1 of 2 shielded rooms with fully rotational gantries or a third treatment room with 2 fixed beams for eye treatment and stereotactic radiosurgery. Courtesy of Jay Flanz, PhD, MGH.

Protons: How User Friendly?

In the past, some proton centers would treat patients only a limited number of months per year because they were laboratory-based facilities; the proton beam was committed for other uses at other times. Newer clinical facilities are treating 52 weeks per year, with operational availability 97–98% of working days. The maintenance is typically performed on weekends. The reliability of most proton equipment remains a little lower than for linear accelerators, and our calculated 'up-time' is approximately 97.5% for the proton facility (personal communication: Jay Flanz, PhD, Massachusetts General Hospital) versus 99% for linear accelerators. However, there is no backup unit available if the cyclotron is down. Maintenance generally requires more time, and may require switching patients to photon therapy to avoid unwanted treatment interruptions. This is particularly problematic for pediatric patients, whose original treatment plans may use only protons

Table 3. Mix of anatomic sites and tumors among 818 patients treated at the Francis H. Burr Proton Therapy Center at MGH between November 2007 and October 2008, the 7th operational year of the facility

Site/tumor	Patients, n	%
CNS	294	36
Eye	157	19
Bone/soft tissue	97	12
Prostate	80	10
Skull base	55	10
Head/neck	53	6.5
Lung	8	1
Liver	4	0.5
Lacrimal	3	0.5
Other	67	8.2
Total	818	

(without a photon component), unlike most adult patients whose plans at our facility often include a component of approximately 20 Gy photons.

The majority of the patients at our facility and at other proton facilities are treated with 3D conformal protons delivered via passive scanning. We did, however, initiate pencil beam scanning at our facility in December 2008, and it is anticipated that over time this will become the dominant mode of proton therapy delivery. It is less costly, more efficient and has a lower rate of neutron scatter in the beam line. It is also the basis for optimal intensity-modulated proton radiation therapy [22].

Figure 8 shows the treatment facility at the MGH. There is a 230-MeV cyclotron, and 3 treatment rooms. Two have 360 degree rotational gantries, and one is a fixed-beam room. The fixed-beam room has 2 stations, an eye line using a degraded 70-MeV beam, and a stereotactic radiosurgery line. There is also an experimental room that is being converted for additional treatment capacity. The treatments are image-guided using orthogonal orthovoltage digital imaging panels, interfaced with a table with 6 degrees of freedom, providing a high degree of accuracy and precision. The stereotactic line has a single scatterer, resulting in a very sharp penumbra for intracranial use.

At Harvard and MGH, proton radiotherapy usage has dramatically increased. The Francis H. Burr Proton Therapy Center at MGH treated its first patient in late 2001, going on to treat about 200 patients in its first year of operation, and increasing to more than 800 patients in both 2008 and 2009, which represents over 12,000 fractions of treatment per year. Nearly 90% of the patients have been adults and 10.5% pediatric. Gantry-based therapy has provided 57% of the treatment courses; fixed-beam eye therapy, 16%; and proton stereotaxis, 27%.

The types of tumors treated are summarized in table 3. The patient mix reflects our particular interest in skull base tumors, sarcomas and CNS tumors, and a significant proportion of the CNS tumors are benign. Many of our patients are on proton therapy clinical research studies. Patients with prostate cancer comprise about 10% of our patient population. Head and neck (primarily paranasal sinus and some nasopharynx) cancers are treated, but this area will be facilitated greatly by the scanned beam rather than the current passively scattered beams. We anticipate treating more locally advanced lung cancers on the randomized phase II study mentioned above as well as more early stage, medically inoperable non-small cell lung cancer, also on clinical research studies. We also anticipate treating more patients with gastrointestinal tumors, as we have open clinical research studies for patients with hepatocellular carcinomas and carcinomas of the pancreas (discussed above). Our pediatric cases reflect the types of pediatric solid tumors that are typically treated by radiotherapy.

Protons: Current Challenges

Motion Management and Beam Uncertainties

Motion management is a particular issue for proton therapy. Dosimetric evaluations in a lung case show that with only a 5-mm setup error, distortion of the therapy isodoses can occur causing the 50% isodose line to nearly reach the gross tumor volume [23]. Respiratory gating is important, and we utilize this for patients with motion greater than 1 cm on 4D CT. Dr. George Chen and his colleagues are currently working on issues of 4D treatment planning [24]. Increasingly, we are trying to understand the effects of motion on the doses delivered. The need for adaptive radiotherapy is substantially greater with protons than with photons. This can be particularly important for treatment of tumors near or within the lung; because of the lower tissue density of the lung, any beam overshoot can be a significant issue.

'Patching' is a technique specific to proton therapy where one stops the distal edge of one beam on the lateral edge of another [25]. It is the workhorse technique used for wrapping the dose around critical structures, though it relies on the penumbra edge of the beam. We may have insufficient knowledge of the distal edge, as discussed above, making scanning beam approaches desirable.

Treatment Efficiency

In our current operation, some of the scattered fields are challenging to deliver, particularly for some of the larger volumes. The apertures are very heavy, and if multiple isocenters are required to treat large tumors, the therapy delivery is very

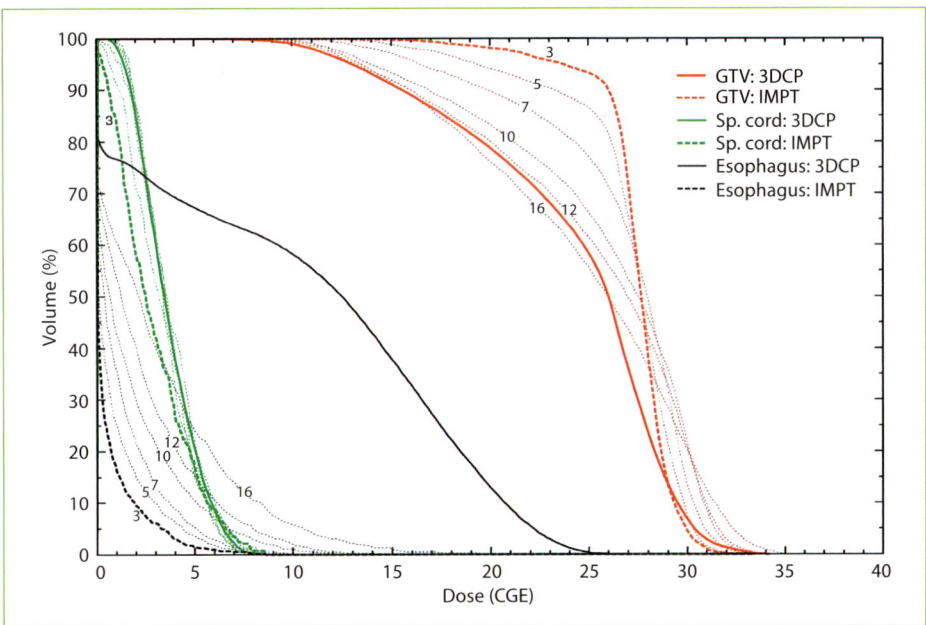

Fig. 9. A treatment planning comparison for 27 GyRBE boost treatment of a paraspinal tumor with 3D conformal protons (3DCP) and intensity-modulated proton therapy with different spot sizes (expressed in sigma from 3 to 16 mm). Gross tumor volume (GTV) coverage with intensity-modulated proton therapy (IMPT) begins to improve with spot sizes less than 12 mm, with similar steps of improvement in normal tissue dose-volume histograms for the spinal cord and esophagus. Courtesy of Thomas Bortfeld, PhD, MGH and Sairos Safai, PhD, MGH.

laborious. As a practical note, each of the brass apertures is very expensive, and the cost to the center last year was more than USD 500,000 for the brass alone.

Scanning Beam Technology

We expect that pencil beam scanning will be a useful tool for some patients. It will lower the neutron doses, especially important for pediatric patients, and it should be cheaper. Scanning beams, as discussed earlier, are still more of a 'crayon beam' than a 'pencil beam' in our own facility at the moment – a little wider than we ultimately hope they will be. As shown in figure 9 and as discussed earlier on p. 457, target conformality for this paraspinal tumor was higher with progressively smaller pencil beam spot sizes. We are looking to develop intensity-modulated protons to give us both distal and proximal dose shaping [20]. We believe this will be very useful for eccentrically based tumors, and will provide dose distributions in the high-dose regions comparable with IMRT, but with substantially less exit dose.

Guidelines for Clinical Practice

- The presumed clinical advantage for proton radiotherapy is due to the reduction in integral dose with protons compared with photons, due to the absence of exit dose beyond the proton Bragg peak.
- The integral dose with protons is approximately 60% lower than that with any external beam photon technique. This is expected to reduce acute and late radiation-associated morbidity, including radiation-associated second malignancies.
- The larger the target, the greater the integral dose advantage for protons.
- Pediatric patients, because of developing normal tissues and anticipated length of remaining life, are likely to have the maximum clinical gain with the use of protons.
- The most common current proton treatment delivery technique is 3D conformal protons. They are range modulated in depth to yield a spread out Bragg peak, and the beam is passively scattered with scattering foils to spread the beam in the x- and y-axes. Lateral field shaping is achieved with custom-milled brass apertures and distal shaping with custom milled Lucite range compensators.
- It is anticipated that magnetic pencil beam scanning will become the dominant mode of proton delivery in the future. This will lower neutron scatter associated with passively scattered beam lines, reduce the need for expensive beam-shaping devices, and allow intensity-modulated proton radiotherapy, which will provide highly conformal dose distributions with less entrance dose and no exit dose compared with intensity-modulated photons (IMRT).
- While impressive treatment results have been reported with protons, their cost is higher than IMRT. Hence, protons should ideally be employed for anatomic sites and tumors not well treated with photons. While protons appear cost-effective for pediatric tumors, their cost-effectiveness for treatment of some adult tumors, such as prostate cancer, is uncertain. Comparative studies have been proposed or are in progress to more rigorously assess their value for a variety of sites, including prostate and locally advanced non-small cell lung cancer.
- The utility of proton therapy will be enhanced by technological developments that reduce its cost. Combinations of 3D protons with IMRT photons may offer improved treatment plans at lower cost than pure proton plans. Hypofractionation appears to be safe and cost-effective in selected cancers such as liver, early-stage non-small cell lung and pancreas.
- Proton treatment plans are more sensitive to variations in tumor size and normal tissue changes over the course of treatment than photon plans, so it is expected that adaptive radiation therapy will be increasingly important for proton therapy.

References

1 Gottschalk B: Passive beam scattering; in DeLaney TF, Kooy HM (eds): Proton and Charged Particle Radiotherapy. Philadelphia, Lippincott Williams & Wilkins, 2008, pp 33–40.
2 Pedroni E: Pencil beam scanning; in DeLaney TF, Kooy HM (eds): Proton and Charged Particle Radiotherapy. Philadelphia, Lippincott Williams & Wilkins, 2008, pp 40–49.
3 DeLuca PM Jr, Wambersie A, Whitmore G: Prescribing, recording, and reporting proton-beam therapy. J ICRU 2007;7:1–210.
4 Goitein M: Radiation Oncology: A Physicist's-Eye View. New York, Springer, 2007.

5 St. Clair WH, Adams JA, Bues M, et al: Advantage of protons compared to conventional X-ray or IMRT in the treatment of a pediatric patient with medulloblastoma. Int J Radiat Oncol Biol Phys 2004;58:727–734.
6 Chung CS, Keating N, Yock T, et al: Comparative analysis of second malignancy risk in patients treated with proton therapy versus conventional photon therapy. Int J Radiat Oncol Biol Phys 2008;72:S8.
7 Torres MA, Chang EL, Mahajan A, et al: Optimal treatment planning for skull base chordoma: photons, protons, or a combination of both? Int J Radiat Oncol Biol Phys 2009;74:1033–1039.
8 Rasch C, Steenbakkers R, van Herk M: Target definition in prostate, head, and neck. Semin Radiat Oncol 2005;15:136–145.
9 Simpson DR, Lawson JD, Nath SK, et al: Utilization of advanced imaging technologies for target delineation in radiation oncology. J Am Coll Radiol 2009;6:876–883.
10 Trofimov A, Nguyen PL, Coen JJ, et al: Radiotherapy treatment of early-stage prostate cancer with IMRT and protons: a treatment planning comparison. Int J Radiat Oncol Biol Phys 2007;18:18.
11 Goitein M, Jermann M: The relative costs of proton and X-ray radiation therapy. Clin Oncol (R Coll Radiol) 2003;15:S37–S50.
12 Hong TS, Ryan D, Blaszkowsky LS, et al: Phase I study of preoperative short course chemoradiation with proton beam therapy and capecitabine for resectable pancreatic ductal adenocarcinoma of the head. Int J Radiat Oncol Biol Phys 2010, E-pub ahead of print.
13 Lundkvist J, Ekman M, Ericsson SR, et al: Cost-effectiveness of proton radiation in the treatment of childhood medulloblastoma. Cancer 2005;103:793–801.
14 Lundkvist J, Ekman M, Ericsson SR, et al: Economic evaluation of proton radiation therapy in the treatment of breast cancer. Radiother Oncol 2005;75:179–185.
15 Konski A, Speier W, Hanlon A, et al: Is proton beam therapy cost effective in the treatment of adenocarcinoma of the prostate? J Clin Oncol 2007;25:3603–3608.
16 Hall EJ: Intensity-modulated radiation therapy, protons, and the risk of second cancers. Int J Radiat Oncol Biol Phys 2006;65:1–7.
17 Paganetti H, Bortfeld T, Delaney TF: Neutron dose in proton radiation therapy: in regard to Eric J. Hall (Int J Radiat Oncol Biol Phys 2006;65: 1–7). Int J Radiat Oncol Biol Phys. 2006;66:1594–1595, author reply 1595.
18 Chang JY, Zhang X, Wang X, et al: Significant reduction of normal tissue dose by proton radiotherapy compared with three-dimensional conformal or intensity-modulated radiation therapy in stage I or stage III non–small-cell lung cancer. Int J Radiat Oncol Biol Phys 2006;65:1087–1096.
19 Zietman AL, DeSilvio ML, Slater JD, et al: Comparison of conventional-dose vs high-dose conformal radiation therapy in clinically localized adenocarcinoma of the prostate: a randomized controlled trial. JAMA 2005;294:1233–1239.
20 Lomax A: Intensity modulation methods for proton radiotherapy. Phys Med Biol 1999;44:185–205.
21 Weber DC, Trofimov AV, Delaney TF, et al: A treatment planning comparison of intensity modulated photon and proton therapy for paraspinal sarcomas. Int J Radiat Oncol Biol Phys 2004;58:1596–1606.
22 Kooy HM, Clasie BM, Lu HM, et al: A case study in proton pencil-beam scanning delivery. Int J Radiat Oncol Biol Phys 2010;76:624–630.
23 Engelsman M, Kooy HM: Target volume dose considerations in proton beam treatment planning for lung tumors. Med Phys 2005;32:3549–3557.
24 Engelsman M, Rietzel E, Kooy HM: Four-dimensional proton treatment planning for lung tumors. Int J Radiat Oncol Biol Phys 2006;64:1589–1595.
25 Kooy HM, Adams JA: Passive scattering; in DeLaney TF, Kooy HM (eds): Proton and Charged Particle Radiotherapy. Philadelphia, Lippincott Williams & Wilkins, 2008, pp 88–98.
26 Gragoudas ES, Munzenrider JE, Lane AM, Collier JM: Eye; in DeLaney TF, Kooy HM (eds): Proton and Charged Particle Radiotherapy. Philadelphia, Lippincott Williams & Wilkins, 2008, pp 151–161.
27 Yock TI, Schneider RJ, Friedmann A, et al: Proton radiotherapy for orbital rhabdomyosarcoma: clinical outcome and a dosimetric comparison with photons. Int J Radiat Oncol Biol Phys 2005;63:1161–1168.
28 Rosenberg AE, Nielsen GP, Keel SB, Renard LG, Fitzek MM, Munzenrider JE, Liebsch NJ: Chondrosarcoma of the base of the skull: a clinicopathologic study of 200 cases with emphasis on its distinction from chordoma. Am J Surg Pathol 1999;23:1370–1378.
29 Slater JD, Rossi CJ Jr, Yonemoto LT, Bush DA, Jabola BR, Levy RP, Grove RI, Preston W, Slater JM: Proton therapy for prostate cancer: the initial Loma Linda University experience. Int J Radiat Oncol Biol Phys 2004;59:348–352.

Thomas F. DeLaney, MD
Francis H. Burr Proton Therapy Center, Massachusetts General Hospital
30 Fruit Street, Boston, MA 02114 (USA)
Tel. +1 617 726 7869, Fax +1 617 724 9532, E-Mail tdelaney@partners.org

Author Index

Abdulrahman, R. 382, 395
Alaly, J. 29

Bastasch, M. 382
Bortfeld, T. 440
Brock, K. 29

Cao, D. 80
Chang, D.T. 412
Cheung, F.W.K. 132
Choy, H. 395
Chu, K.P.-M. 217
Craig, T. 29

David, M.B. 255
Dawson L.A. 196, 315
Deasy, J. 29
DeLaney, T.F. 465
Dieterich, S. 181

Eisbruch, A. 255

Gibbs, I.C. 181

Heinzerling, J. 395
Hittson, W. 382
Hoppe, R. 331

Kavanagh, B.D. 271, 370
Keall, P. 118
Kim, J. 196
King, C. 428
Konski, A.A. 60

Koong, A.C. 412
Kupelian, P. 165, 344

Langen, K. 165
Law, M.Y.Y. 132
Le, Q.-T. 217
Ling, C.C. 132
Loo, B.W., Jr. 271
Low, D. 99

Meyer, J.L. IX, 29, 196, 271, 292, 344, 370, 395
Minn, A.Y. 412
Mohan, R. 440
Moseley, D. 29

Purdy, J.A. 1

Saha, D. 382
Sharpe, M. 29
Shepard, D.M. 80
Star-Lack, J. 132
Steinberg, M.L. 60
Story, M. 382
Swaminath, A. 315

Timmerman, R. 370, 382, 395

Wallner, P.E. 60
White, J.R. 292
Wu, V.W.C. 132

Zakaryan, K. 29

Subject Index

Adaptive radiation therapy
  changing anatomy during radiotherapy course 52, 53
  daily uncertainties 17, 18, 51, 52
  definition 3
  head and neck cancer 233, 234
  megavoltage computed tomography for dose evaluation
    dose accumulation 177–179
    dose recalculation 176, 177
    overview 175, 176
  organ dose distribution variations 17, 18
  planning target volume 50–52
  prospects 53, 54, 212, 213
  prostate cancer
    off-line 363–365
    on-line 364, 366
    overview 359, 360
  rationale 212
Agency for Healthcare Research and Quality, health technology assessment 62, 65
Ambulatory Payment Classification 70

Beamlet, planning 31, 32
Brain tumor, CyberKnife applications 188, 189
Breast cancer, *see also* Breathing motion
  goals of radiation therapy 292, 293
  guidelines for clinical practice 311
  improvement strategies for radiation therapy
    dose homogeneity 294–297
    overview 293, 294
  intensity-modulated radiation therapy planning
    concomitant breast radiotherapy 303, 304
    defining of at-risk organs 300–302
    homogeneity guidelines 302
    motion management 302
    postmastectomy radiotherapy 302, 303
  organ dose minimization
    contralateral breast 299
    heart 297, 298
    lung 298, 299
  partial breast radiotherapy
    intensity-modulated radiation therapy
      advantages 308
      outcomes 309, 310
    patient selection 304, 305
    technique selection 305–308
  surface topology imaging 138, 140
Breathing motion
  cycle 103–105
  four-dimensional computed tomography
    approaches 100, 101
    cone-beam computed tomography 102, 283–285
    prospects 115
    thoracic tumor motion characterization 274–276
  image guidance
    computed tomography in room 283–285
    fluoroscopy for gating verification 283

Breathing motion (continued)
  orthogonal pair planar imaging 282
  overview 281, 282
  tracking 285-287
 impact
  imaging 100-102
  treatment
   dose gradients 105, 106
   interplay effects 106, 107
   tomotherapy 107-109
 management, *see also* Motion correction
  internal target volume 109-111, 277
  maximum intensity projection 110, 111
  respiratory gating
   fluoroscopic simulation 287
   outcomes 290
   overview 111-113, 278, 279
   pancreatic cancer 418-420
   patient positioning 287
   target delineation 288-290
  tracking 113, 114, 279
 tumor motion
  characterization 274-276
  components 272, 274
  reduction
   abdominal compression 280, 281
   breath hold 279, 280
B-spline, automated image registration 49

Calypso System 142
Cancer biology, discovery integration into clinical practice 4, 5
Centers for Medicare and Medicaid Services, *see* Medicare
Charge trapping, cone-beam computed tomography 148-150
Chronic obstructive pulmonary disease, breathing motion management 317
Clinical target volume
 definition 6
 head and neck cancer
  lymph nodes 258, 259
  overview 257
  special considerations 259, 260
 liver cancer 326
 lung cancer 403, 404
 nasopharyngeal carcinoma 234-237
 oropharyngeal carcinoma 245

 pancreatic cancer 322
 proton beam radiation therapy 447, 449
 salivary gland changes 231, 232
Clip box, automated image registration 46, 47
Common Procedural Terminology, payment policy development 68, 69
Comparative effectiveness research, health technology assessment 65, 66
Computed tomography
 breathing motion, *see* Breathing motion
 cone-beam computed tomography, *see* Cone-beam computed tomography
 four-dimensional computed tomography 100-105, 109-112, 114, 115, 274-276
 image registration, *see* Image registration
 megavoltage imaging, *see* Helical tomotherapy, Megavoltage imaging
 on-rails computed tomography 37, 133, 138
 planning overview 8-10
 second cancer risks from daily imaging 22
 tomotherapy imaging 37
Conditional Coverage Strategies 63, 64
Cone-beam computed tomography 133, 143
 four-dimensional computed tomography 102, 149, 150, 283-285
 frequency of imaging 152
 image-guided intervention criteria and protocols 152, 153
 image quality
  charge trapping 148-150
  cone beam geometry 150, 151
  kilovolt images 144
  metal artifact reduction 151
  noise reduction 151
  scattered radiation 147, 148
 imaging volumes, times and doses 145, 146
 intrafraction motion management 153, 154
 liver cancer image-guided radiation therapy 326, 327
 megavolt versus kilovolt imaging 144, 145
 pancreatic cancer image-guided radiation therapy 322-325
 principles 37-39
 reconstruction and data correction 147

Constraints
  planning 30
  stereotactic body radiotherapy 404–409
Cost, *see also* Payment policy development
  overview 5
  proton beam radiation therapy 461, 462, 473–475
Cost-effectiveness
  cost-benefit analysis 72
  cost-effectiveness analysis 72
  cost-minimization analysis 72, 74–76
  cost-utility analysis 73
  prostate cancer radiation therapy example 75–77
  value-based purchasing 73, 74
Couch, movable 127
Coverage with Evidence Development 63, 64
Critical volume model, stereotactic body radiotherapy 388, 390, 405
CyberKnife
  applications
    brain tumor 188, 189
    lung cancer 190
    prospects 192
    prostate cancer 190
    spinal tumor 189
  image guidance and tracking prospects 191, 192
  IRIS collimator 186
  Monte Carlo dose calculation algorithm 184–186
  moving target irradiation 127
  overview 181, 182
  path traversal optimization 183, 184
  prostate cancer stereotactic body radiotherapy 430
  sequential optimization 187
  tumor size 182
  XSight lung tracking 187, 188

Digital tomosynthesis, image guidance 151
Dose calculation
  overview 13
  variability reduction in delivery 198, 199
Dose-response model, plan optimization 10
Dose-volume constraints, planning 30
Dose-volume histogram
  intensity-modulated arc therapy 87, 89–91
  lymphoma 338–340
  prostate cancer 360, 361
  proton beam radiation therapy 447, 449

Electromagnetic monitoring, target localization 125
Erectile dysfunction, prostate cancer stereotactic body radiotherapy outcomes 435, 436
Ergo++, intensity-modulated arc therapy planning 87

Figure-of-Merit, plan optimization 12
Fluence map, planning 32
Fluid flow, automated image registration 48, 49
Fluoroscopy, in-room systems 140

Generalized equivalent uniform dose, planning 30, 31
Geometric uncertainty, *see* Residual geometric uncertainty, Target localization
Gross tumor volume
  definition 6
  head and neck cancer 256, 257
  liver cancer 326
  lung cancer 403
  nasopharyngeal carcinoma 235, 236
  pancreatic cancer 321, 322
  respiratory gating 288, 289

Head and neck cancer
  clinical target volume
    lymph nodes 258, 259
    overview 257
    special considerations 259, 260
  gross tumor volume 256, 257
  guidelines for clinical practice 250, 251, 267, 268
  image-guided radiation therapy
    adaptive radiotherapy 233, 234
    overview 208–210
    setup variability 232
  intensity-modulated radiation therapy
    adaptive replanning 265
    adjunct therapy 265, 266
    case studies 267
    daily setup variability 229

Head and neck cancer (continued)
   dose 264
   intrafraction variability 230, 231
   nasopharyngeal carcinoma
      dose prescription 238, 239
      outcomes 239-242
      stereotactic radiotherapy as boost 242–244
      target volume delineation 234–238
   oropharyngeal carcinoma
      dose prescription 245
      outcomes 246–249
      target volume delineation 245
   planning 264, 265
   positron emission tomography
      accuracy 227
      cervical node delineation 222, 223
      guidance 225
      induction chemotherapy applications 223–225
      registration errors 226, 227
      segmentation approaches 220–222
      target delineation with combined computed tomography 218–220
   prior surgery 262
   recurrence 262
   shoulder immobilization 228, 229
   supraclavicular treatment 262
   swallowing structure preservation 262, 263
   volume changes during fractionated therapy
      salivary gland changes 231, 232
      weight loss 231
 lymphatic drainage 257, 258
 radiation therapy overview 217, 218
 retropharyngeal nodes 260, 261, 263
Health technology assessment
   clinical trials 65
   comparison with Food and Drug Administration evaluation 62
   methodology 62–64
   overview 60, 61
   prospects 66
Heavy ion radiation therapy, delivery techniques 19
Helical tomotherapy
   clinical applications 167
   daily megavoltage-based computed tomography
      benefits 173–175
      deformations 170, 171
      image quality 168
      image registration 168–170
      imaging doses 172, 173
      setup errors 170
   dose evaluation for adaptive radiotherapy
      dose accumulation 177–179
      dose recalculation 176, 177
      overview 175, 176
   system overview 166, 167
HI-ART System 13
Hodgkin's disease, *see* Lymphoma
Hospital Outpatient Prospective Payment System 70

Image-guided radiation therapy
   applications, *see also* specific cancers
      gastrointestinal cancer 211, 212
      head and neck cancer 208–210, 214
      lung cancer 210, 211
      prostate cancer 206–208
   challenges
      dose accumulation 203
      motion
         interfraction motion 204–206
         intrafraction motion 206
         organ motion 203, 204
      organ deformation 202
      planning target volume expansions 202, 203
      prioritizing image guidance 202
   definition 3, 4
   efficiency, workflow and practical integrations 154, 155
   goals 135, 155
   imaging systems, *see* specific techniques
   integration processes 136, 137
   lessons learned 200, 201
   oligofractionation in stereotactic body radiotherapy 156–158
   origins 2
   prospects for study 16, 17, 213–215
   rationale 197, 198
   role in process of radiotherapy 199, 200
   second cancer risks 21
   validation of devices 135, 136

Image registration
   alignment by
      intensity 45
      points 44
      surfaces 44, 45
   automation
      clip box 46, 47
      organ limitation 47, 48
   common surrogates and correspondence between surrogates 43, 44
   daily megavoltage-based computed tomography 168–170
   deformable techniques
      biomechanical approach 49, 50
      B-spline 49
      fluid flow 48, 49
      optical flow 49
      thin-plate spline 49
   overview 41, 42
   steps 42, 43
Informatics systems 23, 24
Intensity-modulated arc therapy
   advantages 81
   applications, *see* specific cancers
   commissioning 92, 93
   comparison with other intensity-modulated radiation therapy techniques 83–85
   delivery systems 85
   historical perspective 81, 82
   quality assurance 93–97
   single arc versus multiple arcs 89–92
   SmartArc 86
   terminology 82
   treatment planning 85–89
   volumetric-modulated arc therapy 82–86
Intensity-modulated proton therapy, prospects 20, 21, 451, 455–458
Intensity-modulated radiation therapy
   accuracy in delivery 7, 8
   applications, *see* specific cancers
   definition 3, 4
   historical perspective 2
   planning
      optimization 10–12
      typical system 30–32
   proton beam radiation therapy comparison
      costs 473–475
      dose 471–473
   quality assurance 15, 16
   rotational approaches 13, 14, 82, 83
   second cancer risks 21
Internal target volume
   breathing motion correction 109–111, 277
   proton beam radiation therapy 448
IRIS collimator, CyberKnife 186

Kilovoltage imaging, target localization 36, 123, 124, 143

Linear accelerator, gimbaled linac for moving target irradiation 128
Linear-quadratic model, cell survival 373, 374, 386, 387
Linear sum objective function, planning 30
Liver cancer, *see also* Breathing motion
   image-guided radiation therapy
      guidelines for clinical practice 328, 329
      image guidance 326, 327
      motion management 325, 326
      overview 316, 317
      planning 326
      simulation 326
   stereotactic body radiotherapy
      clinical studies 422, 423
      fractionation and toxicity 424
      guidelines for clinical practice 426
      indications 425
      rationale 421
      treatment planning 424, 425
Lung cancer, *see also* Breathing motion
   CyberKnife applications 190
   image-guided radiation therapy 210, 211
   motion of tumor 119–121
   proton beam radiation therapy 477, 478
   stereotactic body radiotherapy
      advantages 395, 396
      dose constraints 404–409
      guidelines for clinical practice 410
      historical perspective
         inoperable cancer 398, 399
         operable cancer 399, 400

Lung cancer (continued)
  overview 396–398
  immobilization 400, 401
  motion assessment and management 401–403
  tissue heterogeneity 405, 406
  treatment planning 403, 404
  treatment verification and delivery 407, 408
  XSight lung tracking 187, 188
Lymphoma
  guidelines for clinical practice 342
  history of imaging 331, 332
  radiation therapy planning
    advanced treatment planning 338–340
    dose 337, 338
    field size 337, 341, 342
    positron emission tomography integration 340, 341
  staging
    Hodgkin's disease 332, 333
    non-Hodgkin's lymphoma 333, 334
  treatment response evaluation 334–337, 342

Magnetic resonance imaging
  image registration, see Image registration
  moving target localization 125
  pancreatic cancer breathing motion management 318
  planning 9
  proton beam visualization within patient 460
  radiotherapy guidance 39
MammoSite, breast brachytherapy 306–308
Maximum intensity projection, breathing motion correction 110, 111
Medicare
  Centers for Medicare and Medicaid Services 63, 65, 67–71
  hospital payments 70
  payment policy development 67, 68, 70, 71
Megavoltage electron portal imaging device, target localization 34, 35
Megavoltage imaging
  helical tomotherapy, see Helical tomotherapy
  target localization 34, 35, 124, 142, 143–145, 165
Metastasis, stereotactic body radiotherapy for oligometastasis 379, 380
Monaco, intensity-modulated arc therapy planning 87
Motion correction, see also Breathing motion
  breathing motion, see Breathing motion
  cone-beam computed tomography intrafraction motion management 153, 154
  hitting moving targets 126–128
  image-guided radiation therapy challenges
    interfraction motion 204–206
    intrafraction motion 206
    organ motion 203, 204
  locating moving targets 122–126
  lung tumor 119–121
  overview of technologies 6, 7
  real-time motion correction system components 118, 119
Multileaf collimator
  breathing motion effects 107
  dynamic versus static delivery 32
  leaf position errors 7, 8
  moving target irradiation 127, 128

Nasopharyngeal carcinoma, see Head and neck cancer
National Institute of Health and Clinical Excellence 64
Non-Hodgkin's lymphoma, see Lymphoma

Objective function, planning 30
Oligofractionation, see Stereotactic body radiotherapy
Optical flow, automated image registration 49
Optical tracking, target localization 34, 123, 138, 140
Oropharyngeal carcinoma, see Head and neck cancer

Pancreatic cancer, see also Breathing motion
  chemotherapy outcomes 413
  image-guided radiation therapy
    guidelines for clinical practice 328, 329

image guidance 322–325
motion management 317–319, 324
overview 316, 317
planning 321, 322
simulation 319–321, 324
stereotactic body radiotherapy
clinical trials 413–415
guidelines for clinical practice 425, 426
indications 420
prospects 420, 421
toxicity 415–417
treatment planning
respiratory gating 418–420
target tracking 418
Partial breast radiotherapy, see Breast cancer
Payment policy development, see also Cost-effectiveness
example 67
Medicare mandate 67, 68
socioeconomic context 71, 72
valuation of services 68–70
Picture archiving and communication system 24
Pinnacle, intensity-modulated arc therapy planning 86
Planning target volume, see also Target localization
adaptive radiation therapy 50–52
definition 6
expansions based on image-guided radiation therapy strategy 202, 203
liver cancer 326
lung cancer 403, 404
nasopharyngeal carcinoma 235–237
oropharyngeal carcinoma 245
pancreatic cancer 321, 322, 417
prostate cancer 429, 430
proton beam radiation therapy 447, 449
respiratory gating 289, 290
Positron emission tomography
computed tomography combination 9–11
head and neck cancer
accuracy 227
cervical node delineation 222, 223
guidance 225
induction chemotherapy applications 223–225

registration errors 226, 227
segmentation approaches 220–222
target delineation with combined computed tomography 218–220
image registration, see Image registration
lymphoma
field size 337, 341, 342
historical perspective 331, 332
radiation therapy planning 340, 341
staging 332–334
treatment response evaluation 334–337, 342
proton beam visualization within patient 459
Prostate cancer
CyberKnife applications 190
image-guided radiation therapy
adaptive radiotherapy
overview 359, 360
off-line 363–365
on-line 364, 366
deformable registration 361–363
deformation/rotation of prostate
interfraction 355–357
intrafraction 357, 358
proton therapy 358
dose-volume histogram and patient positioning 360, 361
guidelines for clinical practice 366
interfraction motion
challenges 348–350
daily image guidance importance 347
daily setup accuracy 347
significance 348
intrafraction motion
clinical use of tracking data 351, 352
dosimetric consequences of tracked motion 353, 354
tracking during treatment 350, 351
motion reduction 359
overview 206–208, 345, 346
proton beam radiation therapy 478, 479
stereotactic body radiotherapy
CyberKnife delivery 430
endpoints 431
guidelines for clinical care 437
image-guided therapy 430, 431

Prostate cancer (continued)
    outcome of phase II trial
        endpoints 431
        prostate serum antigen response 433
        toxicity 434–436
    overview 377, 378
    patient selection 429
    rationale 428, 429
    testes incidental dose 431–433
    treatment planning 429, 340

Proton beam radiation therapy
    advantages 466–469
    beam
        active scanning 444–446
        field shaping in passing scattering 446–448
        passive scattering 444, 445
        production 442, 443, 461, 462
    clinical trials
        lung cancer 477, 478
        overview 476, 477
        prostate cancer 478, 479
    cost-effectiveness 475, 476
    delivery techniques 19, 20
    dose comparison with intensity-modulated radiation therapy 471–473
    guidelines for clinical practice 484
    Massachusetts General Hospital instrument 454, 460, 466, 481
    M.D. Anderson Cancer Center instrument 443, 444, 446
    motion management 482, 483
    payment policy development 70, 71
    photon therapy
        combination therapy 470, 471
        cost comparison 473–475
    principles 440–442
    prospects
        beam visualization within patient 458–450
        compact proton accelerator 461, 463
        cost reduction 461, 462
        intensity-modulated proton therapy 20, 21, 451, 455–458
        pencil beam scanning 452–455
        proton linear accelerator 462
        scanning beam technology 483
        wake field accelerator 461, 462
    radiobiology 449, 450
    rationale 466
    second malignancies 470
    spread-out Bragg peak 445, 446, 450, 453, 467
    treatment planning
        normal tissue sparing 470, 471, 479
        target definition 471

Quality assurance
    intensity-modulated arc therapy 93–97
    overview 15, 16

Range compensator, proton beam shaping 446
RapidArc 157
Relative Value Scale Update Committee, payment policy development 68, 69
Renaissance System 15
Residual geometric uncertainty 197
Respiratory gating, see Breathing motion

Salivary gland, volume changes during fractionated radiotherapy 231, 232
Scattered radiation, cone-beam computed tomography 147, 148
Serially functioning tissue, see Stereotactic body radiotherapy
SmartArc, see Intensity-modulated arc therapy
Spinal tumor
    CyberKnife applications 189
    stereotactic body radiotherapy 375–377
Spread-out Bragg peak, proton beam therapy 445, 446, 450, 453, 467
Stereotactic body radiotherapy
    ablative dose delivery
        normal tissue consequences 392, 393
        overview 391
        tumor control dose 393
    advantages 380
    applications, see also Liver cancer, Lung cancer, Pancreatic cancer, Prostate cancer
        growth trends 374, 375
        spinal tumors 375–377
    definition 382, 383
    oligofractionation principles 156–158

prospects
    oligometastasis 379, 380
    protocol standardization 378, 379
radiobiology
    biologically equivalent dose 386, 387
    cell survival curve 373, 374, 384–386
    endothelial apoptosis 372
    organ/tissue function and microscopic architecture 387–390
    overview 371, 383–385
    serially functioning tissue macroscopic architecture 390, 391
    tissue regeneration pathways 372, 373, 384
Stereotaxy, target localization 34

Target localization, *see also* Computed tomography, specific techniques
    Calypso System 142
    comparison of approaches 40, 41
    geometric uncertainty sources 32, 33
    kilovoltage imaging 36, 123, 124
    magnetic resonance imaging 39, 125
    megavoltage imaging 34, 35, 124
    optical tracking 34, 123, 138, 140
    radio-opaque markers 35, 125
    radiofrequency transponders 35, 36
    stereotaxy 34
    technology selection 33, 34
    ultrasound 36, 126, 141
Thin-plate spline, automated image registration 49
TrueBeam system 155

Ultrasound
    image registration, *see* Image registration
    target localization 36, 126, 141

Value, *see* Cost-effectiveness
Volumetric-modulated arc therapy, *see* Intensity-modulated arc therapy

XSight lung tracking 187, 188